NEW COMMERCIAL POLYMERS 2

Related book of interest

New Commercial Polymers 1969−1975
by Hans-Georg Elias
(Gordon and Breach, Science Publishers, Inc., New York, 1977)

New Commercial Polymers 2

Hans-Georg Elias

Michigan Molecular Institute
Midland, Michigan U.S.A.

Friedrich Vohwinkel

Osnabrück College of Technology
Osnabrück
Federal Republic of Germany

Revised and Expanded
English Edition

Translated from the German

GORDON AND BREACH SCIENCE PUBLISHERS
New York • London • Paris • Montreux • Tokyo

Gordon and Breach Science Publishers

P.O. Box 786
Cooper Station
New York, New York, 10276
United States of America

P.O. Box 197
London WC2E 9PX
England

58, rue Lhomond
75005 Paris
France

P.O. Box 161
1820 Montreux 2
Switzerland

14-9 Okubo 3-chome,
Shinjuku-ku,
Tokyo 160
Japan

Originally published in German by Hans-Georg Elias/Friedrich Vohwinkel as "Neue Polymere Werkstoffe für die industrielle Anwendung. 2. Folge" Hanser, Munich 1983

Library of Congress Cataloging-in-Publication Data

Elias, Hans-Georg, 1928-
New commercial polymers 2.

  Rev. translation of: Neue polymere Werkstoffe für die industrielle Anwendung.
  Includes bibliographies and index.
  1. Polymers and polymerization. I. Vohwinkel. II. Title. III. Title:
New commercial polymers two.
QD381.E45513   1986   668.9   85-30244
ISBN 2-88124-078-X

Table of Contents

(References are listed at the end of each chapter)

Foreword from the German Edition

Seven years ago, Hans-Georg Elias published a progress report on the new commercial polymers which were introduced to the market between 1969 and 1975 [1]. The book contained data for over fifty new polymers and thus refuted the often repeated prediction that one could not expect new plastics, elastomers, and fibers with revolutionary physical properties. Since such predictions underestimate human inventiveness, it was not too difficult to foresee a continuous flow of new industrial polymers for the years to come.

This expectation did indeed come true: more than eighty new commercial polymers are described in this book which covers the years 1975 to 1981 with updates for 1982 through 1984. We started to draft the manuscript during the sabbatical of Friedrich Vohwinkel (FV) at the Michigan Molecular Institute. Originally, we intended to deviate from the earlier format by arranging the new polymers according to their use as thermoplastics, thermosets, elastomers, fibers, etc. Several of the new polymers have more than one use, however, which precludes a classification system according to areas of application. Many of the remarks in the foreword of the earlier book are thus equally true for the new report:

"The present book is fundamentally a progress report which is based on references available to us or company brochures kindly sent by many firms. The book is subjective in the sense that no attempt was made to systematically and scientifically search out all relevant literature. Information concerning commercially manufactured synthetic and half-synthetic macromolecular substances is incorporated, without consideration as to their use as thermoplastics, thermosets, elastomers, fibers, coatings, synthetic leather, thickening agents, or others. All those materials are viewed as "commercial" which either went into production or into experimental production or whose commercial sale was announced.

Naturally, the dividing line between "commercial" and "not yet commercial" cannot be sharply drawn. . . . It is also possible that one or another new development was overlooked, because many first announcements are not easily obtainable. Finally, it is often difficult to draw a sharp line between "new" and "already known" structures....

A progress report such as this has to be encyclopedic for the most part, that is to say without correlation between the individual sub-chapters. Several of the new polymers have more than one use, such as an engineering material and as a fiber, which precludes a clas-sification system according to areas of application. Therefore, as a rule, the chemical structure of the chain was chosen as the basis of classification....

All the physical units given in the literature were converted into SI units.... The appendix lists conversion factors from the old to the new units. It also contains lists of abbreviations, common names of physical quantities and the addresses of companies referred to in this book...".

We wish to thank the following individuals for help with the literature search or for other valuable information: Mrs. Mary Reslock (MMI), Professor H. Baumann (Schaum-Chemie Frankenthal), Professor Raymond F. Boyer (MMI), Mr. W. Isler (Ems Industries), Dr. H. Naarmann (BASF), Professor Katsutoshi Nagai (Yamagata University), Dr. S. Schaaf (Ems Industries), Dr. R. Streck (Hüls AG), and Dr. G. Wick (Hoechst AG). We also thank the companies listed in the references for reprints and company brochures.

Hans-Georg Elias
Friedrich Vohwinkel

[1] H.-G. Elias, Neue polymere Werkstoffe 1969-1974, C. Hanser, Munich and Vienna 1975; H.-G. Elias, New Commercial Polymers 1969−1975, Gordon and Breach, New York, London, Paris 1977

1 Introduction

1.1 ECONOMIC DEVELOPMENTS

Nothing is more permanent than change. The truth of this old saying was painfully experienced by the polymer industry during the last decade. The period from 1950 to 1970 saw increases in the world production of natural rubber and natural fibers of 50% each, of synthetic fibers by 500%, of synthetic elastomers by 900%, and of plastics by 1500%. In 1973 OPEC triggered the first oil crisis which caused the world economy to change dramatically. Whereas natural rubber production maintained a slow growth and the production of natural fibers, synthetic fibers, and synthetic elastomers on the whole held up well despite the declines in 1975 and 1980, the production of plastics has stagnated since 1973 and, in addition, took a deep plunge in 1975 (Figure 1-1). The annual world production of plastics dropped from 45.3 Tg/a in 1974 to 36.9 Tg/a in 1975 (1 Tg/a = 1 million metric tons per year). The three leading plastics producing countries, i.e., the United States, the Federal Republic of Germany, and Japan were the chief victims (Table 1-1). In these countries, plastics production fell from 26.8 Tg/a in 1974 to 19.6 Tg/a in 1975, nearly 86% of the total decline in world production between 1974 and 1975.

Although the events of 1975 demonstrate the great influence the three leading Western industrial nations exert on world trade, they hide the fact that these three nations are steadily losing their relative share of world plastics production. This share fell from 66% in 1969 to 50% in 1979 and will probably continue to slide as new plants in OPEC and developing countries come on stream. These new units produce almost exclusively commodity thermoplastics, i.e., PE, PVC, PS, and PP. Due to cheaper feed stocks, production costs in oil exporting countries are likely to be lower than those in oil importing countries. The latter countries, in turn, may become motivated to concentrate more on specialty plastics. Precisely this strategy was announced by several U.S. companies which in the past had mainly dealt with commodity plastics.

Of course, statistics are not always above question and the statis-

1

tics of the United Nations are no exception (Figure 1-1 and Table 1-1). The evaluation of other sources [5] indicates a significantly higher 1978 world production—55.6 Tg/a [6] compared to 37.5 Tg/a [2] (Table 1-2). The statistics of the UN do not include production data of some small countries as well as those of nations under UN boycott, such as Taiwan (Republic of China) and South Africa. Moreover, quoted production figures [2, 6] quite often differ from

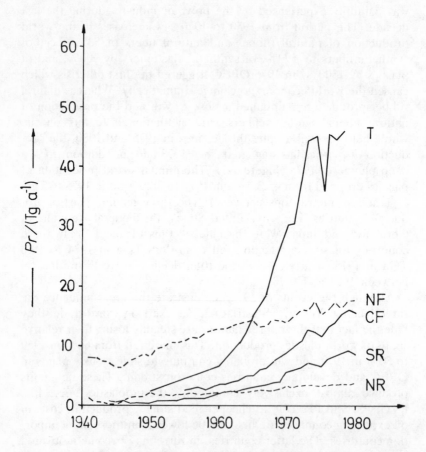

Figure 1-1 Annual world productions, *Pr*, of thermoplastics and thermosets (T), natural fibers (NF), man-made fibers (CF), synthetic rubbers (SR), and natural rubber (NR). 1 Tg/a = 1 million metric tons per year. Data for 1940–1971 from [1], data for 1972 to date from [2].

each other (Table 1-2). Finally, neither set of statistics furnishes data from the People's Republic of China.

The strongly export-oriented countries are all Western European nations. Two of them (Netherlands and Switzerland) export more than they produce, i.e., they export also some of their imports. More than 30% of the plastics production is exported by the Federal Republic of Germany, France, Italy, Great Britain, Sweden, Austria, and Norway, a part of the exports being modified imports. Of the countries outside of Western Europe, only Taiwan has a high export rate; Japan's rate is ca. 20%.

Worldwide, plastics consumption will certainly increase again over the forthcoming years, both in absolute figures and in relation to other materials. One reason is that many countries, including some European ones, have not yet reached the annual high consumption of ca. 90 kg plastics per capita as in the Federal Republic

TABLE 1-1

Development of the annual plastics production in the United States, West-Germany, and Japan (data from [1], except where noted [3]−[5]; data from [4] gives consumption). 1 Gg/a = 1000 metric tons/year.

| Year | World | Production in Gg/a | | | U.S., German, and Japanese production as percent of world production |
		United States	West-Germany	Japan	
1969	27 300	9 909	3 942	4 194	66.1
1970	30 360	9 660	4 321	6 290	66.7
1971	30 460	9 874	4 780	5 125	64.9
1972	37 500	12 046	5 471	5 573	61.5
1973	44 200	13 706	6 434	6 392	60.0
1974	45 300	13 764	6 271	6 722	59.1
1975	36 900	10 206	5 046	4 373	53.1
1976	45 300	13 261	6 444	4 952	54.4
1977	44 400	11 232	6 270	4 977	50.6
1978	46 200	12 276	6 706[5]	5 705	53.5
1979	52 000	15 620[3]	7 255[5]	8 207[5]	59.8
1980	48 000	13 940[3]	6 737[5]	7 518[5]	58.7
1981	48 000	15 000[3]	6 600[5]	7 040[5]	59.7
1982		13 900[3]	6 274[5]	7 135[5]	
1983		16 130[3]			

TABLE 1-2

Production, import, export, and apparent consumption of plastics in 1978. (Data in Gg = 1000 metric tons); estimates in ( ); [n] name-plate capacity; [a] 1977; [b] 1976.

Country	Production [2]	Production [5]	Import [5]	Export [5]	Apparent Consumption [5]
Western Europe					
West-Germany	6755	6752	1743	2651	5844
France	2163	2768	1148	1397	2519
Italy	2467	2780	807	1034	2553
United Kingdom	2765	2650	950	890	2710
Netherlands	1962	1975	750	2275	(450)
Belgium	1103	(1200)			
Spain	1010[a]	1077	178	103	1152
Sweden	596[a]	524	559	384	699
Austria	434	436	294	262	468
Finland	178	355	122	82	437
Yugoslavia	323	293			(536)
Greece	40[a]	(180)	8	8	(180)
Switzerland		114	431	142	403
Norway	146	88	215	63	240
Portugal	78	60[a]	176	7	229
Denmark	50	11	342	35	318
Ireland		0	123	0	123
Eastern Europe					
USSR	3300	3500			
Czechoslovakia	810	810			
East-Germany	762	(750)			
Romania	326	552			
Poland	596	471			
Hungary	213	213			
Bulgaria	120	169[a]			
North America					
United States	12276	15424	252	(1390)	14286
Canada	701	1002	420	128	1294
Mexico	246[a]	237			400
Central and South America					
Brazil	253[a]	1000			1000
Puerto Rico		436[n]			
Argentina	133	103	49	0	152
Venezuela		90[n,a]			
Colombia	19[b]	89[n,a]			
Chile	40	50[n,a]			

TABLE 1-2 (*continued*)

Production, import, export, and apparent consumption of plastics in 1978. (Data in Gg = 1000 metric tons); estimates in ( ); [n] name-plate capacity; [a] 1977; [b] 1976.

Country	Production [2]	Production [5]	Import [5]	Export [5]	Apparent Consumption [5]
		Amounts in Gg/a			
Peru		8[n,a]			
Caribbean/Central America		0	101[b]	0	101[b]
Cuba		0	30	0	30
Ecuador		0	0	0	20
Bolivia		0	5	0	5
Africa					
South Africa		150	100	20	230
Other African Countries		100	430	0	530
Near East					
Israel	36	77	35	20	92
Turkey		65	217	52	230
Iran	18	45	225	—	270
Other Near East Countries	—	320	—		320
Far East					
Japan	5705	6748	193	1403	5538
Taiwan (Rep. of China)		794	61	269	586
South Korea	131	450	210	60	600
India	115	150	50	—	200
ASEAN-Countries	63[a]	140			510
Hongkong					200
People's Rep. of China	?				?
Australia	494	499	116	40	575
Total World	37499	55385			

Caribbean/Central America: Bahamas, Barbados, Bermuda, Costa Rica, Dominican Republic, El Salvador, Haiti, Honduras, Jamaica, Nicaragua, Panama.
ASEAN-Countries: Indonesia, Malaysia, Philippines, Singapore, Thailand.

of Germany, in Sweden, and in Finland (Table 1-3). Some high population countries, e.g., India, consume only 0.4 kg per capita and many African countries even less. Presumably, the expected increase in plastics consumption in the Third World countries will be covered by exports from OPEC countries. Their first plants went on stream between 1979 and 1982 and the production figures are not yet listed in the current statistics.

Although plastics in industrial countries are facing in 1983 the fourth straight year of no-growth, the consumption is likely to grow again because of the well-known advantages of plastics, such as cost benefits, lower energy consumption as compared to other materials, easy processability, low density, and other inherent properties (see also chapter 17). Presently, only 6% of the crude oil is consumed for the synthesis of polymers. The energy costs per production value and the energy consumption for manufacture are lower for plastics than for most other materials (Table 1-4).

TABLE 1-3
Annual Per Capita Plastics Consumption in 1978 (1 Gg/a = 1000 metric tons/year)

	Populations in Millions	App. Consumption in Gg/a	Annual Per Capita Consumption in kg
Western Europe			
West Germany	61	5 844	95
France	53	2 519	48
Italy	54	2 553	47
United Kingdom	56	2 710	48
Netherlands	13	450	35
Spain	34	1 152	34
Sweden	8.2	699	85
Austria	7.5	468	62
Finland	4.7	437	93
Yugoslavia	21	536	26
Greece	8.8	180	20
Switzerland	6.3	404	64
Norway	3.9	240	62
Portugal	8.6	229	27
Denmark	4.9	318	65
Ireland	3.4	123	36
North America			
United States	220	14 286	65
Canada	23	1 294	56
Mexico	48	400	8.3
Central and South America			
Brazil	92	1 000	11
Argentina	23	152	6.6
Caribbean/Central America	21	101	4.8
Cuba	8.6	30	3.5

TABLE 1-3 (*continued*)
Annual Per Capita Plastics Consumption in 1978 (1 Gg/a = 1000 metric tons/year)

	Populations in Millions	App. Consumption in Gg/a	Annual Per Capita Consumption in kg
Ecuador	6.5	20	3.1
Bolivia	4.6	5	1.1
Africa			
South Africa	22	230	10.5
Other African Countries	434	530	1.2
Near East			
Israel	3.2	92	28.8
Turkey	40	230	5.8
Iran	34	270	7.9
Far East			
Japan	112	5 538	49
Taiwan (Rep. of China)	12	586	48
South Korea	35	600	17
India	548	200	0.4
ASEAN-Countries	207	510	2.5
Hongkong	4	200	50
Australia	13.5	575	43

TABLE 1-4
Energy Consumption for the Production of Various Materials (NATO Science Committee)

Material	Density in g/cm^3	Energy Consumption in MJ/kg	kJ/cm^3	Energy Costs / Production Value
Lumber	0.5	4	2	0.1
Plastics	1.1	10	11	0.04
Cement	2.5	9	23	0.5
Paper	1.6	25	40	0.3
Glass	2.5	30−50	75−125	0.3
Magnesium	1.74	80−100	140−175	0.1
Aluminum	2.68	60−170	160−460	0.4
Steel	7.75	25−50	195−390	0.3
Copper	8.96	25−30	225−270	0.05

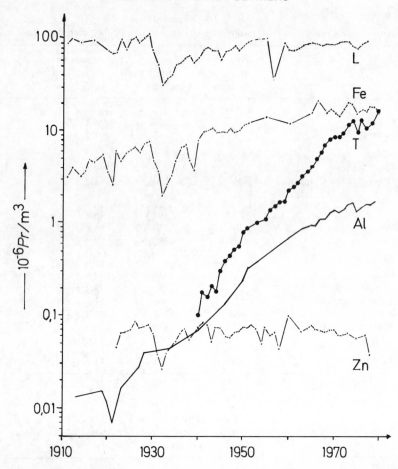

Figure 1-2 Annual United States productions, *Pr*, of lumber (L), raw steel (Fe), thermoplastics (T), aluminum (Al), and zinc (Zn).

Low density is one of the most important properties of plastics, especially for applications in transportation. Materials are sold by weight but used by volume. Quite informative in this respect are the U.S. statistics for the last 70 years (Figure 1-2). Since 1910 the consumption of lumber and zinc remained nearly constant although U.S. population grew from 92 to 230 million during the same time. The production of steel, aluminum and, above all, plastics grew

TABLE 1-5
Apparent Consumption of Plastics in the United States [4, 7] and Production in Japan [5] in the Years 1979–1981; s.b. = see below under "Other"

	US Consumption in Gg/a			Japanese Production in Gg/a		
	1979	1980	1981	1979	1980	1981
Thermoplastics by Addition Polymerization						
Poly(ethylene), low density	3552	3350	3417	1368	1179	1033
, high density	2224	1953	2177	797	681	638
Poly(propylene), incl. copolymers	1770	1647	1774	1023	927	959
Thermoplastic elastomers	139	154	169	s.b.	s.b.	s.b.
Poly(styrene), Standard	1817	1608	1653	694	646	658
, SAN	56	48	51	382	338	380
, ABS	563	441	441			
, mod. PPO	60	53	58	s.b.	s.b.	s.b.
, other (incl. foamable)	410	349	380	151	145	145
Poly(acrylates)	268	235	256	108	104	104
Poly(vinyl chloride), incl. copolymers	2790	2469	2551	1592	1430	1129
Other vinyl polymers	163	150	152	s.b.	s.b.	s.b.
Thermoplastics by Condensation Polymerization						
Polyamides	144	125	132	68	68	69
Polyacetals	48	43	42	s.b.	s.b.	s.b.
Polycarbonates	107	99	111	30	33	33
Polyesters, saturated	409	481	561	s.b.	s.b.	s.b.
Thermosets						
Alkyd Resins	225	210	230	s.b.	s.b.	s.b.
Epoxides	150	144	145	56	59	69
Phenolics	788	701	667	300	304	294
Polyesters, unsaturated	524	440	452	204	182	182
Polyurethanes	856	744	787	191	199	209
Urea and melamine resins	700	589	645	734	716	625
Cellulose derivatives	65	51	49	s.b.	s.b.	s.b.
Specialties (fluoroplastics, poly- (phenylene sulfide), polysulfones, poly(butene) etc.)	40	33	38	s.b.	s.b.	s.b.
Other	—	—	—	511	560	507
Total	17868	16117	16938	8209	7517	7038

faster than the population. The "lighter" materials, i.e., plastics and aluminum, witnessed the biggest growth and plastics production on a volume basis surpassed steel for the first time in 1980.

Changes will, however, occur for the production pattern of plastics. The growth rate of commodity plastics will remain low in highly industrialized countries, partly because of imports from the new OPEC plants, partly because of substitution of commodity plastics by specialty plastics. For example, the Japanese production of PVC and LDPE dropped by 29% and 25%, respectively, between 1979 and 1981. During the same time, and in spite of a general setback in production, the U.S. consumption of specialty plastics such as saturated polyesters (mainly for bottles) and thermoplastic elastomers showed marked increases.

The shift from commodity to specialty plastics is a matter of quality and quantity. In this book, 118 new polymers are reported that were announced between 1975 and 1981, compared to only 64 new polymers previously described [8] for the period from 1969 to 1974. The quoted numbers should not be taken as absolute ones, since some of the new developments probably went unnoticed and the term "new" itself is also somewhat ambiguous. However, the figures indicate an increase of annually introduced new polymers from ca. 13 to ca. 17. They also reveal the shares of individual countries (Table 1-6). The U.S. fell back although its number of 8 new polymers per year remained constant; the Federal Republic of Ger-

TABLE 1-6
New commercial polymers and change of innovation rates per country

	Number of New commercial polymers		New polymers as % of total	
	1969–1974 [8]	1975–1981 [9]	1969–1974	1975–1981
USA	40	56	62	48
Japan	4	25	6	21
West-Germany	10	19	16	16
Other West-European Countries	10	13	16	11
East European Countries	0	5	0	4
Total	64	118	100	100

many kept its relative share of innovations; Japan increased its share of new polymers considerably. It is remarkable also that two Eastern block countries, i.e., the Soviet Union and the Democratic Republic of Germany, appear on the list for the first time.

Whether and how quickly the new polymers will secure a market depends, of course, on many factors such as price, demand, properties, availability of feed stocks, and also on political and environmental aspects. The acrylonitrile resins with barrier properties that were introduced in the early 70's [8] failed to obtain FDA approval for food packaging because the residual monomer was suspected to be a carcinogen. In the U.S. these resins were withdrawn from the market and were quickly replaced by the newly developed bottle-grade poly(ethylene terephthalates). In spite of the general recession, the U.S. trade in saturated polyesters climbed by 37% in 1979–1981. Thermoplastic elastomers also showed a remarkable production increase (Table 1-5). Since this class of polymers offers many structural modifications along with corresponding property changes, one may anticipate quite a number of new products to appear within the next few years. Also, heterocyclic polymers have not been fully exploited with regard to possible chemical structures. Finally, new concepts will presumably be applied to the strategy of synthesizing new polymers; these will be dealt with in chapter 17.

1.2 NOMENCLATURE AND DESIGNATIONS OF POLYMERIC MATERIALS

New polymers appear on the market under all kinds of names. Only in rare cases is the systematic structural name according to the IUPAC nomenclature [10] stated, partly because it is cumbersome to use, partly because companies sometimes attempt to conceal the chemical identity of the new polymer. Thus, we omitted structural names from this book.

In many cases a manufacturer assigns a trivial name to the new polymer. For the sake of the reader's familiarity with those names, we kept to them whenever it seemed practical. Very frequently the new polymer is only known by a trade name. In some cases abbreviations are stated. Those which have been used throughout this book are listed in Table A-5 of the Appendix. The voluminous "complete" list [11] with approximately 700 abbreviations internationally used for thermoplastics, thermosets, elastomers, fibers, and additives was not included.

Indeed, the nomenclature of abbreviations is in bad shape. Identical acronyms are used for different materials, and *vice versa*, different acronyms for identical materials [11]. Not only do every country and each professional organization feel the call of duty to invent their own (and, of course, different) abbreviations, but the subdivisions of the same organization often lack any cooperation. The International Standardization Organization (ISO) is commissioned to standardize scientific and technical terms, test procedures, etc., on an international basis. Therefore, it seems grotesque that the ISO-Commission TC 38 (Textiles) refuses to accept the acronyms of Commission TC 61 (Plastics), whereas Commission TC 45 (Elastomers), of course, maintains its own system. Moreover, some historically developed but non-systematic designations often have to be replaced by more systematic ones. Since prefixes to an abbreviation are prohibited as ruled by ISO 1043, the familiar abbreviation HDPE shall replaced by PE-HD, HIPS by PS-HI, and so on [12].

Substantial progress will be achieved by the introduction of standardized designations for thermoplastic molding materials [12]. In the future, such materials will be characterized through a system with a description and an identity block, the latter consisting of a so-called standard number-block and five data blocks. The description block indicates a molding material and the standard number block points out the corresponding national standard for this material. The first of the five data blocks identifies the thermoplast, e.g., PE, the second block discloses applications (e.g., A = Adhesive) and additives, the third and fourth blocks give the codes for quantitative properties and for fillers and reinforcing materials, respectively, The fifth block is reserved for specifications. It is to be hoped that such a system will be adopted not only for thermoplastic materials but also for other polymeric systems.

Far better than 7 years ago is the position with regard to physical properties. With the exception of a few U.S. companies all others now list property data in SI units. SI units and their conversion into old or Anglo-Saxon units are listed in the Appendix. One notable exception is the permeability of liquids and gases where misleading and non-systematic physical units are still in use [13].

References

[1] Chemical Economics Handbook, Stanford Research Institute, Menlo Park, CA 1974.

[2] Statistical Yearbook 1979–1980, United Nations, New York, NY.

[3] Anon., Chem. Engng. News (June 11, 1984) 39.

[4] Anon., Modern Plastics **59**/1, 77 (1982).

[5] Anon., Kunststoffe **73**/10 (1983).

[6] Data from Kunststoffe **69**, 619 (1979).

[7] Anon., Modern Plastics **58**/1, 67 (1981).

[8] H.-G. Elias, Neue polymere Werkstoffe 1969–1974, Hanser, Munich 1975; New Commercial Polymers 1969–1975, Gordon and Breach, New York 1977.

[9] H.-G. Elias and F. Vohwinkel, this book.

[10] International Union of Pure and Applied Chemistry, Macromolecular Nomenclature Commission, Nomenclature of regular single-strand organic polymers, Macromolecules **6**, 149 (1973); ibid., J. Polymer Sci.-Polym. Letters Ed. **11**, 389 (1973); IUPAC Commission on Macromolecular Nomenclature, Nomenclature for Regular Single-Strand and Quasi-Single-Strand Inorganic and Coordination Polymers, Provisional Recommendations 1980.

[11] H.-G. Elias et. al., Nomenclature Committee, Division of Polymer Chemistry Inc., American Chemical Society, Abbreviations for Thermoplastics, Thermosets, Fibers, Elastomers, and Additives, Polymer News, **9**, 101 (1983); ibid., Part II, **10**, 169 (1984/1985).

[12] K. Wiebusch, Kunststoffe **72**, 168 (1982); see ISO/TC 61/SC 9 N 435 (April 1981).

[13] M. B. Huglin and M. T. Zakaria, Angew. Makromol. Chem. **117**, 1 (1983).

2 Saturated Carbon Chains

2.1 LINEAR LOW DENSITY POLY(ETHYLENE) LLDPE

2.1.1 Structure and synthesis

A property gap exists between the low density poly(ethylenes) (LDPE) produced radically under high pressure and the high density poly(ethylenes) (HDPE) produced under low pressure with organometallic catalysts. This gap has been closed by the new linear poly(ethylenes) with low or medium density (LLDPE or 1-LDPE). The simplified structures of the three poly(ethylenes) can be given as follows:

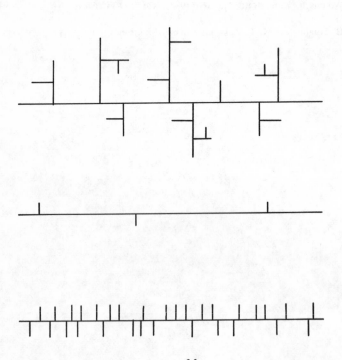

14

The linear HDPE contains less than 5 rather short branches per 1000 carbon atoms, whereas LDPE has between 8 and 40 long and highly developed branches per 1000 chain atoms. The new LLDPE-types are copolymers of ethylene with normally 8−10% of an alpha-olefin such as butene-1, pentene-1, hexene-1 or octene-1. The statistically incorporated comonomers create more and longer branches than in HDPE.

LLDPE is not an entirely new product. The first grades appeared on the market in the sixties but were withdrawn due to problems connected with production and processing. A new generation of LLDPE was introduced around 1978, offering economic advantages based on improved polymerization techniques and on unique properties that were readily accepted by the processors.

At present, there are relatively few producers of LLDPE (Table 2-1). Some companies are still developing their own technology or have obtained a license. Union Carbide Corp. is the leading licensor for its Unipol process (Table 2-2).

LLDPE can be obtained via solution [2], suspension [3, 11], or gas phase [4] polymerization processes. The continuous polymerization with transition metal catalysts is carried out at temperatures between 25 and 300°C and under pressures up to 70 bar, whereas the suspension process works at temperatures between 54 and 80°C and under pressures of about 5 bar [5, 11]. Light hydrocarbons, such as isobutane or propane, serve as diluent. In contrast to the older HDPE processes, no purification step for the diluent recycle stream is necessary [11]. Conversion of ethylene amounts to 50−100% per

TABLE 2-1
LLDPE Producers.

Company	Trade Name	Capacity in t/a	Comonomer	Process
UCC USA	G-Resin	570 000	Butene-1	Gas phase
UCC Canada		80 000	Butene-1	Gas phase
Dow Chemical	Dowlex	350 000	Octene-1, Butene-1	Solution
DuPont Canada	Sclair	112 000	C_5 and higher	Solution
CdF	Lotrex	50 000	Butene-1	High Pressure
DSM	Stamylex	50 000	Octene-1, Butene-1	Solution
Unifos Kemi/UCC		20 000	Butene-1	Gas phase
El Paso/Rexene		10 000	Butene-1	Suspension

TABLE 2-2
LLDPE Technology under development and Licensees for LLDPE
Processes (updated from [1]).

LLDPE Technology under development

Company	Process	Comonomer
USI	Suspension	Butene-1
Phillips	Suspension	Butene-1
Arco	High pressure	Butene-1
Amoco	Gas phase	Butene-1
Cities Service	Gas phase	Butene-1
Dow Chemical	Gas phase	Butene-1
Solvay	Suspension	
DSM	Solution	
Naphthachimie	Gas phase	
BASF	Gas phase	
ICI	Gas phase	
Imhausen Chemie	Gas phase	
Sumitomo	Solution	
Mitsui Petrochemicals	Solution	
Nippon Petrochemicals	?	
Showa Denko	Suspension	
Chem. Werke Hüls	?	
Gulf Oil	?	
Mobil Chemical	UCC	Butene-1
Exxon	UCC	Butene-1
Enesco (Canada)	UCC	Butene-1
Mitsubishi Petrochemicals	UCC	Butene-1
Nippon Unicar	UCC	Butene-1
SABIC [a]/UCC	UCC	Butene-1
SABIC/Exxon	UCC	Butene-1
SABIC/Dow/Mitsubishi	Dow	
ICI Australia	DuPont	C_5 and higher
Idemitsu Petrochemicals	DSM	
Asahi Chemical Industries	DuPont	C_5 and higher
Cities Service	ICI	

[a] SABIC = Saudi Arabian Basic Industries.

reactor pass. The molecular weight is regulated through the polymerization temperature and addition of proton donators, e.g.,
alcohols, carbonic acids, or hydrogen [6, 7].

In the suspension and gas phase processes, the polymerization
temperature is confined to the rather narrow range between 50 to
110°C. The upper temperature limit is given by the equilibrium solu-

tion temperature or the melting temperature of poly(ethylene), the lower temperature limit by the diminishing catalyst activity. Chain terminators are, therefore, needed in these processes. Hydrogen reportedly exerts the best effect [7]. Octene-1 with its high boiling temperature of 121.3°C is not suitable for gas phase processes.

In solution polymerization processes, the molecular weight can be adjusted through the variation of polymerization temperatures. A broad one-phase region exists above the equilibrium solution temperature where poly(ethylene) and aliphatic hydrocarbons are completely miscible. The range is limited towards higher temperatures by the lower critical solution temperature which depends strongly on the chain length of the aliphatic hydrocarbons [8, 9].

Economic advantages result from new, highly active catalysts, e.g., $TiCl_4/Mg(OC_2H_5)_2/AlR_3$ complexes [7] or bis(triphenylsilyl)-chromate/R_2AlOR on aluminum oxide/silicate supports [10].

The new catalysts developed by Montedison have high specific surface areas of $10-100$ m^2/g, ensuring yields of 20 000 g LLDPE per g of catalyst or 1 million g of LLDPE per g of Ti. Catalyst residues in the finished resin are insignificant, about $1-5$ ppm titanium and $10-30$ ppm chlorine [11]. The LLDPE produced with such catalysts in suspension displays highly uniform spherical particles with $400-2000$ μm diameter and a high bulk density of 0.43 g/cm^3. The average particle size for LLDPE from a gas phase polymerization is only 150 μm.

The high catalyst activity requires only low catalyst concentrations and thus eliminates the need for the removal of catalyst residues by after-treatment of the polymers. The production costs for poly-(ethylene) are cut by approximately 20% as compared to conventional suspension polymerization [12]. Gas phase processes are even more economical since they do not require solvent separation, purification, and recycling, monomer stripping, possibly pelletizing, and drying at higher temperatures. The Unipol process offers additional economic advantages through increased reactor capacities of up to 65% [16].

Union Carbide Corp. offers its new Unipol process to licensees for the construction of new plants as well as the conversion of existing low and high pressure polymerization plants. Considerable savings are claimed because of lower investment and energy costs (Table 2-3).

Energy costs of the old Unipol process are only 1/4 compared to

TABLE 2-3
Investments for New Plants or Conversion of Old PE Plants [16]. (1 pound = 454 g).

Polymerization Process	Project	Estimated Investment in US-cents per pound of annual capacity
High pressure/solution	New plant	>20
Low pressure/solution	New plant	>20
Low pressure/gas phase old Unipol process	New plant	ca. 10
High pressure/solution	Conversion from LD to LLDPE with old reactors	<10
Low pressure/gas phase new Unipol process	New plant	ca. 6
ditto	Conversion from LD to LLDPE with old reactors	≥3
ditto	Capacity increase of old Unipol process by 35–65%	ca. 3
ditto	Capacity increase of old Unipol process by 1–35%	≤2

the conventional high pressure process for LDPE. They are reduced to only 1/6 to 1/9 with the new Unipol process [16]. A Unipol plant requires only 1/10 of the area of a high pressure PE plant [7]. On the other hand, all LLDPE plants have additional costs through the use of comonomers; for some LLDPE types up to 19% of comonomers are needed. The solution as well as the gas phase processes are reportedly so flexible as to allow the conversion of the polymerization from LLDPE to HDPE and vice versa in the same plant.

2.1.2 Properties, processing, and applications

With regard to properties, LLDPE not only differs from LDPE but also in its various grades. The properties depend on comonomer type and concentration, average molecular weight, and molecular weight distribution.

The densities of LLDPE are in the range of 0.918–0.94 g/cm^3 depending on type and concentration of the comonomer (Figure 2-1). Ethyl and hexyl groups reduce the densities of the copolymers substantially more than methyl groups [5, 13]. The new, modified

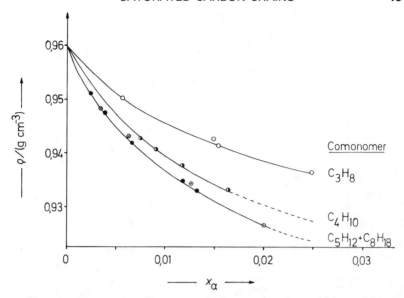

Figure 2-1 Density of copolymers from ethylene and α-olefins as function of the amount of α-olefin units in the copolymer (data of [5]).

catalyst systems achieve a more uniform distribution of the comonomers along the chain than the conventional $TiCl_3/AlR_3$-catalysts. The melting temperatures of the copolymers (Figure 2-2) depend only slightly on the density and tend to correspond with HDPE. The influence of the comonomer, i.e., the length of the side chain, is small. The melting temperatures of the comonomers are about 10–20°C higher than those of LDPE's.

The tensile strengths at yield depend only on the densities and not on the chemical structure, molecular weight or molecular weight distribution (Figure 2-3). In this respect, LLDPE performs like a blend of LDPE and HDPE [5]. The tensile strength at break increases with increasing molecular weight; LLDPE surpasses LDPE. Polymers with broad molecular weight distributions exhibit markedly lower tensile strengths at break than those with narrow molecular weight distributions. The impact tensile strength is lowered with increasing density and increasing molecular weight (Figure 2-4) but the influence of branching is only noticeable at high molecular weights above about 120 000 g/mol. Long chain comonomers yield significantly higher impact tensile strengths than the short chain

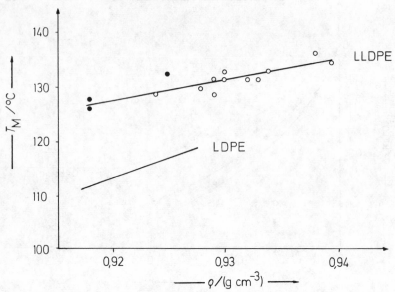

Figure 2-2 Melting temperatures as function of density for copolymers from ethylene and butene-1 (○) or ethylene and hexene-1 (●) as compared to those of free radical polymerized low density poly(ethylene) LDPE (data of [5]).

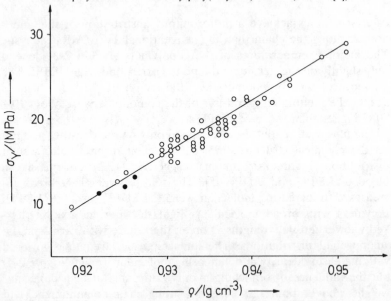

Figure 2-3 Tensile strength at yield of LLDPE (○) and LDPE (●) as function of polymer density (data of [5]).

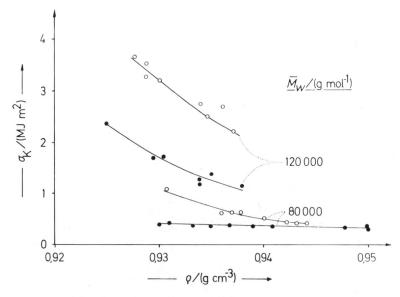

Figure 2-4 Impact tensile strength of LLDPE with various weight average molar masses and short chain (●) or long chain (○) α-olefin comonomers as function of the density of the polymer (data of [5]).

types. The values are comparable to those of high molecular weight HDPE, however, they can only be achieved by polymers with very narrow molecular weight distributions [5]. LLDPE has a markedly narrower molecular weight distribution than LDPE, the $\overline{M}_w/\overline{M}_n$-ratios for LLDPE and LDPE being 3 and 8−10, respectively [14].

The elongation characteristics of LLDPE copolymers are quite different from those of other PEs. Polymers with short side chains display higher elongations at break than polymers with long side chains. The elongations for copolymers with a density of 0.930 g/cm^3 with short or long side chains are 1000% resp. 700−800% compared to LDPE with 600% and HDPE with 500−800%. On impact, the elongation of the polymers decreases with increasing density (Figure 2-5) and longer side chains impart superior elongation characteristics than short chains. Thus the performance of the copolymers is quite different under slow (elongation) or fast (impact tensile strength) deformation [5].

In contrast to LDPE there are only a few grades of LLDPE on the market. They are being offered for film blowing and casting,

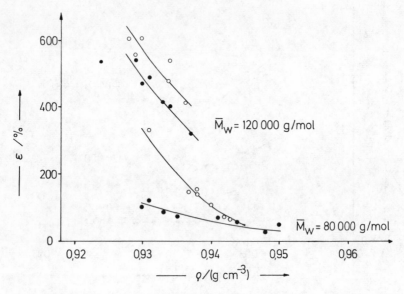

Figure 2-5 Elongation under impact tensile stren for LLDPE with short chain (●) and long chain (○) α-olefin comonomers as function of the density for various weight average molar masses (data of [5]).

injection and rotomolding [15]. Newer developments have led to materials with high transparency, for shrink films, for heavy-duty bags and for extrusion coating [16].

The major portion of LLDPE is used in film blowing which has been the domain of LDPE. Converting a film blowing unit from LDPE to LLDPE requires some alteration of the equipment and of processing. At low shear rates, LLDPE has a lower melt viscosity than LDPE, however, within the range of $200-1000$ s^{-1} normally applied in film blowing, the higher melt viscosity of LLDPE causes a higher back pressure, more friction and a higher motor load. A number of modifications have been recommended [14, 17, 18]: screws with an L/D ratio below 25, no cooling of the feeding zone, feed zone channel depth adjusted to maximum torque and motor power, deeper and/or shorter metering section, larger die gaps, special film cooling systems to improve bubble stability. With these modifications the shear action is reduced and overheating and melt fracture can be avoided. The low melt tension and high melt elasticity of LLDPE allow an increase of the ratio die gap/film thickness

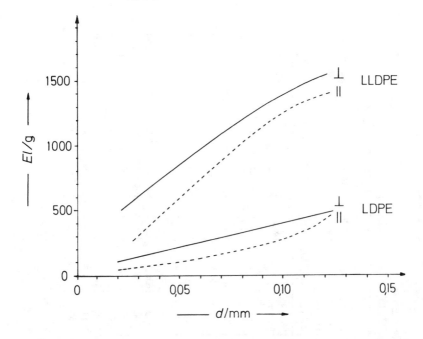

Figure 2-6 Film tear resistance (Elmendorf) in machine direction (‖) and transverse to machine direction (⊥) as function of film thickness for various LLDPE and LDPE films (data of [15]).

up to 250:1, i.e., an fivefold increase compared to LDPE. The LLDPE films are very insensitive to gels and other hard particles, which do not cause bubble breaks and perforations even in thin film gauges. The equally high biaxial elongation properties of LLDPE film ensure an exceptional puncture resistance. With LLDPE the processor has the choice of producing films with improved properties compared to LDPE films with equal thickness, or of maintaining the LDPE film properties by down-gauging LLDPE by 50% (Figure 2-6). At equal thickness, the LLDPE films offer better impact resistance, puncture resistance, higher tensile strength and ultimate elongation than to LDPE films (Table 2-4).

Limited quantities of LLDPE have been available since 1982. It has become a popular practice among film blowing processors to mix LDPE with 15−20% of LLDPE without having to modify the equipment. The small amounts of LLDPE improve the mechanical

TABLE 2-4
Properties of Films of LLDPE, LDPE, and HDPE [16].

Physical Property	Physical Unit	LLDPE Carbide G-Resin 7047	LLDPE Carbide G-Resin 7040	LLDPE Dow Dowlex 2045	High-clarity Grades				High-strength Grades	
					LLDPE[a]		LDPE[b]	HDPE[a]	LLDPE[a]	LDPE[b]
Density	g/cm^3	0.918	0.918	0.919	0.918		0.924	0.950	0.918	0.919
Melt index	g/(10 min)	1.0	2.0	1.0	1.0		2.0	1.0	0.5	0.2
Thickness	mils	1.5	1.5	1.5	1.5	1.0	1.5	1.25	1.5	1.5
	mm	0.038	0.038	0.038	0.038	0.0254	0.038	0.032	0.038	0.038
Impact tensile strength, Dart	g	145	110	190	150	100	90		300	180
Puncture resistance	J/mm	66.8	62.3	62.3	57.9	57.9	13.4		97.9	66.7
Tensile strength										
in machine direction	MPa	40.0	33.8	43.4	36.5	38.6	24.1	43.4	48.3	25.5
transverse to machine direction	MPa	31.7	24.8	34.5	29.6	31.7	19.3	29.0	40.7	22.8
Extensibility										
in machine direction	%	600	690	580	600	580	310	550	590	250
transverse to machine direction	%	760	740	700	700	680	580	650	700	560
Tear Strength (Elmendorf)										
in machine direction	g	248	180	480	210			35	480	120
transverse to machine direction	g	548	465	1050	510			750	1000	130
Turbidity	%	9.0	14.0	14	5.5	5.0	5.0	8.0	9	18
Gloss (Gardner 45°)		70	60	55	72	72	70	60	63	40

[a] Union Carbide, start of production in 1982; [b] commercially available conventional grades.

TABLE 2-5

Comparison of properties of low pressure LLDPE and IDPE with EVAC, LDPE, and EEA/HDPE blends [6].

Physical Property	Physical Unit	LLDPE	EEA/HDPE Blend	EVAC Copolymer	LDPE	IDPE	EEA/HDPE Blend	IDPE
Density	g/cm^3	0.945	0.945	0.933	0.925	0.936	0.934	0.935
Melt index	g/10 min	25	8	1.5	1.5	2.5	5.4	4.1
dto., I_{10}/I_2	—	6.5	10				7.5	7
Ethylacrylate content	%	0	5	0	0	0	5	0
Vinylacetate content	%	0	0	4	0	0	0	0
Yield Strength	MPa	11.2	13.4	12.9	9.7	16.8	14.3	15.7
Tensile strength at break	MPa	9.7	9.5	11.0	10.0	>29.5	8.8	16.6
Elongation at break	%	710	460	500	450	>960	450	790
Flex. (2% Secant)	GPa	0.310	0.426	0.344	0.241	0.524	0.448	0.496
Impact strength (Dart) at −29°C	N	—	—	142	89	222	224	285
Notched impact strength at 25°C	kJ/m	0.454	0.080	—	—	—	0.293	0.166
at −50°C	kJ/m	0.043	0.027	0.059	0.043	0.080	0.064	0.075
Stress corrosion resistance (Bell strip test (F_{50}))	h	20	6.5	50	20	>1000	14	250

and heat sealing properties. Alternatively, film thickness can be reduced.

Injection molded articles from LLDPE show improved toughness, lower brittle temperatures, higher heat deflection temperatures and reduced stress cracking sensitivity [15]. LLDPE and linear low pressure poly(ethylenes) with intermediate densities (IDPE) can compete with EVAC copolymers, blends of LDPE/HDPE, ethylene/ethylacrylate copolymers/HDPE blends, and with other elasticized HDPE compounds (Table 2-5) [6].

Relatively recent applications of LLDPE are rotational molded agricultural and chemical tanks, shipping drums, water storage and holding tanks, waste drums, sports equipment, and toys [19, 20]. Compared to LDPE, rotational molding of LLDPE gives products with 50% higher part stiffness or, alternatively, parts of equal stiffness but 25% less wall thickness. LLDPE can replace crosslinked PE in the rotational molding of large objects, such as canoes or kayaks. Compared to crosslinked PE, LLDPE cuts cycle times and permits reuse of waste material. It is approved by the FDA for contact with food.

Optimistic prognoses claim that LLDPE will eventually replace the use of approximately 70% of conventional LDPE and 20% of HDPE. For the European market, an increase of 39.5% in LLDPE production is forecast for 1980−1990 [11, 16, 21]. Many new production facilities have recently come on stream [22] (see also Table 1-1). Recent developments are described in a number of review and product information articles ([23]−[29]).

The newest development is a "very low density polyethylene" (VLDPE) with a density of ca. 0.900 g/cm^3 [34]. These VLDPE's are available from Union Carbide under the trade name Ucar Flx. The polymers are expected to compete with ethylene/vinyl acetate copolymers; they are recommended for squeeze tubes and bottles, liners, closures, films, hoses, and toys.

References

[1] Anon., Chemical Engineering, (Aug. 24, 1981) page 47.
[2] R. R. Cooper and K. S. Whiteley (ICI), USP. 4133944 (1979).
[3] T. K. Shiomura, Y. T. Kinoshita, and I. Y. Kokue (Mitsui Toatsu Chemicals Inc.), DOS 2714743 (1977).
[4] I. J. Levine and F. J. Karvel (Union Carbide Co.), DOS 2609889 (1976).

[5] W. Payer, W. Wicke, and B. Cornils, Angew. Makromol. Chem. **94** (1981) 49.

[6] J. A. Cook and St. E. Lepper, Modern Plast. **58** (March 1981) 72.

[7] L. L. Boehm, Angew. Makromol. Chem. **89** (1980) 1.

[8] H. Horacek, Makromol. Chem. Suppl. **1** (1975) 415.

[9] F. Hamada, K. Fujisawa, and A. Nakajima, Polym. J. **4** (1973) 316.

[10] W. L. Carrick, R. J. Turbett, F. J. Karol, G. L. Karapinka, A. S. Fox, and R. N. Johnson, J. Polym. Sci. [A-1] **10** (1972) 2609.

[11] C. Cipriani and C. A. Trischman, Jr., Chemical Engineering, (May 17, 1982) 66.

[12] B. Diedrich, Appl. Polym. Symp. **26** (1975) 1.

[13] J. P. Hogan and B. R. Witt, ACS Div. Petr. Chem., Preprints **24** (1979) 2.

[14] Uniforum, Technical Bulletin, Unifos Kemi AB, 1982.

[15] ®Dowlex Polyethylene Resins, product bulletin, Dow Chemical Co., 1981.

[16] R. Martino, Modern Plast. Internat. **12** (April 1982) 44.

[17] Anon., Modern Plast. Internat. **10** (July 1980) 30.

[18] Anon., Modern Plast. Internat. **11** (November 1981) 50.

[19] J. Sneller, Modern Plast. Internat. **10** (October 1980) 55; Modern Plast. **57** (July 1980) 42.

[20] M. C. Restaino, Plastics Engineering **37** (July 1981) 39.

[21] Chem. Systems International, in Chem. Engng. News, (March 22, 1982) 20.

[22] Anon., Kunststoffe 74, 761 (1984).

[23] B. H. Pickover, Adv. Polym. Technol. **3**, 271 (1984).

[24] N. Platzer, Ind. Eng. Chem. Prod. Res. Dev. **22**, 158 (1983).

[25] R. Glaser, Kunststoffe **74**, 543 (1984).

[26] S. Bork, Kunststoffe **74**, 474 (1984).

[27] Anon., Modern Plastics Internat. —, 7 (February 1984).

[28] H. Leder, Kunststoffe **73**, 251 (1983).

[29] Anon., Kunststoffe **73**, 468 (1983).

[30] J. C. W. Chien, J. C. Wu and C. I. Kuo, J. Polym. Sci. Polym. Chem. Ed. **20**, 2019, 2445 (1982).

[31] F. J. Karol, ACS Polymer Preprints **24**/1, 107 (1983).

[32] P. Galli, P. C. Barbe und L. Noristi, Angew. Makromol. Chem. **120**, 73 (1984).

[33] R. P. Quirk, ed., Transition Metal Catalyzed Polymerizations: Alkenes and Dienes, Harwood Academic Publishers, Chur 1983.

[34] Anon., Modern Plastics 13 (October 1984).

2.2 POLY(OLEFIN-CO-VINYL ALCOHOL)

2.2.1 Structure and synthesis

The saponification of copolymers from olefins (mainly ethylene) and vinyl acetate yields copolymers with olefin and vinyl alcohol units. Such copolymers are marketed by the Nippon Synthetic Chemical Industry as ®GL-Resin, by the Kuraray Co. Ltd. as ®EVAL-Resins, by Bayer A. G. ®Levasint, and by Deutsche Solvay as ®Clarene [8].

GL-Resins are engineering plastics [1−3]. A non-reinforced grade GL-N and two glass fiber reinforced grades GL-G 25 and GL-G 40 with glass contents of 25 and 40% respectively are available. Eval-resins are offered as Resin F, as high-flow Resin E and as films. They are used in packaging applications [4, 5]. Levasint is available only as powder designed for fluidized bed coating, more recently also for electrostatic spraying.

For the production of Eval, ethylene and vinyl acetate are polymerized continuously in solution with a conventional radical initiator under pressure to give copolymers with 20−50 mol-% ethylene units. Residual monomers, by-products and the solvent are stripped off. Care has to be taken to avoid the formation of poly(vinyl alcohol) because it is incompatible with Eval and would form fisheyes in films. The saponification is carried out in a tower reactor with a solution of NaOH in methanol to yield a degree of saponification above 99 mol-%. The solution of Eval in a mixture of methanol and water is extruded into cold water, and the porous granulates obtained are washed to remove sodium acetate and other impurities. A capacity of 3000 t/a was reported for the Kuraray plant in 1976 [4].

2.2.2 Properties and applications

GL and Eval resins contain more vinyl alcohol units than Levasint. Densities, stiffness and moduli are higher, while the notched impact resistance is lower compared to Levasint (Table 2-6). When compared with polycarbonates, polyamide 6 and polyacetals, the GL resins demonstrate better tensile and flexural strengths and creep resistance [3].

Transparent Eval films show considerably lower permeabilities for oxygen and water vapor than other films (Table 2-7). These barrier properties are utilized in packaging, e.g., in co-extruded blow-molded containers for food and cosmetics.

Levasint flows better than GL and Eval resins and facilitates film formation when coating hot surfaces. Levasint coatings are elastic and withstand impact tests at −40°C without fracture. No surface damage is observed after 1000 h in a salt spray test, after 4000 h in the Fade-O-meter, or after 8 years of natural weathering in an industrial climate.

TABLE 2-6
Properties of Poly(olefin-co-vinyl alcohols).

Physical Property	Physical Unit	Property values of					
		GI-N	GL-G 25	GL-G 40	Eval F	Eval E	Levasint
Glass fiber content	%	—	25	40	—	—	—
Density	g/cm^3	1.215	1.41	1.52	1.19	1.14	0.97
Melt temperature	°C	186	186	186	180	162	105–108
Melt index (190°C; 2.16 kg)	g/10 min				1.5	5.8	95
Tensile yield strength	MPa	91			74	46	
Tensile break strength	MPa	40	167	212	67	43	13
Elongation at yield	%				5	4	
Elongation at break	%		8.5	8.2	279	313	
Modulus of elasticity	GPa	44.1	9.8	13	3.5	2.6	0.5
Flexural strength	MPa	130	230	300	120		19
Flexural modulus	GPa	3.9	10.3	12.9	3.6		
Impact strength with notch	kJ/m	0.039	0.039	0.039	0.030	0.027	1.3(−40°C)
Hardness (Rockwell R)		120	120	120	87	68	Shore A 95
Linear thermal expansion coefficient $(\times 10^5)$	K^{-1}	4.1	1.6	0.94	11	13	13.7
Vicat temperature	°C	185	186	186			42
Heat distortion temperature (30 min without load)	°C						
Volume resistivity	Ohm × cm	2.9×10^{14}	5.7×10^{15}				10^{15}
Dielectric strength	Ohm	6.6×10^{12}	1.8×10^{15}				
Surface resistivity	MV/m	30	34				41
Relative permittivity	1	6.2	6.0	4.9			3.6
Arc resistance	s	125	125				
Water absorption (24 h)	%	7.5	6.2	6.5	6.5	6.5	0.19

TABLE 2-7
Properties of Plastic Films [4].

Physical Property	Physical Unit		Property values of								
		Eval F	Cello-phane	PVDC/ Cello-phane	PP, biaxially oriented	LDPE	HDPE	PVDC	Poly-ester	Nylon, biaxially oriented
Thickness	μm	15	20	25	20	30	20	30	12	15
Melting temperature	°C	180	—	—	175	105–115	137	150–160	260	223
Tensile break strength in machine direction	MPa	80.4	135.3	72.6	176.6	29.2	25.5	84.3	173.6	213.8
transv. to machine direction	MPa	52.0	82.4	37.3	292.2	20.6	13.6	100	141.2	175.5
Elongation at break in machine direction	%	160	15	24	112	203	311	38	101	64
transv. to machine direction	%	200	42	72	50	497	37	52	58	61
Modulus of elasticity in machine direction	GPa	1.96	3.06	3.14	1.21	0.21	0.76	0.32	2.16	1.61
Moisture regain	%	3.8	15.0	11.4	0.1	0.1	0.1	0.1	0.1	
Moisture permeability (10^{17} P)	m² 24 h 30 μm	50	2030	9.7	5.0	16.3	7.9	10.9	22.1	134
Oxygen permeability (10^{17} P)	cm³ s g^{-1}	0.15	15–7500	1.2	578	2025	1868	7.5–37.5	18	28.5
Haziness	%	1.8	1.5	2–3.1	1.1	12.0	75.5	6.0	2.5	2.0

The oxygen permeability was measured with dry films and the moisture permeability at 40°C and 90% relative humidity. All other data were obtained at 20°C and 65% relative humidity.

GL and Eval resins are hydrophilic whereas Levasint is rather hydrophobic due to its higher content of ethylene units. GL and Eval resins absorb weak acids or alkaline and salt solutions but the mechanical properties are only slightly affected. The resins are attacked by methanol but not by aromatic or halogenated hydrocarbons, ketones, ethers, or higher alcohols. Levasint is also very resistant against inorganic media but is attacked by aromatic and halogenated hydrocarbons, ethers, methanol, and benzyl alcohol but not by greases and fats.

References

[1] H. Kawaguchi and T. Iwanami, Japan Plast. **8**/6 (1974) 6.
[2] H. Kawaguchi and T. Iwanami, Japan Plast. **9**/1 (1975) 11.
[3] H.-G. Elias, New Commercial Polymers, 1969−1975, Gordon and Breach, New York 1977.
[4] H. Iwasaki, K. Sato, K. Akao, and K. Watanabe, Chem. Economy Engng. Rev. **9**/10 (1977) 32.
[5] Eval Resin, Company Literature, Kuraray Co. Ltd.
[6] Levasint, Company Literature, Bayer AG.
[7] H. Blau, Kunststoffber. **23** (1978) 686.
[8] Anon., Kunststoffe **74**, 703 (1984).
[9] H. Burgdörfer, Kunststoffe **74**, 329 (1984).

2.3 POLY(ETHYLENE-CO-ACRYLIC ACID METHYL ESTER)

2.3.1 Structure and synthesis

Since 1976 DuPont has marketed a copolymer under the name of ®Vamac. It contains equimolar portions of ethylene and methyl acrylate and a small amount of a termonomer with a carboxylic group [1, 2].

$$\text{+CH}_2\text{—CH}_2\text{+} / \text{+CH}_2\text{—CH+} / \text{+CHR—CR}^1\text{+} \tag{I}$$
$$\underset{\text{COOCH}_3}{|} \qquad \underset{\text{COOH}}{|}$$

The carboxylic group of the termonomer enables the vulcanization of Vamac by reaction with diamines [4]. Presumably, the polymer is produced by polymerization in emulsion. So far, it has only been available as carbon black filled masterbatch B-124 and as mineral filled grade N-123.

2.3.2 Properties and applications

Masterbatch B-124 contains 24 parts SRF carbon black and additives per hundred parts of polymer. The density is 1.12 g/cm^3 and it has a Mooney viscosity (ML 1 + 4) of 20. Masterbatch N-123 contains 23 parts of silica phr polymer, density is 1.18 g/cm^3 and Mooney viscosity is 30.

Vamac can be crosslinked by primary diamines or faster with hexamethylene diaminecarbamate (®Diac 1, DuPont). The vulcanization by peroxides is confined to compounds free of carbon black. Compounding recipes and the effect of various types and concentrations of carbon black on vulcanizate properties have been described [5].

Vamac was designed to close a property gap in the category of heat and oil resistant elastomers [3, 5, 6]. Elastomers are grouped by ASTM D-2000 and SAE J-200 according to their heat and oil resistance into classes A-H and A-K respectively. Vamac is a EE-EH class elastomer, i.e. it is usable up to 175°C (E) and the volume swell values are between 20% (H) and 80% (E) after 70 h in an aromatic oil (ASTM oil no. 3), depending on compound formulations. With regard to heat resistance, Vamac is only surpassed by the more expensive fluorelastomers and silicones (Figure 2-7) but it displays in general better mechanical properties than those. Isothermal aging shows a 50% loss of elongation at temperature/time [3] of

18 months at 121°C
6 months at 150°C
6 weeks at 170°C
4 weeks at 177°C
10 days   at 191°C
7 days   at 200°C.

The oil resistance of Vamac is lower than that of NBR, acrylic, polysulfide, and fluoroelastomers. Vamac is resistant to hot oil, hydrocarbons, glycols and engine cooling liquids but not to gasoline and brake fluids.

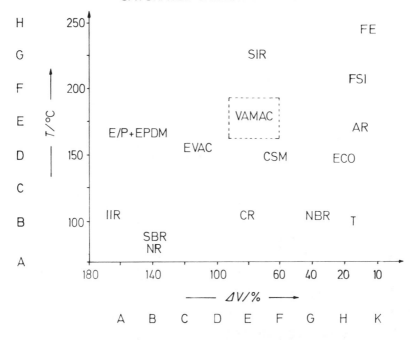

Figure 2-7 Temperature and oil resistance of various elastomers, measured as volume swell. Letters give the SAE-J-200 classification. E/P = ethylene/propylene copolymers, EPDM = terpolymer of ethylene, propylene and a non-conjugated diene, IIR = butyl rubber, SBR = styrene/butadiene rubber, NR = natural rubber, EVAC = copolymer of ethylene and vinylacetate, VAMAC® = ethylene/acrylate elastomer (DuPont), CSM = chlorosulfonated poly(ethylene), CR = poly(chloroprene), NBR = butadiene/acrylonitrile rubber, FE = fluoroelastomers, FSI = fluorosilicones, AR = acrylic ester elastomer, ECO = poly(epichlorohydrine), T = polysulfide elastomer.

Table 2-8 gives typical mechanical properties for black and white vulcanizates. Polar plasticisers such as dioctylsebacate lower the brittle point to −60°C. Vamac has good dampening properties to vibrations and shock impacts over a range from −30°C to 140°C. The saturated carbon backbone and the acrylate structure render Vamac resistant to degradation by weather and ozone.

Vamac has been used in automotive applications [3] for hoses and tubings, seals of all kinds, damping materials, for under-the-hood

TABLE 2-8
Properties of mineral and carbon black filled ®Vamac vulcanizates [1] (Vulcanization: 20 min. at 177°C).

Formulation	Weight-percent	
	Carbon black filled	mineral filled
Vamac B-124 Masterbatch	124	
N-774 SRF Carbon black	35	
Methylenedianiline	1.25	
Diphenylguanidine	4	
Vamac N-123 Masterbatch	123	123
Antioxidant (sterically hindered phenol)	2	2
Calcium carbonate	100	—
Barium sulfate	—	165
N,N'-m-Phenylenedimaleimide	2	2
Dicumylperoxide (40%)	5	5

TABLE 2-8

Properties of mineral and carbon black filled ®Vamac vulcanizates [1] (Vulcanization: 20 min at 177°C).

Physical Properties	Physical Unit	Carbon black filled	Property values for mineral filled	
Tensile strength at break			7.4	7.2
after 7 days at 177°C	MPa	14.0	7.9	8.8
after 3 days at 200°C	MPa	16.5		
	MPa	12.8		
100% modulus	MPa	2.9	4.5[a]	4.7[a]
after 7 days at 177°C	MPa	5.3	7.2	7.6
after 3 days at 200°C	MPa	4.3		
Elongation	%	440	565	540
after 7 days at 177°C	%	240	255	280
after 3 days at 200°C	%	220		
Tear strength C	kN/m	36.2		
Resilience[b]				
Vulcanization 30 min, 177°C	%	35		
After-treatment 4 h, 177°C	%	14		
Brittleness temperature	°C	−48		
Hardness (Durometer (A))		63	58	62
after 7 days at 177°C		66	65	68
after 3 days at 200°C		70		

[a] 200% modulus.
[b] after 70 h at 150°C.

electrical uses and in heat shielding of wire jackets. Burning Vamac develops very little low toxicity smoke and can be used to replace chlorosulfonated poly(ethylene) in cables and electric systems on ships and submarines. Another new Vamac elastomer was described recently [7].

References

[1] J. F. Hagman, R. E. Fuller, W. K. Witsiepe, and R. N. Greene, Rubber Age -/5 (1979).
[2] ®Vamac N-123 and B-124, Company literature, DuPont.
[3] F. V. Bailor, Rubber World -/4 (1979) 42.
[4] T. M. Vial, Acrylic Elastomers, in Encycl. Chem. Technol. **8** (1979), Wiley-Interscience, New York.
[5] H. J. Barager and D. J. Byam, Elastomerics **113**/6 (1981) 36.
[6] G. C. Sweet, Special Purpose Elastomers, in A. Wheelan and K. S. Lee, eds., Developments in Rubber Technology **1**, Appl. Sci. Publ. Ltd, Barking (U.K.) 1979.
[7] Mod. Plast. Internat. **11**/8 (1981) 106.

2.4 OTHER ETHYLENE COPOLYMERS

2.4.1 Structure

In the last few years a number of new ethylene copolymers and modified poly(ethylenes) have appeared on the market in which the producers disclosed only a few details in regard to type of comonomer and modification.

Since 1979, DuPont has marketed 8 different grades of ®CXA-Resins [1, 2]. They are ethylene based polymers with at least two different functional groups [3], presumably EVA terpolymers and quarterpolymers as well as modified polyolefins [2].

Also since 1979 Chemplex has offered three grades of ®Plexar [2, 4]. Plexar 1 and 3 are modified EVA types and Plexar 2 is a modified HD-PE.

Dow Chemical introduced two products named "ELT" (Experimental Liquid Thickeners) XD-30255.02 L and XD-30457.02 L [5]. They are ethylene copolymers containing alkali soluble links in the chain with hydrophilic and hydrophobic groups. The solubility in alkalies is probably due to maleic acid units.

The ®Keldax grades 9100 and 9101 of DuPont are "ethylene inter-polymer compositions" with a high content of inorganic fillers [6].

Mitsui Petrochemical Industries Ltd. offers ®Tafmer copolymers, types A and P, with 3 grades each [7]. They are ethylene-alpha-olefin-copolymers produced with Ziegler-Natta type catalysts [8].

US Industrial Chemicals produces an ethylene-carbon monoxide copolymer in pilot plant quantities. The polymer reportedly is degraded by weathering [8].

2.4.2 Properties and applications

The ®CXA and ®Plexar resins are co-extrudable coupling agents for multi-layer laminates. Mitsui Petrochemical offers similar products under the name of ®Admer-polyolefin-serie [9]. The olefinic coupling agents can bind numerous polymers with different polar and non-polar groups.

Hitherto incompatible polymer combinations, such as PE/Nylons, PP/PS, SB/PS/PE, Barex nitrile resin/PE or PP, are now possible with the aid of the new coupling materials. They can replace the formerly used and more expensive PCTFE or PVDC. Thinwalled laminates with barrier properties, such as films, thermoformable semi-finished goods and blow molded bottles, are produced by co-extrusion at reasonable costs [2].

First applications of such laminates are FDA approved containers for food and medical products. A laminated film of Nylon/Plexar/PE serves as hydrocarbon barrier to prevent pentane losses from packed expandable poly(styrene).

Dow's "ELT" Experimental Liquid Thickeners are milky white emulsions, with 20% solids, pH 3.5−5.5, Brookfield viscosities (No. 2 at 100 rpm) 2−25 mPa × s and densities of 1.03 and 1.12 g/cm^3 [5]. In alkaline solution, they associate equally well with the inorganic and organic components of latex paints because of the simultaneous action of hydrophilic and hydrophobic groups in the polymer [10]. The thickeners are suitable for most of the vinyl/acryl-, acryl-, EVAC-, PVC- and styrene-butadiene-latices. Contrary to the natural thickeners, such as modified cellulose, guar gum or xanthane, the "ELT" polymers show only a moderate viscosity increase in water. ELT XD-30457.02 L is designed for high viscous, high-shear systems, whereas the less expensive ELT XD-30255.02 L yields lower viscosities under high shear action.

TABLE 2-9

CXA and Plexar Resins. Properties, Processing Temperatures, and Application [1, 2, 4].

Resin	Density in g/cm³	Melt index in g/10 min	Vicat Temperature in °C	Processing Temperature in °C	Compatibilizing Agent for the following substrates
CXA 1025	1.00	35	35	200–225	ABS, PAN, Acrylates, HIPS, PC, PET, SAN, SMA, PVAC
CXA 1064	0.944	20	42	200–225	HIPS, PP, PET, SAN, SMA
CXA 1104	0.955	6.0	43	215–225	EVA, HIPS, Ionomers, PC, PE, PET, PP, PVDC, SAN
CXA 1202	0.935	3.5	61		EVA, Ionomers, PE, PP
CXA 2002	0.942	10	48	288–316	EVA, Ionomers, Nylon, PE, PP
CXA 2022	0.939	35	55	288–316	same as CXA 2002
CXA 3095	0.937	2.3	82	215–225	EVA, Nylon, PE, PVA Copolymers
CXA 3101	0.948	3.5	55	215–225	EVA, Ionomers, Nylon, PC, PE, PET, PP, PS, PVDC
Plexar 1	0.93	1.0		204–227	Nylon, PE and its copolymers, PVAC,
Plexar 2	0.96	0.85		232–249	tin free steel, copper, aluminium
Plexar 3	0.93	3.0		221–232	paper, wood, glass

TABLE 2-10
Properties of ®Keldax and ®Tafmer

Physical Properties	Physical Unit	Property values			
		Keldax 9100[1]	Keldax 9101[1]	Tafmer A-4085	Tafmer P-0180
Density	g/cm^3	1.67	1.83	0.88	0.88
Melt index	g/10 min	2.4	2.5	4.0	5.0
Ash content	%	65	72.5	—	—
Moisture content	%	0.15	0.15	—	—
Tensile strength	MPa	4.2	4.6	—	—
Elongation at break	%	185	70	800	700
Flexural modulus					
at 23°C	MPa	79	101	—	—
at −20°C	MPa	450	475	—	—
Impact strength					
at 23°C	kJ/m^2	99	101	—	—
at −20°C	kJ/m^2	71	19	—	—
Surface hardness JISA		—	—	85	44
Vicat temperature	°C	43	44	54	not measurable
Heat distortion temperature at 455 kPa	°C	−21	−11	—	—
Brittleness temperature	°C	—	—	<−70	<−70
Volume resistivity	Ohm · cm	—	—	≥10^{18}	≥10^{17}
Loss factor		—	—	1.3×10^{-4}	2×10^{-4}
Relative permittivity		—	—	2.1	2.3
Breakdown voltage	kV/mm	—	—	55	48

The properties of the thermoplastic ethylene containing polymers ®Keldax and ®Tafmer are compiled in Table 2-10.

The highly filled black ®Keldax grades are processed on conventional machinery. Keldax was designed for use as sound barrier and dampening material. Applications include pressed sheets, injection molded, extruded or vacuum-formed sound-absorbing parts in automotives, snow-mobiles, agricultural equipment, office machinery, and industrial equipment.

®Tafmer A is a copolymer of low crystallinity and flexibility compared to EVAC or plasticized PVC, whereas ®Tafmer P is a noncrystalline thermoplastic elastomer. Tafmer P is used exclusively for elasticizations of polyolefins and EVAC. Tafmer A is processed as such and serves also as modifier.

Tafmer A provides high transparency and gloss and has a very low brittleness temperature compared to other polymers. Tafmer A is fabricated, together with PE, to laminates and laminated films. Tafmer P is recommended for the elasticization of highly filled or fiber reinforced poly(propylenes).

DuPont has recently announced the commercialization of random copolymers from ethylene and methacrylic acid, trade named Nucrel® [11]. Current grades contain up to 15% methacrylic acid. The good flow characteristics and resiliency allow Nucrel to compete with EVAC.

Dow Chemical introduced a new ethylene/acrylic acid copolymer (EAA 469) for extrusion coating which is claimed to be competitive with ionomer resins and superior to ethylene/methacrylic acid polymers, conventional EAA, and EVAC [12].

References

[1] CXA, Bulletin DuPont, March 1981.
[2] Anon., Modern Plastics Internat., **10**, 56 (September 1980).
[3] P. S. Blatz, DuPont, personal communication (23 June 1981).
[4] Plexar, Bulletin, Chemplex Co.
[5] "ELT" Experimental Liquid Thickeners XD-30255.02 L and XD-30457.02 L, Bulletin, Dow Chemical Co., 1979.
[6] Keldax 9100 and 9101 Sound Barrier Resins, Bulletin, DuPont, March 1981.
[7] Tafmer, Bulletin, Mitsui Petrochemical Industries.
[8] Anon., J. Commerce (15 December 1980) 8.
[9] Hj. Saechtling, Kunststoffe, **69**, 910 (1979).
[10] N. Sarkar and R. L. Lalk, J. Paint Technol., **46**/590, 29 (1974).
[11] Anon., Modern Plastics Internat. (April 1983) 54.
[12] Anon., Modern Plastics Internat. (January 1984) 47.

2.5 POLY(P-METHYLSTYRENE) AND COPOLYMERS

2.5.1 p-Methylstyrene

Since early 1981 Mobil Chemical Co. has marketed p-methylstyrene under the name of Mobil PMS Monomer; in addition, poly(p-methylstyrene) and several copolymers are being developed [1]. There are various incentives for producing p-methylstyrene from toluene. Toluene costs less than benzene (Figure 2-8), and the demand for benzene exceeds by far the demand for toluene. Benzene is being used for a number of important intermediate products such as styrene, cumene/phenol, cyclohexane, nitrobenzene/aniline and maleic acid. In the USA, ca. 65% of the toluene needed by the chemical industry is dealkylated to benzene. Monsanto attempted to utilize excess toluene in a process to make styrene. Toluene is oxidized by oxygen to stilbene which is reacted with ethylene to yield styrene:

(2-1)

$$2\ C_6H_5CH_3 + O_2 \longrightarrow C_6H_5CH{=}CHC_6H_5 + 2\ H_2O$$

$$C_6H_5CH{=}CHC_6H_5 + CH_2{=}CH_2 \longrightarrow 2\ C_6H_5CH{=}CH_2$$

Pure p-methylstyrene has not been available before in commercial quantities. A mixture of ca. 33% para and 67% meta-methylstyrene is being used as vinyltoluene. The preparation, properties and applications of vinyltoluene have been described [2, 3]. The synthesis is analogous to that of styrene by Friedel-Crafts alkylation of toluene with ethylene and $AlCl_3$ as catalyst to give a crude mixture of 11.9% para, 19.3% meta and 3.8% ortho isomers of ethyltoluene besides 14.2% of poly(ethyltoluenes) [3]. The o-ethyltoluene is separated and recycled to the alkylation unit because it would be cyclized to indene and indane in the dehydrogenation step. The isolation of p-ethyltoluene by distillation is not economical because the boiling points are very close (p-ethyltoluene b.p. 162°C, m.p. −62.3°C, m-ethyltoluene b.p. 161.3°C, m.p. −95.5°C) and an outlet for m-ethyltoluene would have to be found. Presumably, the dehydrogenation is a gas-phase oxy-dehydrogenation with gases rich in

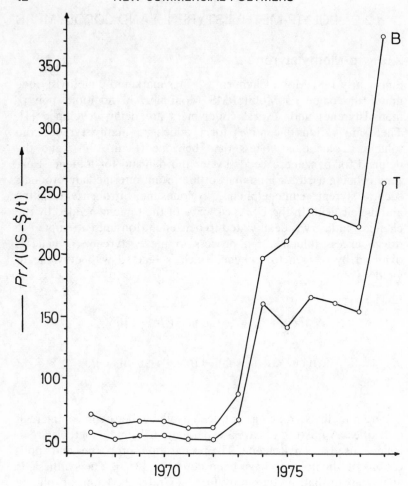

Figure 2-8 Development of benzene (B) and toluene (T) prices between 1967 and 1979.

oxygen or COS [4, 5]. Catalytic one-step processes to oxy-alkylate aromatics with ethylene and oxygen are being developed [6].

Mobil supplies only polymerization-grade p-methylstyrene (Table 2-11). The monomer has a higher boiling point than styrene, the flash point corresponds to that of vinyltoluene. Polymerization activity is only slightly different from styrene, thus, Mobil recom-

TABLE 2-11

Specifications of p-methylstyrene, styrene and vinyltoluene

Physical Properties	Physical unit	Property values for		
		Mobil PMS	Styrene	Vinyltoluene
Vinyl content	%	99.7	99.7	99.2
Isomer distribution				
para	%	97	—	33
meta	%	3	—	66.7
ortho	%	0	—	0.3
Polymer content	ppm	0	0	0
Carbonyl content	ppm	10	10	10
Peroxides content	ppm	5	5	5
Inhibitor content	ppm	25	12	50
Chloride content	ppm	6	10	5
Sulfur content	ppm	25	25	25
Polymerization enthalpy	kJ/mol	62−71	71	62−71
Copolymerization parameters				
Q-value		1.27	1.00	1.06
e-value		−0.98	−0.80	−0.78
Vaporization enthalpy at 25°C	J/g	411	428	
Specific heat capacity at 25°C	$J\,g^{-1}\,K^{-1}$	1.73	1.71	
Refractive index n_D		1.5408	1.5440	1.5395
Density at 25°C	g/cm^3	0.892	0.902	0.893
Viscosity at 25°C	mPa·s	0.79	0.72	0.78
Surface tension	N/m^2	34	32	31
Boiling temperature	°C	170	145	168
Freezing temperature	°C	−34	−31	−77
Volume contraction at polymerization	%	12	14	12
Vapor pressure at 0°C	kPa		0.13	
at 20°C	kPa		0.53	0.13
at 40°C	kPa	0.53	1.86	0.53
at 60°C	kPa	1.86	5.05	1.73
at 80°C	kPa	4.66	12.0	4.66
at 100°C	kPa	10.6	25.4	11.0
at 120°C	kPa	22.6	48.9	23.4
at 140°C	kPa	43.2	87.9	45.4
at 160°C	kPa	77.1	148.2	81.9
Flash point	°C	60	31	60
Self ignition temp.	°C	515	490	575
Explosion limits in air				
at room temp.	Vol.-%		1.1−6.1	
at 100°C	Vol.-%	1.1−5.3		
at 130°C	Vol.-%			0.8−11.0

mends the PMS-Monomer in applications where normally styrene or vinyltoluene is used. The replacement of styrene by PMS-Monomer yields products with lower densities, higher heat deflection temperatures and, in some cases, improved flow.

PMS-Monomer can replace vinyltoluene in plain or reinforced polyester materials [1]. In unfilled polyesters higher heat deflection temperatures and somewhat better mechanical properties are achieved with PMS, whereas in reinforced materials the use of PMS instead of vinyltoluene yields no marked improvement in properties (Table 2-12).

2.5.2 Co- and terpolymers with p-methylstyrene

In analogy to styrene polymers, Mobil has developed the corresponding p-methylstyrene polymers under the designations Poly-PMS (Poly(p-methylstyrene)), PMSAN (Poly(p-methylstyrene-co-acrylonitrile)) and ABPMS (Poly(acrylonitrile-co-butadiene-co-p-methylstyrene)) [1]. Available types of Poly-PMS include a general-purpose, a high-heat, and a high-impact grade and two products with 10 and 25% glass fiber reinforcement.

Compared with the corresponding styrene polymers (Tables 2-13 and 2-14), the p-methylstyrene polymers display lower densities, higher Vicat and heat deflection temperatures and increased hardness. The polymers can be processed like styrene polymers. At the same melt index, p-methylstyrene polymers flow better than styrene polymers. A shorter cycle time results because p-methylstyrene polymers can be removed sooner from the mold due to their higher heat resistance. The application of p-methylstyrene polymers instead of styrene polymers is recommended in cases where an improved flow and a higher heat resistance are required. The toxicology of the new products has not been fully examined as yet. At present it is not advisable to use the new products in contact with food.

Mobil Chemical Co. began the commercial production of p-methylstyrene in a 16000 t/year plant in 1982. The plant was originally built by American Hoechst for the manufacture of styrene. It is now jointly owned by American Hoechst and Mobil Chemical [7, 8]. For additional literature on p-methylstyrene and its polymers see [9–12].

TABLE 2-12

Properties of unfilled and glass-fiber reinforced polyester resins cured with p-methylstyrene (PMS) or vinyltoluene (VT).

Compounding recipe	unfilled resin	reinforced resin
Polyester } premix	74%	70%
PMS or VT } premix	25%	25%
Benzoylperoxide } premix	1%	—
Premix		40.0%
Glass fiber		33.0%
Clay		25.0%
Zinc stearate		1.5%
Benzoylperoxide		0.5%

Properties	Physical unit	Property values of unfilled polyester resin with PMS	Property values of unfilled polyester resin with VT	Property values of reinforced polyester resin with PMS	Property values of reinforced polyester resin with VT
Density	g/cm^3			1.71	1.70
Tensile strength at break	MPa	43.4	38.6	33.8	33.8
Flexural strength	MPa	82.1	80.7	90.3	96.5
Flexural modulus	GPa	2.73	2.73	9.17	8.96
Tensile modulus	GPa	2.55	2.52		
Notched impact strength	kJ/m			0.48	0.53
Barcol hardness		36	35	56	53
Heat distortion temperature	°C	107	88	270	270
Dielectric strength	MV/m			12.8	11.4
Mold shrinkage	%			0.19	0.16

TABLE 2-13

Properties of poly(p-methylstyrenes) and poly(styrenes).

Physical Properties	Physical units	Property values for				Glass fiber reinforced			
		Mobil Poly-PMS-Resin	Mobil Poly(styrene) 224EP	Mobil Poly-PMS-Resin High-Heat	Mobil Poly(styrene) 110 S	PMS 10%	PMS 25%	PS 10%	PS 25%
Density	g/cm^3	1.01	1.05	1.01	1.05				
Melt index (200°C, 50 N)	g/10 min	12	12	2.2	2.2				
Tensile strength at break	MPa	45	45	52	54	59	66	60	67
Elongation	%	3.0	3.0	3.0	3.0	2.8	2.5	2.6	2.2
Tensile modulus	GPa			2.20	2.48				
Flexural strength	MPa					93	101	91	101
Flexural modulus	GPa			2.90	3.38				
Notched impact strength	kJ/m	0.016	0.016	0.016	0.016	0.032	0.048	0.032	0.043
Hardness (Rockwell M)		80	70	80	75				
Vicat temperature	°C	110	89	117	107				
Heat distortion temperature at 1.81 MPa	°C	90	85	94	91	102	109	100	106

TABLE 2-14
Properties of various PMS and PS grades.

Physical Properties	Physical unit	Property values of					
		Mobil Poly-PMS High-Impact Resin	Mobil High-Impact poly(styrene) 444	Mobil PMSAN Resin	Mobil SAN 120	Mobil ABPMS Resin	Mobil ABS 808 K
Density	g/cm³	1.01	1.04	1.04	1.07	1.01	1.03
Melt index (200°C, 50 N)	g/10 min	6.4	5.8			5.3	
" " (230°C, 50 N)	g/10 min						
Tensile strength at yield	MPa	34	32	21	21	36	35
Elongation at break	%	20	30	68	55	12	11
Tensile modulus	GPa	1.72	1.70	2.42	2.57	1.57	1.62
Flexural strength	MPa			82	97		
Flexural modulus	GPa			3.14	3.32		
Notched impact strength at 23°C	kJ/m	0.091	0.107	0.016	0.021	0.262	0.320
at -40°C	kJ/m					0.128	0.107
Hardness (Rockwell)		L70	L65	R122	R122	R98	R93
Vicat temperature	°C	102	98	112	108	106	103

References

[1] Mobil Chemical Co., product information and technical publications: Mobil PMS Monomer, Mobil PMS Monomer Molding Compound, Bulk Molding Compound, Mine Bolt Adhesive Compound, Mobil Poly-PMS Resin, Mobil Poly-PMS Impact Grade Resin, Mobil PMSAN Resin, Mobil ABPMS Resin, Mobil Poly-PMS Resin Glass-Reinforced.

[2] R. H. Boundy and R. F. Boyer, Styrene. Reinhold Publishing Corp., New York 1952.

[3] K. E. Coulter, H. Kehde and B. F. Hiscock, Styrene and Related Monomers, in E. C. Leonard (ed.), Vinyl and Diene Monomers, Part 2, Wiley-Interscience, New York 1971; C. A. Brighton, G. Pritchard, and G. A. Skinner, Styrene Polymers, Applied Science Publishers Ltd., London 1979.

[4] USP 3 399 243 (1968), D. E. Boswell (to Mobil Oil).

[5] USP 3 787 517 (1974) and 3 875 252 (1975), W. Haag and J. N. Miale (to Mobil Oil).

[6] K. Weissermel and H. J. Arpe, Industrielle Organische Chemie, Verlag Chemie, Weinheim 1976; Industrial Organic Chemistry, Verlag Chemie, New York 1978.

[7] Anon., Modern Plastics (July 1982) 14.

[8] Anon., Chem. Engng. News (May 31, 1982) 20.

[9] Anon., A better styrene goes commercial, Chem. Week (February 17, 1982) 42.

[10] C. L. Myers, ACS Org. Coat. Appl. Polym. Sci. **46** (1982) 302.

[11] Anon., Chem. Engng. News (May 24, 1981) 15.

[12] W. Kaeding, W. Young, L. Brewster and A. G. Prapas, Para-methylstyrene, Chemtech **12** (1982) 556.

2.6   POLY(TRIBROMOSTYRENE)

2.6.1   Structure and synthesis

Ferro Corp., Chemical Division, offers poly(tribromostyrene) under the trade name ®Pyro-Chek 68 PB [1, 2]. The structure has not been disclosed; however, it was stated that the bromine is attached to the aromatic ring. 2,4,6-Tribromostyrene [3] and 2,4,5-tribromostyrene [4] have been described in the literature.

The preparation of poly(2,4,6-tribromostyrene) starts from (2-bromoethyl)-2,4,6-tribromobenzene. The reaction of an alcoholic solution with NaOH and azobisisobutyronitrile leads directly to the polymer [3]:

(2-2)

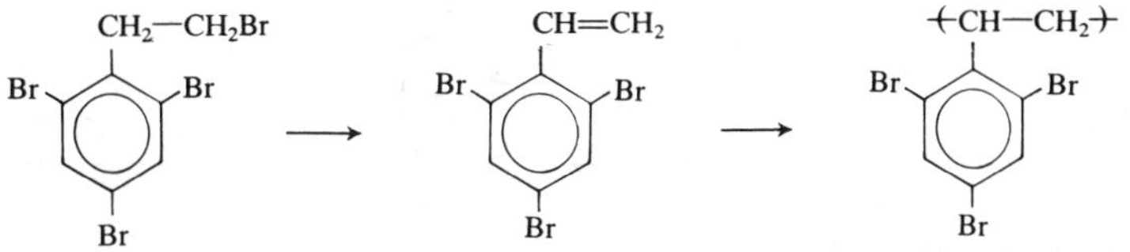

The 2,4,5-tribromostyrene is obtained according to the general equation

(2-3)

$$Br_nC_6H_{5-n}CH_2CH_2Br \longrightarrow Br_nC_6H_{5-n}CH=CH_2 + HBr$$

by eliminating HBr in the gas phase with steam at $280-470°C$ or in alcoholic solution in the presence of peroxides [4].

2.6.2 Properties and applications

The polymer, density 2.8 g/cm^3, softening point 220°C, contains 68% bromine which agrees with the calculated value for poly(tribromostyrene). A thermographic analysis at a heating rate of 40°C/minute shows weight losses of 1% at 340°C, 10% at 408°C, and 50% at 435°C. The polymer has a low toxicity. The acute oral dose LD_{50} for rats is >15 g/kg, the acute dermal dose LD_{50} for rabbits is >3 g/kg. The product has been approved by the EPA.

PyroChek 68 PB is used in combination with antimony trioxide as flame retardant. In poly(butyleneterephthalate) the material tends to bloom out slightly after 500 h heating at 120°C in an air circulation oven, whereas decabromodiphenylene oxide shows a moderate migration after only 5 h at 120°C. No blooming out of PyroChek 68 PB in PBT is observed after 650 h heating at 65°C. A flame rating of V-O (UL-94) in 0.8 mm PBT specimen is reached by 14% Pyro-Chek 68 PB together with 4.7% Sb_2O_3. The recommended concentrations of PyroChek 68 PB and Sb_2O_3 are listed in Table 2-15.

TABLE 2-15
Recommended concentrations of Pyro-Chek 68 PB and antimony trioxide.

	Recommended concentration in %	
Polymer	Pyro-Chek 68 PB	Antimony trioxide
Poly(butyleneterephthalate)	10−16	3− 6
Poly(ethyleneterephthalate)	9−15	3− 5
Nylon	10−15	3− 5
Crosslinked poly(ethylene)	25−40	8−14
ABS	15−20	4− 6
High-impact poly(styrene)	12−15	3− 5
Unsaturated polyester	12−15	3− 5
Poly(ethylene)	25−40	8−14
Poly(propylene)	25−40	8−14
Epoxy resins	20−28	4−10

PyroChek 68 PB costs 1.50 $/lb. The prices of competitive materials are 1.50 $/lb for decabromodiphenyloxide, 2 $/lb for Dechlorane (Hooker), 2.05 $/lb for Saytex BT-93 (Saytech Inc.), and 2.25 $/lb for the polymeric Firemaster TSA [1].

References

[1] A. v. Hassell, Plastics Technology (July 1980) 71.
[2] Pyro-Chek 68 PB, Technical information Ferro Corp., Chemical Division.
[3] Ger. P. 1 570 395 (1969), Chem. Fabrik Kalk, Köln.
[4] Ger. P. 2 505 807 (1974).

2.7 NEW STYRENE-BASED THERMOPLASTIC AND HIGH-IMPACT ENGINEERING PLASTICS

2.7.1 Introduction

ABS plastics based on acrylonitrile, butadiene and styrene have been known for more than 25 years. The polymers combine high-impact resistance with good stiffness, high dimensional stability under heat, excellent surface quality, chemical resistance, superior stress cracking resistance and perfect thermoplastic processability. They allow various post-treatments, e.g. metallizing. These properties and a reasonable price led to many applications. The consumption in West Europe alone was more than 1 million tons in 1980. Recent developments in ABS polymers have been summarized in several review papers [1, 2].

The disadvantages of ABS polymers are the low flame resistance, the poor weathering resistance, the lack of transparency and, for many applications, the insufficient dimensional stability at elevated temperatures. The flame resistance of ABS can be improved by addition of PVC or rather large quantities (15−20%) of flame retardants, however, the additives lower the toughness considerably.

Transparent ABS types are obtained when the grafting is carried out in the presence of methylmethacrylate. Such MABS or MBS polymers have been on the market since 1964, e.g. under the name of ®XT-Polymer.

The replacement of the light sensitive butadiene units by other elastifying components leads to polymers with improved weather resistance. Such components are acrylic elastomers (ASA polymers), chlorinated poly(ethylenes) (ACS polymers) and EP or EPDM elastomers (AES polymers).

The ASA types were introduced in 1969. More recent commercial products are the ®ACS-Resins (Showa Denko, 1976), the AES types such as ®Rovel (Uniroyal, 1977−1980; 1984 sold to Dow Chemical), the SAA SC 1000 series (Stauffer, 1978; 1984 sold to General Electric Plastics), the Dow XP 5272 Resins (1978) and ®Cadon (Monsanto, 1981) which is an ABS type with improved heat resistance.

Table 2-16 lists old and new thermoplastic, high-impact resistant engineering plastics based on styrene, stating trade names, producers, grades supplied, and hard and elasticizing components of the products.

Available grades of the new products are:

®ACS-Resins   NF-920 (stiff/high-impact)
               NF-960 (high-impact)
               NF-760 (high-heat)
               NF-1060 (high-flow)

®AES Rovel   401 (for sheets)

501 (for profiles)
701 (injection molding)

®1000 (Stauffer)   SCC 1001 (Teflon filled)
SCC 1002 (high-impact)
SCC 1004 (injection molding/extrusion)

®Cadon   112 (high-impact, medium heat resistant)
127 (high-impact, high-heat resistant)
135-PG ⎱ high-impact, high-heat resistant,
155-PG ⎰ for metal plating
180 (extrusion-grade; high impact, high heat
resistance)

2.7.2   Structure and synthesis

The preparation of the older polymers ABS [3], AAS, and MABS/ MBS [4] have been described in detail.

ACS is obtained by grafting styrene and acrylonitrile onto a chlorinated poly(ethylene) elastomer and then compounding with SAN [5].

®Rovel consists of a SAN matrix and a dispersed elasticizing component which is a Uniroyal proprietary saturated olefin elastomer, presumably an EP copolymer [6, 7, 8]. Other AES products are obtained by grafting SAN with EPDM terpolymers, the diene being 5-ethylidene-2-norbornene [2]. In 1984, the Rovel product line was purchased by the Dow Chemical Co.

The ®1000 polymer series of Stauffer are reportedly not ASA-polymers, in spite of their composition of styrene, acrylonitrile and acrylic esters, but so-called multi-polymers. No details about the production method have been given [9].

It is well known that terpolymers prepared from styrene, acrylonitrile and maleic anhydride display glass transition temperatures above 122°C [10]. As expected, ®Cadon and Dow's Resin XP 5272, both based on styrene/maleic anhydride, show high heat resistance. For both products, the elasticizing component has not been revealed. Since Cadon is not recommended for out-door use, it is likely to contain butadiene based elastomers [11]. Other styrene/ maleic acid copolymers are marketed by Arco Chemical under the name ®Dylarc 350 and 700 [13] and by Dainippon Ink and Chemical under the name Ryurex A-15®.

TABLE 2-16

Thermoplastic and impact resistant engineering plastics based on styrene.

Polymer Type	Trade name	Company	Composition and basic units									
			"hard component"					"soft" elasticizing component				
			S	A	MMA	MA	B	ACE	EP	CPE	AE	U
SB	many	many	X				X					
BDS Block	K-Resin	Phillips	X				X					
ABS	many	many	X	X			X					
ASA	Luran S	BASF	X	X				X				
	AAS-Resin	Hitachi	X	X				X				
MABS/MBS	XT	Am. Cyanamid	X	X	X		X					
	Terluran XR2802	BASF	X	X	X		X					
	Cyrolite	Rohm & Haas	X		X		X					
	Cyclolac CIT	Borg-Warner	X		X		X					
	Metacrylene	Ugine Kuhlmann	X		X		X					
	Toyolac 900	Toray Ind.	X		X		X					
	Sicoflex	Mazzucchelli	X		X		X					
AES	Rovel	Uniroyal	X	X					X			
		Japan Synth. Rubber	X	X					X			
	Hostyren XS	Hoechst	X	X					X			
	Novodur AES	Bayer	X	X					X			
ACS	ACS Resin	Showa Denko	X	X						X		
SAA	SC 1000 Serie	Stauffer	X	X							X	
SMA	Cadon	Monsanto	X			X						X
	Resin XP 5272	Dow	X			X						X
	Dylarc	Arco	X			X	unknown					

S   Styrene
A   Acrylonitrile
MMA   Methacrylic acid methylester
MA   Maleic acid anhydride
B   Butadiene
ACE   Acrylic ester elastomer
EP   Ethylene propylene co or terpolymers
CPE   Chlorinated poly(ethylene) elastomer
AE   Acrylic ester
U   Elasticizing component of unknown composition

2.7.3 Properties and applications

The properties of older and recent commercial polymers are listed in Table 2-17.

®K-Resin is very notch sensitive. The common notched impact test demonstrates only low values comparable to those from unmodified poly(styrene), however, improved toughness results after orientation [12]. The other mechanical properties are similar to impact resistant poly(styrenes). K-Resins possess excellent transparency. The material is FDA approved, it is used for deep-drawn blow molded food containers. K-Resins can replace cellulose esters, rigid PVC and oriented polyester films.

The flammabilty of ABS is moderately reduced by the addition of PVC, however, the mechanical properties, in particular the notched impact resistance, are lowered considerably. The best flame rating (UL 94 V-O) is observed in ACS-Resins which simultaneously maintain good property balance comparable to ABS but better weather resistance. ACS-Resins are used where an inherently high flame resistance is required, such as for TV cabinets, electric office machines, projectors, lamp sockets, fire detecting devices, etc.

Similar good weather resistance is demonstrated by polymers containing olefin elastomers, e.g., Rovel (see Figure 2-9) [6]. Rovel has weather resistance and mechanical properties which are generally comparable to polycarbonate but Rovel costs less. Due to rheological properties resembling ABS, the two polymers can be co-extruded with Rovel forming the top layer. Applications are roofings for recreational vehicles, sidings for boats, outboard engine covers, and traffic signs. Rovel substitution for rigid PVC by Rovel in window frames is being explored.

The black pigmented polymers were tested in a so-called chip impact test developed by Uniroyal. Hereby, the weathered side of the unnotched specimens ($2.54 \times 1.27 \times 0.17$ cm) is exposed to the striking pendulum. The test is reportedly very sensitive to the presence of microcracks which are formed by UV irradiation in the elasticizing component of ABS or other products [6].

SCC 1004 combines high heat deflection temperatures with high impact resistance. It is intended to use the product as a blend with PVC for sidings. Compared to PVC alone, the blend is expected to have an improved weather resistance and to retain its color and gloss over a period of many years, especially when applied in dark

TABLE 2-17

Properties of thermoplastic and impact resistant engineering plastics based on styrene.

Physical Properties	Phys. unit	ABS General Purpose type [5]	ABS/PVC [5]	ASA BASF Luran S 757 R	Property values of BDS Phillips K-Resin KR 03 [12] [5]	ACS Showa Denko NF-960 [5]	AES Uniroyal Rovel 701 [6]	SAA Stauffer SCC 1004 [9]	SMA Monsanto Cadon 135-PG [11]
Density	g/cm^3	1.04	1.21	1.07	1.04	1.16	1.02	1.06	1.07
Tensile strength at break	MPa	40.2	39.2	51.0	27.6	39.2	41.4	29	30.4
Elongation at break	%	20	30	15		50		40	20
Tensile modulus	GPa	2.06	1.67		1.38	2.06	2.07	1.1	2.26
Flexural strength	MPa	49.1	44.1	78.5	46.9	49.1	61.4	47	55.2
Flexural modulus	GPa	2.2	1.6	2.6	1.55	2.2	2.0	1.5	2.14
Notched impact strength	kJ/m	0.29	0.15		0.02	0.12	0.32	0.48	0.15
Hardness (Rockwell R)		106	106	108	72	102	102	95	102
Heat distortion temperature (at 1.81MPa)	°C	87	70	88	71	89	92	100	116
Volume resistivity	Ohm × cm	1.1×10^{16}		1.0×10^{14}		7.3×10^{15}		4.1×10^{13}	
Surface resistivity	Ohm			1.0×10^{13}		2.7×10^{15}			
Dielectric strength	MV/m			22	11.8	25.4			
Dielectric constant		2.81	2.87	3.4		3.05			
Dissipation factor				0.02					
Arc resistance	s	100				80			
Water absorption (24 h)	%	0.3	0.2	0.5	0.09	0.2	0.09	0.31	
Oxygen index	%	18.4	28.5			28.5			
Flammability (UL)		94 HB	94V-1			94V-0		94 HB	
Transparency		translucent	translucent	transparent	transparent	translucent	translucent	translucent	

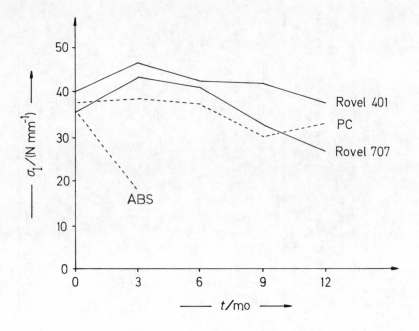

Figure 2-9   Effect of weathering in Florida on impact resistance of Rovel®, ABS, and polycarbonate.

TABLE 2-18
Notched impact strength and heat distortion temperatures of ®Cadon grades.

Physical Properties	Physical unit	Property value of		
		Cadon 112	Cadon 127	Cadon 135-PG
Notched impact resistance	kJ/m	0.25	0.23	0.15
Heat distortion temperature (at 1.81 MPa)				
without annealing	°C	106	112	106
after annealing	°C	116	120	116
metallized				
without annealing	°C			150
after annealing	°C			133
Vicat temperature	°C	125	127	

shades. Other outdoor uses aim at the same applications as cited for AES products.

The best combination of high heat deflection temperature and high impact resistance is represented by maleic anhydride containing polymers, e.g., ®Cadon [11]. With annealing or metal plating, the heat deflection temperature can be further increased (Table 2-18).

Cadon is recommended for applications in the automotive sector, where manufactured parts are exposed to high temperatures in hot paint drying equipment or in metallizing baths. Here Rovel is superior to special grades of ABS as well as to PPO and ABS/PC blends. Competitive materials are modified polysulfones, e.g., ®Mindel (UCC).

In 1981, the following prices were quoted: ABS 0.95−1.11 $/lb, Rovel 1.05−1.11 $/lb, Cadon 1.05−1.11 $/lb and SCC 1004 0.80−0.85 $/lb. The SCC 1004 is aimed to replace rigid PVC in sidings rather than to substitute ABS; PVC costs ca. 0.30 $/lb.

References

[1] H. Jenne, Kunststoffe **66** (1976) 581.

[2] K. Schneider, Kunststoffe **70** (1980) 617.

[3] A. Lebovits, Encycl. Polym. Sci. Technol., Vol. **1** (1964) 441.

[4] T. O. Purcell, Jr., Encycl. Polym. Sci. Technol., Suppl. 1 (1976) 307.

[5] M. Ogawa and S. Kokury, Brit. Plast. **42** (1969) 42. ®ACS-Resins, Technical Information, Showa Denko.

[6] Anon., Plast. Technol. (February 1980) 24.

[7] Anon., Plast. Engng. (March 1980) 69.

[8] J. F. Wefer, ACS Div. Org. Coat. Plast. Chem., Preprints **44** (1981) 376.

[9] ®SCC 1004, Technical Information, Stauffer. Anon., Modern Plast. **46** (June 1979) 56.

[10] A. W. Hanson and R. L. Zimmermann, Ind. Eng. Chem. **49** (1957) 1803.

[11] ®Cadon, Technical Information, Monsanto. Anon., Chemical Week (May 13, 1981) 23; Anon., Modern Plast. Intern. (July 1981) 21.

[12] L. M. Fodor, A. G. Kitchen, and C. C. Biard, in R. D. Deanin (ed.): New Industrial Polymers, ACS Symp. Ser. **4**, Amer. Chem. Soc., Washington 1974, 37.

[13] Anon., Modern Plast. Intern. (July 1981) 62.

[14] Anon., Kunststoffe **74**, 300 (1984).

[15] H. Jenne, Kunststoffe **74**, 551 (1984).

2.8   MODIFICATION OF POLY(VINYL CHLORIDE)

2.8.1   Structure

The portion of rigid PVC in total PVC processing increased from 25% in 1960 to over 60% in 1980 [1]. Processing of PVC is considered to be difficult; also the toughness of the material is insufficient for many applications.

The processability of PVC can be improved by copolymerization of vinyl chloride (VC) with ethylene, propylene, long-chain vinyl esters, acrylic esters, or vinyl acetate. However, the copolymers display comparatively low thermal stabilities and heat deflection temperatures and, further, they are more expensive than PVC because of the high portion of comonomers needed. It is, therefore, preferred to incorporate polymeric processing aids which in small concentrations facilitate the plasticization of PVC. Recent commercial products (Table 2-19) are poly(alpha-methylstyrene), MMA, and acrylate containing copolymers and poly(vinyl chloride-co-ethylene) with high contents of vinyl acetate.

Previously, blends of PVC with NBR or chlorinated poly(ethylene), copolymers of VC/acrylic esters, and VC/fumaric acid, as well as graft copolymers of ABS, MBS, and AMBS have been used as impact modifiers. The poor aging properties of butadiene compounds exclude them from out-door uses. On the other hand, there is an increasing demand for weather and impact resistant PVC types for window frames and sidings. New impact modifiers (Table 2-19) are modified polyacrylates and ethylene/vinyl acetate copolymers with a high vinyl acetate content. To improve the compatibility with PVC, the modifiers can be grafted with VC.

Furthermore, ready-to-use impact resistant PVC types which have already been modified by the producers are commercially available. They are produced by grafting an acrylic compound onto PVC (Lonza B 6805), of VC onto EVAC (Levapren, Vestolit Bau, Vinnol K, Pantalast) or PVC onto EPDM.

Modifiers and the modification of PVC have been reviewed in the literature [1−6].

More recently, transparent or translucent PVC types which have been developed for roofings, are available as ®Hostalit Z and ®Hostapren by Hoechst, as ®Sinticlad by ICI and as ®Benvic by Solvay [7].

TABLE 2-19
Modification of PVC.

Trade name	Producer	Polymer type	Concentration of modifier in phr PVC	Reference (2.8)
Processing aids				
Amoco Resin 18	Amoco Chemicals	Poly(alpha-methylstyrene)	2–6	[8]
Diakon APA 1	ICI	MMA-ethylacrylate-copolymer	1–3	[9, 10]
P 210-D	Richardson	Styrene-acrylic-multipolymer	1–2	[21]
SCC-7149	Stauffer	Modified polyacrylate		[11]
VAE 611	Wacker-Chemie	VAc-ethylene-copolymer (60% VAC)	1.5	[3, 12]
Impact modifiers				
Durastrength 200	M & T Chemicals	Modified poly(acrylate)	0.6–4[1] / 4–10[2]	[13, 14]
Elvaloy 837 P	DuPont	EVAC copolymer (28% VAC)	4–6	[5, 21]
Elvaloy 838 P	DuPont	EVAC terpolymer	4–6	[5]
Pantalast W	Pantasote	VC grafted onto EVAC (12% EVAC)		[15, 16]
Vinnol K 550/79	Wacker-Chemie	50 VC grafted onto 50 EVAC (45% VAC)	10–20	[12, 17]
Vinuran KR 3820	BASF	polymeric acrylester	5–15	[18]
Vynathene	U.S.I.	EVAC with 40–60% VAC	4–7	[5, 19, 21]
Ready-to-use impact PVC grades modified by the producer				
Lonza B 6805	Lonza	VC grafted onto poly(acrylate)		[6]
Pantalast L	Pantasote	50% VC grafted onto 50% EVA		[15, 16]
Rucodur 1400	Hooker/Ruco Div.	PVC/EPDM graft copolymers		[20]
1706				[20]
1803				[20]
Vinnol K 505/68	Wacker-Chemie	5% EVAC grafted onto 95% PVC		[12, 17]
Vestolit BAU	Chem. Werke Hüls	6% EVAC grafted onto 94% PVC		[26]

1) as processing aid.
2) as impact modifier.

2.8.2   Properties and applications

2.8.2.1   Processing aids

Poly(alpha-methylstyrene) has been marketed by Amoco since 1976 under the designation ®Resin 18. The grades available are Resin 18-210, 18-240 and 18-290 with molecular weights of 685, 790 and 960, respectively, and softening points of 99, 118 and 143°C, respectively. Resin 18 causes a reduced melt viscosity in PVC and allows either a higher production rate or a higher load of fillers at reduced concentrations of stabilizers. Resin 18 is recommended for both rigid and plasticized PVC to see service in sidings, in pipes with high content of fillers, and in injection molded articles. Resin 18 costs 0.34 $/lb and thus is a low-priced modifier.

®Diakon APA 1 (ICI) is a copolymer of MMA with 10% ethyl acrylate, molecular weight above 1 million, glass transition temperature 91°C, refractive index 1.49. Diakon accelerates the gelation rate of powdered PVC compounds, prevents melt fracture in injection molding, eliminates scaling in extrusion processes, and improves gloss and surface quality. Diakon is approved for applications in contact with food. Diakon costs 1.32 £/kg.

®P-210-D is a so-called styrene-acrylic-multipolymer. Like Diakon, it is supplied in the form of microballoons which can be pneumatically transported. The price is 0.48 $/lb.

®SCC-7149 is a non-hygroscopic "acrylic-bearing" polymer with a granular size distribution matching that of suspension PVC. It is used to improve the surface quality of transparent PVC articles. The FDA approved material costs 0.73 $/lb.

The ®VAE-Polymers (Wacker-Chemie) with 60−70% VAc can be blended with PVC in any ratio. Transparent compounds show no white fracture under load. Transparent films are obtained from a solution of the mixed polymers. An equivalent compatibility is not achieved by EVA polymers with a lower VAc content. VAE 611 with 60% VAc, density 1.04 g/cm^3, melt index <1 (at 190°C), Mooney-viscosity 40 (ML 4 at 100°C), reduces the plasticization time of impact grade PVC compounds at concentrations of 1.5 phr. In extrusion processes, gloss is improved and less streakiness is observed. Articles deep-drawn from sheets show a more accurate reproduction of contours and a more uniform wall thickness; moreover, a greater drawing depth can be achieved.

2.8.2.2 Impact modifiers

®Durastrength 200 (M&T Chemicals) is supplied with the following specifications: density 1.09 g/cm³, bulk density 0.35 g/cm³, volatiles <1.2%, tensile strength 6.2−8.3 MPa, elongation 75−120%, notched impact resistance 0.96 kJ/m. At low concentrations it reduces the melt viscosity of PVC and increases the out-put thus acting as a processing aid. At higher concentrations, it also increases the impact resistance, even at low temperatures, and it improves the weather resistance of PVC compounds. In order to obtain equivalent impact resistance at room temperature, only 3 parts of Durastrength 200 are required compared to 5 parts of a conventional polyacrylate or 6 parts of a chlorinated PE. Contrary to CPE or polyacrylates, Durastrength allows a reduction of the TiO_2 content in PVC compounds by 20−30% without sacrificing notched impact resistance. PVC compounds containing Durastrength 200 are used in the manufacture of window frames and sidings. Durastrength 200 costs 1.20 $/lb.

The ®Elvaloy-EVA-types with 28% VAc have molecular weights of ca. 1 million, which is considerably higher than other EVA types with molecular weights of about 125 000.

Elvaloy's are modified EVAC polymers. The modification imparts improved weather resistance. With 4−6 phr of Elvaloy in PVC sidings, similar good results are obtained as with 6−9 phr of poly-(acrylates). The addition of Elvaloy leads to matted surfaces and helps to save matting pigments. Prices are 1.10−1.25 $/lb, depending on grades.

®Pantalast W (Pantasote Inc.) is produced by grafting VC onto EVAC in suspension. The grafted copolymer with 12% EVAC has the following specification: density 1.35 g/cm³, relative viscosity (1% in cyclohexanone at 25°C) 2.26, K-value 67, volatiles 0.5%. Pantalast imparts excellent impact resistance at low temperatures and good thermal stability at elevated temperatures.

®Vinnol K 550/79 (Wacker-Chemie) is a suspension polymer which is obtained by grafting 50 parts VC onto 50 parts of the EVAC copolymer ®Levapren. The data are: density 1.17 g/cm³, bulk density 0.71 g/cm³, intrinsic viscosity 153 cm³/g, K-value 79, chlorine 28.5%, volatiles <0.3%. The free-flowing material can be pneumatically transported. Addition of 10−20 parts of Vinnol K 550/79 to 80−90 parts of PVC yields low temperature impact resistant products with high light and weathering fastness. Applications are window frames and other articles for out-door use.

®Vinuran KR 3820 (BASF) is a free flowing powder based on acrylic esters which is added to PVC in amounts of 5–15%. In blends with rigid PVC, the product imparts high transparency combined with high impact and weather resistance. Applications are transparent covers for sport halls, green-houses, hotbeds, light transmitting panels and decorative articles.

The ®Vynathene EVA types (U.S.I. Chemicals) contain 40–60% VAC. Incorporated calcium stearate prevents the free flowing powder from agglomerating. Addition of 5–7 parts of Vynathene to PVC imparts similar good impact and weather resistance and color fastness as with 6–8 parts of the conventional poly(acrylates).

2.8.2.3 Ready-to-use high impact PVC's

This class of PVC polymers consists of two groups. The first group consists of PVC polymers to which impact modifiers are added after polymerization. Almost all PVC producers have been offering such products for several years, e.g., ®Hostalit Z, HM and H (Hoechst) [22].

The second group consists of PVC graft polymers (Table 2-20). B 6805 (Lonza) is a graft polymer of VC on dispersed polyacrylates with a K-value of 68. A transparent grade is available under the name EMA [6].

®Pantalast L is a suspension polymer which is obtained by grafting of VC onto EVA [16]. The granular material, containing 50% EVA, K-value 78, demonstrates a high elongation; its properties resemble a thermoplastic elastomer. Pantalast can be used as is or in PVC blends where it causes increased hardness. Applications include translucent and flexible articles, cable sheathing and parts for the shoe industry. Pantalast L costs 0.99 $/lb.

More recently, several other types of PVC have been marketed which can clearly be classified as thermoplastic elastomers (see chapter 4.3). These polymers are "XEL" (Kanegafuchi Chemical Industry Co.) and "Sumiflex" (Sumitomo Bakelite Co. Ltd.).

The ®Rucodur polymers (Hooker Chemical and Plastics) contain EPDM as low-priced, weather and impact resistant component to which VC is grafted. Rucodur 1400 is recommended for injection molding, Rucodur 1700 for extrusion and Rucodur 1803 for blow

TABLE 2-20

Properties of ready-to-use impact PVC grades for exterior applications.

Physical Properties	Physical unit		Property values of					
		Lonza B 6805	Pantalast L	Rucodur 1400	Rucodur 1706	Rucodur 1803	Vinnol K505/68	Vestolit BAU
Density	g/cm³	1.36	1.18	1.34	1.42	1.32	1.40	
Tensile strength at yield	MPa	50	14.5	51.8	48.5	45	37	>45
Tensile modulus	GPa			2.89	2.78	2.37	2.50	
Elongation	%	40	250	3	3	3	60	<5
Flexural strength	MPa			96	76	77.3	80	
Flexural modulus	GPa	2.75		2.82	2.90	2.76	2.40	>2.5
Notched impact strength (Izod)								
at 23°C	kJ/m			0.91	1.28	1.01		
at −29°C	kJ/m			0.064	0.08	0.064		
Notched impact strength (Sharpy)								
at 23°C	kJ/m²	30					>20	>30
at 0°C	kJ/m²						>8	>8
at −40°C	kJ/m²						>5	
Notched impact tensile strength	kJ/m²	250		819	884	709	170	170
Hardness (Rockwell R)				100	103	98		
Hardness (Shore D)		78					78	78
Heat distortion temperature (at 1.81 MPa)	°C			70	70	70		
Vicat temperature	°C	80					76–82	>76
Brittleness temperature	°C		−55					
Dielectric constant (10⁶ Hz)				2.75	3.00	2.72	3.0	
Dielectric strength	MV/m		43	16	14.7	16.9	28	
Dissipation factor 10⁶ Hz				0.019	0.0165	0.0197	0.0185	
Flammability 1/16″ (UL 94)				V-0	V-0	V-0		
Oxygen index	%		23.5	41	44.1	40		

molding. The flame retardant Rucodur polymers are a priceworthy alternative to flame retardant ABS and PPO polymers. Rucodur costs 0.55 $/lb.

®Vinnol K 505/68 (Wacker-Chemie) contains 5% EVAC (®Levapren with 45% VAc) onto which VC is grafted. The graft polymer. K-value 68, is used as such in extrusion for manufacturing impact resistant window frames. Vinnol K 510/68 with 10% EVA yields products with even higher impact resistance, however, it is mostly used in blends with PVC homopolymers [17].

A similar product is ®Vestolit BAU (Chem. Werke Hüls) which contains 6% of grafted EVA (®Levapren). Green, red, orange and blue colored grades have been commercially available since 1979. Applications are weather resistant window frames and sidings. With 5% and 8% EVAC, impact strengths of 40 and 50 mJ/mm^2 are obtained. Higher concentrations of EVAC give no further improvement. Prolonged homogenization on a roll compounding mill reduces impact resistance to a level of $10-15$ mJ/mm^2. This reduction occurs the sooner the smaller the concentrations of impact modifiers are. The initial honeycomb-like structure with EVAC cementing the primary PVC particles together undergoes a phase inversion when subjected to conditions of high temperature, pressure and shear. This mechanism is considered to apply generally to any impact grade PVC system with two phases [25].

References

[1] K. H. Miehl, Kunststoffe **70**/10 (1980) 591.

[2] P. Förster and M. Herner, Kunststoffe **69** (1979) 146.

[3] K. Adler and K. P. Paul, Kunststoffe **70** (1980) 411.

[4] J. T. Lutz, Polym.-Plast. Technol. Eng. **11**/1 (1978) 55; ACS Org. Coat. Plast. Chem. **46** (1982) 635.

[5] Anon., Modern Plastics Intern. (September 1981) 40.

[6] H.-G. Elias, Neue polymere Werkstoffe 1969–1974, C. Hanser Verlag, München 1975; New Commercial Polymers 1969–1975, Gordon and Breach, New York 1977.

[7] Anon., Modern Plastics Intern. (March 1981) 44.

[8] ®Amoco Resin 18, Technical Information, Amoco Chemicals, Bulletin No. R-23, R-25 and R-28 (1976).

[9] ®Diakon APA 1, Technical Information, ICI, December 1979.

[10] R. W. Gould and J. M. Player, Kunststoffe **69**/7 (1979) 393.

[11] Anon., Modern Plastics Intern. (September 1981) 39 and 72.

[12] ®Wacker VAE, Technical Information Wacker Chemie GmbH, April 1981.

[13] ®Durastrength 200 Impact Modifier, Technical Information, M&T Chemicals Inc., 1979.

[14] A. Stoloff, Plastics Engineering (July 1979) 29.

[15] ®Pantalast, Technical Information, Pantasote Inc., 1980.

[16] A. D. Varcnelli, L. Weintraub and St. Pearson, Paper SPE 38th Annual Technical Conference, May 1980.

[17] ®Vinnol K and Vinnol K 550/79, Technical Information, Wacker-Chemie GmbH, 1979 and 1977.

[18] ®Vinnuran KR 3820, Technical Information, BASF A. G., Sept. 1979.

[19] ®Vynathene VAE Copolymers, Technical Information, U.S.I. Chemicals, 1981.

[20] ®Rucodur, Technical Information, Hooker Chemicals & Plastics Corp., Ruco Division.

[21] Anon., Modern Plastics Intern. (December 1978) 54.
    ®Hostalit Z, H and HM, Technical Information, Hoechst A.G., December 1976.

[23] G. Hundertmark and G. Menzel, Der Lichtbogen **195**/3 (1979) 10.

[24] G. Menzel, G. Hundertmark and A. Polte, Kunststoffe **67** (1977) 339.

[25] D. Fleischer, J. Brandrup and O. Heinzmann, Kunststoffe **67** (1977) 312.

[26] ®Vestolit BAU, Technical Information, Chem. Werke Hüls A.G., September 1981.

2.9 MODIFIED VINYLESTER RESINS

2.9.1 Structure and synthesis

Recently, several companies started to market thermosetting resins based on acrylates and methacrylates: ®Diacryl (Akzo) [1], ®Derakane Vinyl Ester (Dow Chemical) [2, 3], and ®Spilac (Showa High Polymer Co. Ltd.) [4, 5].

Diacryl (I) is prepared by esterification of the reaction product of ethylene oxide and bisphenol-A with methacrylic acid. The Derakane Vinyl Esters (II) are obtained by reacting methacrylic acid, epichlorohydrin, and bisphenol-A. The flame retardant grades Derakane 510 A and 510 N contain bromine which is presumbably introduced by reacting tetrabromo-bisphenol-A in lieu of bisphenol-A. The production of Spilac starts from diallylidenepentaerythritol (III) which is derived from the reaction of pentaerythritol with acrolein (eq. 2-4). Compound (III) can be reacted with glycols (eq. 2-5) or with acids in the presence of Lewis acids (eq. 2-6). Spilac (V) is obtained from the reaction product (IV) by transesterification with acrylic or methacrylic acid or by esterification with those acids. Another route leads via eq. (2-6) using acrylic or methacrylic acid.

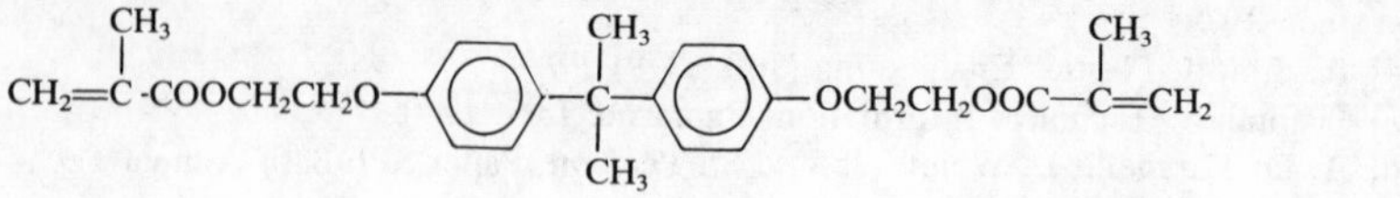

$$CH_2=\overset{\underset{\displaystyle CH_3}{|}}{C}\text{-}COOCH_2CH_2O-\!\!\!\!\bigcirc\!\!\!\!-\overset{\underset{\displaystyle CH_3}{|}}{\underset{\displaystyle CH_3}{C}}-\!\!\!\!\bigcirc\!\!\!\!-OCH_2CH_2OOC-\overset{\underset{\displaystyle CH_3}{|}}{C}=CH_2 \qquad (I)$$

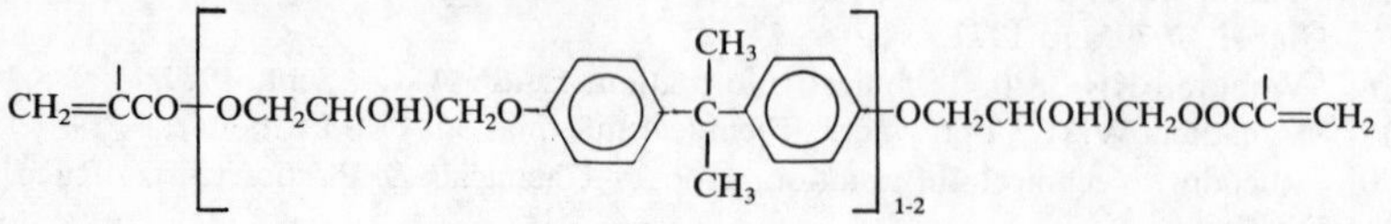

$$CH_2=CCO\left[OCH_2CH(OH)CH_2O-\!\!\!\!\bigcirc\!\!\!\!-\overset{\underset{\displaystyle CH_3}{|}}{\underset{\displaystyle CH_3}{C}}-\!\!\!\!\bigcirc\!\!\!\!-OCH_2CH(OH)CH_2OOC\right]_{1\text{-}2}C=CH_2 \quad (II)$$

(2-4)

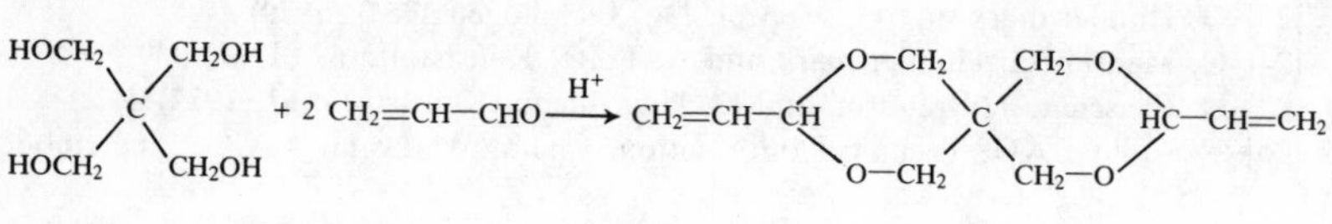

$$\overset{HOCH_2}{\underset{HOCH_2}{}}\!\!\!>\!\!C\!\!<\!\!\overset{CH_2OH}{\underset{CH_2OH}{}} + 2\ CH_2=CH-CHO \xrightarrow{H^+} CH_2=CH-CH \overset{O-CH_2}{\underset{O-CH_2}{}}\!\!\!>\!\!C\!\!<\!\!\overset{CH_2-O}{\underset{CH_2-O}{}}\!\!\!HC-CH=CH_2 \quad (III)$$

(2-5)

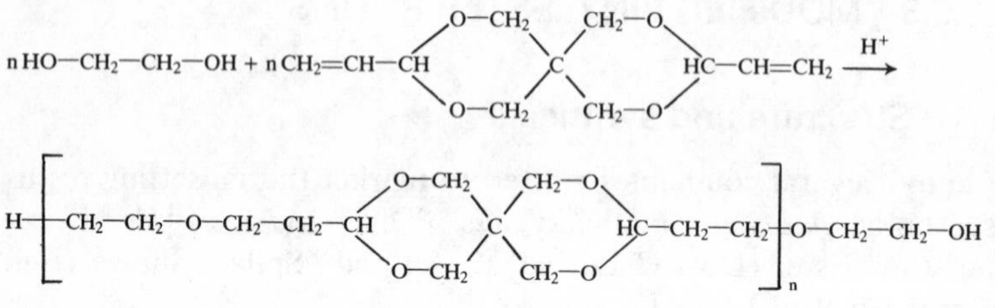

$$n\ HO-CH_2-CH_2-OH + n\ CH_2=CH-CH \overset{O-CH_2}{\underset{O-CH_2}{}}\!\!\!>\!\!C\!\!<\!\!\overset{CH_2-O}{\underset{CH_2-O}{}}\!\!\!HC-CH=CH_2 \xrightarrow{H^+}$$

$$H\left[CH_2-CH_2-O-CH_2-CH_2-CH \overset{O-CH_2}{\underset{O-CH_2}{}}\!\!\!>\!\!C\!\!<\!\!\overset{CH_2-O}{\underset{CH_2-O}{}}\!\!\!HC-CH_2-CH_2\right]_n O-CH_2-CH_2-OH \quad (IV)$$

(2-6)

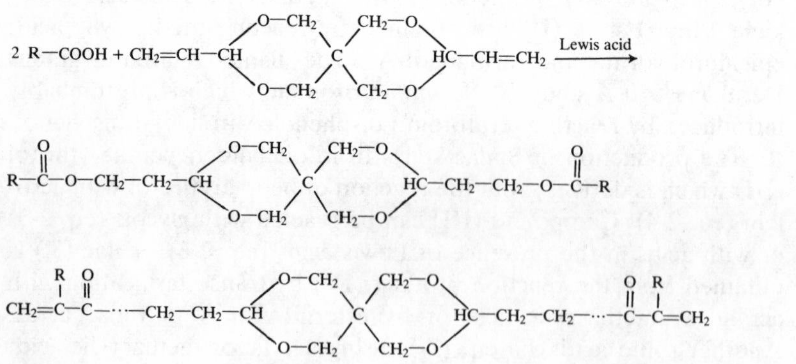

$$2\ R-COOH + CH_2=CH-CH \overset{O-CH_2}{\underset{O-CH_2}{}}\!\!\!>\!\!C\!\!<\!\!\overset{CH_2-O}{\underset{CH_2-O}{}}\!\!\!HC-CH=CH_2 \xrightarrow{Lewis\ acid}$$

$$R-\overset{\overset{\displaystyle O}{\|}}{C}-O-CH_2-CH_2-CH \overset{O-CH_2}{\underset{O-CH_2}{}}\!\!\!>\!\!C\!\!<\!\!\overset{CH_2-O}{\underset{CH_2-O}{}}\!\!\!HC-CH_2-CH_2-O-\overset{\overset{\displaystyle O}{\|}}{C}-R$$

$$CH_2=\overset{\overset{\displaystyle R}{|}}{\underset{}{C}}-\overset{\overset{\displaystyle O}{\|}}{C}\cdots CH_2-CH_2-CH \overset{O-CH_2}{\underset{O-CH_2}{}}\!\!\!>\!\!C\!\!<\!\!\overset{CH_2-O}{\underset{CH_2-O}{}}\!\!\!HC-CH_2-CH_2\cdots \overset{\overset{\displaystyle O}{\|}}{C}-\overset{\overset{\displaystyle R}{|}}{C}=CH_2 \quad (V)$$

The ®Silmar Resin S-808 (Vistron, Silmar Division) is based on a modified bisphenol-A polyester [6]. Further details about the structure of the resin have not been revealed, possibly, structures similar to (I) or (II) exist.

For ®Atlac 580 (ICI) the structure (VI) has been disclosed [7, 8] which indicates internal and terminal unsaturated groups.

U = Urethane groups   (VI)
R = Hydrogen or alkyl groups

Presumably, the preparation is carried out by reacting a bisphenol-A fumaric acid polyester containing terminal bisphenol-A groups with diisocyanate and glycol. The resulting product with terminal hydroxy groups is then esterified with acrylic or methacrylic acid.

The resins are supplied in solutions of styrene or vinyltoluene. Solvent-free resins are also available. The Spilac resins were designed to be used without styrene.

The above resins are commonly described as vinyl resins although the hitherto known structures represent acrylic or methacrylic resins.

2.9.2   Properties and applications

All resins yield thermosets and thus compete with unsaturated polyesters. Usually they are cured with peroxide/cobalt naphthenate systems with the exception of the Spilac resins which allow for peroxide-free curing after adding cobalt compounds.

Derakane Resins 411 and 411 C are recommended for filament winding, rotational casting and low pressure molding processes. The cured resins show high corrosion resistance. Resin 510 A is also corrosion resistant and in addition has an improved flame resistance. Resin 470 exhibits superior solvent, oxidation, corrosion, and thermal resistance. Resin 510N combines the advantages of Resin 470 and Resin 510 A. Resin 790 containing 50−65% chopped glass fibers is used as prepreg for SMC (Sheet Molding Compound) and HSMC (High Strength Molding Compound), particularly in the automotive industry [3].

The Spilac resins demonstrate exceptional good weatherability due to the absence of aromatic groups. They have low viscosities and high tracking resistance. Grades with improved heat resistance and flexural strength are available; these are recommended for fiber reinforced and SMC materials. Special grades are offered for binders and adhesives, as well as for water soluble or dispersable systems.

®Atlac displays high corrosion resistance and good mechanical properties at elevated temperatures.

Properties of unfilled cured Derakane resins have been described [9]. Properties of laminates from modified vinyl resins are summarized in Table 2-21.

Properties of Derakane HSMC materials have been reported [3]. Static and dynamic mechanical properties in correlation with glass and calcium carbonate content were determined in a temperature range of −30 to 150°C. Sheets of 35.6 cm × 71.1 cm were cured 3 minutes at 149°C under 3.5 bar pressure. Molding materials reinforced with 10% glass and 65% calcium-carbonate demonstrate excellent properties at 150°C: tensile strength 105 MPa, flexural strength 140 MPa, and flexural modulus 9.5 GPa. The materials are designed to replace metals in the automotive industry. Precautions must be taken to adjust materials with different coefficients of thermal expansion. Table 2-22 compares properties of reinforced Derakane 790 with those of metals [3].

References

[1] H. Saechtling, Kunststoffe **65** (1975) 836.
[2] ®Derakane Vinyl Ester Resins, Technical Information, Dow Chemical Co.
[3] J. H. Enos, R. L. Erratt, E. Francis and R. E. Thomas, Polymer Composites **2** (1981) 53.
[4] E. Takiyama, T. Hanyuda and T. Sugimoto, Jap. Plastics **9**/2 (1975) 6, 29.
[5] E. Takiyama, Plastics Age **21** (1975) 93.
[6] ®Silmar, Technical Information, Vistron Corp., Silmar Division.
[7] ®Atlac, Technical Information, ICI United States, Inc.
[8] R. J. Lewandowski, E. C. Ford, Jr., D. M. Longenecker, A. J. Restaino and J. P. Burns, SPE 30th Ann. Techn. Conf. (1975) Papers Sect. 6-B, 1.
[9] H.-G. Elias, Neue polymere Werkstoffe 1969−1974, Carl Hanser Verlag, München 1975; New Commercial Polymers 1969−1975, Gordon and Breach, New York 1977.

TABLE 2-21

Mechanical properties of laminates from modified vinyl ester resins.

Physical Properties	Physical unit	Derakane 411	Derakane 470	Property values of Derakane 510A	Derakane 510N	Atlac 580-05
Styrene content of resin	%	45	45	40	36	47
Glass content	%	40	40	40	40	35
Tensile strength at break						
at room temperature	MPa	142.7	117.9	113.1	144.8	104.8
at 93°C	MPa	150.3	119.3	134.5		110
at 149°C	MPa	53.1	99.3			52
Tensile modulus						
at room temperature	GPa	12.0	10.6	10.3	9.5	7.2
at 93°C	GPa	10.3	10.6	9.0		
at 149°C	GPa		6.1			
Flexural strength						
at room temperature	MPa	204.1	157.2	164.1	172.4	179.6
at 93°C	MPa	188.9	168.2	165.5		172
at 149°C	MPa	22.1	97.2			31
Flexural modulus						
at room temperature	GPa	7.1	8.2	7.6	7.9	7.6
at 93°C	GPa	5.9	7.0	6.7		5.2
at 149°C	GPa	3.6	2.6			1.2

Laminates were prepared by hand lay-on. Laminates with Derakane resins were composed of screen, two mats, roving mat, mat, roving mat, mat; laminates with Atlac resins were composed of screen, three mats, roving mat, mat.

TABLE 2-22

Coefficient of linear thermal expansion, α, of Derakane 790 molding resins at various temperatures as compared to metals.

Material	$10^6 \, \alpha/K^{-1}$ at			
	$-40°C$	$-30°C$	$71°C$	$150°C$
Resin 35%/Glass 60%	16.6	16.5	12.9	12.2
Resin 25%/Glass 50%/CaCO$_3$ 25%	17.0	16.9	13.4	11.7
Resin 25%/Glass 40%/CaCO$_3$ 35%	17.4	17.1	13.5	12.4
Steel		11	12	13
Aluminum		22	23	24

2.10 POLY(ISOBUTYLENE-CO-MALEIC ANHYDRIDE)

Poly(isobutylene-co-maleic anhydride) is a developmental polymer supplied by The Humphrey Chemical Co. under the designation Poly(isobutenyl-succinic anhydride).

Maleic anhydride undergoes a fast reaction with low carbon alpha olefins at 70°C under pressure. The reactivity of the olefins decreases in the order isobutylene, propylene, ethylene, butene-1 [1]. Poly(isobutylene-co-maleic anhydride) is obtained in good yields after 1 h at 65°C by redox polymerization with benzoyl peroxide and p-toluene sulfonic acid [2].

The copolymer contains alternate units, giving the structure:

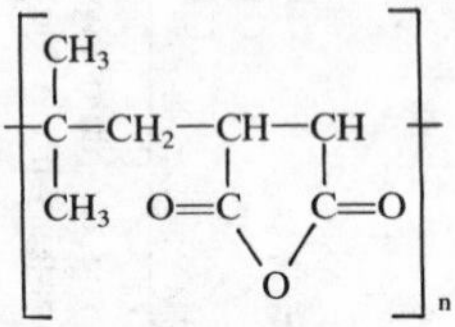

$$\left[\begin{array}{c} CH_3 \\ | \\ -C-CH_2-CH-CH- \\ | \qquad | \quad | \\ CH_3 \quad O=C \quad C=O \\ \diagdown \diagup \\ O \end{array} \right]_n \qquad VII$$

The copolymer supplied by Humphrey under code K-66 is a brown viscous liquid with a molecular weight of 1018 and acid number 98–115 [3]. The product is recommended as corrosion inhibitor, hardener for epoxy resins and as additive in detergents for cleaning metal surfaces. Prices are $ 16.92/kg and 4.80 $/kg for quantities of 19 and 210 liters, respectively.

References

[1] H. P. Frank, Makromol. Chem. **114** (1968) 113.
[2] H. P. Frank and W. Lang (Österr. Stickstoffwerke), Brit. P. 1047768 (1966).
[3] The Humphrey Chemical Co., Technical Information, October 1978 and June 1981.

2.11   POLY(1-OCTADECENE-CO-MALEIC ANHYDRIDE)

2.11.1   Structure and synthesis

As early as 1971, Gulf Oil Chemicals offered copolymers of alpha-olefins and maleic anhydride. The developmental products were designated PA-6, PA-10, PA-14 and PA-18 whereby the figures indicated the number of the olefinic carbon atoms. Several years later the Advanced Materials Division of Gulf started the production of poly(1-octadecene-co-maleic anhydride) under the name Gulf PA-18/Polyanhydride Resin.

Gulf produces 1-octadecene from ethylene through the Ziegler step growth reaction with aluminum triethyl at 150−200°C and ethylene pressure of 140−280 bar. The product spectrum consists of C_4 to C_{30} olefins from which the volatile alpha-olefins are separated by fractional distillation.

The polymerization of 1-octadecene with maleic anhydride and di-tert-butyl peroxide as initiator is carried out either in bulk [1] or in xylene under reflux [2]. The resulting copolymer contains alternate units of 1-octadecene and maleic anhydride in 1:1 molar ratio.

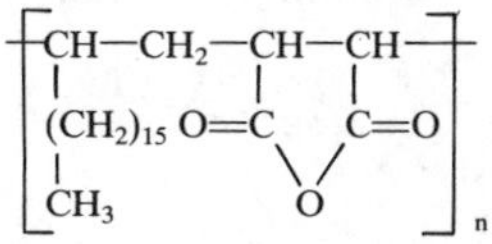

VIII

2.11.2   Properties and applications

PA-18 is a white powder, density 0.97 g/cm^3, molecular weight ca. 50000 g/mol, inherent viscosity (5.0 g PA-18 in 100 ml methylisobutyl ketone at 25°C) 10−13 ml/g, mp 110−120°C, melt viscosity at 150°C and 160°C 20 and 8 Pa·s, respectively. PA-18 shows a good thermal stability, weight losses are 1% at 250°C, 3% at 300°C, 10%

at 350°C, and 23% at 400°C. Volatiles are <1% and residual monomers <3%. The copolymer contains 15–23% of maleic anhydride units corresponding to 3.10–3.25 mmol/g; in addition, it has 5–10% of maleic acid units causing the rather high acid number of 270–310 [3].

PA-18 is very reactive due to its anhydride groups. The reactions lead to a number of technical products. PA-18 competes with other olefin/maleic anhydride copolymers, such as ethylene-based EMA-Resins (Monsanto) and isobutylene-maleic anhydride copolymers (Kuraray Co. and The Humphrey Chemical Co.).

Its hydrophilic and hydrophobic properties render PA-18 soluble in aqueous alkalines, aromatic, cycloaliphatic, and chlorinated hydrocarbons, esters, and ketones. Methanol and propanol yield soluble esterification products. Disodium and potassium salts are obtained by treating PA-18 with 30% aqueous solutions of NaOH or KOH at temperatures above 85°C. The salts are precipitated by methanol. Addition of polyvalent cations to the aqueous salt solutions yields insoluble precipitates. On acidification the free acid is recovered:

(2-7)

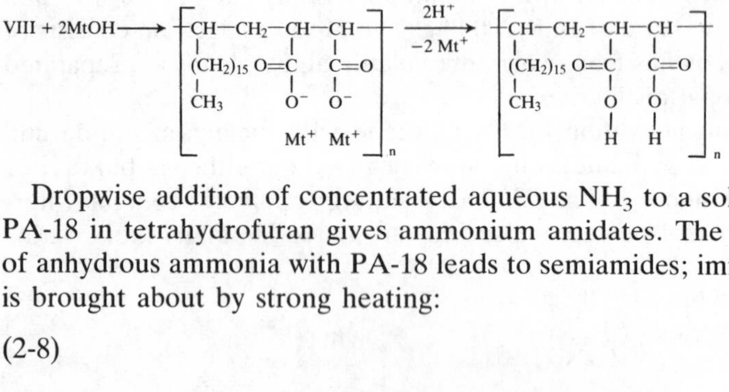

Dropwise addition of concentrated aqueous NH_3 to a solution of PA-18 in tetrahydrofuran gives ammonium amidates. The reaction of anhydrous ammonia with PA-18 leads to semiamides; imidization is brought about by strong heating:

(2-8)

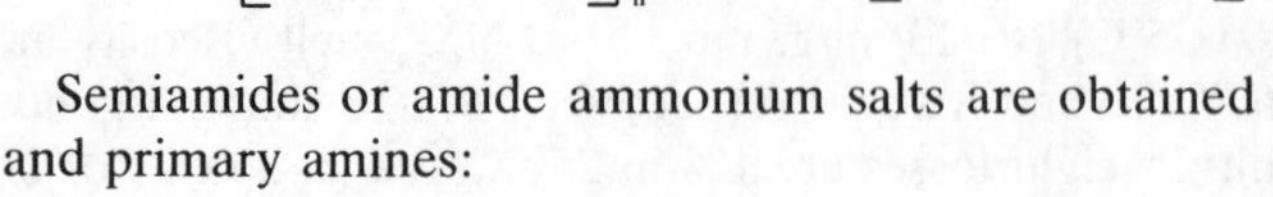

Semiamides or amide ammonium salts are obtained from PA-18 and primary amines:

(2-9)

$$\begin{bmatrix} CH-CH_2-CH-CH \\ | \quad\quad | \quad | \\ (CH_2)_{15} \quad O{=}C \quad C{=}O \\ | \quad\quad | \quad | \\ CH_3 \quad\quad O^- \quad NH \\ \quad\quad\quad + \quad | \\ \quad\quad\quad NH_3 \quad R \\ \quad\quad\quad | \\ \quad\quad\quad R \end{bmatrix}_n$$

VIII $\xrightarrow{+2RNH_2}$

$\xrightarrow{+RNH_2}$

$$\begin{bmatrix} CH-CH_2-CH-CH \\ | \quad\quad | \quad | \\ (CH_2)_{15} \quad O{=}C \quad C{=}O \\ | \quad\quad | \quad | \\ CH_3 \quad\quad OH \quad NH \\ \quad\quad\quad\quad | \\ \quad\quad\quad\quad R \end{bmatrix}_n$$

Semiesters are prepared by dropwise addition of the calculated amount of alcohol to a solution of PA-18 in methyl-iso-butyl-ketone containing sulfuric acid or p-toluene sulfonic acid as catalysts, whereas excess alcohol and removal of water leads to the formation of esters:

(2-10)

$$\begin{bmatrix} CH-CH_2-CH-CH \\ | \quad\quad | \quad | \\ (CH_2)_{15} \quad O{=}C \quad C{=}O \\ | \quad\quad | \quad | \\ CH_3 \quad\quad OR' \quad OH \end{bmatrix}_n$$

VIII $\xrightarrow{+R'OH}$

$\xrightarrow{+2R'OH}$

$$\begin{bmatrix} CH-CH_2-CH-CH \\ | \quad\quad | \quad | \\ (CH_2)_{15} \quad O{=}C \quad C{=}O \\ | \quad\quad | \quad | \\ CH_3 \quad\quad OR' \quad OR' \end{bmatrix}_n$$

PA-18 is an effective hardener for epoxy metal adhesives and in resin systems which contain epoxy or hydroxy groups. Thus a cross-linked polymer is obtained by reacting PA-18 with a polymeric polyol, e.g. poly(vinyl alkohol). At temperatures above 80°C the crosslinks dissociate and the material becomes thermoplastic. Slow cooling restores the network and products insoluble in acetone are formed. Crosslinked materials are also obtained by heating dry PA-18 with IIA-group oxides or hydroxides, e.g. CaO or Ca(OH)$_2$.

Recommended applications for PA-18 are as release agents, for paper coating, corrosion protection, as thickeners, chelate formers, adhesives, hardeners for epoxy resins, additives in basic dyestuffs to improve light fastness, dispersing agents, and hydrophobation of clay [3].

The prices for PA-18 range from 6 $/lb for small amounts down to 2.20 $/lb for delivery of more than 20 000 lb (9100 kg).

References

[1] Brit. P. 1 245 879 (1971), to Mobil Oil Co.
[2] Brit. P. 1 121 464 (1968), to Monsanto Co.
[3] Gulf PA-18/Polyanhydride Resin, Technical Information Gulf Oil Chemicals, Advanced Materials Division.

2.12 POLY(DIVINYLETHER-CO-MALEIC ANHYDRIDE)

Alternating cyclopolymers from divinylether (I) and maleic anhydride (II) are known as Pyran copolymers (Hercules Co.) or DIVEMA. The polymers with the presumed structure III are obtained by radical copolymerization of divinylether and maleic anhydride through inter- and intramolecular reaction steps [1]:

(2-11)

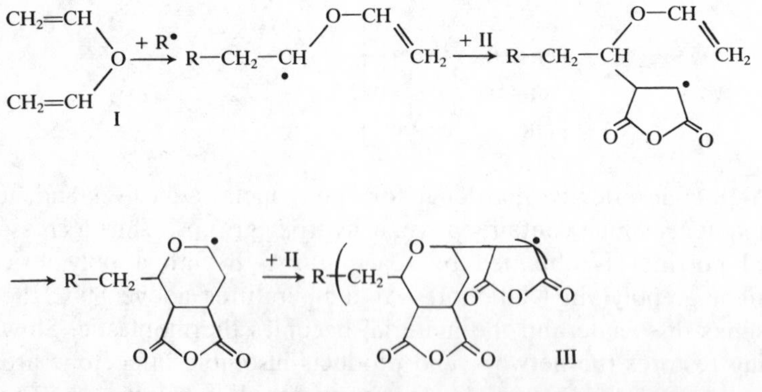

The polymer contains units of divinylether (as part of a six-membered ring) and maleic anhydride in 1:2 molar ratio, however, formation of five-membered ether ring structures cannot be excluded.

After hydrolysis and neutralization, DIVEMA shows biological activity, e.g. tumor inhibition, induction of interferon formation, inhibition of viral RNA dependent DNA polymerase (reverse transcriptase) and activation of macrophages in stimulation of immune reactions in test animals. Furthermore, DIVEMA demonstrates bacteriostatic, fungistatic and anti-arthritic properties [2]. Some of the biological activities are strongly influenced by molecular weight [3]. DIVEMA [4] and related polymers [3] are presently being tested for their clinical usefulness [5, 6].

References

[1] G. B. Butler, J. Macromol. Sci.-Chem. A-5 (1971) 219.
[2] D. S. Breslow, Pure Appl. Chem. **46** (1976) 103.
[3] G. B. Butler, in L. G. Donaruma, R. M. Ottenbrite and O. Vogl (Ed.), Anionic Polymeric Drugs, Wiley, New York 1980, 49; R. Takatsuka and R. M. Ottenbrite, Org. Coat. Plast. **46** (1982) 310.
[4] D. S. Breslow, ACS Polymer Preprints **22** (1981) 24.
[5] G. B. Butler, J. Macromol. Sci.-Rev. Macromol. Chem. Phys. C **22**, 89 (1982/3).
[6] J. R. Baldwin, R. A. Carrano, A. R. Imondi, J. D. Iuliucci and L. M. Hagerman, Polym. Mat. Sci. Engng. **51**, 136 (1984).

2.13 POLY(PHENYLENES)

2.13.1 Structure and synthesis

Poly(p-phenylene) with the structure

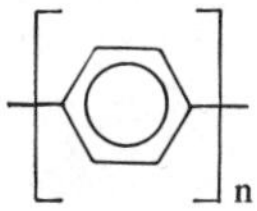

I

was the first commercial material of this class, already available in 1967 from Eitel-McCollough under the name ®Eimac 221. It was prepared by the oxidative cationic coupling of benzene [1].

Low-molecular weight, soluble, and fusible poly(phenylenes) were commercially produced in 1968 by Monsanto via oxidative cationic polymerization of terphenyl isomers. The melt process, optionally carried out in the presence of other compounds such as biphenyl or biphenyl/phenol, resulted in the formation of prepolymers which were soluble in chloroform, dichloroethane, or chlorobenzene. These solutions, containing m-benzenedisulfonyl chloride, p-toluene sulfonic acid, sulfuryl chloride, or boron fluoride etherate with Cu_2Cl_2 as catalysts, served to fabricate asbestos or graphite prepregs. After evaporation of the solvent, heating at 300°C under 35−70 bar yielded an insoluble and infusible polymer. The Hughes Aircraft Co. promoted the development of these materials.

Poly(phenylenes) were recommended for use as semiconductors, photoconductors, solid lubricants, insulators, pigments, ablatives, stabilizers to heat and light, laminate binders, graphite fiber precursors, and ion exchange resins. Preparation, properties, and processing of the earlier poly(phenylenes) have been extensively described [1−6].

In 1974 Hercules, Inc., introduced a new family of crosslinkable oligophenylenes [7−9] under the designation Hercules ®H-Resins. The branched prepolymers have the simplified structure given below.

The preparation of H-Resins follows the synthetic route proposed by Korshak [11]. Reactive oligophenylenes with terminal acetylene groups are obtained by copolycyclotrimerization of p-diethynylbenzene (DEB) with phenylacetylene or by cyclotrimerization of DEB [10] in aromatic solvents with $AlR_3/TiCl_4$, $NiCl_2/P(C_6H_5)_3/NaBH_4$,

®H−Resin
Prepolymer

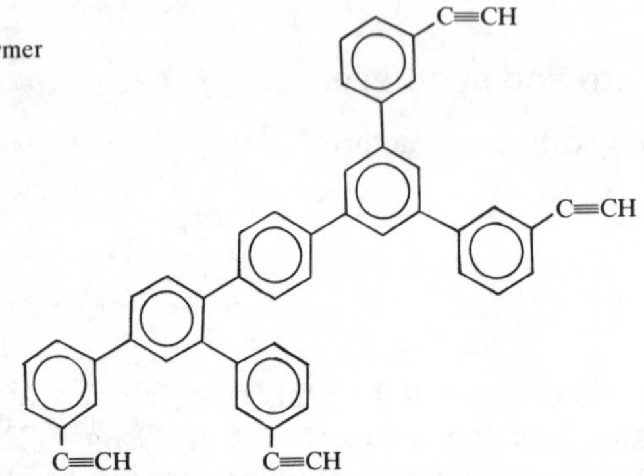

II

$Ni(C_5H_7O_2)_2/P(C_6H_5)_3$, or $CoBr_2/(C_6H_5O)_3P$ as catalysts.

Soluble prepolymers, representing the so-called B-stage resins, are obtained from DEB with $TiCl_4/Al(C_2H_5)_2Cl$ in 75% yield at 86% monomer conversion. Higher conversions result in gelation. The prepolymers still contain considerable amounts of unreacted DEB. The number and weight average molecular weights of the monomer-free prepolymer after 86% monomer conversion are $\overline{M}_n$ = 2300 g/mol and $\overline{M}_w$ = 8060 g/mol [10].

2.13.2   Processing, properties, and application of ®H-Resins

The B-stage prepolymers are soluble in aromatic and halogenated hydrocarbons, ketones, and cyclic ethers to give solutions with more than 50% solids. The prepolymers are insoluble in water, alcohols, and aliphatic hydrocarbons.

Hercules Inc. supplies 3 grades of H-Resins: H-A 43 for laminates and molding processes, H-112 for solution coating, and H-132 for powder coating [9].

Curing of H-A 43 starts after 10 minutes of heating at 130°C. Common molding cycles are 2−5 minutes at 160−170°C [8]. Post-curing at 230°C improves mechanical properties and solvent resistance, hence H-Resins can be processed at lower temperatures than many reactive poly(imides), poly(amide-imides), poly(sulfones), or poly(phenylenesulfide). The initial melt viscosity of H-A 43 at 130°C is 100 Pa·s, sufficiently low to take filler loads up to 90% of glass fiber, talcum, silica, or graphite.

No volatile products are generated in curing. The cured resins are hard but not porous. They exhibit good mechanical properties (Table 2-23). The continuous service temperature is 200−300°C in air and 400°C in an oxygen-free atmosphere. At 500°C slow carbonization begins.

Laminates are fabricated by treating glass or graphite fibers with a 50% solution of the prepolymer in methyl ethyl ketone. Laminates with 60% E-glass demonstrate flexural strengths of 414−552 MPa and flexural moduli of 21−34 GPa. Unidirectional laminates with 60% carbon fiber display a flexural strength of 1379 MPa and a flexural modulus of 110 GPa at room temperature; at 230°C more than 90% of these values are retained [8].

Corrosion resistant coatings are obtained from H-112 and H-132 [9]. Metal surface adhesion is enhanced by primers such as epoxides, phenolic resins, PPO, poly(sulfones), or poly(imides). For applications at temperatures of above 250°C only poly(imides) are suitable as primer. The coatings withstand heating in air at 200−220°C for more than 10 000 hours. They are resistant for up to 300 h against the attack of 50% sulfuric acid at 90°C, trichloroethylene at 90°C, methyl isobutyl ketone at 120°C, and toluene at 110°C. Neither 168 h in boiling water nor 500 h at 38°C in a salt spray test, nor in an atmosphere with 100% relative humidity show any effect.

TABLE 2-23
Properties of unfilled and filled/cured ®H-Resins [8, 10].

Physical Properties	Physical unit	H-Resin unfilled	Property value for H-Resin filled with		
			talcum 50%	silica 50%	asbestos 30%
Density, uncured	g/cm³	1.35			
cured	g/cm³	1.145			
Flexural strength					
at 23°C	MPa	34−69	48−69	41−83	55−83
at 250°C	MPa		48−62	62	
at 350°C	MPa		48−55	46	
at 360°C	MPa	34−55			
Flexural modulus					
at 23°C	GPa	4.83	9.65	10.3−12.4	10.3−12.4
at 250°C	GPa		6.27	5.86	
at 300°C	GPa	3.45			
at 350°C	GPa		6.76	4.62	
Barco hardness		85			
Taber abrasion C 17					
(1000 g load)	g	0.0015			
Oxygen index	%	>55			
Specific heat					
at 23°C	J/g	1.22			
at 125°C	J/g	1.45			
Thermal conductivity	W m⁻¹ K⁻¹	2.72×10^{-3}			
Enthalpy at cure	J/g	max. 837			
Volume resistivity	Ohm × cm	10^{17}			
Dissipation factor					
60 Hz at 23°C		0.002			
60 Hz at 100°C		0.0009			

H-resins are used for corrosion resistant coatings, e.g. in pipelines, for pumps, for geothermal drilling operations, in sodium and lithium high-temperature batteries, as non-inflammable coatings for electric and electronic parts and as anti-friction resins. However, it was hinted that Hercules Inc. may shut down its production of H-resins [12].

References

[1] P. Kovacic and F. W. Koch, Polyphenylenes, in Encycl. Polym. Sci. Technol. **11** (1969) 380.

[2] G. K. Noren and J. K. Stille, Polyphenylenes, J. Polym. Sci.-Macromol. Revs. **D 5** (1971) 285.

[3] J. G. Speight, P. Kovacic and F. W. Koch, J. Macromol. Sci.-Rev. Macromol. Chem. **C 5** (1971) 295.

[4] N. Bilow and L. J. Miller, J. Macromol. Sci. Chem. **A 3** (1969) 501.

[5] G. Ensor, Brit. Polym. J. **2** (1970) 264.

[6] H.-G. Elias, in Ullmanns Encycl. Techn. Chem. 4th ed., **15** (1978) 424.

[7] L. C. Cessna, Jr., and H. Jabloner, J. Elast. Plast. **6** (1974) 103.

[8] T. M. Bednarski, J. H. DelNero, R. H. Mayer and J. A. Hagan, SPE Ann. Techn. Papers **21** (1975) 90.

[9] J. French, ACS Coat. Plast. Prepr. **35/2** (1976) 72.

[10] J. Jabloner and L. C. Cessna, Jr., ACS Polymer Prepr. **17** (1976) 169.

[11] V. V. Korshak et al., ACS Polymer Prepr. **16/1** (1975) 328.

[12] Dr. E. Vandenberg, Hercules Inc., personal commun., May 1981.

2.14 HYDROGENATED NITRILE RUBBERS

Nitrile rubbers (copolymers from butadiene and acrylonitrile) can be selectively hydrogenated at the carbon-carbon double bond [1, 2] under improvement of their heat stabilities. The hydrogenated polymers have also better mechanical properties in the cold. Since they are also more swellable than the starting materials, their nitrile content has to be increased relative to conventional nitrile rubbers. Bayer introduced recently a hydrogenated nitrile rubber under the trade name Therban® [3].

The Therban polymers are practically saturated and can no longer be vulcanized by sulfur but rather with peroxides or high energy radiation. The polymers possess excellent mechanical properties, high resistance against abrasion, excellent aging behavior, and a good resistance against various oils [3−5].

References

[1] Brit. P. 1558491 (3 January 1980), Belg. P. 845775 (2 March 1977); D. Oppelt, H. Schuster, J. Thörmer and R. Braden, to Bayer AG.

[2] A. H. Weinstein, Rubber Chem. Technol. **57**, 203 (1984).

[3] J. Thörmer, G. Marwede and B. Buding, Kautsch. Gummi Kunststoffe **36**, 269 (1983).

[4] J. Mirza, J. Thörmer und H. Buding, Rubber Chem. Technol. **57**, 413 (1984).

[5] W. Hofmann, Kunststoffe **74**, 613 (1984).

3 Unsaturated carbon chains

3.1 1,2–1,4-POLY(BUTADIENES) OR "VINYL-POLYBUTADIENES"

3.1.1 Structure and synthesis

Poly(butadienes) with various ratios of 1,4- to 1,2-addition of buta-diene units have been termed "Vinyl-Polybutadienes". Variables including catalyst systems, solvents, and polymerization tempera-ture exert an influence on the degree of 1,2-addition.

Chemische Werke Hüls uses Li alkyls as catalysts in solution poly-merization [1–6]. Addition of Lewis bases provides a simple way to adjust the ratio of 1,2- to 1,4-linked butadiene units in a controlled manner. Besides altering the microstructure, the macrostructure, i.e., molecular weight distribution and long chain branching, and the processability can be influenced. These have been achieved by applying chain coupling techniques to the "living polymers" present in the polymerization medium, at elevated temperatures and under adiabatic process conditions [7].

Ube Industries, Ltd., employs a two-stage polymerization of buta-diene with two catalyst systems [8, 9]. First, a typical cis-catalyst, e.g., cobalt octanoate/diethyl aluminum chloride, is used to carry the polymerization to the desired conversion. Then, a 1,2-system is created through the addition of carbon disulfide and aluminum trialkyl. This procedure gives only small amounts of block polymers; the main reaction product is a mixture of cis-1,4- and syndiotactic 1,2-poly(butadienes). However, the small portion of block polymers allows the coupling of the otherwise incompatible homopolymer phases [10]. The poly(butadiene) with partially segmental vinyl groups is supplied by Ube under the designation ®Ubepol VC PBD.

3.1.2 Properties and applications

On a series of poly(butadienes) which all have approximately the same raw Mooney viscosity of 45–47, weight average molecular weights of 247 000–374 000 g/mol and number average molecular weights of 79 400−125 000 g/mol, the influence of vinyl group content in the range of 1−65% has been investigated by Chem. Werke Hüls [6], see Table 3-1. Processability of such poly(butadienes) depends less on their microstructure but is dominated by the macrostructure, e.g. molecular weight distribution [5, 6]. Increasing the content of vinyl groups causes no enhanced crosslinking activity, which is independent of the various sulfurous curing agents applied. Hence, crosslinking occurs mainly via the double bonds in the backbone chain. With peroxides the vulcanization rate increases with increasing vinyl content, however, this is attributed to be more a function of augmented tertiary carbon atoms. Increasing the content of vinyl groups exerts no effect on vulcanizates in hot air aging. Apparently, the pending double bonds initiate no cyclization reactions when heated in the presence of oxygen. Also, there is no interdependency between vinyl content and tensile strength in the temperature range of −20° to 70°C except at very high deformation rates. Here tensile strength at break is lowered with increasing vinyl content.

Increasing the vinyl content results in a considerable rise of glass transition temperatures T_g due to a reduction in thermal mobility of the chain segments. Pure cis-1,4-poly(butadiene) has a T_g of −100°C; increasing the vinyl content to 70% leads to a T_g of ca. −35°C and to poor low-temperature properties. Swelling in oil, gas diffusion and resilience decrease with increasing vinyl content in the temperature range of −20° to 70°C. Surprisingly, a lower heat-build-up is observed in spite of reduced elasticity. This effect becomes more obvious the tougher the chosen test conditions are. It is explained by the excellent reversion resistance of polymers with high vinyl content at elevated temperatures, because the increasing vinyl content diminishes the number of double bonds in the main chain and thus minimizes the likelihood of such side reactions. The greater reversion resistance allows higher vulcanization temperatures and leads to shorter and more economic heating cycles.

A correlation exists between glass temperature, abrasion resistance and wet grip performance. With respect to their abrasion/wet grip property balance vinyl-poly(butadienes) with 35% and 55% vinyl correspond to the performance of SBR/cis-BR compounds and E-SBR, respectively. With vinyl-poly(butadienes) in passenger tires it is possible to get both improved wet safety and reduced rolling resistance. Similar results are obtained with the new solution SBR

TABLE 3-1

Properties of vinyl-poly(butadienes) compared to cis-1,4-BR (Ubepol BR 150).

Physical Properties	Physical unit	Ubepol VC PBD[1]	Ubepol BR 150[1]	Property values of				
				Hüls I[2]	Hüls II[2]	Hüls III[2]	Hüls IV[2]	Hüls V[2]
Microstructure								
1,2 (vinyl)	%	8	1	1	10	35	53	65
cis	%	91	98	2	37	26	18	15
trans	%	1	1	97	53	39	29	20
Compound viscosity ML-4		50	43	70	68	64	71	65
Tensile strength	MPa	21	21	13.9	12.2	11.9	13.6	12.5
Elongation at break	%	390	540	472	457	415	421	435
Modulus 300%	MPa	16.1	9.9	7.4	7.1	7.6	7.5	7.3
Hardness, Shore A		70	60	62	62	61	62	62
Rebound resilience 20°C	%			50	47	45	42	40
70°C	%			55	52	49	47	47
Tear resistance	kN/m	45	40					
Crack propagation from 2 to 15 mm	cycles	141 000	3800					
Goodrich flexometer test (45 lbs, 4.45 mm, 100°C, 60 min)	°C			60	52	41	33	31

[1] Recipe (phr): rubber 100.0, HAF carbon black (N330) 50.0, aromatic oil 10.0, ZnO 5.0, stearic acid 2.0, phenyl-beta-naphthyl-amine 1.0, sulfur 1.5.
Cure 40 min at 140°C.

[2] Recipe (phr): rubber 100.0, HAF carbon black 50, HAR oil 5, 4010 NA 1, ZnO 3, stearic acid 2, CBS 0.75, sulfur 2.

types, see chapter 3.2. The low heat-build-up and the high reversion resistance of vinyl-poly(butadienes) allow blending with NR for use in truck tire treads. First road tests with truck tires made from NR/vinyl-poly(butadiene) with 50% 1,2 have shown that such blends lead to lower tread temperatures [6]. The new vinyl-poly(butadienes) and solution SBR types of Chem. Werke Hüls still bear preliminary experimental designations and are collectively tagged "BUNA VI" [11].

In Ube's ®Ubepol VC PBD the vinyl blocks form finely dispersed crystalline domains securing a higher green strength in raw compounds. With vulcanization, a stress induced crystallization occurs producing considerable improvement in moduli, split tear resistance, and flexural crack resistance. In particular, dynamic crack growth is reduced compared to conventional BR vulcanizates (Table 3-1). Ubepol VC PBD has been developed primarily for use in tires.

References

[1] K. H. Nordsiek, Kautschuk und Gummi—Kunststoffe 25 (1972) 87.

[2] K. H. Nordsiek, PRT Polymer Age 4/9 (1973).

[3] K. H. Nordsiek, Der Lichtbogen (Magazin Chem. Werke Hüls) September 1980, 10.

[4] K. H. Nordsiek and K. H. Kiepert, Kautschuk, Gummi, Kunststoffe 33/4 (1980) 251.

[5] J. Markert, Kautschuk, Gummi, Kunststoffe 34/4 (1981) 269.

[6] K. H. Nordsiek and K. M. Kiepert, Internat. Rubber Conf., Harrogate/England, June 10, 1981; Kautschuk, Gummi Kunststoffe 35/5 (1982) 371.

[7] N. Sommer, Kautschuk, Gummi, Kunststoffe 28 (1975) 131.

[8] M. Kono, H. Ishikawa and M. Takayanagi, IISRP Conf., Hongkong, April 1978; Ger. Offen. 1 933 620 (1969) and Ger. Offen. 2 163 542 (1971), to Ube Industries Ltd.

[9] Anon., Japan Chemical Week (January 1, 1981) 5.

[10] J. Witte, Angew. Makromol. Chem. 94 (1981) 119.

[11] Dr. Hesse (Chem. Werke Hüls), personal communication, May 7, 1982.

3.2 COPOLYMERS OF STYRENE AND 1,2–1,4-POLY(BUTADIENES) ("SOLUTION SBR")

3.2.1 Structure and synthesis

Selected variation of micro- and macrostructure allows preparation of new SBR types with improved property characteristics. With regard to microstructure, the ratio of 1,2-/1,4-addition, cis/trans ratio and styrene content are variables, while those in macrostructure are average molecular weight, molecular weight distribution and long chain branching. Target of the development has been a rubber for tires which both reduces rolling resistance and thus fuel consumption and maintains or possibly improves wet grip.

It has been known that elastomers with low glass transition temperatures T_g, e.g. cis-BR, demonstrate high resilience and favorable wear resistance but low wet grip properties. Elastomers with higher T_g, e.g. SBR 1516 and 1712, display comparatively low resilience and wear resistance but excellent wet grip. All synthetic elastomers have high inherent rolling resistance, whereas natural rubber exhibits low rolling resistance but inferior wet grip.

Shell International Chemical Co. in cooperation with Dunlop UK has developed a new SBR for tire tread compounds. Developmental quantities have been available from Shell since early 1982 under the designation ®Cariflex SSCP 901 [1-3]. Production plants are scheduled for start up in 1983 by Shell in the Netherlands and by Sumitomo Rubber Industries in Japan [1].

The new SBR is produced in solution with Li alkyl catalysts. Process variables with effects exerted on polymer micro- and macrostructure include the sequence of monomers addition, the initiator type, the solvent system, the polymerization temperature, structure modifiers and coupling agents of various functionality [4]. Random SBR copolymers are obtained in apolar solvents in the presence of so-called randomizers which also effect an increased degree of 1,2 addition. Typical structure modifiers are ethers and amines. The maximum vinyl levels that can be realized with randomizers are: diethyleneglycol methylether (Diglyme) 90%, THF 60%, diethylether 40%, tetramethylenediamine 70% and triethylamine 20%. The vinyl content is reduced with increasing diglyme concentration and with decreasing polymerization temperature in the range of 40–60°C. The glass transition temperature T_g is increased with increasing vinyl and styrene contents. For a given T_g value there exist numerous SBR compositions with different ratios of vinyl/styrene.

The linear SBR polymers possess excellent dynamic-mechanical properties which lead to tires with low rolling resistance. The lithium catalysts, however, produce polymers with very narrow molecular weight distribution thus impairing processability. Polymers with bi-, tri- or polymodal molecular weight distributions can be obtained by reaction of living polymers with coupling agents of various functionality. The resulting polymers have linear structures or contain short and long chain branchings. In contrast to linear polymers, the viscosity of branched polymers shows pronounced dependency on shear. At low shear rates cold flow is minimal, at high shear rates the viscosity decreases with increasing degree of branching and thus facilitates loading with fillers are processability [4].

The Chemische Werke Hüls has also been active in developing improved solution SBR types [5, 6]. Proceeding from SBR 1712 the stepwise substitution of phenylic by vinylic sidegroups results in polymers with significantly lower tensile strengths, however, heat-build-up is drastically reduced while wear and grip properties are fully maintained. Copolymers with less than 10% styrene units no longer show a tendency to reversion. The new Vinyl-SBR and solution SBR types of Chemische Werke Hüls are still experimental products classified as "BUNA VI".

3.2.2 Properties and applications

Table 3-2 gives a comparison of properties of ®Cariflex SSCP 901 with a conventional SBR grade S-1502 [3] and with other SBR, solution SBR and Vinyl-SBR types with different microstructures [5, 6].

Shell emphazises that the superior properties of its Cariflex SSCP 901 become only evident in tire road tests rather than in laboratory tests [7]. Compared to other elastomers Cariflex SSCP 901 exhibits both improved wet grip and reduced rolling resistance and thus fuel consumption (Figure 3-1). Energy losses of 6% in a car running at a constant speed of 80 km/h are due to tire rolling resistance. Rolling resistance can be reduced when light-weight tires are used or if hysteresis losses are minimized resulting in lower heat-build-up. Hysteresis losses depend on tire construction and tire tread formulation. Wet grip performance is improved, first as the rebound resilience of the tread compound decreases, and second as the glass transition temperature of the polymer or tread compound increases [2].

Figure 3-2 demonstrates the developments which up to now have been achieved in wet grip and fuel consumption improvements [2].

TABLE 3-2

Property value comparison of ®Cariflex SSCP with E-SBR, solution-SBR, and vinyl-SBR.

Physical Properties	Physical unit	Cariflex SSCP 901[1]	SBR S-1502[1]	Propety values of			
				E-SBR[2]	L-SBR-I[2]	L-SBR-II[2]	Vinyl-BR[2]
Oil content	%	—	—	37.5	37.5	37.5	37.5
Styrene content	%	23.5	23.5	23.5	25	10	—
Cis content	%	20	10	10	34	24	20
Trans content	%	30	75	70	44	31	25
Vinyl content	%	50	15	20	22	45	55
Mooney viscosity ML1 + 4 (100°C)	—	45	50				
Density	g/cm^3	0.93	0.93				
Glass transition temperature	°C	−40	−55	−50	−50	−50	−50
Molar mass distribution M_w/M_n	—	—	—	7.6	5.5	4.1	5.4
Vulcanizate properties							
Tensile strength	MPa	22	24	19.3	19.0	15.5	14.0
Elongation	%	425	400	577	540	485	494
Modulus 300%	MPa	13	14	8.4	8.7	9.0	8.1
Hardness, Shore A		61	62	62	61	60	60
Tear resistance	kN/m	50	52				
Rebound resilience 23 and 22/75	%	42	52	28/45	29/46	31/47	32/47
Heat build-up (over 38°C)	°C	34	33				
Relative wear, index 12 000 km				100	99	98	102
Relative wet grip performance				100	98	100	99

[1] Recipe (phr): polymer 100, carbon black (N339) 45, ZnO 5, stearic acid 3, Santoflex-13 1, Santocure MOR 1, sulfur 2. Cure at 145°C until t$_{90}$ for SSCP 901 30 min, for SBR S-1502 32 min.

[2] Recipe (phr): polymer 137.50, carbon black (N339) 80.00, aromatic oil 12.00, stearic acid 1.00, ZnO 3.00, 4010 NA 1.00, Vulkacit CZ 1.10, Vulkacit D 0.25, sulfur 1.90.

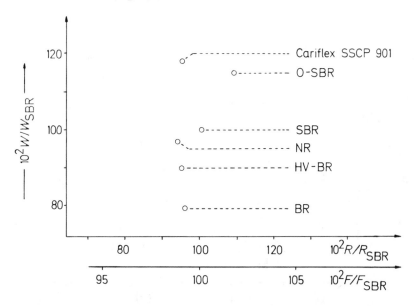

Figure 3-1 Road performance of tires made from Cariflex SSCP and other elastomers in comparison with a conventional styrene-butadiene rubber with 23% styrene units [3]. W/W_{SBR} = relative rating of wet grip, R/R_{SBR} = relative rating of rolling resistance, F/F_{SBR} = relative rating of fuel consumption. BR = Butadiene rubber, HV-BR = Butadiene rubber with high vinyl content, NR = Natural rubber, SBR = Styrene-butadiene rubber with 23% styrene units, O-SBR = SBR highly extended with mineral oil.

A bias ply tire with NR tread as it was being used in 1955 serves as a point of reference. The bias ply tire with SBR tread of 1962 led to improved wet grip performance but at a sacrifice of fuel consumption. The radial tire with textile belt of 1967 showed improved wet grip and lower fuel consumption; it was further improved by steel cord radial tires in 1971 and by constructional modifications of the steel belt in 1979. The road tests carried out by Dunlop with Cariflex SSCP 901 in steel belt tires resulted in a fuel saving of 5.2%. Presently, 60 000 passenger car tires fabricated by Dunlop are being road tested in England, France, Germany, Japan and in the U.S.A. The performance of Cariflex SSCP 901 in truck tires will also be examined.

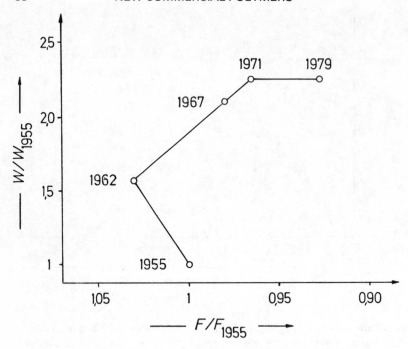

Figure 3-2 Changes in wet grip performance W/W_{1955} and fuel consumption F/F_{1955} in relation to the year 1955. Improvements stem from the introduction of radial tires, changes in tread composition and/or fiber cord type [2].

References

[1] R. Grace, Europ. Rubber Journal (Nov. 1981) 4.
[2] R. Bond, Shell Polymers **6**/1 (1982) 6.
[3] ®Cariflex SSCP 901, Technical Information, Koninklijke/Shell Laboratorium, Amsterdam, January 1982.
[4] P. Luijk, Kautschuk, Gummi, Kunststoffe **34**/3 (1981) 191.
[5] K. H. Nordsiek and K. H. Kiepert, Kautschuk, Gummi, Kunststoffe **33**/4 (1980) 251.
[6] J. Markert, Kautschuk, Gummi, Kunststoffe **34**/4 (1981) 269.
[7] W. Jacobsen (Deutsche Shell Chemie GmbH), personal communication, February 5, 1982.

3.3 POLY(NORBORNENE)

3.3.1 Structure and synthesis

The C_5-cut obtained in naphtha steam cracking processes contains ca. 15% by weight of cyclopentadiene (I) and dicyclopentadiene (II), which nearly amounts to the content of isoprene. Several processes have been developed in order to utilize the ample supply of I or II, e.g., in the preparation of poly(pentenamer), in bisdiene-polymers and as a comonomer in EPDM [1, 2]. The Diels-Alder reaction of I or II with ethylene yields bicyclo [2, 2, 1] heptene-2 or norbornene (III):

(3-1)

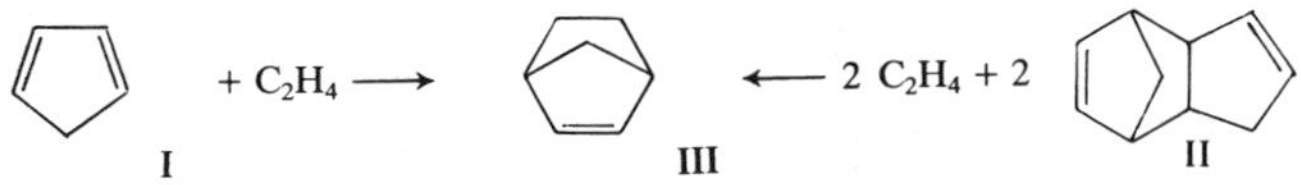

Norbornene (III) undergoes polymerization to give high-melting crystalline products or amorphous polymers with low softening points [3]. The polymerization of III with $WCl_6/AlR_3/I_2$ [4] or ruthenium [5, 6] or iridium [6] catalysts proceeds under ring opening to poly(norbornene) (IV) with the double bond in cis or trans position:

(3-2)

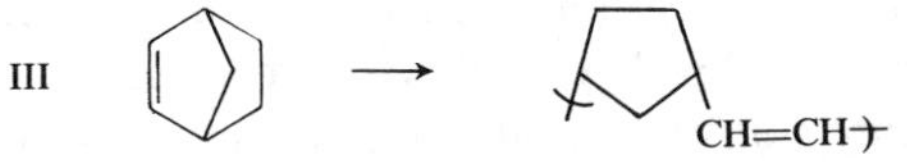

Poly(norbornene) with a molecular weight of ca. 2 million is produced by CdF Charbonnages de France under the name ®Norsorex. It is supplied as powder or as masterbatch with 150 phr oil.

Copolymers of norbornene with other monomers are thermoplastic resins, marketed as Telene® by the B. F. Goodrich Chemical Group [11].

3.3.2 Properties and applications

Poly(norbornene) is partially crystalline; glass transition temperatures of 35°C [7, 8] and 45−47°C [4] as well as melt temperatures of 170−190°C have been reported. Thermal characteristics fall between those of typical thermoplasts and elastomers.

Norsorex is the first elastomer which has been specially designed for powder technology. It is a white, free flowing powder, particle size ca. 0.3 mm, bulk density 0.30 g/cm^3. It contains no fillers and no release agents. The porous powder absorbs up to 400 parts of mineral oil phr polymer which lowers the glass transition temperature to a range between −45°C and −60°C [3, 7]. Incorporation of plasticizers should be done in kneaders since working on a roll mill would destroy the porous structure of the powder. Formulation of compounds, vulcanization and vulcanizate properties have been described [3, 7−10]. Preferably naphthenic and aromatic plasticizers are used to obtain a hardness in the range of 10−80° Shore A. The effect of various active fillers on tensile strength, elongation and hardness is shown in Figure 3-3. The low temperature brittleness depends on the type and amount of plasticizer used (Table 3-3). The vulcanization with accelerators and sulfur is carried out at temperatures of 140−200°C by the common techniques such as compression, transfer, injection, autoclave, LCM or UHF.

Typical formulations with Norsorex contain only ca. 20% of poly-(norbornene). The polymer is not resistant to ozone due to its double bond. Addition of antiozonants and 20−30 parts of EPDM yield ozone resistant compounds, however, EPDM impairs the aging characteristics (Table 3-4). Norsorex vulcanizates have a continuous service temperature of 80°C. They are resistant to the attack of water, alkali, non-oxidizing acids, fats and vegetable oils, the resistance to mineral oils is on a level with poly(chloroprene).

Norsorex is used in the automotive and building industry as a soft sealing material. Non-vulcanized compounds with high loads of fillers serve as noise and sound damping materials, sealants, caulkings and interlayers. In heavy-duty tires Norsorex can be substitute for foamed or compact poly(urethanes): An inexpensive paste compounded from 150 parts of Norsorex and 250−300 parts of medium viscosity naphthenic or aromatic plasticizers is poured into the tire where it soon forms a strong, elastic gel.

Norsorex readily absorbs oil and for this purpose is supplied by CdF under the trade name ®Norsopol [10]. The absorbing capacity of one part of polymer is 15 parts of toluene, or 10 parts of aromatic, 4 parts of naphthenic or 2 parts of paraffinic oils, all parts being by weight.

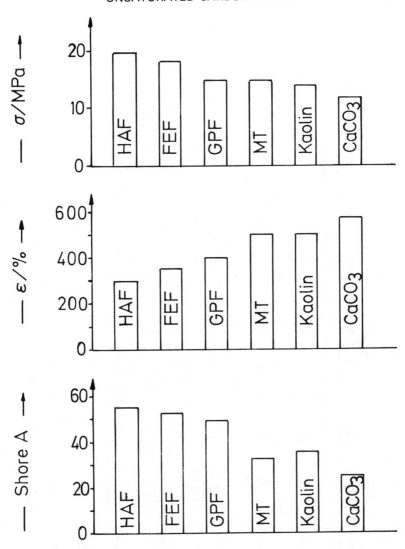

Figure 3-3 Effect of different active fillers on Norsorex tensile strength σ, elongation ε and Shore A hardness [8]. Formulation: 100 pt Norsorex, 180 pt low-viscosity aromatic oil, 20 pt paraffinic oil, 200 pt filler, 5 pt zinc oxide, 1 pt stearic acid, 5 pt CBS and 1.5 pt sulfur. $CaCO_3$ = limestone; HAF, FEF, GPF, MT = different carbon black types.

TABLE 3-3

Properties of ®Norsorex vulcanizates (containing 200 parts of FEF carbon black) as function of plasticizer type and amount [8].

Plasticizer	Amount of plasticizer per 100 parts of polymer and 200 parts of carbon black									
aromatic oil, high viscosity	180									
aromatic oil, low viscosity		180								
naphthenic oil			180							
paraffinic oil				180						
dioctylphthalate					160	160	160	160	90	180
dioctylsebacate					20	20			90	
dioctyladipate							20	20		40

Physical Properties	Physical unit	Property values									
Tensile strength	MPa	17.5	17.5	16.9	15.4	17.2	17.7	17.5	17.6	17.0	14.0
Elongation	%	200	300	180	170	310	310	320	300	260	380
Hardness, Shore A	—	68	64	66	67	65	64	65	65	66	45
Brittleness temperature	°C	-10	-30	-42	-49	-34	-38	-39	-38	-38	-52

TABLE 3-4
Properties and aging characteristics of ®Norsorex and ®Norsorex/EPDM vulcanizates [8].

Formulation	*Compound A*	*Compound B*
Norsorex	100	100
EPDM (Vistalon 6505)	—	20
Naphthenic oil	220	220
MT carbon black	100	100
p-Phenylenediamine	5	—
Antilux 600	1.5	—
TE 75 processing aid	1.5	—
Zinc oxide	5	5
Stearic acid	1	1
TMTD	1.5	1.5
Vulkacit I	1.5	1.5
Sulfasan R	1.5	1.5
Tellurac	0.8	0.8
DOTG	1	1
Vulkacit NPV	0.5	0.5

Physical Properties	Physical unit	Original A	Original B	Compound A 7 days/70°C	Compound A 4 days/100°C	Compound B 7 days/70°C	Compound B 4 days/100°C
Hardness Shore A		20	20	23	23	24	23
Tensile strength	MPa	12.0	10.0	10.1	9.9	9.3	7.9
Modulus 100%	MPa	0.3	0.3	0.4	0.4	0.4	0.5
Modulus 200%	MPa	0.7	0.8	1.3	1.5	1.5	1.9
Modulus 300%	MPa	1.8	2.1	2.9	3.1	3.5	4.1
Elongation	%	620	550	520	500	450	410
Resilience	%	15	15[1)]				

[1)] after 22 h at 70°C.
Curing of compounds A and B: 50 s at 195°C.

References

[1] H.-G. Elias, Polymer News **3**/2 (1976) 90.

[2] M. Fefer and A. B. Small, Cyclopentadiene and Dicyclopentadiene, in Kirk-Othmer, Encycl. Chem. Technol., volume 7 (1979), Wiley & Sons, New York.

[3] R. F. Ohm and T. M. Vial, J. Elastomers Plast. **10** (1978) 150.

[4] J. C. Muller, Germ. P. 1 961 743 (1970), to Soc. Chimique des Charbonnages de France.

[5] J. Vergne, L. Solaux, J. C. Robinat and P. Lacroix, USP 3 676 390 (1972), Brit. P. 1 230 597, FrAD 94 571, all to CdF.

[6] L. Porri, R. Rossi, P. Diversi and A. Lucherini, Makromol. Chem. **175** (1974) 3097.

[7] C. Stein and A. Marbach, Rev. Gen. Caout. Plast. **52**/1-2 (1975) 71; Plast. Mod. Elastomers **26**/9 (1974) 78.

[8] H.-D. Manger, Kautschuk, Gummi, Kunststoffe **32**/8 (1979) 572.

[9] P. D. Nelson and R. F. Ohm, Elastomerics **111**/1 (1979) 28.

[10] R. F. Ohm, Chemtech **10** (1980) 183.

[11] Anon., Ind. Res. Dev. (October 1982) 104.

3.4 POLY(OCTENAMER)

3.4.1 Structure and synthesis

Poly(octenamer) is prepared by ring opening polymerization of cyclooctene which is available by partial hydrogenation of the butadiene dimer, cyclooctadiene:

(3-3)

The polymerization is carried out in solution with $RAlCl_2/WCl_6$ [1], with $RAlCl_2/WCl_6/C_2H_5OH$ [2] or in bulk with $RAlCl_2/WF_6$ [3]. Hexene-1 acts as chain stopper, the cis content is adjusted via the Al/W-mol ratio [3]. Several companies have investigated the production of poly(octenamer), in particular Chemische Werke Hüls, Goodyear and Montedison. The latter two have concentrated on polymers with cis contents of $75-80\%$ [3, 4], whereas Hüls has developed a polymer with 80% trans which was the first one to appear on the market in 1980.

The melting points of the pure cis- and trans-polyoctenamer are 38°C (extrapolated) and 73°C, respectively [3, 5]. The high crystallization rate of the cis polymer decreases with decreasing cis content. Polymers with high cis content are difficult to process on roll mills [4].

3.4.2 Properties and applications

The trans-poly(octenamer) or TOR is supplied by Hüls under the name ®Vestenamer 8012. The number indicates a trans content of 80% and an intrinsic viscosity of 120 ml/g. Some properties of TOR are listed in Table 3-5.

The viscosity and rheological characteristics of TOR are unique (Figure 3-4) [7]. At room temperature, TOR is extremely high viscous and hard but it becomes easy flowing above 60°C. The low viscosity of TOR at elevated temperatures renders it difficult to process. Hence TOR is preferably recommended for blending with other elastomers, e.g. NR, SBR, BR, IIR, NBR, CR, and EPDM. Relatively small amounts of TOR improve the green strength and the dimensional stability of compounds at room temperature (Figure 3-5), whereas at higher temperatures TOR containing compounds

TABLE 3-5
Properties of trans-poly(octenamer) [6, 7].

Physical Properties	Physical unit	Property value
Density	g/cm^3	0.91
Defo hardness at 40°C		>20 000
Defo hardness at 50°C		8 000
Mooney viscosity (ML 4) at 60°C		12
Mooney viscosity (ML 4) at 100°C		5
Crystallinity	%	33
Melting temperature	°C	55
Glass transition temperature	°C	−65
Decomposition temperature	°C	250
Ignition temperature	°C	>400
Inherent viscosity (0.1 g in 100 ml toluene at 25°C)	ml/g	120
Molecular weight	g/mol	60 000
cis/trans ratio		20:80
Ash content	%	0.05

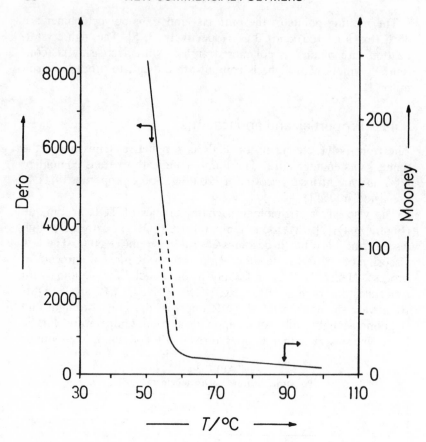

Figure 3-4 Defo hardness and Mooney-viscosity (ML 1 + 4) of poly(octenamer) Vestenamer 8012 with molecular weight of ca. 60 000 g/mol vs. temperature [7].

are softer and display significantly reduced viscosities compared to TOR-free formulations. Thus higher output and lowered energy consumption are achieved (Figure 3-6) [7]. TOR can be vulcanized with sulfur, peroxides, phenolic resins, quinone dioxime, and other crosslinking systems.

Because of the relatively low molecular weight of TOR one would expect that the mechanical properties of TOR blends with other rubbers would become poorer. Surprisingly, some properties are improved; this is attributed to the partial macrocyclic structure of

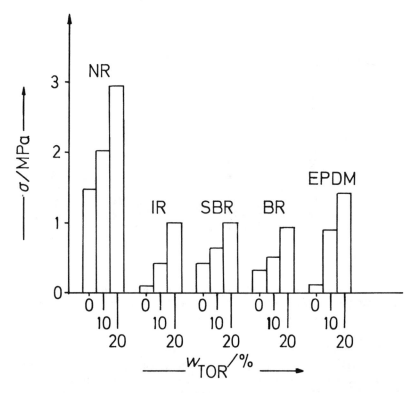

Figure 3-5   Green strength of different elastomers containing 0, 10 or 20% poly-(octenamer) [7]. Formulation: 100 pt polymer, 5 pt zinc oxide RS, 1 pt stearic acid, 50 pt carbon black N 550.

TOR. With TOR, improvements in abrasion resistance, hysteresis loss, compression set, Shore hardness and tensile strength have been achieved. Table 3-6 lists the shift in properties of SBR/TOR blends with TOR ranging from 0 to 40% [6]. The relatively slow vulcanization rate of TOR requires higher concentrations of accelerators.

TOR is recommended for extrusion and injection molding to improve flow ability, particularly in blends of high hardness, as well as, in all processes where high green strength and dimensional stability of blends are required. The good flow properties of TOR allow low pressures at vulcanization and facilitates bonding, e.g., in calendering, doubling, and in the production of rubber rollers and tire carcasses.

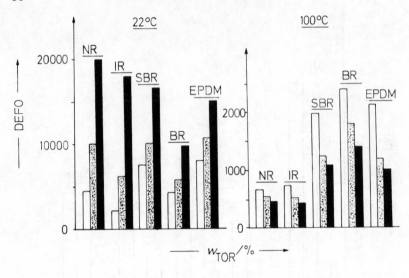

Figure 3-6   Defo hardness of different elastomers containing 0 (□), 10 (▦) or 20 wt-% (■) poly(octenamer) [7] at 22° and 100°C. Formulation: 100 pt polymer, 5 pt zinc oxide RS, 1 pt stearic acid, 50 pt carbon black N 550.

References

[1] N. Calderon, E. A. Ofstead and W. A. Judy, J. Polymer Sci. A-1 5 (1967) 2209; N. Calderon, J. Macromol Sci.-Revs. Macromol. Chem. C 7 (1972) 105.

[2] K. W. Scott, N. Calderon, E. A. Ofstead, W. A. Judy and J. P. Ward, in W. F. Gorham, (Ed.) Addition and Condensation Polymerization Processes, Adv. Chem. Ser. 91, Amer. Chem. Soc., Washington D.C., 1969, 399.

[3] G. Gianotti, A. Capizzi and L. DelGiudice, Rubber Chem. Technol. 49 (1976) 170.

[4] G. Dall'Asta, Rubber Chem. Technol. 47 (1974) 511.

[5] N. Calderon and M. C. Morris, J. Polymer Sci. A-2 5 (1967) 1283.

[6] ®Vestenamer 8012, Technical Information, Chem. Werke Hüls.

[7] A. Dräxler, Der Lichtbogen 39/198 (1980) 4 (Company periodical of Chemische Werke Hüls); A. Dräxler, Kautschuk, Gummi, Kunststoffe 34/3 (1981) 185.

3.5   BUTYL RUBBER WITH CONJUGATED DOUBLE BONDS

3.5.1   Structure and synthesis

The copolymer of isobutylene with a small amount of isoprene,

TABLE 3-6

Property value comparison of belt conveyor repair compounds of SBR with various proportions of TOR [6].

Compound formulation	Composition of compound no.			
	1	2	3	4
SBR 1500	100.0	90.0	80.0	60.0
Vestenamer 8012	—	10.0	20.0	40.0
Zinc oxide RS	5.0	5.0	5.0	5.0
Stearic acid	2.0	2.0	2.0	2.0
Highly-aromatic oil	2.0	3.0	5.0	6.0
Carbon black N 234	50.0	50.0	50.0	50.0
Vulkanox 4010 NA	2.0	2.0	2.0	2.0
Antilux	2.0	2.0	2.0	2.0
Sulfur	1.8	1.8	1.8	1.8
Vulkacit CZ	1.0	1.1	1.2	1.3
Vulkacit D	0.3	0.3	0.4	0.5
Compound viscosity ML (1 + 4) 100°C	71	63	54	41
Vulkameter 150°C t_{10} min	7.4	7.7	7.6	7.0
t_{90} min	15.6	15.3	14.6	14.9

Cure 20 min at 150°C

Physical Properties	Physical unit	Property values of compound no.			
		1	2	3	4
Tensile strength at break	MPa	26.0	24.1	21.4	18.9
Elongation at break	%	519	492	443	394
Modulus, 100%	MPa	2.3	2.6	3.1	3.9
Modulus, 300%	MPa	11.8	12.2	12.7	13.6
Permanent elongation	%	14	15	17	27
Shore A hardness at 22°C	—	67	69	74	79
Shore A hardness at 75°C	—	65	65	66	67
Compression set at 22°C	%	35	36	34	36
Compression set at 75°C	%	46	47	47	47
Density	g/cm^3	1.14	1.14	1.13	1.12
Abrasion (DIN 53516)	mm^3	90	81	84	73
Split tear resistance	N/mm	11.8	12.3	13.7	13.8

Oven ageing: 7 days at 100°C

Physical Properties	Physical unit	Property values of compound no.			
		1	2	3	4
Tensile strength at break	MPa	16.3	14.9	15.4	14.9
Elongation at break	%	170	160	164	160
Modulus, 100%	MPa	7.8	7.8	8.0	8.3
Shore A hardness at 22°C	—	81	81	82	84

known as butyl rubber (IIR), can be chlorinated and brominated [1]. Chlorobutyl rubber has been available since 1960 from Exxon Chemicals Co. under the designations Chlorobutyl 1066 and 1068 [2]. Polysar Ltd. has commercially produced its Chlorobutyl 1240 since 1979 [3]. Polysar is also the sole producer of brominated butyl rubber, marketed as Bromobutyl X 2, with plants in Canada and Belgium [1]. Exxon Chemicals Co. introduced a developmental polymer under the name CDB (Conjugated Diene Butyl) in 1981 [4, 5].

The synthesis of CDB starts with halogenated butyl rubber. After halogenation of butyl rubber I with chlorine or bromine [1], most of the halogen is in allyl position to the double bond of the isoprene unit (II, III, eq. 3-4). The dehydrohalogenation produces a second double bond and yields a diene with conjugated double bonds (IV, V, eq. 3-5). The catalyzed dehydrohalogenation to CDB is carried out with soluble zinc salts of carbonic acids [6] or with an insoluble copper (I)-oxide/aluminum oxide catalyst [7]. Chlorination of butyl rubber is known to be highly regioselective. More than 90% of the product consists of the moiety III [6]. There are altogether six different configurations possible for CDB arising from dehydrohalogenation: IV with exo-cis and exo-trans structure, and V with cis-cis, cis-trans, trans-trans and trans-cis structures.

(3-4)

I

$$\sim CH_2-\underset{\underset{CH_3}{|}}{\overset{\overset{CH_3}{|}}{C}}-CH_2-\overset{\overset{CH_3}{|}}{C}=CH-CH_2-CH_2-\underset{\underset{CH_3}{|}}{\overset{\overset{CH_3}{|}}{C}}\sim \qquad \xrightarrow[-\,HCl]{+\,Cl_2}$$

II

$$\sim CH_2-\underset{\underset{CH_3}{|}}{\overset{\overset{CH_3}{|}}{C}}-CH_2-\underset{\underset{Cl}{|}}{\overset{\overset{CH_2}{||}}{C}}-CH-CH_2-CH_2-\underset{\underset{CH_3}{|}}{\overset{\overset{CH_3}{|}}{C}}\sim$$

III

$$\sim CH_2-\overset{\overset{CH_3}{|}}{C}-CH=\overset{\overset{CH_3}{|}}{C}-\underset{\underset{Cl}{|}}{CH}-CH_2-CH_2-\underset{\underset{CH_3}{|}}{\overset{\overset{CH_3}{|}}{C}}\sim$$

(3-5)

$$II + III \xrightarrow{- HCl}$$

IV

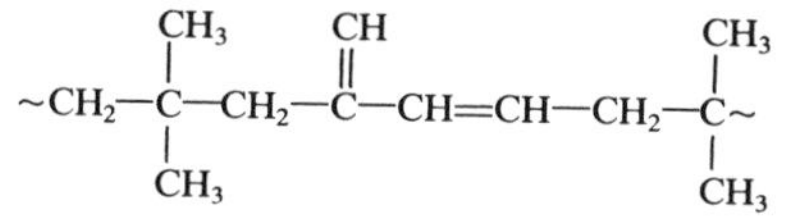

~CH$_2$—C(CH$_3$)(CH$_3$)—CH$_2$—C(=CH)—CH=CH—CH$_2$—C(CH$_3$)~

V

~CH$_2$—C(CH$_3$)(CH$_3$)—CH=C(CH$_2$)—CH=CH—CH$_2$—C(CH$_3$)~

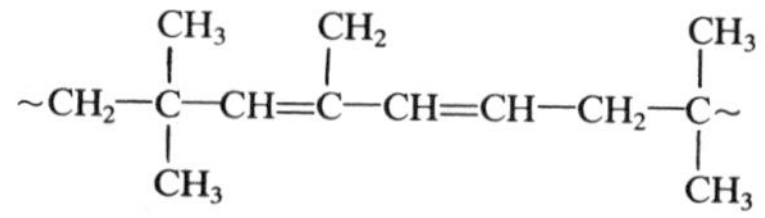

Owing to its conjugated double bonds, CDB is a polyfunctional macromolecule accessible to many reactions [7–13]. Self-vulcanization, grafting, Diels-Alder reactions and vulcanization with sulfur have been investigated.

3.5.2 Properties and applications

CDB self-vulcanizes at elevated temperatures via dimerization of the double bonds. The activation energy for this reaction is appr. 75 kJ/mol for temperatures between 154 and 177°C.

The Mooney viscosity increases by 6 points after one year storage at 127°C. The shelf life is improved by reduction of the diene content or by addition of plasticizing oils. CBD should be compounded at temperatures lower than 150°C in order to avoid premature self-vulcanization.

Carbon black filled CDB compounds vulcanize with sulfur and accelerators much faster than butyl rubber. CDB binds more carbon black than butyl, chlorobutyl or SBR rubber. The mechanical properties of CDB vulcanizates are similar to those of NR and SBR; the resistance against ozone is however much higher. CDB can also be vulcanized with halogenated formaldehyde resins or quinone dioxime derivatives.

CDB undergoes free radical polymerization with mono- and multifunctional monomers under grafting and cross-linking [10], with high rates due to the Trommsdorf-Norrish effect. No reaction occurs with butyl rubber under the same conditions since chain propagation is prevented by chain transfer to the isolated double bonds of butyl rubber. The graft polymers of CDB with styrene or methacry-

TABLE 3-7
Property values of sulfur cured CDB, NR, and SBR elastomers.

Formulation (phr):

Polymer	100
Zinc oxide	5
Stearic acid	1
HAF carbon black	50
Sulfur	2
Santocure NS-1	1

Physical Properties	Physical unit	Property values of					
		CDE		NR		SBR	
after curing at 153°C	min	15	30	15	30	15	30
Tensile strength at break	MPa	16.7	20.4	25.7	24.4	16.4	25.8
Modulus 300%	MPa	6.2	13.1	16.7	16.6	5.7	17.5
Elongation	%	680	465	440	430	690	460
Hardness, Shore A		52	57	65	65	62	68
Ozone resistance (20% elongation, 100 ppm ozone, 38°C) until appearance of cracks	h	126	203	4	4	2	4

lic acid are crosslinked, tack free, and surprisingly transparent. A vulcanization by grafting is also possible with maleic anhydride, alkyl acrylates, alkyl methacrylates, acrylonitrile, divinylbenzene, vinylpyridine and other polyfunctional monomers. CDB can also be crosslinked with monomers like trimethylolpropane trimethacrylate or 1,6-hexanediol dimethacrylate on irradiation with high energy electron beams.

Applications recommended for CDB graft polymers include coatings, encapsulating compounds for electrical parts, adhesives, unfilled and reinforced molding materials, semi-conductive and oil resistant compounds for extrusion, sulfur-free vulcanizates for ampules and tubes for medical, use as well as injection molding materials. CDB has been used as an elasticizing component in polyester composites [12, 13]. To comply with the tendency to facilitate handling of unsaturated polyesters and to automate processing, SMC and BMC materials have been developed. They exhibit improved elasticity, impact resistance, moldability, dimensional stability and surface quality. With the so-called Low-Profile-Resins containing PVAC or PMMA, the dimensional stability and surface quality can be improved, however, the increased absorption of solvents and water is disadvantageous. Addition of CDB to SMC materials secures the desired low-profile properties while stiffness is maintained and impact resistance is improved (Table 3-8). Electron beam crosslinking of CDB/HDPE (50/50) blends yields elasticized crosslinked PE's whose mechanical properties largely surpasses those of a crosslinked butyl rubber/HDPE-blend at temperatures above 150°C. Such blends of CDB are applied in shrink insulating material, cables and films [14].

TABLE 3-8
Property comparison of prepregs (SMC) based on CDB and on a commercial Low-Profile-Resin.

Formulation	Composition in wt.-%	
	CDB	LP-Resin
Polyester resin (Propyleneglycol/maleic anhydride) 62%	15.17	16.83
Axel Int-20E	0.41	—
Black pigment	0.06	—
PVAc resin (40%)	1.93	11.22
CDB resin (30%)	10.48	—
Zinc stearate	1.10	1.12
tert.-Butylperbenzoate	0.36	0.29
Activator solution	0.55	—
Calcium carbonate	33.1	42.08
ASP-400 clay	8.28	—
Calcium oxide	0.30	0.31
Water	0.03	0.03
OCF 951 glass fiber 25.4 mm long	28.14	28.00
Cure 2 min at 150°C		

Physical Properties	Physical unit	Property values of	
		CDB-Resin	LP-Resin
Tensile strength at break	MPa	78.1	76.5
Tensile modulus	GPa	11.7	14.1
Elongation	%	1.45	1.11
Flexural strength	MPa	179.3	186.2
Flexural modulus	GPa	11.4	14.6
Impact strength	Nm	28.9	18.2
Surface roughness (profilometer)	μm	<50	100−150
Barcol hardness		42	51
Expansion	%	0.42	0.20

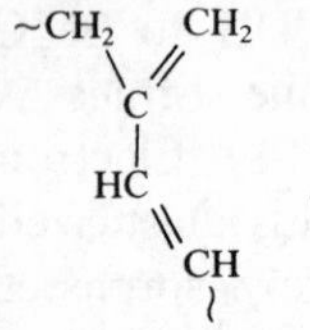

~CH₂   CH₂   CH₃   CH
  \   //   \   //
   C   VI   C   VII
   | |
  HC   exo, trans HC   trans, trans
   \\   \\
    CH   CH

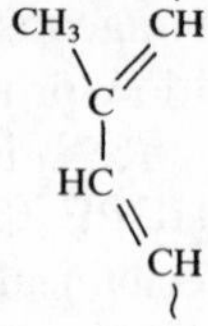

The exo-trans (VI) and trans-trans (VII) diene structures of CDB should be readily accessible to Diels-Alder reactions. UV and IR investigations proved that ca. 50% of the 1,2−1,4 mol-% conjugated double bonds present can react with dienophils. Effective dienophilic compounds are mono- and polyfunctional acrylic esters, acrylamides, methacrylamides, maleimides, maleic anhydride and unsaturated polyesters.

The Diels-Alder reactions of VI with maleic anhydride or acrylic esters yield polymers with cyclohexene structures VIII and IX, respectively, in the backbone chain:

(3-6)

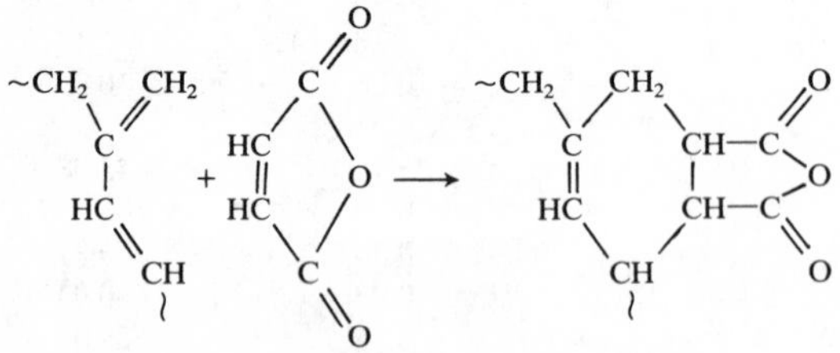

$$
\sim CH_2 \; CH_2 \quad + \quad \text{(maleic anhydride)} \quad \longrightarrow \quad \sim CH_2 \; CH_2 \quad VIII
$$

$$
\sim CH_2 \; CH_2 \quad + \quad CH_2{=}CH{-}C({=}O){-}OR \quad \longrightarrow \quad \sim CH_2 \; CH_2 \quad IX
$$

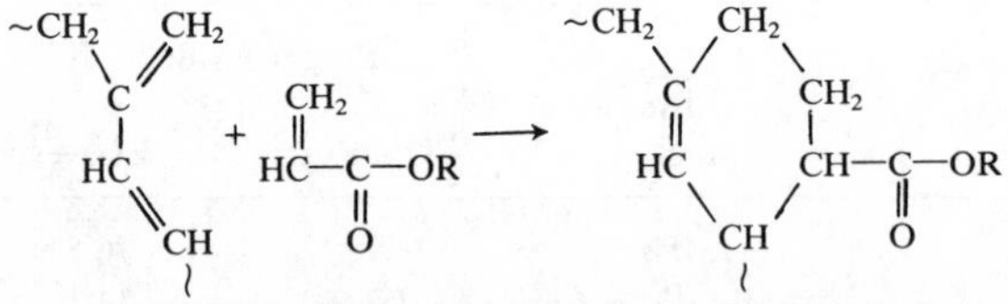

The activation energy of 50 kJ/mol for the Diels-Alder reaction between CDB and trimethylolpropanetriacrylate is only half the activation energy for accelerator-sulfur-systems. Shorter cure times at elevated temperatures together with postcuring at low temperatures allow reduced energy consumptions. The CDB-Diels-Alder-vulcanizates demonstrate high damping characteristics, low water vapor and air permeabilities, and good resistance to heat, oxidation, ozone and polar solvents. Their oven heat aging performance is superior to EPDM, Butyl (Table 3-9), and Chlorobutyl elastomers. Because the double bond is positioned within the ring structure, the ozone resistance of CDB-Diels-Alder-vulcanizates surpasses sulfur-cured vulcanizates in which the double bonds remain in the backbone chain. The time for the first cracks to appear at 38°C and 100 ppm ozone is more than 580 h for CDB-vulcanizates compared to 200 h for sulfur-cured vulcanizates. The resistance to superheated steam is very good despite the presence of polar groups. The application of CDB-Diels-Alder-vulcanizates in tires is being investigated [5].

References

[1] H.-G. Elias, Neue polymere Werkstoffe 1969–1974, Carl Hanser Verlag, München 1975.

[2] Chlorobutyl Rubber, Technical Information, Exxon Chemicals Co.

[3] Polysar Chlorobutyl 1240, Technical Information, Polysar Ltd.

[4] I. J. Gardner, personal communication, May 19, 1981.

[5] I. J. Gardner, F. P. Baldwin and J. V. Fusco, Paper presented at the Plastics and Rubber Institute Polymer Conference, Harrogate, 1981.

[6] F. P. Baldwin and A. Malatesta, USP 3 816 271 (1974); F. P. Baldwin, USP 3 775 387 (1973).

[7] I. J. Gardner, USP 4 145 492 (1979) and USP 4 211 855 (1980).

[8] F. P. Baldwin, I. J. Gardner, A. Malatesta and J. A. Rae, "Polyisobutylene Based Elastomers Containing Conjugated Diene Butyl", 108th Meeting of the ACS Rubber Division, New Orleans, October 1975.

[9] I. J. Gardner and F. P. Baldwin, "Interpenetrating Network from Isobutylene Copolymers Containing Conjugated Diene Butyl", 108th Meeting of the ACS Rubber Division, New Orleans, October 1975.

[10] F. P. Baldwin and I. J. Gardner, in S. S. Labana (Ed.), Chemistry and Properties of Crosslinked Polymers, Academic Press, New York 1977, 273; I. J. Gardner, USP 4 208 491 (1980).

[11] I. J. Gardner, "Elastomeric Toughening of Thermoset Resins", SPE 32th Meeting Washington, D.C., 1977.

[12] I. J. Gardner and J. V. Fusco, "High Impact Polyester Composites", SPE 33th Meeting, Washington, D.C., 1978.

[13] J. V. Fusco, I. J. Gardner and L. Gursky, "Conjugated Diene Butyl as an Elastomeric Additive in Fiber Glass Reinforced Polyester Composites", 118th Meeting of the ACS Rubber Division Detroit, 1980.

TABLE 3-9
Aging of CDB-Diels-Alder, EPDM, and Butyl rubber vulcanizates.

Formulation	Composition in phr			
	EPDM	Butyl	Butyl	CDB
Vistalon 4608	100.0	—	—	—
Butyl 4608 EPDM	—	100.0	100.0	—
CDB	—	—	—	100
HAF carbon black	36.0	36.0	36.0	36.0
Super Multiflex	52.0	52.0	52.0	52.0
Stearic acid	—	1.0	1.0	2.0
Sunpar 2280	7.5	7.5	7.5	7.5
Zinc oxide	—	5.0	5.0	—
Agerite Resin D	2.0	2.0	—	2.0
Dicup 40 KE	7.0	—	—	—
Triallylcyanurate	2.0	—	—	—
Resin SP-1056	—	10.0	—	—
Lead oxide	—	—	10.0	—
Quinonedioximidibenzoate	—	—	6.0	—
Trimethylolpropanetriacrylate	—	—	—	1.75
Cure at 160°C, min	20	60	20	30

Physical Properties	Physical unit	Property values of		
		EPDM	Butyl	Butyl	CDB
Tensile strength at break	MPa	16.2	9.9	10.8	11.5
after 7 days at 170°C	MPa	7.0	4.9	melts	6.8
after 14 days at 170°C	MPa	brittle	degraded	—	3.9
Elongation	%	375	680	455	535
after 7 days at 170°C	%	115	50	melts	210
after 14 days at 170°C	%	brittle	degraded	—	210
Hardness, Shore A	—	65	73	59	52
after 7 days at 170°C	—	79	56	melts	65
after 14 days at 170°C	—	brittle	degraded	—	60

3.6 POLY(1-HEXENE-CO-4/5-METHYL-1,4-HEXADIENE)

3.6.1 Structure and synthesis

Poly(1-hexene-co-4/5-methyl-1,4-hexadiene) is prepared by copolymerization of 1-hexene with a mixture of 4- and 5-methyl-1,4-hexadiene. The polymer is offered by Goodyear Tire & Rubber Co. under the designation ®Hexsyn-Rubber.

A mixture of 4- and 5-methyl-1,4-hexadiene is obtained by co-dimerization of isoprene and ethylene in a metathesis reaction:

(3-7)

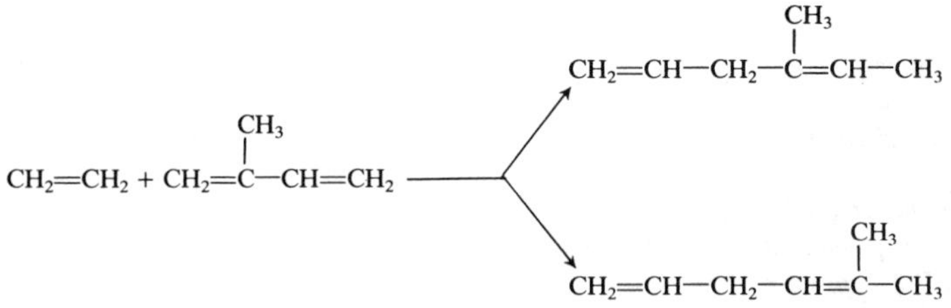

$$CH_2{=}CH{-}CH_2{-}\underset{\overset{|}{CH_3}}{C}{=}CH{-}CH_3$$

$$CH_2{=}CH_2 + CH_2{=}\underset{\overset{|}{CH_3}}{C}{-}CH{=}CH_2 \longrightarrow$$

$$CH_2{=}CH{-}CH_2{-}CH{=}\underset{\overset{|}{CH_3}}{C}{-}CH_3$$

The copolymerization of 1-hexene with a mixture of 4- and 5-methyl-1,4-hexadiene is carried out in hydrocarbon solution with $(C_2H_5)_3Al/TiCl_4/VCl_4$ or $(C_2H_5)_3Al/alpha\text{-}TiCl_3$ as catalysts [1]. Vulcanizates with good mechanical properties are only obtained when the methyl-1,4-hexadiene isomer mixture contains at least 15 mol-% of 5-methyl-1,4-hexadiene. The monomer feed mixture contains 70−98 mol-% of 1-hexene.

3.6.2 Properties and applications

The curing of Hexsyn is similar to that of EPDM. The vulcanizate has a relatively low modulus and shows the high oxidation and ozone resistance that is expected of a largely saturated hydrocarbon. Compared to other elastomers, Hexsyn exhibits a far superior flexing fatigue life under dynamic load (Table 3-10). The resistance to hydrolysis is excellent. Properties of Hexsyn are summarized in Table 3-11 and 3-12 [2].

The particular high flexing fatigue life of Hexsyn leads to applications as diaphragms in membrane pumps for artificial hearts [2] and in ®Bion artifical finger joints marketed by Lord Corp. [3].

The application in implantates requires a blood-compatible surface. First, the compound (Table 3-11) is prepared in a Banbury internal mixer and on mill rolls in accordance with ASTM D 3182. All processing equipment is run first with pure natural rubber for the purpose of cleaning and "rinsing". Only cleaned air is used in the working area. The compound is poured into molds pre-heated to

TABLE 3-10

Flexing fatigue life of elastomers under dynamic load in accordance with ASTM D 430 DeMattia test.

Elastomer	Millions of flexing cycles until damage
Silicone rubber	0.8
Natural rubber	3−4
Styrene-butadiene rubber	3−4
Neoprene	6−8
Poly(propyleneoxide) rubber	10
Ethylene-propylene-diene-terpolymer	14−15
®Biomer Poly(urethane) rubber	18
®Hexsyn	200*

* no damage.

TABLE 3-11

Mechanical properties of ®Hexsyn [2]. Compound: 100 pt polymer, 50 pt ISAF carbon black, 2 pt zinc oxide, 2 pt stearic acid, 1 pt tetramethylthiuram disulfide, 0.5 pt 2-mercaptobenzthiazole and 2 pt sulfur. Vulcanization at 154°C.

Physical Properties	Physical unit	Property value
Tensile strength at break	MPa	11.7
Elongation at break	%	450
Modulus, 100%	MPa	1.5
Modulus, 300%	MPa	6.7
Compression set after 100% elongation	%	5
Tear resistance ASTM D 624 A	$N\ m^{-1}$	219
Tear resistance, Instron	$N\ m^{-1}$	6.0

TABLE 3-12

Mechanical property changes of Hexsyn (cured as stated in Table 3-11) after aging at 70°C and 100% relative humidity.

Physical Property	Physical unit	Property values after time in days			
		0	60	90	120
Tensile strength at break	MPa	11.7	12.1	13.1	13.8
Elongation at break	%	450	500	410	420
Modulus (100%)	MPa	1.5	1.4	1.7	1.7

154°C and is then kept in the molding press for 15 min under a load of 1.36 t. The time applied represents only half of the necessary cure time. The diaphragm is removed from the mold. A mixture of a 10% Hexsyn solution in toluene and powdered sodium chloride (sieve portions used are >250 μm and <190 μm) is applied in three layers to the blood side of the diaphragm surface. After drying in vacuum, the vulcanization is continued to completion. The extraction of the salt by water leaves a porous layer of 100 μm thickness with a pore size of 10−50 μm. Unreacted sulfur and excess curing agents are removed by 6 h extraction with a mixture of 30 parts toluene and 70 parts acetone.

The so-called "biolizing", i.e., the formation of a blood-compatible surface, is carried out by soaking the porous layer with a 5% gelatin solution, drying and post-treatment with a 0.5% glutaricaldehyde solution. The gelatin crosslinks and becomes insoluble. In vivo animal tests with artifical hearts and Hexsyn diaphragms treated as above showed practically no blood clots or other deposits within 5 months [2].

Artificial finger limbs produced by Lord Corp. consist of TiAl6V4 alloy for the stiff parts and joints of ®Bion elastomer. The single step vulcanization of Bion and the cure of the adhesive bonding between the metal and elastomer parts are carried out simultaneously in a heated press [3].

References

[1] J. Lal and P. H. Sandstrom, USP 3 933 769 (January 20, 1976) and USP 3 991 262 (November 9, 1976), both to Goodyear Tire & Rubber Company.
[2] R. J. Kiraly, R. Arconti, D. Hillegas, H. Harasaki and Y. Nose; Trans. Amer. Soc. Art. Intern. Organs **23** (1977) 127.
[3] Lord Corp., Biomedical Group, Technical Information, May 1978.

4 Thermoplastic elastomers

4.1 OVERVIEW

4.1.1 Introduction

Thermoplastic elastomers (TPE's), sometimes also called elasto-plastics, plastomers, or thermolastics, are not irreversibly chemically crosslinked like conventional elastomers, but physically crosslinked. The reversible physical crosslinking results from a separation in 2 phases. The molecular architecture of TPE's is based on either block, graft or segmented copolymers composed of the two structural units A and B. The A and B sequences are mutually incompatible and form locally seperated regions. With a well designed molecular architecture, the domains of the "hard" A segments act as physical crosslinks dispersed in the continuous matrix of the "soft" B segments. Only such B segments are appropriate in which the service temperature remains above the glass transition temperature. The glass transition temperature of amorphous or the melt temperature of crystalline A segments, however, must be above the service temperature in order to sustain the crosslinking effect of the A domains. Hence, at the service temperature, TPEs perform like elastomers. The A segments attain mobility above their characteristic transition temperatures and the elastoplasts become processable like thermoplasts [1].

Ionomers are polymers which have a number of properties and applications in common with TPEs. In ionomers, the reversible crosslinking is created by coordination of polyions around a metal ion. The reversible crosslinks dissociate at the processing temperature and render the material processable as common thermoplasts. Ionomers perform like slightly crosslinked thermoplasts at their service temperatures. Because of some features in common with TPEs, ionomers will be dealt with in the same chapter of this book.

The first PUR TPE was introduced in 1958, styrene-butadiene-block TPE in 1965, its second generation with enhanced properties in 1972, polyolefin blends in 1972, polyester block TPE also in 1972, and 1,2-poly(butadiene) in 1975. Of more recent origin are some

TPE graft copolymers, elastomeric polyesteramides, polyetherami-des, modified PVC types and a new ionomer based on sulfonated EPDM.

The syntheses and properties of the older TPEs have been de-scribed in several books and review articles; see the recent literature since 1975 [1−14]. Styrene-diene block copolymers [4, 15, 16], poly-olefin based TPE [17], polyurethane block TPE [18, 19], elastomeric polyesters [20] and ionomers [21−23] have been reviewed in detail.

Producers and trade-names of commercial TPE's are listed in Table 4-1 to Table 4-4. A variety of other TPE's has been devel-oped, however, they are considered developmental products and are not yet commercial. They include graft polymers of poly(pivalolac-tone) onto EPDM [24, 25], onto polydienes [26], onto poly(iso-butylene) [26] and onto polyacrylate elastomers [27]. Polysiloxanes as a "soft" phase have been combined with various "hard" phases such as polycarbonates [28], poly(alpha-methylstyrene) [2, 29], poly-(styrene) [29], and polysulfones [30]. The grafting of poly(dimethyl-siloxane) containing small portions of unsaturated groups onto poly-(ethylene) or EVAC yields two-phase slightly crosslinked polymers which are processable like TPE [31].

The thermoplastic elastomers are enjoying a period of expanding market development. In the USA, ca. 85 000 t of styrene-based TPE, 27 000 t of PUR TPE, 19 000 t of polyolefin-based TPE and 6 000 t of polyester TPE were consumed in 1979 [32]. Other sources report a consumption of 219 000 t of TPE in the USA for 1979 and predict an annual increase of 17.5% for the period to 1985 [33]. The consumption of TPE's in West Europe climbed from 5 500 t in 1970 to 45 000 t in 1976 [9].

4.1.2   Comparison of typical TPE properties

The most obvious advantage of thermoplastic elastomers over vul-canized elastomers is the ease of processing by conventional thermo-plastic techniques. With TPE, mastication, incorporation of curing agents on roller mills, and in some cases any compounding, and of course any vulcanization step are omitted. Detailed cost comparison of each processing step prove the superior economics of TPE [34]. TPEs benefit from increasing energy costs because increasing vul-canization costs for conventional elastomers give TPE an additional cost advantage. The possibility of reprocessing scrap from the

TABLE 4-1

Thermoplastic elastomers based on styrene and urethane.

Chemical composition	Trade name	Company
Styrene block copolymers		
Linear styrene-butadiene-styrene triblock copolymers	Kraton (in Europe; Cariflex)	Shell
Linear styrene-isoprene-styrene triblock copolymers	Kraton	Shell
Linear styrene-ethylene-butylene styrene block copolymers	Kraton G-Series	Shell
based on Kraton G for cable	Elexar	Shell
Radial, star, or teleblock copolymers from styrene-butadiene-styrene	Solprene	Phillips
ditto	Europrene Sol T	Anic (under license from Phillips)
Styrene butadiene star copolymers connected by divinylbenzene	Macromer	Arco
Urethane segment or block copolymers		
Polyester or polyether/polyol with excess of MDI to prepolymer, followed by chain extension with glycol	Caprolan	Elastogran
	Con-Than	Continental Gummi
	Cyanopren	Amer. Cyanamid
	Cytor	Amer. Cyanmid
	Daltomold	ICI Europa
	Desmopan	Bayer
	Disogrin	Disogrin Ind.
	Elastollan	Elastogran
	Estane	B. F. Goodrich
	Jecothan	Dunlop UK
	Orthane	Ohio Rubber
	Paraprene	Nippon Polyurethane
	Pandex	Dainippon Ink
	Pellethane	Upjohn (now Dow)
	Q-thane	K. J. Quinn
	Roylar	Uniroyal
	Rucothane	Ruco Div. Hooker
	Simputhan	C. Freudenberg
	Texin	Mobay
	Experim. Elastomers E 913, E 914	Mobay

majority of the TPEs is still another advantage.

The most obvious disadvantages of TPE compared to vulcanized elastomers are poor compression sets, a rather high hardness, poor

TABLE 4-2
Thermoplastic elastomers based on poly(olefins).

Chemical Composition	Trade name	Company
EP/EPDM Elastomer/poly(olefin) blends		
<50% PP + EP	Dutral	Montedison
>50% PP + EP + inorganic fillers	Moplen	Montedison
LDPE-Olefin elastomer	HC-Series	Heisler Compounding
	Polytrope	A. Schulman
	Prolastic	Prolastomer Inc.
PP/EP-Elastomer	Renflex	REN-Plastic (Ciba-Geigy)
	Keltan TP	DSM
	Propathene	ICI
	Vestolene EM	Hüls
	Vistaflex	Exxon
Poly(olefin)/EPDM	Somel	DuPont
Non-cryst. ethylene/alpha-olefin copolymer	Tafmer P	Mitsui Petrochem.
	Tersyn	Copolymer Rubber & Chem.
60−65% EP-Elastomer + 35−40% high mol. wt. PP, pre-crosslinked	TPR	Uniroyal (TPR purchased by Reichhold Chemicals in 1983)
EPDM-Blend	Uniprene	ISR
EPDM/Poly(olefin) blend	Vestopren TP	Hüls
	Telcar	Tecnor Apex (formerly B. F. Goodrich)
	Levaflex EP	Bayer
	Santoprene	Monsanto
	Ferra-Flex	Ferro

resistance to oil and hydrocarbons, and relatively low service temperatures. The comparatively high prices hamper TPE in gaining broad acceptance in markets that are being served by the conventional elastomers. Although some of these disadvantages can be overcome with certain TPE types, none of the present TPEs exhibit the broad property balance required in high performance rubber goods as specified in ASTM D-2000/SAE-J-200 which are met by conventional elastomers such as EPDM, CR, CSM and NBR (Table 4-5).

TABLE 4-3
Various other thermoplastic elastomers.

Chemical composition	Trade name	Company
Poly(butyleneterephthalate-co-(multi(butyleneoxy))terephthalate)	Hytrel	DuPont
Poly(etherester) based on terephthalic acid, butanediol and poly(tetrahydrofuran) or poly(propylene oxide)	Arnitel	Akzo
Co(polyester)	Pelprene	Toyobo Co. Ltd.
Poly(esteramide) with soft segments from aliphatic poly(esters) and hard segments from aromatic poly(amides)		Upjohn (now Dow)
Poly(esteramide) from laurolactame, 1,10-decanedicarbonic acid and α,ω-dihydroxy(poly-(tetrahydrofuran))	Vestamid X	Hüls
Poly(etheramide) from aliphatic C_{12}-poly(amide) hard segments and poly(ether) soft blocks	Ely 1256	Ems Chemistry
Elastomer modified nylon-6 types	KL 1-2310 KL 1-2310/2	Bayer
1,2-Poly(butadiene) with soft amorphous and hard crystalline segments	JSR RB	Japan Synthetic Rubber
Trans-1,4-poly(isoprene)	Trans-Pip	Polysar
Modified PVC TPE	Sumiflex	Sumitomo Bakelite Co.
Pre-crosslinked PVC	Kanevinyl XEL	Kanegafuchi Chem. Industry Co.
Poly(ester-ether-amide)	Pebax	Ato Chimie

The styrene based TPE's meet the requirements with regard to tensile strength and hardness but not in oil resistance, heat aging and compression set. SBS and SIS types are not recommended for continuous use above 65°C whereas some SEBS grades reputedly have an upper service temperature limit of 120°C. The poor weathering resistance of SBS and SIS is due to the butadiene (SBS) and isoprene units (SIS). This limitation is removed in SEBS types with saturated center blocks of poly(ethylene-co-butylene).

The polyolefin based TPEs meet all requirements with regard to

TABLE 4-4
Thermoplastic Elastomers based on graft polymers and ionomers.

Chemical composition	Trade name	Company
Graft polymers		
Graft polymer of butyl rubber on LDPE or HDPE	ET-Polymers	Asea Kabel Sweden, formerly Allied Chemical
50/50 Graft polymer of vinylchloride on poly(ethylene-co-vinylacetate)	Pantalast L	Pantasote
Graft polymer of EPDM on PVC	Rucodur 1900	Ruco Division, Hooker Chem.
Graft polymer of styrene and acrylonitrile on saturated acryl rubber	AAS-Polymers	Hitachi
Ionomers		
Poly(ethylene-co-methacrylic acid), crosslinked by metal ions	Surlyn	DuPont
Zinc salt of a sulfonated poly-(ethylene-co-propylene-co-5-ethylidene-2-norbornene) EPDM	Thionic	Uniroyal in license from Exxon
Poly(ethylene-co-acrylic acid)	Dow EEA	Dow Chemical

hardness, tensile strength and aging at moderate temperatures, however, compression set and oil resistance are poor. On the other hand, special grades with improved compression set have low tensile strength.

The copolyester TPEs generally are harder than the other TPEs and exceed the values of the ASTM D-2000 specification. Their advantages are the high service temperature and the excellent resistance to oil and hydrocarbons. Drawbacks are the high water absorption, the sensitivity to saponification and the high price. Some of the new polyamide TPEs have lower hardnesses than copolyester TPE at an otherwise equivalent balance of properties but also at equivalent high prices.

The PUR TPEs have rather low service temperatures and are sensitive to hydrolytic attack, here the poly(ether-urethanes) perform better than the poly(ester-urethanes) whereas the latter are less sensitive to oxidation and heat aging. In general, PUR TPEs meet the requirements of ASTM D-2000 except compression set.

In the following chapters, the recent developments of TPE are described with the exception of PUR and styrene based TPEs.

TABLE 4-5
Property values of unfilled thermoplastic elastomers (TPE's).

Physical Property	Physical Unit	Property values of							
		SBS	SIS	SEBS	Poly-(olefins)	Poly-coesters	Poly-amides	Polyester urethanes	Polyether urethanes
Density	g/cm³	0.94–0.95	0.9	0.901–0.907	0.87–0.95	1.17–1.25	1.01–1.15	1.15–1.24	1.10–1.20
Hardness (Shore A/D)	—	62A–90A	34A–29A	75A	54A–70D	40D–74D	60A–67D	70A–70D	70A–55D
Tensile strength	MPa	19–32	9–21	31–34	3–27	20–45	12–50	20–76	20–62
Elongation at break	%	700–880	1050–1300	500	150–1100	350–800	200–680	200–700	390–600
Modulus 100%	%								
300%	%	2.7–3.4	0.6–1.0	4.8–5.5	6–14	6–28	14–18	3–21	3–10
Compression Set (22 h/70°C)	%				40–92	29–38	32–70	25–55	25–40
Tear strength (Die C)	kN/m	19–30		71–80	10–144	98–260	60	78–100	87–156
Rebound resilience	%	60–70			30–40	40–62	60	40–50	
Abrasion (Taber C17)	mg/1000					3–28	4	2–15	
(Taber H 18)	cm³/1000				40–180	64–160	55–180		
Brittleness temperature	°C	>–73		–48/–86	–5/–69	<–70		–45/–80	
Continuous service temperature	°C	≤65	≤65	≤120	120–150	125–175		≤80	
Price (1981)	US-$/lb	0.65–1.90	0.65–1.90	0.65–1.90	0.95–1.40	2.35–2.60	3.00–3.50	1.80–2.25	2.25–2.60

For PUR TPE, only a moderate growth is expected for the 80's because of the advancement of the competitive RIM technology and the replacement of PUR TPE by other and less costly TPEs. The development of SBS, SIS and SEBS types seems to have reached a final stage although some modified grades are still being marketed. After 1975, the Phillips Petroleum Co. introduced the SBS ®Solprene grades 315 P, 416, 420, 422, 423 and 481, all with radial structure, the SIS grade 418, and a carboxylated SB block copolymer ®Solprene 312 to be used as elasticizer in SMC and BMC [35]. On the other hand, Phillips decided to sell or phase out its Solprene thermoplastic elastomers by the end of 1982 [36].

References

[1] H.-G. Elias, Makromoleküle, 4th edition, Hüthig & Wepf, Basel-Heidelberg 1981; Macromolecules, Plenum, New York, 2nd ed. 1984

[2] A. H. Ward, T. C. Kendrick and J. C. Saam, in N. A. J. Platzer (Ed.) Copolymers, Polyblends and Composites, (Advances in Chemistry Series **142**) American Chemical Society, Washington D.C., 1975.

[3] J. A. Manson and L. H. Sperling, Polymer Blends and Composites, Plenum Press, New York 1976.

[4] H. G. Elias, New Commercial Polymers 1969−1975, Gordon and Breach Science Publ., New York 1977.

[5] A. Noshay and J. E. McGrath, Block Copolymers, Academic Press, New York 1977.

[6] S. Cooper and G. M. Estes (Ed.), Multiphase Polymers (Advances in Chemistry Series **176**), American Chemical Society, Washington D.C. 1979.

[7] E. N. Kresge, Rubbery Thermoplastic Blends, in D. R. Paul and S. Newman (Ed.) Polymer Blends, Academic Press, New York, vol. 2, 1978.

[8] G. Wegner, Kautschuk und Gummi, Kunststoffe **31**/2 (1978) 67.

[9] K. v. Henten, Kautschuk und Gummi, Kunststoffe **31**/6 (1978) 426.

[10] B. M. Walker, (Ed.), Handbook of Thermoplastic Elastomers, Van Nostrand Reinhold, New York 1979.

[11] A. F. Finelli, R. A. Marshall and D. A. Chung. Thermoplastic Elastomers, Encycl. Chem. Technol., Vol. **8**, J. Wiley & Sons, New York 1979.

[12] H.-G. Elias, Kunststoffe **70** (1980) 705.

[13] P. Dreyfuss and L. J. Fetters, Elastomeric Block Polymers, Rubber Chem. Technol. **53** (1980) 728.

[14] A. D. Thorn, Thermoplastic Elastomers, RAPRA, Shawbury, U.K., 1980.

[15] M. Morton, Styrene-Diene Block Copolymers, Encycl. Polym. Sci. Technol., Vol. **15**, J. Wiley & Sons, New York 1971.

[16] T. Thorstadt and J. Vandendael, Kunststoffe **66** (1976) 405.

[17] J. B. Titus, New Plastics, Plastec Report R48, Plastics Technical Evaluation Center, Dover, New Jersey 1977.

[18] S. Cooper, J. C. West and R. W. Seymour, Polyurethane Block Polymers,

Encycl. Polym. Sci. Technol., Suppl. Vol. 1, 52, J. Wiley & Sons, New York 1976.

[19] W. Goyert and H. Hespe, Kunststoffe **68** (1978) 819.

[20] R. J. Cella, Elastomeric Polyesters, Encycl. Polym. Sci. Technol., Suppl. Vol. **2**, 485, J. Wiley & Sons, New York 1977.

[21] L. Holliday, (Ed.), Ionic Polymers, Halsted Press, J. Wiley & Sons, New York 1975.

[22] H. G. Elias, Kunststoffe — mit besonderen Eigenschaften, Ullmanns Encycl. der Techn. Chemie, Band **15**, 421, Verlag Chemie, Weinheim 1978.

[23] A. Eisenberg, (Ed.), Ions in Polymers (Advances in Chemistry Series **187**), American Chemical Society, Washington D.C. 1980.

[24] S. A. Sundet, R. C. Thamm, J. M. Meyer, W. H. Buck, S. W. Caywood, P. M. Subramanian and B. C. Anderson, Macromolecules **9** (1976) 37.

[25] R. C. Thamm and W. H. Buck, J. Polym. Sci. **16** (1978) 539.

[26] W. H. Starkley, R. P. Foss and J. F. Harris Jr., ACS Organic Coating and Plastics Chemistry **37**/1 (1977) 560.

[27] S. W. Craywood, Rubber Chem. Technol. **50** (1977) 127.

[28] R. P. Kambour, in S. L. Aggarwal, (Ed.), Block Polymers, Plenum Press, New York 1970.

[29] J. C. Saam, A. Howard and F. W. G. Fearon, J. Inst. Rubber Ind. **7**/2 (April 1973) 69.

[30] A. Noshay, M. Matzner and C. N. Merriam, J. Polym. Sci. [A-1] **9** (1971), 3147.

[31] J. R. Falender, S. E. Lindsey and J. C. Saam, Polym. Eng. Sci. **16** (1976) 54.

[32] Anonym, Modern Plast. **57** (November 1980) 54.

[33] P. A. DePaolo, Rubber World (April 1980) 43; DePaolo, Walker Report on Thermoplastic Elastomers in the U.S., Prolastomer Research and Consulting Div., Cheshire, CT.

[34] L. E. Fithian, Rubber World **184**/6 (1981) 34.

[35] D. S. Hall (Phillips Petroleum Co.), personal communication, June 8, 1981.

[36] Anon., Modern Plast. Intern. (July 1982) 61.

4.2 THERMOPLASTIC POLYOLEFIN ELASTOMERS

4.2.1 Structure and synthesis

The majority of thermoplastic polyolefin elastomers (TPO) are blends of isotactic poly(propylene) (PP) and ethylene/propylene rubber (EPR and/or EPDM). They are either composed of relatively high portions of EPR (60–65%) together with a high melt index (6–10) PP, or less EPR together with a high molecular weight PP. Apparently, partially crystalline EPR and EPDM with long poly-(ethylene) segments are preferred. Unlike the conventionally curable EPR and EPDM types which should have the greatest possible random distribution of sequences, the ethylene units in TPO are preferrably kept in long sequences to attain microcrystalline PE regions. At a sufficient segment length, the microcrystals act as physical crosslinks; this effect is different from those in intertwined structures or entanglements. The sequence length of the crystalline PE is controlled to such an extent that its melt temperature is above 80°C. Under certain polymerization conditions, copolymers containing more than 65% of ethylene yield PE sequences capable of crystallisation [1]. Some TPOs, e.g. the recently introduced ®Telcar and ®TPR grades, contain partially chemically crosslinked olefin/ elastomer blends. The products have improved compression sets but processability is reduced [2–6].

The first commercial TPO was introduced in 1972 by Uniroyal under the name TPR®; this line was purchased by Reichhold Chemicals in 1983. Within two years, it was followed by products of DuPont (®Somel), B. F. Goodrich (®Telcar), Hercules (®Profax) and Exxon (®Vistoflex), however, by 1980, these TPOs had either been withdrawn or the product line was sold to compounders. Today, smaller US companies, e.g., A. Schulman, Cooke Division of Reichhold, Heisler, RenPlastics (Ciba/Geigy) and Prolastomer, nowadays also produce TPO blends, in part according to customer specifications [7]. The Tecnor Apex Co. acquired the Telcar line from B. F. Goodrich.

Since 1975, a number of new TPO producers have appeared, on the market. They include Bayer (®Levaflex EP), Chemische Werke Hüls (®Vestopren TP), Copolymer Rubber & Chemicals Corp. (®Tersyn), DSM (®Keltan TP), Ferro (®Ferroflex in the USA, ®Ferra-Flex in Europe), Hoechst (®Hostalen PP Elastomer Blends), ISR (®Uniprene), Mitsui Petrochemical (®Tafmer P), Monsanto (®Santoprene), and Montedison (®Dutral and ®Moplen).

Tersyn, Keltan TP, Uniprene, Dutral and Moplen are TPO blends. Hoechst supplies PP/elastomer blends Hostalen PPK VP 1018, PPN VP 1009 and PP LP 290, as well as a PP block copolymer PPR 1042 [8, 9]. Hüls has not disclosed whether its Vestopren TP grades are produced by blending or grafting [10]. Ferro's TPO is said to be based on "an interpenetrating matrix system bridging crystalline and amorphous phases to obtain the desired rubber-like characteristics through third components" [29]. Tafmer P is a non-crystalline ethylene-alpha-olefin copolymer [11]. The nature of the alpha-olefin comonomer has not been disclosed.

The ®ET-Polymers are also TPOs; they were produced by Allied Chemical [12] and are now supplied by Asea Kabel in Sweden [13]. ET-Polymers are graft polymers of butyl rubber onto LDPE or HDPE. Grafting is achieved by phenolic resins, e.g. with a brominated poly(methylolphenol). Asea has marketed the grades XT 6001 (HDPE/IIR), CT 6001 (LDPE/IIR) and two semi-conducting grades.

®Santoprene TPO of Monsanto [14] is based on the new concept of the so-called "dynamic vulcanization" [15]. The process involves the blending of a conventional rubber with a crystalline poly(olefin). On mastication, mechanical degradation leads to chain scission and the generation of macroradicals. These macroradicals cause a cross-linking of the rubbery domains, leading to gel particles with a diameter of appr. 1 μm. The new concept has been applied to EPDM/PP, EPDM/PE [16, 17], to NBR/nylon [18] and to 11 other rubbers in combination with 9 other thermoplastics resulting in 75 thermoplastic elastomers [19, 20]. ®Santoprene is based on EPDM/ PP and is produced by mixing EPDM, PP, zinc oxide, stearic acid, tetramethylthiuram disulfide, 2-benzothiazyl disulfude and sulfur at 180−190°C [17]. Unlike the older processes of Exxon and Uniroyal [2−4] which lead to partially crosslinked materials, the dynamic vulcanizates yield completely crosslinked elastomers dispersed in a matrix.

4.2.2 Properties and applications

TPO properties range from soft rubbery to very stiff rubbery-plastic characteristics. Shore hardnesses range from 55A to 96A and from 14D to 65D (Table 4-6). As the elastomer characteristics of TPO increase, melt index, Vicat temperature, Shore hardness, tensile strength at yield and tensile modulus decrease [10]. Tensile strength values of about 20 MPa are attainable; they can be improved by adding talc or chopped fiberglass strands, e.g. in Vestopren TP 2083 TG. Filler loads of 20% are generally used, however, TPOs filled up to 85% with talc, chalk or mica have also been reported, e.g. for the new HC-8000 series of Heisler Compounding, Div. of Container Corp. of America [21]. Some recently introduced unfilled TPOs, e.g. the stiffer grades of the RenFlex series [22], reportedly have a tensile strength of about 27 MPa at 850% elongation. TPO reinforced with carbon black exhibits a tensile strength of up to 30 MPa at elongations of 800%.

TABLE 4-6

Property values of thermoplastic poly(olefin) elastomers (TPO) ([10, 25, 28]).

Physical property	Physical Unit	Keltan TP		Physical property values of Vestopren TP					Levaflex EP		
		0300	0602	2047	1056	2084	2053	2083TG	360	370	690S
Density	g/cm^3	0.89	0.89	0.90	0.90	0.90	0.90	1.10	0.89	0.89	1.04
Hardness Shore A	—	30	58	75	88	50	57	65	60	71	92
Shore D	—										
Tensile strength	MPa	10.8	19.5	8.8	7.5	18.8	15.3	25.9	3.5	5.2	8.6
Tensile strength at yield	MPa		21	5.4	7.0	19.1	21.3	25.8			
Elongation at break	%	1100	420	537	250	120	159	59	340	340	500
Modulus of elasticity	MPa				109	536	809	1752			
Compression set (22 h/70°C; at 23°C)	%								59	61	
Notched impact strength Izod at −40°C	J/m	n.b.	1600	n.b.	680	75	67	30	n.b.	n.b.	n.b.
at 23°C	J/m	n.b.	n.b.	n.b.	n.b.	682	491	102	n.b.		
Vicat temperature	°C	55	128		49	118	140	145			

Vestopren 2083 TG is filled with talcum and glass fibers; Levaflex EP 690 S contains light fillers.
n.b. = no break.

Compression sets of TPO generally range from 40 to 92% and thus are much higher than the 10−60% usually obtainable with conventionally cured elastomers [23]. With the new TPO ®TPR 078 a compression set of only 25% at 70°C is achieved [24].

TPOs have a service temperature range from −45 to 125°C, for some grades an upper use temperature of about 150°C is claimed, but at this temperature only 50% of room temperature tensile strength is retained. The heat deflection temperature is at least 30°C higher when compared to plasticized PVC. The resistance of TPO to UV, ozone and weathering is excellent. The tensile strength at break decreases from 12.5 to only 11.9 MPa after 4 000 h in a Xenotest 150 whilst the elongation increases from 140 to 200% [10].

The resistance of TPO to acids, bases and aqueous salt solutions is excellent, but it is only fairly resistant to polar organic solvents. Strong swelling is observed with fuel, oil, or aromatic and halogenated solvents. The degree of swelling increases as the elastomeric properties of TPO increase. In hot mineral oils the majority of TPO grades disintegrate completely, however, the partially and fully pre-crosslinked TPO grades are largely resistant to oils and hydrocarbons.

These products include TRP 8302 and TRP 8392 of Uniroyal and Santoprene of Monsanto.

In Santoprene TPO, the particle diameter of the crosslinked rubber ranges from 1 to more than 80 µm. Tensile strength and elongation increase as the rubber particle size decreases; at 1−1.5 µm diameter, tensile strengths of 25 MPa and elongations of 500% are obtained. The tensile strength also depends on the crosslinking density and the PP content. Below 30 pts of PP phr EPDM tensile strength remains below 15 MPa, however, a level of high tensile strength of 25−28 MPa is achieved with 50 to about 250 pts PP phr EPDM [17]. Santoprene was developed for applications hitherto served by vulcanizates of EPDM, CR and chlorosulfonated PE (CSM). The heat aging properties of Santoprene at 125 and 150°C are superior to those of CR, EPDM and CSM (Table 4-7). EPDM is not resistant to 125°C hot oil, whereas Santoprene under such conditions demonstrates swelling and retention of tensile strength as do CR and CSM. Although EPDM, CR and CSM all show initially lower compression sets than Santoprene after 22 h at 100°C, these values increase markedly after prolonged heating. Thus, after 70 h at 100°C, the compression sets of the softer Santoprene grades are

TABLE 4-7

Comparison of property values of Santoprene-TPO with those of CR, EPDM and CSM vulcanizates ([14, 15]).

Physical Property	Physical Unit	Property values Santoprene				CR	EPDM	CSM
		201–73	201–80	201–87	203–40			
Density	g/cm³	0.98	0.97	0.96	0.95	1.42	1.21	1.44
Hardness (Shore)	—	A 73	A 80	A 87	D 40	A 80	A 80	A 78
Tensile strength	MPa	7.6	9.7	15.2	18.2	18.1	14.5	17.8
Elongation at break	%	375	400	530	600			
Modulus 100%	MPa	3.2	4.6	6.9	8.6	ca.6	ca.6	ca.8
Tear strength at 25°C	kN/m	26	36	53	74	40	32	37
at 100°C	kN/m	11	14	26	42	19	16	17
Compression set								
22 h at 25°C	%	25	27	35	39	31	45	36
22 h at 100°C	%	33	39	52	65	43	61	50
70 h at 100°C	%	36	43	53	77			
Brittleness temperature	°C	<–60	<–60	–60	–56	–34	–52	–41
After aging at 125°C, 30 days								
Strength remaining	%	120	115	110	110	36	60	70
Elongation remaining	%	90	80	90	90			
Modulus 100% remaining	%	120	115	115	120			
After aging at 150°C, 15 days								
Strength remaining	%	82	80	77	75	0	30	0
After immersion in ASTM 0:1 No. 3								
125°C, 1 week								
Volume increase	%	78	65	47	42	47	n.r.	36
Strength remaining	%	60	62	70	75	63	n.r.	68

n.r. = not resistant.

considerably lower than those of EPDM, CR and CSM. Santoprene cannot be compounded with conventional fillers or extenders, however, it is processable by extrusion at 190−230°C and by injection molding at 170−190°C on conventional machinery.

The majority of TPOs can be injection and compression molded, extruded, vacuum formed, blow molded, embossed and painted. Applications are in the automotive industry (bumpers, body extensions, mud flaps, spoilers, dashboard panels, steering wheels, seat panels), in wires and cables, in applicances, hoses, tubings, and sheets, and in the footwear industry, e.g. as soling material and in ski boots. TPO has replaced rigid PVC in road marking posts [25]. Due to the excellent electrical properties and the very good aging characteristics of TPO, increased usage is expected in the cable industry [7, 13, 26]. The Asea Kabel Co. has marketed semi-conductive ®ET-Polymers with a volume resistivity of less than 10^2 Ohm × cm; the normal value of TPO is about 10^{16} Ohm × cm. The attempt to achieve an equivalent semi-conductivity with conventional LDPE or HDPE requires such high loadings of carbon black that the PE compounds would become very brittle and barely processable [27].

References

[1] G. Kerrutt, Kautschuk, Gummi, Kunststoffe **26** (1973) 54.

[2] A. M. Gessler (Esso Research), US-Pat. 3 073 954 (1962).

[3] W. K. Fischer (Uniroyal), US-Pat. 3 758 693 (1973), US-Pat. 3 835 201 (1974), US-Pat. 3 862 106 (1975).

[4] H. L. Morris (Uniroyal), US-Pat. 4 031 169 (1977).

[5] Anonym, Modern Plastics (October 1978) 44.

[6] K. van Henten, Kautschuk, Gummi, Kunststoffe **31** (1978) 426.

[7] P. A. DePaolo, Rubber World (April 1980) 43.

[8] ®Hostalen PP-Elastomerblends, Hostalen PP Service, Technische Kunststoffe 12, Hoechst A. G., Techn. Information, Oct. 1978.

[9] G. Heufer, Kunststoffe **68** (1978) 145.

[10] ®Vestopren TP, Chem. Werke Hüls, Technical Information, April 1977 und March 1978.

[11] ®Tafmer, Mitsui Petrochemicals Ind. Ltd., Technical Information.

[12] B. M. Walker, (Ed.), Handbook of Thermoplastic Elastomers, van Nostrand Reinhold Co., New York 1979.

[13] ®ET-Polymer, Asea Kabel AB, Technical Information.

[14] ®Santoprene Thermoplastic Rubber, Monsanto, Technical Information and Data Sheets No. 1−13 (1981−1982).

[15] G. E. O'Connor and M. A. Fath, Rubber World **185** (January 1982) 26.

[16] A. Y. Coran, B. Das and R. P. Patel (Uniroyal), US-Pat. 4 130 535 (1978).

[17] A. Y. Coran u. R. Patel, Rubber Chem. Technol. **53** (1980) 141.

[18] dgl., Rubber Chem. Technol. **53** (1980) 781.

[19] dgl., Rubber Chem. Technol. **54** (1981) 54.

[20] dgl., Rubber Chem. Technol. **55** (1982) 536.

[21] Anonym, Modern Plastics (November 1980) 55.

[22] ®Ren-Flex, Ciba-Geigy, REN Thermoplastic Department, Technical Information 1980.

[23] A. D. Thorn, Thermoplastic Elastomers, Rubber and Plastics Research Association of Great Britain, Shawbury, England, 1980.

[24] C. C. Ho and J. R. Johnson, Elastomerics **111** (Juni 1979) 19.

[25] ®Keltan TP, DSM, Technical Information, April 1981.

[26] ®TPR Rubber Wire and Cable Selection Guide, Uniroyal, Technical Information.

[27] Polymer Compounds you did'nt know existed—from a supplier you've never even heard of, Asea Compound, Asea Kabel AB, Technical Information.

[28] ®Levaflex EP, Bayer A. G., Technical Information, March 1982.

[29] Anon., Modern Plast. Intern. (April 1982) 16.

4.3 THERMOPLASTIC POLY(VINYL CHLORIDE) ELASTOMERS

4.3.1 Structure and synthesis

Since 1980, Kanegafuchi Chemical Industry has produced a PVC-TPE with a monthly capacity of 600 t [1]. The two grades offered, i.e ®Kanevinyl XEL-A and XEL-B, are described as "pre-cross-linked PVC suspension polymers" [2].

Pantasote has marketed a graft copolymer of 50% vinyl chloride onto 50% EVA under the trade-name ®Pantalast-L; it is prepared by free radical polymerization in suspension [3, 4].

®Rucodur 1900 of Ruco Division, Hooker Chemical & Plastics Corp., is a graft copolymer of vinyl chloride grafted onto EPDM [5].

Sumitomo Bakelite Co. Ltd has marketed six grades of ®Sumiflex which has been described as a "modified PVC-TPE" [6].

The Sekisui Chemical Industries Ltd has recently commercialized a flexible PVC which contains no plasticizer. The polymer is prepared from vinyl chloride, ethylene and a termonomer of undisclosed composition [7].

TABLE 4-8
Property values of thermoplastic PVC-Elastomers ([2, 4, 6]).

Physical property	Physical unit	Kanevinyl		Pantalast L	Sumiflex		Weich-PVC[1]
		XEL-A	XEL-B		K 500C	K 760	
Density	g/cm³			1.16	1.25	1.30	
Tensile strength at break	MPa	16.2	11.3	14.5	13.2	14.7	12.3
Elongation	%	370	240	250	310	260	410
Modulus 100%	MPa	4.6	4.1	8.6	3.6	5.8	
Hardness (Shore A)	—	66	66	82	58	66	50
Heat distortion JIS K-6723	%				6	7	27
Compression Set (22 h/70°C)	%	49	46		47	47	65
Heat aging (100 h/120°C) Break strength remaining	%			100	110	99	115
Elongation remaining	%			100	92	100	90
Brittleness temperature	°C			−55	−48	−40	−50

[1] Sumikon VM-1225 C-90 of Sumitomo Bakelite.

4.3.2 Properties and applications

Thermoplastic PVC elastomers (PVC-TPE) contain no plasticizers. They exhibit properties which range from those of plasticized PVC to elastomer-modified PVC types (see chapter 2.3 and Table 4-8).

Kanegafuchi's XEL grades are superior to PVC homopolymers with regard to heat deflection temperature and compression set [2]. The partial crosslinking yields a gel. The gel portion which is insoluble in tetrahydrofuran amounts to 50%; the soluble portions of XEL-A and XEL-B have molecular weights of 2000 and 3000 g/mol and K-values of 169 and 219, respectively.

The advantages of Pantalast over PVC reportedly are lower flammability and less smoke emission, better properties at low and elevated temperatures, improved weathering resistance, and less abrasion [4].

The impact resistance of Rucodur is superior to PVC. The claimed advantages of Sumiflex over PVC, elastomer-modified PVC, EVA and EPR are improved compression sets, higher resilience, and better mechanical properties at elevated temperatures. In comparison to some other rubbers, Sumiflex offers enhanced resistance to weathering, ozone and mineral oils.

None of the PVC TPE's contains plasticizers, hence no migration or exudation occurs as with plasticized PVC which makes PVC TPE a strong candidate for applications in the medical field. Other applications include jacketing for wires and cables, and use in the footwear industry, in the automotive industry, e.g., bumper covers, in films and in sidings for exterior use.

PVC TPE costs more than PVC; e.g. the price of Pantalast is 0.99 US-$/lb.

References

[1] Anon., Japan Chemical Week (March 13, 1980) 1.
[2] ®Kanevinyl XEL-A and XEL-B, Kanegafuchi Chemical Industry Co., Ltd., 1981.
[3] A. D. Varenelli, L. Weintraub and S. Pearson, Paper presented at the SPE 38th ANTEC, May 5, 1980.
[4] ®Pantalast L, Pantasote Inc., Film and Compound Division, Technical Information, September 1980.
[5] Anon., Plastics Technol. **10** (1980) 41; ®Rucodur, Ruco Division, Hooker Chemicals & Plastics Corp., Technical Information.
[6] ®Sumiflex Modified PVC Thermoplastic Elastomer, Sumitomo Bakelite Co., Ltd., Technical Information 1981.
[7] Anon., Kunststoffe **73**/1 (1983) 51.

4.4 THERMOPLASTIC POLYESTER ELASTOMERS

4.4.1 Structure and synthesis

®Hytrel of DuPont is a poly[butyleneterephthalate-co-(multi-butylene-oxy)-terephthalate]; it became commercially available in 1972 and has been described in detail [1–5]. The Toyobo Co. Ltd, a recent competitor, produces copolyester-TPE based on butylene-glycol-terephthalate under the name ®Pelprene with an annual capacity of 1800 t.

In 1977, Akzo NV marketed a series of thermoplastic copolyesters under the trade name ®Arnitel [6]. They are based on butanediol, terephthalic acid and poly(tetrahydrofuran) [7]. Hytrel and Arnitel are copoly(ether-esters) with soft amorphous poly(alkylene-ether)-terephthalate blocks and hard crystalline tetramethylenegly-col-terephthalate blocks:

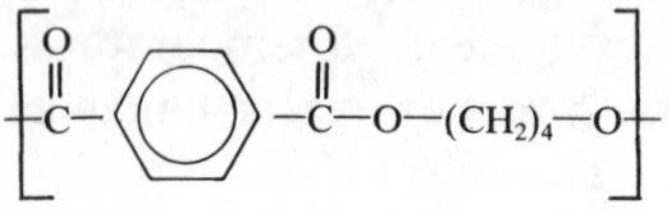

4 GT hard segment 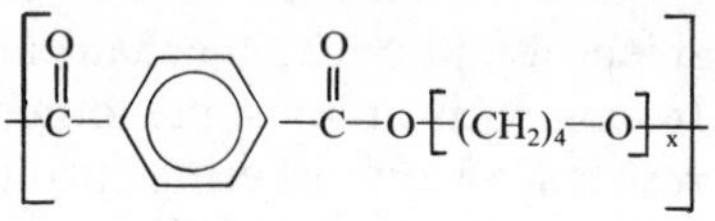   PTMEGT soft segment

The copoly(ether-ester) elastomers are prepared by transesterification from terephthalic ester, poly(alkylene ether) glycol and a short-chained diol, e.g. 1,4-butanediol.

The effect of different poly(alkylene ether)glycols, e.g. poly(tetramethylene ether)glycol (PTMEG), poly(ethyleneoxide)glycol (PEG), poly(propylene oxide)glycol (PPG) and ethylene oxide capped PPG (EOPPG), in copolymers with 4 GT hard segments has been studied [8]. Copolymers with PTMEG exhibit the highest tensile and tear strengths. Copolymers with PEG suffer from strong swelling in hot water and hydrolyze rapidly. The elongation at break is virtually independent of the nature of the polyglycol. In PTMEG/4GT copolymers, the hard terephthalic ester segments can be partially replaced by esters of isophthalic or sebacic acid. Such copolymers show increased tensile and tear strengths without sacrifice of hardness, modulus or low temperature flexibility [9].

4.4.2   Properties and applications

Presently, DuPont offers Hytrel injection molding and extrusion grades 4056, 5526 and 5556 with a good balance of properties for low and elevated temperature applications, the heat-stabilized Hytrel 5555 HS and the non-staining Hytrel 6346 for use at elevated temperatures.

Akzo has marketed Arnitel injection molding grades EL 550, EL 630, UV-stabilized EL 550 L, EL 630-L and EL 740, extrusion grades EM 400, EM 460, EM 550, EM 630 and EM 740, UV-stabilized EM 550 L and EM 630 L, ER 460 for rotational molding, and the powder ER 550 [6, 10]. The first digits in the Hytrel and Arnitel grades specify the Shore hardness (Table 4-9).

The elastomeric polyether-esters exhibit high notched impact strength at low temperatures, high flexibility, tear strength and abrasion resistance. The combination of high modulus, high stiffness and good recovery from deformation under tensile and compression strains allow applications as an engineering plastic. Polyester TPEs have higher upper service temperatures than TPEs based on styrene, olefins and urethanes. At elevated temperatures, unstabilized co-polyether-esters are susceptible to oxidative attack on the polyether backbone. They are also degraded by UV light. Properly stabilized grades, however, showed no loss of tensile strength after 1000 h at 120°C in air or after 1000 h in the UV Xeno-Tester 450 [6]. The new Arnitel EL 700 can be used up to 150°C. About 50% of tensile strength is retained after more than 1000 h at 135°C or 850 h at 150°C [6].

The elastomeric polyether-esters demonstrate excellent resistance to bases and acids of moderate concentrations, to aliphatic hydrocarbons, mineral oils and hydraulic fluids; however, they swell in acetone, ethanol, in mixtures of benzene and methanol, in ethylacetate, and in aromatic hydrocarbons. Copolymers with less than 60% hard segments are heavily attacked by chlorinated hydrocarbons, some grades dissolve in methylenechloride. Arnitel 740 reputedly is resistant to these solvents. Hydrolytic stability can be improved by incorporation of poly(carbodiimides). DuPont has marketed masterbatches of Hytrel with 20% of poly(carbodiimide) under the designations HTG 3820 and 10 MS.

Reported applications for elastomeric polyether-esters are hoses and tubings for hydraulic and brake fluids, low speed tires, snow-

TABLE 4-9

Property values of thermoplastic copoly(etherester) elastomers ([3, 6]).

Physical Property	Physical Unit	Property values of Arnitel					Hytrel	
		EM 400	EM 460	EL 550	EL 630	EL 740	4055	6345
Hardness (Shore D)	—	40	46	55	63	74	40	63
Density	g/cm³	1.17	1.18	1.20	1.23	1.27	1.17	1.22
Melting temperature	C	195	185	202	213	221	168	206
Yield strength	MPa			17	23	36		
Tear strength (Graves)	kN/m	98	125	152	176	260		
Tensile strength at break	MPa	20.5	21.5	32	38	45	25.5	39.3
Elongation at break	%	700	714	570	540	360	450	350
Flexural strength	MPa			12	20			
Flexural modulus	MPa	53	85	185	330	830	48	345
Abrasion resistance (Taber CS 17)	mg/1000	29	38	27	28	17	3	8
Compression Set (22 h/70°C)	%						60	
Notched impact strength (Izod)								
at 20°C	J/m			n.b.	n.b.	90	107	107
at 0°C	J/m					45		
at −20°C	J/m			n.b.	36	35		
at −40°C	J/m			n.b.	17	30		30
Vicat temperature (10 N)	°C			178	194	207	112	184
Brittleness temperature	°C			<−70	<−70		<−70	<−70
Heat distortion temperature (1.81 MPa)	°C			49	57	48		
Water absorption (24 h)	%			0.18	0.12	0.06	0.06	0.3
Volume resistivity	Ohm × cm			2.5×10^{12}	1.2×10^{13}	8.8×10^{15}	2.3×10^{13}	1.4×10^{13}
Surface resistivity	Ohm			7.8×10^{13}	2.5×10^{14}	8.8×10^{12}		
Dielectric strength	MV/m			51	50	>53	35.5	27
Relative permittivity (10³ Hz)	—			4.3	4.1	3.35	6.0	5.0

n.b. = no break.

mobile tracks, sealings, gaskets and packings, damping elements, jackets for electrical cable connectors, transmission belts, a cork substitute, telephone retractable cords, pipe bells, innersoles, and rotomolded corrosion-resistant covers for automotive steering gears [11]. Arnitel EL 740 with a shore hardness of 74 D is said to replace other engineering plastics such as PA 6, PA 11 and PA 12 [6].

References

[1] H.-G. Elias, Neue polymere Werkstoffe 1969−1974, Carl Hanser Verlag, München 1975; New Commercial Polymers 1969−1975, Gordon and Breach, New York 1977.

[2] R. J. Cella, Elastomeric Polyesters, Encycl. Polym. Sci. Technol., Suppl. Vol. **2**, 485, J. Wiley & Sons, New York 1977.

[3] B. M. Walker (Ed.), Handbook of Thermoplastic Elastomers, Van Nostrand Reinhold, New York 1979.

[4] P. Dreyfuss and L. J. Fettes, Rubber Chem. Technol. **53** (1980) 728.

[5] J. D. Ryan, ACS Organic Coatings and Plastics Chem. **44** (1981) 307.

[6] ®Arnitel, Akzo Plastics bv, Technical Information, July 1977 and December 1978; ®Arnitel EM 400 and 460, Akzo Plastics bv, Technical Information 1980.

[7] B. F. Steggerda, Gummi, Asbest, Kunststoffe **30** (1977) 428.

[8] J. R. Wolfe, ACS Rubber Div. Meeting, Chicago, May 1977.

[9] W. K. Witsiepe, in N. A. J. Platzer (ed.), Polymerization Reactions and New Polymers (Advances in Chemistry Series 129), American Chemical Society, Washington, DC, 1973.

[10] Anon., Modern Plast. Intern. **8**/7 (July 1978) 22.

[11] Anon., Modern Plastics **57**/11 (November 1980) 54.

4.5 THERMOPLASTIC POLYAMIDE ELASTOMERS

4.5.1 Structure and synthesis

Thermoplastic polyamide elastomers are block copolymers of hard polyamide segments with soft polyether and/or polyester segments.

In 1979, Chemische Werke Hüls introduced copolyether-amides under the trade name ®Vestamid with the grades E62L (formerly X 4018, X 3808), E52L (X 3912), E47L (X 3978, X 3933), E40L (X 4138, X 4006), and E33Lw (X 4083) [1] (Table 4-10). Also, black permanent antistatic grades, heat and light stabilized and flame-retardant grades are commercially available [2].

TABLE 4-10
Property values of thermoplastic poly(amide) elastomers ([1, 2, 6, 14, 16]).

Physical property	Physical unit	ELY 1256	Property values of Vestamid					Upjohn	
			E 62 L (X4018)	E 52 L (X3912)	E 47 L (X3978)	E 40 L (X 4138)	E 33 LW (X4083)	90A	55D
Density	g/cm³	1.01	1.02	1.01	1.01	1.01	1.04	1.15	
Yield strength	MPa	20	24						
Elongation at yield	%	45	37						
Tensile strength at break	MPa	30	30	23	18	13	9	38.8	33.9
Elongation at break	%	300	200	200	200	200	200	465	390
E Modul	MPa	250	360	200	140	80	70		
Notched impact strength (−40°C)	kJ/m²	10	5	30	n.b.	n.b.	n.b.		
Hardness (Shore)	—	56D	62D	52D	47D	40D	33D	90A	55D
Vicat temperature (10 N)	°C	150							
Water absorption (23°C/50% RF)	%	0.50							
Volume resistivity	Ohm × cm	$5 \cdot 10^{12}$							
Dielectric strength	kV/mm	40							
Surface resistivity	Ohm	10^{12}							
Relative permittivity (50 Hz)	—	5							

Copolyether-amides are synthesized by copolymerization of lactams with $C \geq 10$ (I), a α,ω-dihydroxy(polytetrahydrofuran) with molecular weights between 160–3000 (II) and a dicarbonic acid (III), where the ratio of I:(II + III) ranges from 30:70 to 98:2, and the OH- and COOH-groups in II and III are applied in equimolar amounts. The polycondensation at 270°C is conducted in the presence of 2–30% of water, based on I [3, 4], e.g.:

(4-1)

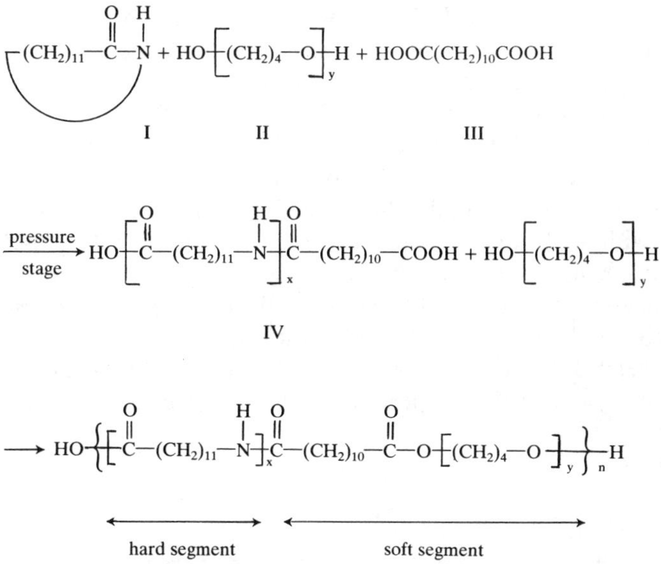

$$-(CH_2)_{11}-\overset{O}{\underset{\|}{C}}-\overset{H}{\underset{|}{N}} + HO-\left[(CH_2)_4-O\right]_y H + HOOC(CH_2)_{10}COOH$$

I   II   III

$$\xrightarrow[\text{stage}]{\text{pressure}} HO\left[\overset{O}{\underset{\|}{C}}-(CH_2)_{11}-\overset{H}{\underset{|}{N}}\overset{O}{\underset{\|}{C}}-(CH_2)_{10}-COOH\right]_x + HO\left[(CH_2)_4-O\right]_y H$$

IV

$$\longrightarrow HO\left\{\left[\overset{O}{\underset{\|}{C}}-(CH_2)_{11}-\overset{H}{\underset{|}{N}}\right]_x \overset{O}{\underset{\|}{C}}-(CH_2)_{10}-\overset{O}{\underset{\|}{C}}-O\left[(CH_2)_4-O\right]_y\right\}_n H$$

hard segment   soft segment

First, an oligoamide (IV) is formed by reacting the starting materials at a pressure of 19 bar and using 1,10-decanedicarbonic acid as chain regulator. The polycondensation to high molecular weight copolyether-amides is conducted in a "one-pot-method" without applying vacuum. The polycondensation rate is markedly increased with dialkyltin oxides as catalysts. Unlike polyester homopolymers, the high molecular weight copolymers can be prepared under normal pressure, because their equilibrium water concentration is much higher at the desired average molecular weight of about 20 000 g/mol [4]. Other components for the preparation of copolyether-

ester-amide elastomers are laurolactam, adipic acid and 1,4-bis-
(hydroxymethyl)cyclohexane [5]. In the elastomeric copolyether-
amides, the hard poly(laurolactam) and the soft poly(tetrahydrofurane) blocks are linked through ester groups.

In 1979, Ems-Chemie A. G. (formerly Emser-Werke A. G.)
introduced ®Grilamid ELY 60 (formerly ®ELY 1256) which is composed of hard PA 12 and soft polyether blocks but contains no
ester groups [6, 7]. ELY 60 is also commercially available as self-
extinguishing grade free of halogens and phosphorus.

In 1981, ATO Chimie marketed 8 grades of its ®PEBAX series
(Table 4-11). The name is derived from *Poly-Ether-Block-A*mide.
The copolyether-ester-amides are prepared by polycondensation of
polyamides containing terminal carboxy groups with linear or
branched α, ω-dihydroxy(polyethers). PA 6, PA 66, PA 11, PA 12,
PA 6, 11 and PA 6, 12 can serve as polyamide block. Presumably,
PA 11 is preferred because of the available feed stocks already
being used for ®Rilsan. In the first reaction step, an oligomeric
polyamide with molecular weights between 300 and 15 000 g/mol
and two carboxylic groups per mol is prepared. Adipic acid is the
preferred acidic component. In the second step, the oligomeric
polyamide is condensed at 260°C under vacuum with polyethers of
molecular weights from 100 to 6 000 g/mol, e.g. poly(ethylenegly-
col), poly(propyleneglycol) or preferably with poly(tetramethylene-
glycol) [8]:

(4-2)

$$\text{HO}-\underset{\underset{O}{\|}}{C}-\text{PA}-\underset{\underset{O}{\|}}{C}-\text{OH} + \text{HO}-\text{PE}-\text{OH} \rightleftharpoons \text{HO}-\left[\underset{\underset{O}{\|}}{C}-\text{PA}-\underset{\underset{O}{\|}}{C}-\text{O}-\text{PE}\right]_n \text{OH} + \text{H}_2\text{O}$$

PA = polyamide block, PE = polyether block

Titanate esters Ti(OR)_4 are used as catalysts. Esters of zirconium
or hafnium reportedly yield products with lighter color [9]. The
hydrolytic stability of copolyether-ester-amides is enhanced when
the number of free carboxylic groups is reduced. To this end, one
adds oligomeric polyamides which contain only one terminal carboxylic group, e.g. products that are available from the reaction
with stearic acid or other monocarbonic acids [10]. The preparation

TABLE 4-11
Property values of ®PEBAX thermoplastic poly(ether-amide) Elastomers ([19, 20]).

Physical properties	Physical unit	5533 SN 00	4033 SN 00	3533 SN 00	2533 SN 00	6312 MN 00	5512 MN 00	5562 MN 00	4011 RN 00
Density	g/cm^3	1.01	1.01	1.01	1.01	1.11	1.10	1.06	1.14
Melting temperature	°C	168	168	162	160	195	190	120	190
Hardness (Shore D)		55	40	80	70	63	55	55	40
Melt index (235°C/1 kg)	g/10 min	8	7	7	6	10	25		10
Tensile strength at break	MPa	33	33	29	29	42	49	51	38
Tensile strength at yield	MPa	22				19	15	14	8
Elongation at yield	%	28				23	26	27	21
Elongation at break	%	510	620	650	680	410	530	510	530
Flexural modulus	MPa	200	105	29.5	20	410	223	210	110
Stiffness in torsion									
at 20°C	MPa	85	25	15	15	110	84	81	35
at −40°C	MPa	190	120	47	45	410	350	320	190
Flexural strength	MPa	10	6	2	1	19	13	12	6
Abrasion resistance (Taber H18, 100 μ/min 1000 g)	mg	55	55	55	55	120	180	170	140
Vicat temperature (1 daN)	°C	156	145	82	63	186	168	105	160
Linear thermal expansion coefficient (×10^5) between −40 to +140°C	K^{-1}	23	22	22	21	16.5	19	21	13
Surface resistivity	Ohm	10^{13}	10^{13}	10^{13}	10^{13}	10^{13}	10^{13}	10^{12}	10^6
Water absorption (24 h)	%	1.2	1.2	1.2	1.2	6.4	6.1	3.5	119
Water absorption to saturation	%	0.5	0.5	0.5	0.5	2.8	2.4	1.3	4.5

Injection molded specimens conditioned 15 days at 23°C and 50% RF. No break at impact tests and notched impact tests at 20°C and −40°C.

of oligomeric, ω,ω'-bifunctional or monofunctional polyamides 11 with different molecular weights as model compounds for the esterification with mono-alcohols [11], their polycondensation with various ω,ω'-dihydroxy-poly(oxyethylenes) [12] and the polycondensation kinetics [13] have been described.

Upjohn developed elastomeric polyester-amides of the general structure $(CO-R-CO-NH-Ar-NH)_n$ where R represents $(CH_2)_x$ or $m-C_6H_4$ and Ar is $C_6H_4CH_2C_6H_4$ or toluylene. The polymer forming reaction was the condensation of an aromatic diisocyanate and a dicarboxylic acid with the liberation of carbon dioxide. Carboxylic acid terminated soft segment prepolymers were prepared by the esterification of commercial polyester diols such as hexamethylene adipate (Hooker) or tetramethylene azelate (Emery) with a 2:1 excess of aliphatic dicarboxylic acid (eq. 4-3). The polymerization was carried out at elevated temperatures, e.g. at 100–160°C in tetramethylenesulfone, by adding an aromatic diisocyanate (MDI and/or TDI) to the solution of the acid terminated prepolymer and additional aliphatic dicarboxylic acid (eq. 4-4). The content of hard segments and the Shore hardness was controlled by adding this additional acid as a hard segment chain extender [14, 15, 17]. In addition to the aliphatic dicarbonic acids, aromatic dicarbonic acids can also be used, e.g. isophthalic acid [16].

(4-3)

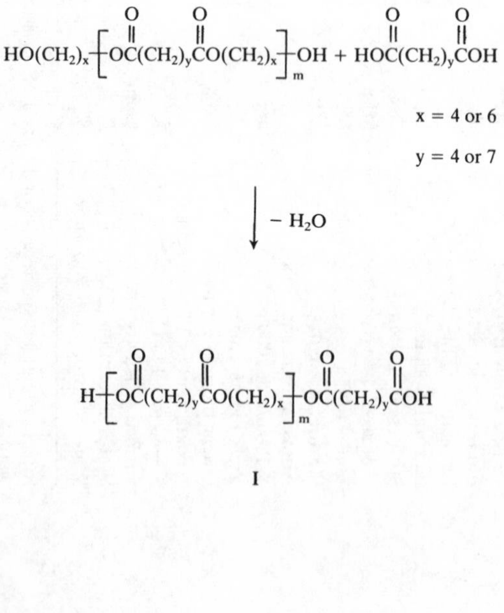

$$HO(CH_2)_x\left[OC(CH_2)_y CO(CH_2)_x\right]_m OH + HOC(CH_2)_y COH$$

x = 4 or 6

y = 4 or 7

$$\downarrow -H_2O$$

$$H\left[OC(CH_2)_y CO(CH_2)_x\right]_m OC(CH_2)_y COH$$

I

$$\mathrm{I} + n\,HO\overset{\overset{\displaystyle O}{\parallel}}{C}(CH_2)_y COH + (n+1)\,OCN\!-\!\!\bigcirc\!\!-CH_2\!-\!\!\bigcirc\!\!-NCO$$

$$\xrightarrow{-\,CO_2}$$

$$\left[\overset{\overset{\displaystyle O}{\parallel}}{C}(CH_2)_y C\!-\!O(CH_2)_x O\overset{\overset{\displaystyle O}{\parallel}}{C}(CH_2)_y\overset{\overset{\displaystyle O}{\parallel}}{C}\!-\!NH\right]_m\!\!-\!\!\bigcirc\!\!-CH_2\!-\!\!\bigcirc\!\!-NH\!-\!\!\left[\overset{\overset{\displaystyle O}{\parallel}}{C}(CH_2)_y\overset{\overset{\displaystyle O}{\parallel}}{C}NH\!-\!\!\bigcirc\!\!-CH_2\!-\!\!\bigcirc\!\!-NH\right]_n$$

(4-4)

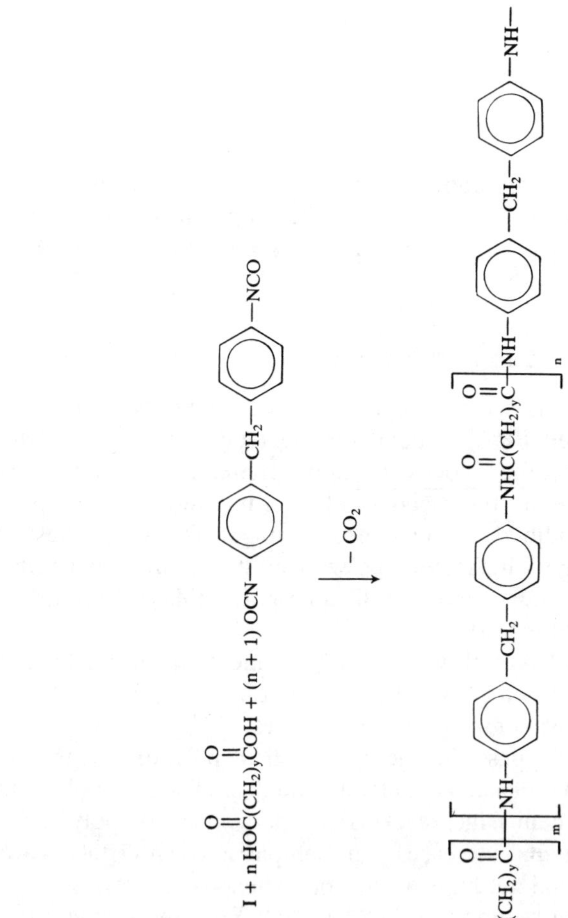

The reaction rate in the second step is accelerated by employing phospholenes or derivatives of phospholene-1-oxide as catalysts, e.g., 1,3-dimethyl-3-phospholene-1-oxide (V) [18]:

CH₃

V

O CH₃

Akzo Plastics has announced that it will also enter the nylon block-copolymer field. The three initial development grades, trade-named ®Arnetal, will have Shore hardnesses of 42, 52 and 74 D [22].

4.5.2 Properties and applications

The properties of elastomeric polyether-amides and polyether-ester-amides, abbreviated PEA, depend on the type and length of the polyamide and polyether blocks. By appropriate selection and control of these segments, the mechanical, thermal and chemical properties can be modified within a broad range. The properties of PEA cover a range which overlaps with elastomer-modified polyamides on the one hand, and with flexible polyamides on the other side (see also chapter 9.6).

The density of PEA with C_{11} and C_{12} segments from PA 11 and PA 12 or 1.10-decanedicarboxylic acid is $1.01-1.02$ g/cm^3, those with PA 6 or PA 66 segments have densities of $1.11-1.15$ g/cm^3.

In the Vestamid types, the incorporation of polyether segments results in an internal plasticizing effect. The flexibility of copolymers with poly(lauro lactam) increases as the percentage of poly(tetra-hydrofuran) segments increases. In comparison with poly(lauro lactam), the torsional modulus of the copolymers is reduced through internal plasticizing by about a factor of ten. With increasing ether content, the glass transition temperature is lowered to below $-60°$C. The melt temperature of the poly(lauro lactam) segments in the copolymers is lowered as the molar concentration of the soft component increases. Thus, the melting point is reduced from an initial value of 179°C for the pure poly(lauro lactam) to about 135°C for a copolymer with 30 mol-% of the soft component [OC-(CH₂)₁₀COORO]. The depression of the melt temperature is 1.5°C per mol-% of the co-component. The properties of the Vestamid series can be varied over a broad range by controlling the polyamide segment length and by variation of the molar ratio of 1,10-decanedicarboxylic acid/lauro lactam [4].

In the ®PEBAX series, the melt temperature ranges from 120 to 195°C depending on the type and length of the polyamide blocks (Table 4-11). By appropriate selection of the polyether blocks, water absorption can range from 1,2 to more than 100% which also affects the antistatic properties [19]. Flexibility (Figure 4-1) and Shore hardness (Figure 4-2) are controlled by the weight ratio of polyamide to polyether blocks ranging from 80:20 to 20:80. The tensile strength at yield is progressively diminished and flexibility is increased as the percentage of polyether is increased [20, 21]. The Shore hardnesses from 63 D to 60 A cover the gap between thermoplastics and elastomers. PEA-TPEs are applied in such cases where PUR TPEs at high Shore hardnesses (above 55 D) and polyester-TPEs at low Shore hardnesses (below 40 D) show a loss in performance.

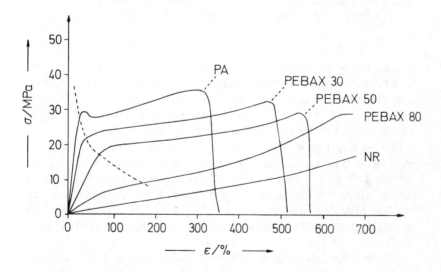

Figure 4-1 Stress/strain curves for polyamidethers with 30, 50 or 80% polyether as compared to those of a polyamide (PA) and natural rubber (NR). Broken line indicates lower yield value [19, 20].

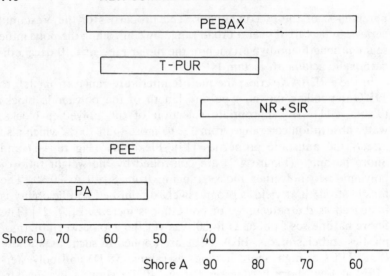

Figure 4-2 Shore hardness of PEBAX polyetheramides compared to other thermoplastics and elastomers [19, 20]. T-PUR = thermoplastic polyurethanes, NR = natural rubber, SIR = silicone rubber, PEE polyetherester, PA = polyamides.

Elastomeric polyether-amides contain no plasticizers. Also, they demonstrate high toughness and notched impact strength at low temperatures, the flexibility varies only slightly over a temperature range from −40 to 80°C (Figure 4-3). They have good abrasion resistance, good resilience, and high fatigue resistance.

For the developmental ®Arnetal of Akzo the following properties have been reported for the grade with Shore hardness 52 D [22]: density 1.07 g/cm^3, yield stress 12.5 MPa, elongation at yield 25%, tensile strength 40 MPa, elongation at break 500%, flexural modulus 180 MPa, stiffness factor at −20/+20°C 2,0, Charpy notched impact test at −40°C giving no break, and melting point (DSC) 210°C.

All copolyether-amides are susceptible to thermo-oxidative attack. Heating in air at 120−150°C for 3 days degrades unstabilized copolyether-amides resulting in almost complete loss of strength [4]. Relatively high concentrations (1−2%) of antioxidants are required. Copolyether-amides are also susceptible to the attack by organic

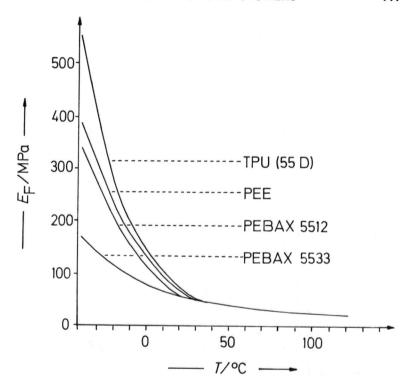

Figure 4-3 Temperature dependence of flexural modulus, E_F, of a thermoplastic polyurethane TPU (55 D) with a Shore hardness of 55 D as compared to a polyetherester, PEE, and two polyamid ethers, PEBAX [19, 20].

solvents. The values for the absorption to saturation are 10% in gasoline, 8% in mineral oil, 13% in methylenechloride, 5% in acetone and 8% in ethanol [6, 7]. Properly stabilized copolyesteramides have superior thermal and aging resistance over polyester and PUR TPEs. The upper service temperature of PUR TPE with Shore hardness 90 A is about 100°C, for copolyester TPE with Shore 40 D and 55 D 107 and 149°C, respectively, and for annealed copolyester-amides at short-term exposures 175°C, and for long-term exposures 150°C [14] (see Table 4-12). Both the elastomeric copolyester-amides and copolyether-amides swell in organic solvents. For copolyester-amides, weight increases in toluene were 30% for the grade with Shore 90 A and 13% for the 55 D grade,

TABLE 4-12

Influence of temperature on the properties of thermoplastic copoly(ester-amide) elastomers ([14, 17]).

Properties at test temperature	Physical unit	Upjohn PA-Elastomer with Shore hardness 90 A				Upjohn PA-Elastomer with Shore hardness 55 D			
Test temperature	°C	22	100	150	175	22	100	150	175
Tensile modulus at									
50% extension	MPa	11.4	8.8	6.8	5.9	15.5	6.9	4.5	3.3
100% extension	MPa	14.0	10.4	7.9	6.6	18.4	8.6	5.4	4.0
300% extension	MPa	23.2	13.1			28.3	13.3	7.7	5.4
Tensile strength at break	MPa	38.8	21.7	9.4	7.6	33.9	19.6	8.8	5.7
Elongation at break	%	465	500	250	235	390	535	430	375
Compression set (22 h/70°C)	%	32				70			
Data* after heat aging (120 h/150°C)									
Tensile modulus at									
50% elongation	%	130	107	118	106	108	123	102	86
100% elongation	%	125	101	109	101	103	108	87	77
300% elongation	%	132	101	136	102	104	95	77	82
Tensile strength at break	%	121	160			100	101	104	93
Elongation at break	%	99	116	198	134	95	99	116	109

* in % of the data before heat aging.

however, mechanical properties reportedly are only slightly affected [14].

Elastomeric polyamides are processed by injection molding, extrusion, extrusion blow molding and rotational molding techniques and can also be formed into monofilaments. Conventional machinery is used with screws similar to the processing of PA, polyether-esters and PUR. The screw profiles generally match those in PA processing. But unlike PA, the molds should have short sprues as in the processing of PUR [20, 21].

Recommended uses are for ski boots, ski spoilers, soles, balls, hoses and tubings, gaskets, pump membranes, fittings, damping elements, catheters and gasoline tank caps. The flame-retardant grades are used in cable and wire jacketings; the permanent antistatic grades see service in safety areas.

The elastomeric polyamides are relatively expensive. For example, PEBAX costs 3.50 US-$/lb compared to only 0.80 US-$/lb for rubbers used for similar applications. However, finished articles made of elastomeric polyamides may cost 10−20% less than those made of rubber, because their density is lower by about 25%, the wall thickness is lower by about 50%, and the molding cycles can be shortened by about 30 s [19].

References

[1] ®Vestamid X 3808, X 3933, X 4066, Chemische Werke Hüls A. G. Technical Information, April 1979 and August 1981.
[2] ®Vestamid X 3967, X 4005, X 4062, Chemische Werke Hüls A. G., Technical Information, April 1979 and August 1981.
[3] K. Burzin, S. Mumcu, R. Felfmann, R. Feinauer and H. Jadamus (Chem. Werke Hüls), Ger. Offen. 2712897 (1978) and Ger. Offen. 2716004 (1977).
[4] S. Mumcu, K. Burzin, R. Feldmann and R. Feinauer, Angew. Makromol. Chem. **74** (1978) 49.
[5] K. Burzin and R. Feldmann (Chem. Werke Hüls), Ger. Offen. 2542467 (1977).
[6] ®Grilamid ELY 60 (formerly ®ELY-1256), Ems-Chemie A. G., Technical Information; Ger. Offen. 3006961.
[7] S. Schaaf, Swiss Plastics **2**/5 (1980) 47.
[8] G. Deleens, P. Foy et al. (ATO Chimie), Ger. Offen. 2523991 (1975), Fr. P. 2418913 (1974), 2531828 (1975), 2623399 (1976) and 2628848 (1976).
[9] ditto, Ger. Offen. 2837687 (1979).
[10] ditto, Ger. Offen. 2802989 (1977) and Fr. P. 2378058 (1978).
[11] G. Deleens, P. Foy and E. Maréchal, Europ. Polym. J. **13** (1977) 337.
[12] ditto, Europ. Polym. J. **13** (1977) 343.
[13] ditto, Europ. Polym. J. **13** (1977) 353.

[14] R. G. Nelb, A. T. Chen, W. J. Farrissey, Jr., and K. B. Onder, SPE 39th ANTEC, Boston, May 1981, ANTEC Preprints (1981) 421.

[15] A. T. Chen, W. J. Farrissey and R. G. Nelb (Upjohn Co.), USP 4129715 (1978).

[16] K. B. Onder (Upjohn Co.), USP 4087481 (1978) and Ger. Offen. 2756605 (1978).

[17] J. T. Chapin, B. K. Onder and W. J. Farrissey, Jr., ACS Polymer Preprints **21**/2 (1980) 130.

[18] K. B. Onder and C. P. Smith (Upjohn Co.), USP 4156065 (1979) and Ger. Offen. 2801701 (1978).

[19] ®PEBAX, ATO Chimie, Technical Information 1981; G. Deleens, 39th ANTEC, Boston, May 1981; SPE **27** (1981) 419.

[20] Anon., Industrie-Anzeiger **104**/4 (January 13, 1982).

[21] Anon., Kunststoffe **72** (1982) 282; A. Sternfield, Modern Plast. Intern. (February 1982) 22.

[22] A. Sternfield, Modern Plast. Intern. (February 1982) 23.

4.6 IONOMERS

4.6.1 Poly(ethylene-co-acrylic acid)

4.6.1.1 Structure and synthesis

Poly(ethylene-co-acrylic acid) or EAA block or graft polymers have been known since the '50s [1]. Recently, random EEA copolymers have been marketed by Dow Chemical [2−4]. Ethylene undergoes a free radical polymerization with acrylic acid (AA) under high pressure and yields highly branched EAA polymers. The commercial products contain 3.5, 8.0 and 20.0% of AA units. The hydrogen bonds between neighboring carboxylic groups exert a crosslinking effect at room temperature. Allied has also commercialized a new EAA copolymer under the designation A-C Copolymer 540 [5].

4.6.1.2 Properties and applications

The EAA copolymers combine excellent adhesion to various metallic and non-metallic substrates with high toughness, tear strength, and good stiffness (Table 4-13).

The density of EAA copolymers increases with increasing content of AA and reaches 0.96 g/cm^3 at 20% AA. The tensile yield strength has a maximum between 10 and 15% AA. The falling Dart impact strength of EAA films is superior to LDPE by the factor of 3−4,

TABLE 4-13
Property values of EAA resins ([3, 4]).

Physical Property	Physical unit	Property values of			
		EAA 435	EAA 449	EAA 459	XO-2375.33
Acrylic acid content	%	3.5	8.0	8.0	20
Density	g/cm^3	0.925	0.932	0.932	0.96
Melt index	g/10 min	11.0	5.5	9.0	300
Tensile strength at yield	MPa	9.3	9.7	9.3	6.2
Tensile strength at break	MPa	14.1	19.0	17.6	11.0
Elongation at break	%	650	650	650	300
Vicat temperature	°C	86	85	82	50
Brittleness temperature	°C	−73	−73	−73	−20
Shore hardness D	—	47	49	48	54

however, it is inferior to copolymers of ethylene with acrylic or vinyl esters. The thermal properties are equivalent to those of ®Surlyn-A ionomers. The Vicat temperature of EAA is 82−86°C, and the brittleness temperature is −73°C. The Dow developmental product XO 2375.33 has an exceptional high melt index, and its properties differ significantly from those of the other EAA grades.

With increasing content of AA, the resistance to chemical attack decreases while resistance to stress cracking increases. EAA copolymers demonstrate better resistance to fats and oils than LDPE. They have FDA approval for direct contact with food. Compared to LDPE films of identical thickness, in EAA the permeabilities of oxygen and nitrogen are lowered, and the permeability of water is increased as the AA percentage in EAA increases.

EAA copolymers exhibit excellent adhesion to metals, glass, poly(ethylene) and chlorinated poly(ethylene). Other plastics adhere to EAA only when primers are applied or the surface is oxidized prior to bonding. The initial good adhesion of EAA to copper is almost completely destroyed on aging in humid climate, whereas composites of EAA with aluminum or tin coated steel are not affected under such conditions. The bonding strength of EAA/metal laminates increases sharply as the percentage of AA in EAA increases.

EAA copolymers can be processed like LDPE, however, the carboxylic groups cause some corrosion. Temper-hardened steel performs well at AA contents of up to 8% and melt temperatures of less than 177°C; at higher AA content, nickel or chrome plated surfaces are required.

Extrusion coated Al/EAA laminates are used in heat-sealable flexible packaging materials, e.g. for tooth paste tubes and for moist face refreshing tissues. Blow extruded EAA films are used in heat-activated adhesive tapes and in tough packaging films.

Based on its product XO-2375.33, Dow manufactures an aqueous ammonia-neutralized dispersion under the designation XD-60899 EAA Dispersion [6]. The dispersion has 25% solids, pH 9−10, viscosity 0,1−0,2 Pa × s (100−200 cp) at 25°C; the minimum film formation temperature is 30°C. On drying, the dispersion gives flexible, glossy and glass-clear films with a density of 0.96 g/cm^3, high tensile yield and tear strengths, elongation 400%, Shore hardness 50 D, melt temperature 90°C, brittle temperature −22°C, and water absorption ≤1%. The films have low gas permeabilities, they can be sealed at 83°C and crosslinked to enhance adhesive strength. Applications are coatings, laminates and bondings with paper, metal foils, nylon, poly(ethylene) and polyesters.

The A-C Copolymer 540 of Allied is recommended as a processing aid in sheet extrusion and injection molding of PA [5]. It acts as lubricant and improves mold release, surface gloss and, in some cases, also mechanical properties.

The new A-C 201 is a low molecular weight ionomer commercially available from Allied Fibers and Plastics Co. [7]. The product is compatible with ABS, PS, nylon, PC and thermoplastic polyesters. When used at 15−20% of the pigment weight, excellent dispersion quality in color concentrates is reported. In addition, carbon black at 60−65%, flame retardants at 80%, and TiO$_2$ at 80−85% concentrations have been successfully dispersed in A-C 201 as the sole carrier. As a processing aid, it helps to lower processing temperatures, increase output, and reduce scrap. It is also recommended as a flush vehicle.

In 1978, A-C 504 was reported to cost 0.58−0.62 US-$/lb.

References

[1] M. C. Young, Plast. Eng. **31**/8 (August 1975) 52.
[2] H.-G. Elias, in Ullmanns Encykl. Techn. Chem., Verlag Chemie, 4th edition (1978), volume **15**, 421.
[3] Dow EAA Copolymers, Dow Chemical Co., Technical Information 1974 and 1976.

[4] XO-2375.33, Dow Chemical Co., Technical Information 1979.
[5] Anon., Modern Plast. Intern. **8**/12 (December 1978) 54.
[6] XD-60899, Dow Chemical Co., Technical Information 1979.
[7] Anon., Modern Plast. Intern. (August 1982) 52.

4.6.2 Poly(ethylene-co-propylene-co-ethylidene norbornyl sulfonate) ("Sulfonated EPDM")

4.6.2.1 Structure and synthesis

Exxon developed an ionomer based on sulfonated EPDM named "Thionic". It was licensed to Uniroyal which in 1981 introduced the two commercial grades IE-1025 and IE-2590.

The EPDM used as starting material contains 55% ethylene and 5% ethylidene norbornyl groups; it is sulfonated with acetyl sulfate and then reacted with zinc salts [1−5]:

(4-5)

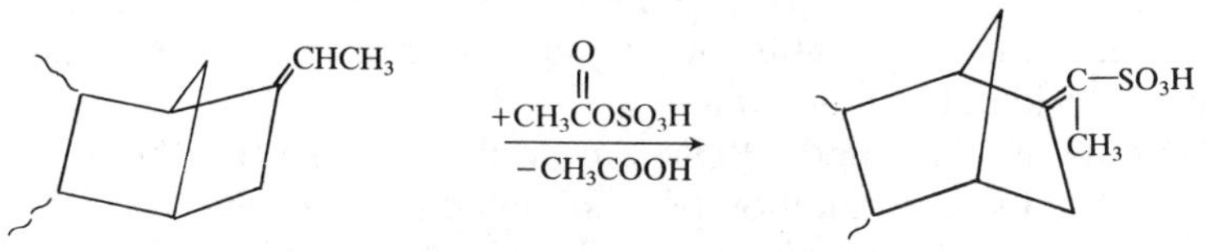

The reaction occurs at the exocyclic double bond of the ethylidene-norbornyl group and presumably yields an intermediate vinylsulfone derivative and finally its zinc salt.

For the synthesis of Thionic, EPDM is dissolved in an aliphatic hydrocarbon, preferably hexane, and is then sulfonated with acetyl sulfate at room temperature. The reaction is terminated after about 30 min by addition of alcohol or an aqueous solution of zinc hydroxide. By another procedure, the solution of the sulfonated EPDM is fed into hot water, the solvent is stripped off by flash distillation, and the product is then reacted with zinc compounds [5]. By employing a much simpler process technology, molten EPDM is continuously sulfonated in a corrosion-resistant extruder with a mixture of sulfuric acid and acetic anhydride at 85°C [6]. After adding molten zinc stearate to the sulfonated polymer, the components are homogeneously blended in a static Kenics mixer at 93°C.

4.6.2.2 Properties and applications

The free sulfonic acid groups of sulfonated EPDM show little association; the material is soluble in aliphatic and aromatic solvents, has poor strength and exhibits a tendency to degradation and crosslinking. The incorporation of metal compounds results in thermally stable polymers. The properties depend on the crosslinking density which is a function of the degree of sulfonation and the nature of the counter ion. Salts have been prepared from sulfonated EPDM with NH_4, Li, Na, Cs, Mg, Ca, Ba, Co, Hg, Pb and Zn. The zinc salts have by far the lowest melt viscosities of the salts investigated [5, 8].

The melt viscosity of zinc salts increases as the degree of sulfonation increases, however, the EPDM type, its termonomer content and the termonomer distribution exert additional effects. Tensile strength also increases as the degree of sulfonation increases (Table 4-14), but again, is dependent on EPDM type; maxima are observed at sulfonate levels ranging from 30 to nearly 50 mmol/100 g polymer. Likewise, elongation and the 300% modulus increase as the sulfonation degree increases. The water absorption of zinc and lead salts is less than 10% after 300 h immersion at 50°C; other salts absorb considerably more water.

The salts of sulfonated EPDM are insoluble in hydrocarbons but dissolve readily after addition of small amounts of alcohols or other polar solvents. The solutions exhibit viscosity anomalies. The viscosity passes through a minimum at about 40°C and then increases markedly up to 100°C [7]. This isoviscosity effect occurs over broad

TABLE 4-14
Property values of sulfonated EPDM ("Thionic") ([5]).

Physical Property	Physical Unit	Property data for a sulfonate content (mmol/(100g polymer)) of			
		20	20	40	40
Mooney viscosity (ML 1 + 8, at 100°C)	—	22.3	46.7	22.9	48.4
Melt index (100°C; 1.72 MPa)	g/10 min	0.61	0.04	0.18	0.005
Tensile strength at 25°C	MPa	6.7	20.4	2.1	15.6
at 70°C	MPa	2.8	2.9	0.5	3.4
Elongation at 25°C	%	630	470	740	390
at 70°C	%	410	420	275	300

temperature ranges and is mechanistically different from the behavior of conventional polymer solutions. This behavior is explained by a simple equilibrium involving solvated ion pairs, and is an example of a general phenomenon [7].

The melt viscosity of the neat zinc salts is too high for injection molding or extrusion processing. The melt viscosity is drastically reduced by addition of about 20 phr zinc stearate. Additional advantages hereby achieved are improved tensile strengths at 25°C and 70°C, and a reduced water absorption [9, 10].

Such modified zinc salts yield compounds which can be processed like thermoplastic elastomers. Uniroyal's grades IE-1025 and IE-2590 contain 1.1 and 2.7 wt-%, respectively, of ionic groups. The products are supplied as powders which readily mix with extender oils, fillers and other polymers. Filler loadings of up to 70−80% are attainable. The addition of 100 phr extender oils still yields free flowing powders [11]. Compound densities range from 0.95 to 1.25 g/cm^3, tensile strengths from 3.5 to 17.2 MPa, and elongations from 300 to 900%. The compounds demonstrate high toughness at low temperatures, good resistance to weathering and ozone, high abrasion resistance, and good heat-aging characteristics.

The melts show superior mechanical stability over those of other thermoplastic elastomers and hence favors processing by extrusion, calendering, and blow molding. The thermal stability of the melt is equally good and allows short-term temperatures of up to 270°C. In general, processing temperatures range from 100 to 230°C.

First applications include calendered sheets and films for inner linings of swimming pools and waste dumps, flat roof covers, solings, hoses, fabric coatings, and solvent cements [11−13].

Furthermore, foams with open and closed cell structures can be prepared [14]. Their properties are: densities 100−800 g/l, brittle temperature −70°C, tensile strength 0.41−2.8 MPa, elongation 240−320%, 100% modulus 0.24−1.8 MPa, and compression sets 20−60% at 23°C.

Uniroyal sells its products at 1.35 US-$/lb. Compound material costs can be lowered by high filler loadings down to 0.50−0.85 US-$/lb [11].

References

[1] N. H. Canter (Esso Res. & Eng.), USP 3 642 728 (1972).

[2] C. P. O'Farrell and G. E. Serniuk (Esso), USP 3 836 511 (1974).

[3] H. S. Makowski, R. D. Lundberg and G. Singhal (Esso), USP 3 870 841 (1975).

[4] R. D. Lundberg, H. S. Makowski and L. Westerman (Esso), USP 4 014 847 (1977).

[5] H. S. Makowski, R. D. Lundberg, L. Westerman and J. Bock, ACS Polym. Prepr. **19**/2 (1978) 292; ditto, in A. Eisenberg (Ed.), Ions in Polymers (Adv. Chem. Ser. **187**), Amer. Chem. Soc., Washington, D.C., 1980.

[6] B. Siadat, R. D. Lundberg and R. W. Lenz, Polym. Eng. Sci. **20** (1980) 530.

[7] R. D. Lundberg and H. S. Makowski, J. Polym. Sci.-Polym. Phys. Ed. **18** (1980) 1821; R. D. Lundberg, J. Appl. Polym. Sci. **27** (1982) 4623.

[8] H. S. Makowski and R. D. Lundberg, ACS Polym. Prepr. **21**/1 (1980) 304.

[9] K. B. Wagener and I. Duydeyani, ACS Polym. Prepr. **21**/1 (1980) 2.

[10] P. K. Agarwal, H. S. Makoswki and R. D. Lundberg, Macromolecules **13** (1980) 1679.

[11] Anon., Plastics Technol. (March 27, 1981) 17.

[12] Anon., The Journal of Commerce (March 11, 1981) 5.

[13] Anon., Modern Plast. (May 1981) 80.

[14] D. Brenner and R. D. Lundberg, Rubber Chem. Technol. **50** (1977) 437.

4.7   1,2-SYNDIOTACTIC POLY(BUTADIENE)

4.7.1   Structure and synthesis

The solution polymerization of butadiene-1,3 with the catalyst system $CoHal_2$/ligand/AlR_3/H_2O yields poly(butadiene) with ca. 90% 1,2-structure; of the 1,2 structure 51−66% form syndiotactic and 49−34% form heterotactic triads. The molecular weights are greater than 100 000 g/mol. Such polymers have been produced by the Japan Synthetic Rubber Co. In the USA and Canada, Uniroyal held production and sales licenses until the end of 1981; these were transferred to JSR America in 1982.

4.7.2   Properties and applications

The properties of the 3 products which are now commercially available differ from those of the older developmental products [1, 2]. The mechanical properties of syndiotactic poly(butadienes), abbreviated SBD, are in between those of typical elastomers and typical thermoplastic elastomers (Table 4-15). Nowadays, SBD is generally classified as a thermoplastic elastomer.

TABLE 4-15
Property values of 1,2-syndiotactic poly(butadiene) [3].

Physical Property	Physical Unit	Property values of		
		RB-810 (XP2625)	RB-820 (XP2626)	RB-830 (XP2627)
Density	g/cm^3	0.901	0.906	0.909
Crystallinity	%	18	25	29
Amount of 1,2 units	%	90	92	93
Melt index (150°C/2.16)	g/10 min	3	3	2
Vicat temperature	°C	39	52	66
Melt temperature (DSC)	°C	75	80	90
Glass transition temperature (DSC)	°C	−30	−25	−17
Brittleness temperature	°C	−40	−37	−35
Tensile strength	MPa	6.4	10.3	13.2
300% Modul	MPa	3.9	5.9	7.8
Elongation at break	%	750	700	670
Hardness (Shore D)		25	34	41
Transparency	%	91	91	91
Gas permeability				
Carbon dioxide	0.1 mm/		36	
Oxygen	(m^2 × 24 h × atm)		6.9	
Ethylene oxide			320	
Water vapor	(g × 0.1 mm)/ (m^2 × 24 h × atm)		98	

SBD is processable like LDPE; furthermore, crosslinking with peroxides or sulfur (via the pendant vinyl groups) is also possible and leads to elastomers or thermosets.

Primary applications are transparent packaging films with outstanding tear resistance and high permeabilities for gases and water vapor. Owing to the high ultimate elongation, SBD films are not pierced by sharp objects. The high permeability for ethylene oxide allows the rapid sterilization of packed food and medical parts. The crosslinking of films, e.g. in film wrapping of poultry, by electron beam technique occurs much faster than with EVA films. SBD is also used for cable insulation and in injection molding, e.g. for solings.

SBD is compatible with poly(olefins) and with elastomers based on butadiene or olefins. The addition of 10−15% SBD to non-oriented PP improves the film tensile strength without sacrifice of transparency. Films fabricated from such blends cost less than oriented PP films. In some plastics, impact strength is improved by the addition of SBD. SBD is a likely candidate for use as elasticizing agent in thermosets; it takes part in the crosslinking reaction of some matrix resins.

SBD costs 0.85−0.95 US-$/lb; as with EPDM, compound costs can be reduced by extenders and fillers [4].

References

[1] S. I. Kimura, N. Shiraishi, S. Yanagisawa and M. Abe, Polymer-Plast. Technol. Engngn. **5** (1975) 83.
[2] H.-G. Elias, Neue polymere Werkstoffe 1969−1974, Carl Hanser Verlag, München 1975; New Commercial Polymers 1969−1975, Gordon and Breach, New York 1977.
[3] 1,2-Syndiotactic Polybutadiene, Uniroyal Chemical Co., Technical Information 1981.
[4] Anon., Plastics Technol. **27** (March 1981) 17.

5 Fluoropolymers

5.1 INTRODUCTION

Since publication of our previous book [1], the still fast developing field of fluoropolymers has been reviewed with regard to thermoplastic materials [2−4], fluorine containing elastomers [4−9], and fluoropolymer films [10]. Since 1974 also a number of new fluoropolymers have been marketed, some of which are still developmental products.

Recently introduced thermoplastic fluoropolymers are ®CM-1 of Allied Chemical Corp. (now Allied Corp.) and a membrane filter material based on a copolymer manufactured by Asahi Glass Co., Ltd. ®Teflon EPE of DuPont has been tested since 1980, but it is as yet not a semi-commercial product [11]. Also not yet commercial is poly(trifluoromethylstyrene) which has excellent optical properties [12]. Hoechst introduced Hostaflon TFB, a terpolymer from tetrafluoroethylene, hexafluoropropylene and vinylidene fluoride and also a copolymer on basis tetrafluoroethylene and perfluoroalkoxy compounds [13].

New fluoroelastomers are ®Aflas of Asahi Glass Co., the ®Viton types GH, GLT, and GF of DuPont, ®Tecnoflon XHF of Montedison, and Fasil developed by the U.S. Air Force. Xenox® is a tetrafluoroethylene/propylene copolymer [14]. For a review see [15].

References

[1] H. G. Elias, Neue polymere Werkstoffe 1969−1974, Carl Hanser Verlag, München 1975; New Commercial Polymers 1969−1975, Gordon and Breach, New York 1977.
[2] A. B. Robertson and E. C. Lupton, Jr., Fluorinated Plastics, in Encycl. Polymer Sci. Technol., suppl. vol. 1 (1976) 260.
[3] Several authors and articles in Encycl. Chem. Technol. 11 (1980) 1 and following pages.
[4] H. Fitz, Kunststoffe 70 (1980) 659.
[5] D. A. Stivers, Fluoroelastomers, The Vanderbilt Rubber Handbook, R. T. Vanderbilt Co., New York 1978, 244.
[6] J. D. MacLachlan, Polym. Plast. Technol. Engng. 11 (1978) 41.

[7] A. C. West and A. G. Holcomb, Fluorinated Elastomers, Encycl. Chem. Technol. **8** (1979) 500.

[8] G. C. Sweet, in A. Whelan and K. S. Lee (ed.) Developments in Rubber Technology, Applied Science Publishers Ltd., Barking (UK) 1979, 45.

[9] S. Geri and C. Lagana, Kautschuk, Gummi, Kunststoffe **33/1** (1980) 9.

[10] H. Fitz, Kunststoffe **70** (1980) 27.

[11] N. B. Lamb, Fluoropolymers Div., DuPont, pers. communication, June 18, 1982.

[12] B. Bömer and H. Hagemann, Angew. Makromol. Chem. **109–110**, 285 (1982).

[13] H. Fitz, Kunststoffe **74**, 586 (1984).

[14] D. Hall, Rubber Chem. Technol. **56**, 1148 (1983).

[15] L. D. Albin, Rubber Chem. Technol. **55**, 902 (1982).

5.2 POLY(HEXAFLUOROISOBUTYLENE-CO-VINYLIDENEFLUORIDE)

5.2.1 Structure and synthesis

In 1975, Allied Chemical Corp. introduced a thermoplastic poly-(hexafluoroisobutylene-co-vinylidenefluoride) under the designation ®CM-1 [1]. The polymer is available in limited quantities for testing purposes by customers [2–4], however, it was still not fully commercialized by 1981 [5].

The copolymer is prepared by free radical suspension or emulsion polymerization in aqueous phase from hexafluoroisobutylene I and vinylidenefluoride II. Starting with liquid I cooled to 5°C, II is continuously added to keep pressures between 10 and 20 bar at 20°C.

CM-1 is an alternating copolymer with head to tail arrangement of units III. There is no evidence of CH_2CF_2 blocks being present. As structure III illustrates, the two trifluoromethyl groups shield the protons effectively from thermal and oxidative attacks.

$CH_2{=}C(CF_3)_2$ I

$CH_2{=}CF_2$ II

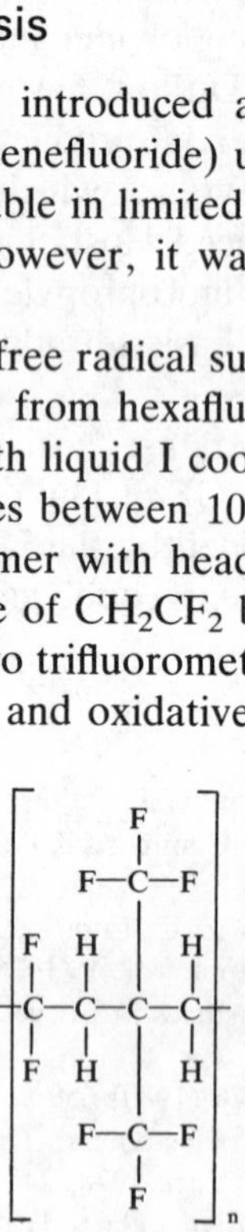

III

X-ray diffraction studies show the conformation of a 2_1-helix with a 118° angle between the C-C chain atoms [1]. Optical and electron microscopy show formation of spherulites in III.

5.2.2 Properties and applications

CM-1 can be melt processed by transfer molding at 350°C or by injection molding at 370−380°C. Powder coatings are obtained by fluidized bed, flame spraying or electrostatic spraying techniques.

CM-1 competes with poly(tetrafluoroethylene) (Table 5-1). It possesses a similar melt temperature as PTFE but a lower density. CM-1 has superior tensile modulus and tensile strength but lower notched impact resistance than PTFE. The upper service temperature limit is 20°C above PTFE, moreover, the coefficient of linear thermal expansion is smaller and less dependent on temperature.

CM-1 is superior to PTFE and to other fluoropolymers with regard to surface hardness, wear and scratch resistance. CM-1 is not attacked by acids, alkalis, halogens, metal salt solutions, hydrogen peroxide, aliphatic and aromatic hydrocarbons, alcohols and halogenated hydrocarbons. Acetic acid, ketones and esters cause swelling of CM-1, the degree of swelling is higher when compared to PTFE but less compared to poly(vinylidenechloride). CM-1 shows good adhesion to metals which makes it a suitable lining material for reaction vessels.

CM-1 is an engineering plastic. Its creep resistance is by far superior to that of PTFE or FEP; the deformations observed after 200 h at 100°C were [1]:

> FEP (under 5.2 MPa)   30%
> PTFE (under 6.9 MPa)   12%
> CM-1 (under 13.8 MPa) 1.5%.

The torsion modulus at elevated temperatures significantly surpasses that of FEP, PTFE and poly(ethylene-co-chlorotrifluoroethylene) (Figure 5-1). Electrical properties of CM-1 are similar to those of PTFE. At 415°C, significant weight loss occurs on heating in air or nitrogen.

The critical surface tension of CM-1 is $19.3 \cdot 10^{-3}$ N/m and is close to that of PTFE with $18.8 \cdot 10^{-3}$ N/m but much lower compared to poly(ethylene-co-chlorotrifluoroethylene) with $31.2 \cdot 10^{-3}$ N/m. This allows CM-1 to be used as an antiadhesive coating, e.g. for frying

TABLE 5-1
Property values of fluoroplastics ®CM-1 and PTFE.

Physical Property	Physical Unit	Property values of ®CM-1	PTFE
Density	g/cm³	1.88	2.17
Melt temperature	°C	327	327
Melt index	g/10 min	1–3	—
Mold shrinkage (linear)	—	0.020	—
Tensile strength at break			
at 23°C	MPa	37.9	13.8–34.5
at 200°C	MPa	20.7	3.5
at 300°C	MPa	15.9	—
Modulus of elasticity			
at 23°C	GPa	3.79	0.38
at 200°C	GPa	0.76	0.10
Elongation			
at 23°C	%	2	200
at 200°C	%	200	300
at 300°C	%	220	>300
Flexural strength (23°C)	MPa	36.6	
Flexural modulus (23°C)	GPa	4.48	0.62
Impact strength with notch	kJ/m	0.02	0.16
Hardness (Rockwell R)	—	115	16
Heat distortion temperature (1.81 MPa)	°C	220	56
Continuous service temperature	°C	280	260
Flammability	UL 94	V–0	V–0
Limiting oxygen index	%	60	>95
Expansion coefficient (×10⁵)			
between −45 and 24°C	K⁻¹	3.7	8.5
between 24 and 60°C	K⁻¹	4.1	12.0
between 60 and 149°C	K⁻¹	4.4	14.0
between 149 and 204°C	K⁻¹	4.7	17.0
Water absorption	%	0.01	—

pans, injection molds or in the treatment of roller surfaces. Other recommended applications include parts under mechanical strain, e.g. bearings and rollers, a matrix for fiber reinforced composites to be used at high temperatures in aggressive environments, or linings and coatings for temperatures up to 280°C.

The most obvious disadvantages of CM-1 are the relatively arduous processing, the poor notched impact resistance and the sensitivity to the attack of some organic solvents.

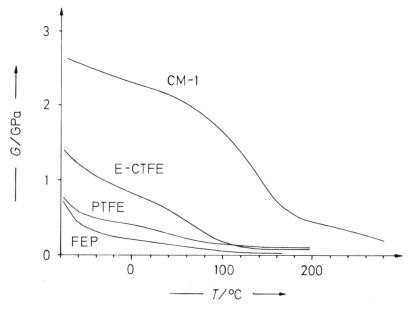

Figure 5-1 Temperature dependence of the shear modulus G of various fluoropolymers. CM-1 = copolymer from hexafluoroisobutene and vinylidene fluoride, E-CTFE = copolymer from ethylene and chlorotrifluoroethylene, PTFE = poly(tetrafluoroethylene), FEP = poly(tetrafluoroethylene-co-hexafluoropropylene).

References

[1] P. S. Minhas and F. Petrucelli, Plastics Engn. **33**/3 (1977) 60.
[2] J. B. Titus, New Plastics, Plastec Report R48, Plastics Technical Evaluation Center, Picatinny Arsenal, Dover, New Jersey 1977.
[3] H.-G. Elias, New Commercial Polymers 1969–1975, Gordon and Breach, New York 1977.
[4] H.-G. Elias, Polymer News **4** (1977) 78.
[5] K. S. Culley, Allied Chemical, personal communication, April 21, 1981.

5.3 POLY(TETRAFLUOROETHYLENE-CO- -CARBALKOXY-PERFLUOROALKOXY-VINYL ETHER)

5.3.1 Structure and synthesis

The Asahi Glass Co. Ltd. in Yokohama produces copolymers from tetrafluoroethylene (I) and ω-carbalkoxy-perfluoroalkoxy-vinylether (II) [1, 2]. Examples of vinylethers represented by the general formula (II) are the compounds (III) and (IV).

I $CF_2{=}CF_2$

II $CF_2{=}CF$
        |
        $(O{-}CF_2{-}CF)_m{-}O{-}(CF_2)_n{-}C{-}OR$
                    | ‖
                    CF_3 O

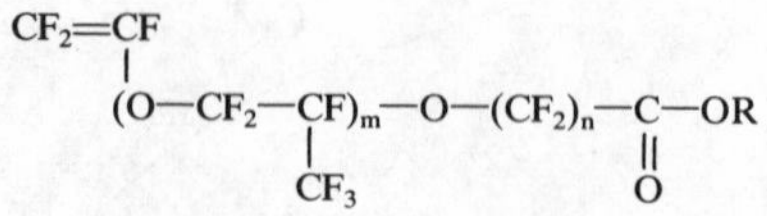

m = 0 oder 1; n = 1 – 5; R = Alkyl

III $CF_2{=}CF$
         |
         $O{-}(CF_2)_3{-}C{-}O{-}CH_3$
                       ‖
                       O

IV $CF_2{=}CF$
        |
        $O{-}CF_2{-}CF{-}O{-}(CF_2)_3{-}C{-}O{-}CH_3$
                    | ‖
                    CF_3 O

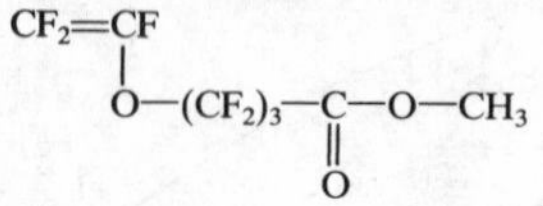

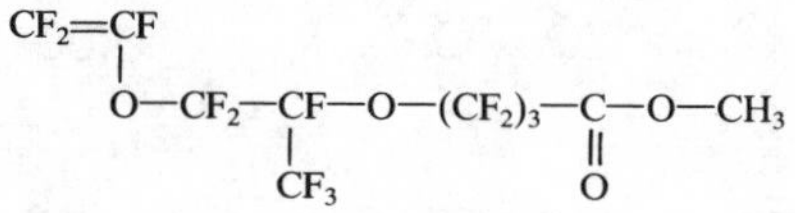

The monomer (III) is prepared from I by a multi-step reaction. Monomer (IV) is obtained by a similar reaction whereby presumably hexafluoropropene is used instead of tetrafluoroethylene in the fourth step of the reaction sequence leading to (III):

(5-1)

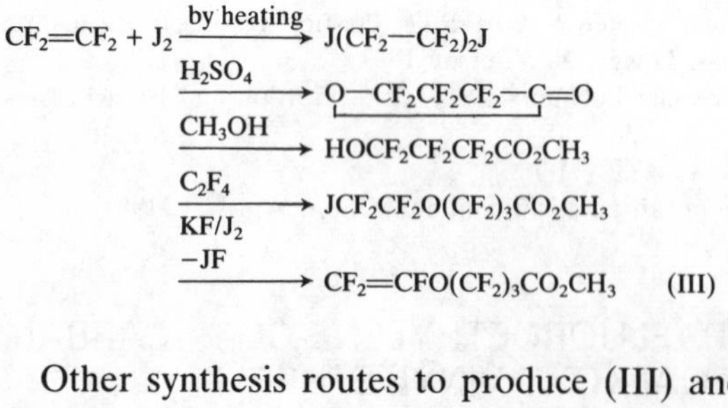

$CF_2{=}CF_2 + J_2$ $\xrightarrow{\text{by heating}}$ $J(CF_2{-}CF_2)_2J$

$\xrightarrow{\text{H}_2\text{SO}_4}$ $O{-}CF_2CF_2CF_2{-}C{=}O$

$\xrightarrow{\text{CH}_3\text{OH}}$ $HOCF_2CF_2CF_2CO_2CH_3$

$\xrightarrow[\text{KF/J}_2]{\text{C}_2\text{F}_4}$ $JCF_2CF_2O(CF_2)_3CO_2CH_3$

$\xrightarrow{-\text{JF}}$ $CF_2{=}CFO(CF_2)_3CO_2CH_3$ (III)

Other synthesis routes to produce (III) and (IV) start from perfluorosuccinic acid fluoride and perfluoro-γ-butyrolactone, respectively [1]. These syntheses presumably require the use of perfluoroepoxides which is state of art in the production of perfluorovinylether monomers needed for ®Teflon PFA, ®Kalrez and ®Nafion.

5.3.2 Properties and applications

Copolymers of I with III and I with IV are random copolymers with molecular weights of above 10 000 g/mol. Their crystallinity decreases as vinyl ester content increases. The glass transition temperatures are 10–15°C, the decomposition starts around 320°C.

The copolymers can be processed to films by extrusion or compression molding. The melt viscosity between 230 and 250°C reaches about 10^3 Pa × s. The copolymers in the shape of films are treated with caustic soda resulting in a complete saponification. The structure of the copolymer sodium salts is similar to that of ®Nafion:

Asahi Glass
Copolymers

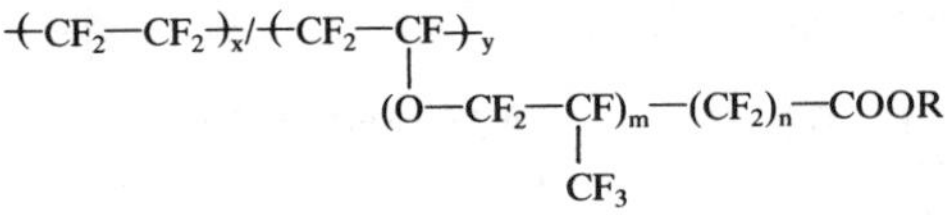

$$+CF_2-CF_2+_x/+CF_2-CF+_y$$
$$(O-CF_2-CF)_m-(CF_2)_n-COOR$$
$$CF_3$$

®Nafion

$$+CF_2-CF_2+_x/+CF_2-CF+_y$$
$$O-(CF-CF-O)_n-(CF_2)_2-SO_3H)$$
$$CF_3$$

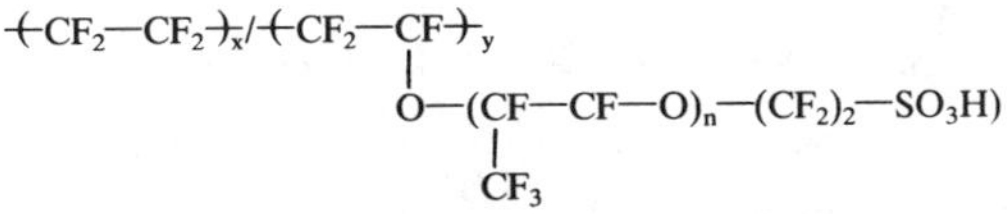

The Asahi copolymers have superior strength to Nafion. The strength can be further enhanced by reinforcement with woven PTFE fabrics [3, 4].

TABLE 5-2
Comparison of properties of Asahi copolymers with ®Nafion of DuPont.

Polymer	Tensile strength at break in MPa	Elongation in %	Ion exchange capacity in mmol/(g dry resin)
Copolymer I with 15–20 mol-% III			
as ester	167	300	—
as sodium salt	265	50	1.1–1.3
Copolymer I with IV (as sodium salt)	—	—	1.1–1.5
®Nafion (with 25% water)	21	150	0.7–0.9

Membranes manufactured from the Asahi copolymers are perm-selective; they are permeable only to cations, not to anions. The membranes find increasing use in electrochemical processes, e.g. in the electrolysis of sodium chloride solution. A chlor-alkali electrolysis reportedly worked more than 700 h with Asahi membranes, cell voltage was 3.5 V, current efficiency up to 95%, and caustic soda concentration reached more than 35%. The transport number of sodium ions in caustic soda solutions with a concentration range of 20−40% was 0.9. The electrolysis plant is reported to consume less energy than the conventional process operating with mercury cells.

References

[1] H. Ukihashi, ACS Polymer Preprints **20**/1 (1979) 195.
[2] H. Ukihashi, Chemtech **10** (February 1980) 118.
[3] H.-G. Elias, Neue polymere Werkstoffe 1969−1974, Carl Hanser Verlag, München 1975, 60; New Commercial Polymers 1969−1975, Gordon and Breach, New York 1977.
[4] W. Grot, Chem.-Ing. Technik **44** (1972) 167.

Supplement

[5] S. C. Stinson, Chem. Engng. News (March 15, 1982) 22.
[6] A. D. Eisenberg and H. Yeager (ed.), Perfluorinated Ionomer Membranes, ACS Symp. Ser. **180**, Amer. Chem. Soc., Washington, D.C. 1982.

5.4 PEROXIDE-CURABLE FLUOROELASTOMERS

5.4.1 Structure and synthesis

Between 1976 and 1980, DuPont expanded its ®Viton-A series (copolymers of vinylidenefluoride/hexafluoropropene) and ®Viton-B series (terpolymers of vinylidenefluoride/hexafluoropropene/tetrafluoroethylene) by three new polymers belonging to the ®Viton-G series [1−5]. The three new peroxide-curable fluoroelastomers are designated ®Viton GH, GLT and GF [6]. They are quaterpolymers [7]. GH and GF are poly(vinylidenefluoride-co-tetrafluoroethylene-co-hexafluoropropylenes), GLT is a poly(vinylidenefluoride-co-tetrafluoroethylene-co-perfluoromethyl-vinylether). The composition and the nature of the fourth monomer that enables the peroxide cure has not been disclosed.

The 3M Co. produces a developmental fluoroelastomer named ®Fluorel FKM 4826 from vinylidenefluoride, hexafluoropropylene and a cure site monomer for enhanced peroxide curability; it contains 69% fluorine by weight and is crosslinked using an organic peroxide/triallyl isocyanurate cure system [8].

5.4.2 Properties and applications

®Viton GH was developed for processing by extrusion and for continuous vulcanization, the GLT type is intended for applications at low temperatures, and the GF type shows improved resistance to inorganic and organic media.

The recommended peroxides are 2,5-dimethyl-2,5-bis(tert-butylperoxy)hexane (®Luperco 101-XL) for the curing of Viton GLT and GF and 2,5-dimethyl-2,5-bis(tert-butylperoxyhexyne) (®Luperco 130 XL) for Viton GH. The peroxide curability is enhanced by activators such as triallyl isocyanurate. Inorganic acid acceptors such as $Mg(OH)_2$ or $Ca(OH)_2$ are useful in absorbing any traces of hydrogen fluoride generated during the curing process. Lead oxide is used when the vulcanizate is required to demonstrate low swelling in water and good aging characteristics.

Viton GF can also be cured with aromatic dihydroxy compounds or with diamines. Peroxide cured samples which are subsequently over cured (postcured) in art at 232°C for 12 to 24 h exhibit improved tensile strength and compression sets. The vulcanizates show good ageing characteristics at elevated temperatures (see Table 5-3).

Peroxide-cured fluoroelastomers offer higher resistance to hot water and steam than their conventional counterparts which were cured by diamines or bisphenol compounds. The latter contain bonds which are susceptible to hydrolytic attack. Another advantage offered by peroxide curing is that it can be carried out under mild pressures or even without pressure, e.g. by extrusion of profiles into vulcanization baths, or in recirculating hot air ovens. No water is generated during peroxide curing. Thus, unpressurized processing results in vulcanizates without spongy appearance.

Applications of Viton GH include extruded profiles, hose and sealings where resistance to hot water, steam and aqueous acids is

TABLE 5-3
Property values of peroxide vulcanizable fluoroelastomers.

Physical property	Physical Unit	Property values of Viton GH	Viton GLT	Viton GF
Density of raw elastomer	g/cm³	1.86	1.78	1.91
Mooney viscosity (121°C; ML-10)	—	90	90	60
Formulation				
Elastomer	—	100	100	100
MT Carbon black	phr	30	30	30
Lead oxide	phr	3	—	3
Calcium hydroxide	phr	—	4	—
Diak No. 7	phr	3	4	3
Luperco	phr	3	4	3
Mooney Scorch of mixture (121°C, M5)	—[1]	60	47	40 [3]
Properties of vulcanized elastomer				
Tensile strength at break	MPa	14.5 10.0 [2]	18.8 10.4 [2]	17.6 12.8
Elongation	%	180 200	185 170	210 240
Modulus (100%)	MPa	5.3 4.0	7.2 4.8	5.4 3.4
Hardness (durometer A)	—	77 75	67 63	76 74
Compression set [4]				
70 h, 200°C	%	41	21–30 [5]	38
70 h, 232°C	%	54	26–53	63
Brittleness temperature	°C	-42	-51	-49

1) Vulcanization in the press (10 min at 177°C), followed by 24 h post-curing in the oven at 232°C.
2) 70 h aging at 275°C.
3) 3 days aging at 275°C.
4) ASTM D395, method B (O-rings 25.4 mm × 3.5 mm).
5) The lower values for the compression set apply to a formulation with 10 phr MT carbon black and 20 phr Austin carbon black instead of 30 phr MT carbon black.

required. Viton GLT imparts excellent mechanical properties at low temperatures. Viton GF vulcanizates show outstanding resistance to hot organic and aqueous media (Table 5-4).

TABLE 5-4
Swelling and compression of ®Vitron GF vulcanizates in different solvents.

Solvent	Time in weeks	Temperature in °C	Swelling in %	Compression set in %
Water	3	162	5	82
Water vapor	3	162	2	65
Hydrochloric acid (37%)	1	70	6	43
Nitric acid (70%)	1	70	12	—
Ethylene glycol/water (50/50)	3	162	2	79
Mobil Jet Oil II	-	200	—	29
Mobil Jet Oil II	12	200	—	76
Reference Fuel C	1	70	—	18

References

[1] J. B. Finlay, A. Hallenbeck and J. D. MacLachlan, J. Elastomers Plast. **10** (January 1978) 3.
[2] J. E. Alexander and H. Omura, ACS 110th Meeting Rubber Division, San Francisco 1976, Abstract in Rubber Chem. Technol. **50** (1977) 417.
[3] L. F. Pelosi and E. T. Hackett, ibid. 418.
[4] J. D. MacLachlan and A. Hallenbeck, ibid. 419.
[5] J. G. Bauerle and J. B. Finlay, ACS 117th Meeting Rubber Division, Las Vegas 1980, Abstract in Rubber Chem. Technol. **53** (1980) 1262.
[6] ®Viton GH, GLT and GF, Technical Information, DuPont.
[7] A. C. West and A. G. Holcomb, Fluorinated Elastomers, Encycl. Chem. Technol., vol. **8** (1979) 500.
[8] R. R. Campbell, D. A. Stivers and R. E. Kolb, Rubber Chem. Technol. **55** (1982) 1137.

Supplement

[9] D. Apotheker, J. B. Finlay, P. J. Krusic and A. L. Logothetis, "Curing of Fluoroelastomers by Peroxide", Rubber Chem. Technol. **55** (1982) 1004.
[10] J. D. Eddy and R. P. Kane, "New Fluoroelastomers with Outstanding Resistance to Harsh Fluids", ACS 120th Meeting Rubber Division, Cleveland 1981.
[11] D. L. Tabb and J. B. Finlay, "A New Gelled Fluoroelastomer", Rubber chem. Technol. **55** (1982) 1152.

5.5 POLY(TETRAFLUOROETHYLENE-CO-PROPYLENE)

5.5.1 Structure and synthesis

The Asahi Glass Co. Ltd developed an elastomer based on poly-(tetrafluoroethylene-co-propylene) which is marketed in Japan under the trade name ®Aflas by Asahi and in the USA by Xenox Inc., Houston.

Aflas is produced by copolymerization of tetrafluoroethylene with propylene at 25°C using a modified persulfate redox system as initiator [1] and ammoniumperfluorooctanate as emulsifier. The reactivity ratios determined by the Fineman-Ross method were $r_{C_2F_4} = 0.05$ and $r_{C_3H_6} = 0.10$ [2].

Copolymerization is also achieved by γ-irradiation with ^{60}Co at −78°C [3]. Recently, this process was improved by the Japan Atomic Energy Research Institute (JAERI) [4]. The emulsion polymerization is carried out at 30−50°C with a ^{60}Co radiation dose of $3 \cdot 10^5$ rad/h. The copolymer obtained has molecular weights of up to 180 000 g/mol. The JAERI process is open to licensing.

5.5.2 Properties and applications

In the Aflas copolymers, the C_2F_4/C_3H_6 molar ratio is about 55/45 with ca. 70% of the C_2F_4 units alternating with the C_3H_6 units forming tetrads. The C_3H_6 units in the C_3H_6 sequences display atactic structures [2]. The nearly amorphous copolymer has a density of 1.55 g/cm^3 and a glass transition temperature of −2°C. The weight average molecular weight can be regulated by polymerization conditions within the range of 100 000−180 000 g/mol; the $\overline{M}_w/\overline{M}_n$-ratio was found to be between 3 and 5 [5].

The commercially available Aflas grades 100 and 150 have Mooney viscosities (ML 1 + 10 at 100°C) of 85 and 100, respectively [6]. Aflas can be cured only with peroxides and the curing is enhanced by activators. A combination of α,α′-bis(tert.-butylperoxy)-p-diisopropylbenzene with triallyl isocyanurate yields vulcanizates with good mechanical properties and high resistance to thermal and chemical attacks. Generally, curing is carried out in two steps, e.g. first under pressure at 160°C succeeded by postcuring in an air oven at 160°C for 1 h, then for another hour at 180°C, and finally for 2 h at 200°C.

Carbon black reinforced vulcanizates of Aflas 100 and 150 demonstrate good mechanical properties (Table 3-5). The upper service temperature limit is around 200°C. About 50% of the initial tensile strength is retained after 60 days at 230°C, or 14 days at 260°C or 9 days at 287°C. The lowest service temperature limit is −40°C, however, Aflas is not recommended for applications under dynamic stress at temperatures below 0°C. The compression set after 30 days at 200°C is 62% for Aflas 100 and 50% for Aflas 150. Ozone concentrations of up to 100 ppm have no effect on the properties of Aflas. The volume resistivity is higher than that of silicon, SBR, CPE and fluoroelastomers. The permeability to nitrogen is lower than that of silicon, NR and SBR elastomers but higher compared to fluoro- and epichlorohydrine-elastomers. Aflas vulcanizates exhibit good resistance to inorganic and to some organic media (Table 5-6), however, they are only moderately resistant to solvents such as acetone, benzene, chloroform or fuel B.

Aflas 100 is compression molded to O-rings, gaskets, seals, diaphragms and rolls. The higher Mooney grade 150 is processed by transfer molding and is extruded to profiles and tubes. Aflas has been used in hot, corrosive media [7], especially in oil fields [8], and for linings [9].

TABLE 5-5
Property values of ®Aflas vulcanizates.

Physical property	Physical unit	Property value of	
		Aflas 100	Aflas 150
Density	g/cm^3	1.60	1.60
Tensile strength at break	MPa	17.7−21.6	14.7−17.7
Elongation	%	200−350	200−350
Modulus (100%)	MPa	2.5−3.4	2.5−3.4
Hardness (Shore A)	—	70	70
Brittleness temperature	°C	−40	
Retraction temperature (TR-10)	°C	3	
Volume resistivity	ohm·cm	3×10^{16}	3×10^{16}
Relative permittivity (1 kHz)	—	6	
Permeability of gases			
Nitrogen	cm^3 s g^{-1}	5.3×10^{-13}	
Oxygen	cm^3 s g^{-1}	17.3×10^{-13}	
Carbon dioxide	cm^3 s g^{-1}	21.8×10^{-13}	

Formulation: Aflas 100; α,α'-bis(t-butylperoxy)diisopropylbenzene 1 phr; triallyl-isocyanurate 5 phr; MT carbon black 35 phr. Vulcanization: 30 min at 160°C in the press followed by oven curing 1 h at 160°C (Aflas 100) or 180°C (Aflas 150) and 2 h at 200°C.

TABLE 5-6
Resistance of ®Aflas vulcanizates against chemicals.

Solvent	Temperature in °C	Time in days	Volume increase in %	Remaining tensile strength at break in %	Elongation in %	Hardness (relative change)
Sulfuric acid	100	3	4.4	99	101	−3
Fuming nitric acid	25	7	10	42	126	−7
Nitric acid	25	7	1.3	94	95	−1
Hydrochloric acid	25	7	0	100	107	−1
Sodium hydroxide	100	3	1.1	101	116	−1
Ammonia	70	3	0.5	82	116	−1
Hydrogen peroxide (30%)	100	7	−1.1	105	99	0
Sodium hypochlorite (10%)	100	7	1.1	100	95	−1
Water vapor	160	7	0	112	84	−2
Methanol	25	7	0.2			
Acetic acid	25	7	71			
Acetone	25	7	50.3			
Chlorohydrine	25	7	0			
Toluene	25	7	41			
Nitrobenzene	25	7	5.6			
n-Hexane	25	7	24			
Chloroform	25	7	112			
Carbontetrachloride	25	7	86			
Trichloroethylene	25	7	95			
ASTM Oil No. 3	25	7	1			
ASTM Oil No. 3	175	3	15			
Fuel B	25	7	55			

References

[1] HH. Kojima, M. Hisasue and G. Kojima, 8th Intern. Symp. Fluorine Chem., Kyoto 1976.
[2] G. Kojima, H. Kojima and Y. Tabata, Rubber Chem. Technol. **50** (1977) 403.
[3] Y. Tabata, K. Ishigure and H. Sobue, J. Polym. Sci. **A2** (1964) 2235.
[4] Anon., Chem. Engineering **86**/3 (March 1979) 103.
[5] G. Kojima and H. Wachi, Rubber Chem. Technol. **51** (1978) 940.
[6] ®Aflas, Technical Information Asahi Glass Co., Ltd., Tokyo.
[7] D. Hall, Rubber Chem. Technol. **56**, 1148 (1983).
[8] D. Hall, Rubber Chem. Technol. **57**, 402 (1984).
[9] M. Morozumi, Rubber Chem. Technol. **57**, 416 (1984).

5.6 POLY(FLUOROALKYLARYLENESILOXAN-YLENE)

5.6.1 Structure and synthesis

The Wright Patterson Air Force Base laboratories of the U.S. Air Force developed a new thermally stable elastomer based on poly-(fluoroalkylarylenesiloxanylene) (I) from which the code designation FASIL is derived [1−5].

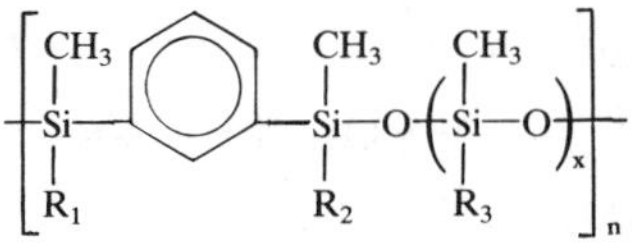

$$(I)$$

$R_1 = R_2 = R_3 = CH_3$ or $CF_3CH_2CH_2$ $x = 0, 1$ or 2

Of the many polymers represented by the general structure (I), the poly(m-phenylene-1,3,5,7-tetrakis(3,3,3-trifluoropropyl)tetrasiloxanylene) (II) demonstrated the best balance of properties.

The synthesis of II starts with commercially available raw materials. First, 1,3-bis[ethoxymethyl-3,3,3-trifluoropropyl)silyl]benzene (IV) is prepared by in situ Grignard reaction from m-dibromobenzene, magnesium and diethoxymethyl(3,3,3-trifluoropropyl)silane (III). Without isolating the product, it is reacted with ice water and potassium dihydrogenphosphate and then with aqueous NaOH to give 1,3-bis-hydroxymethyl(3,3,3-trifluoropropyl)silylbenzene (V) in a yield of 91.5%. It is a clear oil with a boiling point of 140−142°C

(5-2)

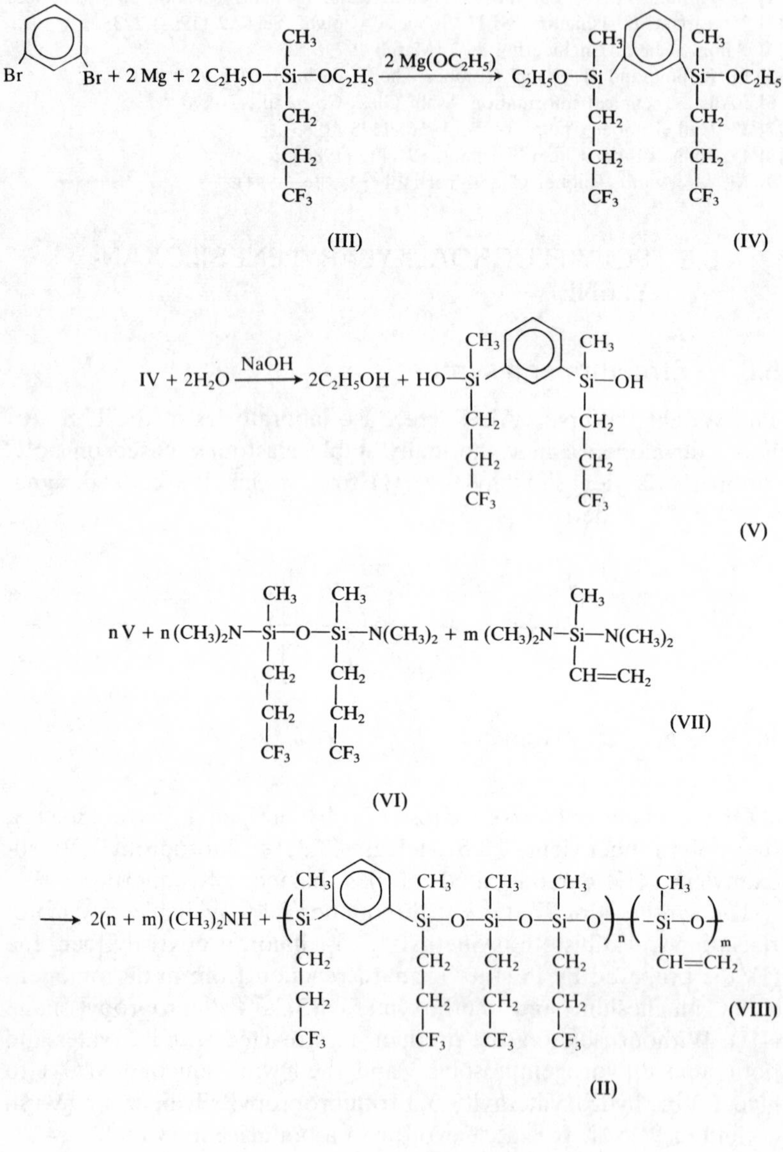

(III)   (IV)

$$IV + 2H_2O \xrightarrow{\text{NaOH}} 2C_2H_5OH + \text{(V)}$$

(V)

$$nV + n(CH_3)_2N-\overset{CH_3}{\underset{\underset{CF_3}{\overset{|}{\underset{|}{CH_2}}}}{\underset{|}{Si}}}-O-\overset{CH_3}{\underset{\underset{CF_3}{\overset{|}{CH_2}}}{\underset{|}{Si}}}-N(CH_3)_2 + m(CH_3)_2N-\overset{CH_3}{\underset{CH=CH_2}{\underset{|}{Si}}}-N(CH_3)_2$$

(VI)   (VII)

$$\longrightarrow 2(n+m)(CH_3)_2NH + \text{(VIII)}$$

(VIII)

(II)

at 0.1 Torr. The compound (V) is reacted with equimolar amounts of 1,3-dimethyl-1,3-bis(3,3,3-trifluoropropyl)-1,3-bis(dimethylamino)silane (VI) in toluene at 110°C. The copolymerization with small amounts of methyl-vinyl-bis(dimethylamino)silane (VII) leads to the incorporation of vinyl groups which are required for curing with peroxides. The reaction mixture is hydrolysed by addition of water after the evolution of dimethylamine has ceased. The copolymer obtained consists of (II) with 3 mol-% units of (VIII).

5.6.2 Properties and applications

The polymer is a very viscous oil with the following properties: molecular weight 3700 g/mol, solubility in toluene, limiting viscosity number 2 ml/g, glass transition temperature −49°C [5], weight loss of 25% at 441°C in vacuum of 10^{-5} Torr at a heating rate of 5°C/min [4].

The polymer is cured with di-tert.-butylcumylperoxide under pressure at 170°C. Vulcanizates have an upper continuous service temperature of 260°C and are flexibel down to −54°C. They demonstrate good adhesion to titanium and aluminum and cause no stress corrosion in titanium. The vulcanizates exhibit high resistance to JP-4 fuel and to hydrolysis; heating for a period of 7 days at 204°C causes no depolymerization.

This balance of properties has not been achieved by any of the other polymers used as sealing compounds in aircraft fuel tanks. For example, sealings manufactured from polysulfide vulcanizates have an upper continuous service temperature of only 120°C. Sealing compounds in supersonic aircraft operating at speeds of Mach 2−3 must survive temperatures in the range of 150−315°C in contact with fuel [6]. Copolymers based on II/VIII are used by the U.S. Air Force.

References

[1] D. C. Bonner, K. C. Chen and H. Rosenberg, ACS Polymer Preprints **17/2** (1976) 372.

[2] H. Rosenberg and E. W. Choe, ACS Coating and Plastics Preprints **37/1** (1977) 166.

[3] I. J. Goldfarb, E. W. Choe and H. Rosenberg, ACS Coatings and Plastics Preprints **37/1** (1977) 172.

[4] H. Rosenberg and B. D. Nahlovsky, ACS Polymer Preprints **19**/2 (1978) 625.

[5] H. Rosenberg and E. W. Choe, ACS Organic Coating and Plastics Preprints **40**/1 (1979) 792; H. Rosenberg and E. W. Choe, in A. May (ed.) Resins for Aerospace, ACS Symp. Ser. **132**, Amer. Chem. Soc., Washington D.C. 1980.

[6] W. E. Anspach, Chemtech **5** (December 1975) 752.

6 Polyethers and related polymers

6.1 PHENOL-ARALKYL-RESINS

6.1.1 Structure and synthesis

Since 1978 the English company, Advanced Resins, holding exclusive license rights, has produced a series of phenol-aralkyl-resins trade named ®Xylok [1] which formerly were supplied by Albright & Wilson. The Xylok grades 209, 214, 225 P, and 237 are commercially available. Some are further developed or supplemented products of the older grades 210, 211, 225, and EX 53 [2, 4].

The prepolymers are synthesized by condensation of aralkylethers with phenol at elevated temperatures and in the presence of Friedel-Crafts catalysts, e.g., from α,α'-dimethoxy-p-xylene and phenol [3, 4]:

(6-1)

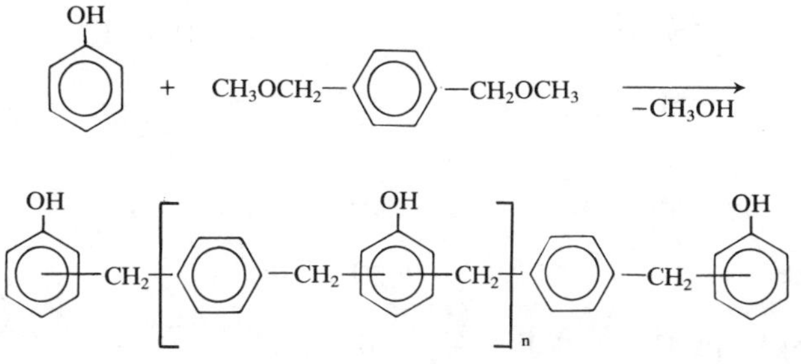

The average degree of polycondensation n is only 1.6 with a distribution between n = 0 and n = 6.

6.1.2 Properties and applications

Xylok 209 is supplied in 2-ethoxyethanol solution with 50% solids, viscosity 250−600 cSt at 25°C, and flash point 48°C. The solution also contains the hardener hexamethylenetetramine. Xylok 214 is supplied in alcoholic solution with 55% solids, kinematic viscosity 2−5 cm²/s, and flash point 13°C. This solution contains no hardener and hexamethylentetramine has to be added. Xylok 237 is a 2-component system comprising the resin solution (with 60% solids, kinematic viscosity 2.5−6 cm²/s, flash point 3°C) and the solution of an epoxy hardener (with 90% solids, kinematic viscosity 1−3 cm²/s, flash point 3°C) dissolved in methylethylketone. Xylok 225 P is a powdery mixture of 100 parts prepolymer and 12.5 parts hexamethylenetetramine. The mixture melts at 85−105°C.

Xylok 209 and 237 are used as matrices in composites manufactured from glass, graphite, asbestos, and aramid fibers. Prepregs show good storage stability and can be processed like phenol or epoxy resins. Table 6-1 illustrates properties of composites manufactured from Xylok 209 and 237.

TABLE 6-1

Property values of hardened laminates of phenol-aralkyl resins with fabrics of glass, asbestos or Nomex 411.

| Physical property | Physical unit | Property values of | | | |
| | | Xylok 209 with | | Xylok 237 with | |
		Glass fabrics	Asbestos fabrics	Glass fabrics	Nomex 411 fabrics
Density	g/cm³	1.77	1.65	1.80	1.26
Tensile strength					
at 20°C	MPa	434	133	—	—
at 250°C	MPa	301	—	—	—
Flexural strength					
at 20°C	MPa	—	188	552	178
at 180°C	MPa	—	—	448	—
Hardness (Rockwell)	—	M120	R120	—	—
Heat deformation					
temperature (1.81 MPa)	°C	>330	325	—	—
Relative permittivity					
at 50 Hz	—	—	—	5.0	—
at 1 MHz	—	4.77	—	—	0.02
Dissipation factor					
at 50 Hz	—	0.011	—	0.006	—
at 1 MHz	—	—	—	—	3.9
Dielectric strength	MV/m	27.6−33.5	—	—	—
Water absorption	mg	5.2	93	—	—
	%	—	—	0.14	0.16
Flammability	UL-94	VE-0	—	—	—
Limiting oxygen index	%	69.2	—	28	—

Xylok 209 imparts an upper continuous service temperature of 250°C to glass or asbestos fiber reinforced laminates. The aging properties are superior to some other thermosets (Figure 6-1 and 6-2). The properties of glass fiber reinforced Xylok 209 laminates correspond to dielectric materials of class "H". Applications include coatings, wire enamels, and dip or vacuum impregnating of electric components. Blades of rotary multivane compressors are manufactured from Xylok 209 reinforced with asbestos fabric. Their service life is reported to reach 10 000 h at 180°C.

Glass fiber reinforced Xylok 209 laminates exhibit excellent resistance to chemical attack and hence are used in chemical pumps. After 100 h exposure at 90°C, the remaining strength is 79% in 98% sulfuric acid, 51% in 37% hydrochloric acid, 75% in 10% sodium hydroxide, 61% in saturated sodium hydroxide, and 92% in dimethylformamide.

The mechanical and thermal properties of phenol-resol-resins are

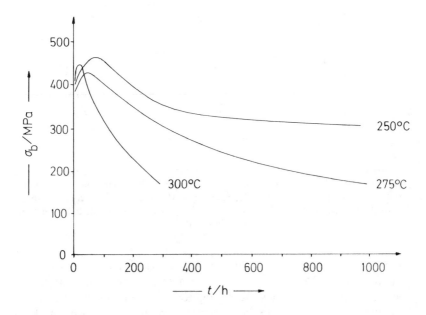

Figure 6-1   Flexural strengths of laminates from Xylok 209 and glass fiber fabrics at 250°C after aging at various times and temperatures [1].

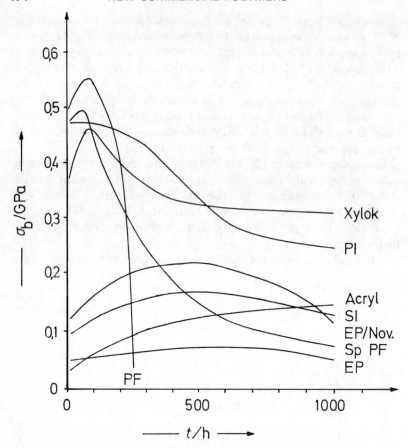

Figure 6-2 Time dependence of flexural strengths of a Xylok 209/glass fiber fabric laminate (Xylok) as compared to other polymers. Aging and measuring temperatures both at 250°C. [1]. PI = polyimide, Acryl = acrylic resins, SI = silicon, EP/Nov = epoxide/novolac blend, Sp PF = special phenol/formaldehyde resin, EP = epoxide, PF = phenol/formaldehyde resin.

improved by adding up to 20% of Xylok 214 (Figure 6-3, 6-4). The addition of Xylok 214 reduces the water absorption and improves wet insulation properties. Such upgraded phenolic resins see service in glass fiber reinforced laminates, oil filters to be used at high engine temperatures, baking enamels, coil coating, grinding wheels binder, foams, and stoving laquers. In lamp capping cements, Xylok

214 replaces the less thermally stable phenolic resins; the service life of common household bulbs is nearly doubled [1, 4].

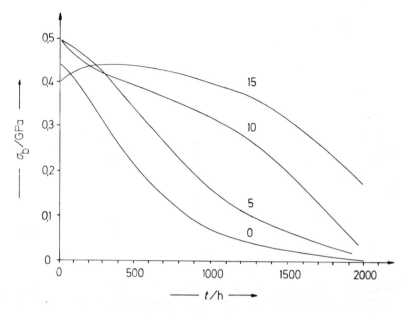

Figure 6-3 Flexural strengths of a laminate from glass fiber fabrics and a resol with 0, 5, 10, or 15% Xylok 214 [1]. The flexural strengths were measured at 180°C after an aging at 180°C for the times given.

Xylok 225 P is used per se or in mixtures with up to 25% of phenolic-novolak resins as binder. Applications include nickel-coated diamond or boron nitride abrasive grinding wheels and cutting tools, brake linings, matrices for carbon fiber reinforced plastics, pump bearings, and rotor blades for compressoors.

Epoxy-cured Xylok 237 serves as matrix in composites with glass fabric, carbon fiber, Nomex, Kapton, Kevlar, mica, and asbestos. Non-annealed glass cloth laminates demonstrate significantly higher strength at 180°C than annealed anhydride cured epoxy, epoxy/novolak, or silicone systems (Figure 6-5). Epoxy-cured Xylok 237 is particularly suited to the manufacture of thick-walled components because no volatiles are released during the curing process. Copper-clad Xylok 237 is used in printed circuits. Contrary to most of the epoxides, Xylok 237 meets the requirements of class "H" (180°C)

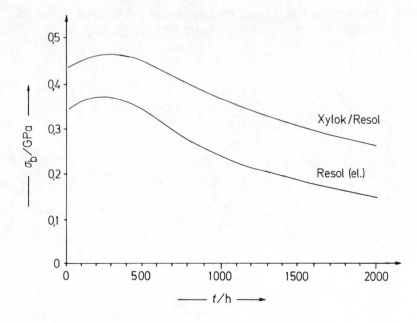

Figure 6-4 Flexural strengths of laminates from a glass fiber fabric and a resol and a resol/Xylok 214 blend [1]. Measurements at 200°C after aging at 200°C for the times given.

insulating materials as specified in the BS and IEC specifications. The temperature index, i.e., the time required to produce a 50% reduction in strength, is 20 000 h at 180°C. With Xylok 237, only one curing step is needed in manufacturing carbon fiber reinforced composites for use in the aircraft industry [5]. Molded parts and laminates reinforced with Nomex 411 are commercially available under the designations Etronax and N-Etronit-XY, respectively [6, 7]. They are approved for application in extra high voltage transformers of classes "H" (180°C) and "C" (220°C). Gas compressor blades operating at high temperatures which were manufactured from such laminates demonstrated prolonged service life times and less wear compared to blades made from glass fiber laminates.

The phenol-aralkyl-resins were designed for long term applications in the temperature range from 165 to 230°C. Processing reportedly is easier and more economical compared to polyimides, poly(quinoxalines), poly(benzimidazoles), and other polymers

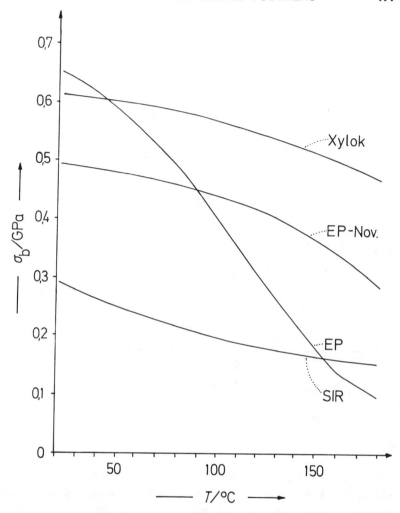

Figure 6-5 Temperature dependence of flexural strengths of laminates from glass fiber fabrics and various resins [1]. EP-Nov = epoxide/novolac, EP = epoxide, SIR = silicon.

which are used between 250 and 315°C. Thus, they compete with the modified high-heat phenolic resin grades. Recent products of this category include among others ®Bakelite B 1040 (Bakelite GmbH), ®Norsophen 1200 (CdF), 4000 F series (Fiberite Corp.), GF 700 series (General Electric), and MX 582 (General Electric).

References

[1] ®Xylok, Technical Information, Advanced Resins Ltd., Cardiff.

[2] H.-G. Elias, New Commercial Polymers 1969—1975, Gordon and Breach Science Publ., New York 1977.

[3] G. I. Harris, A. G. Edwards, and B. G. Huckstep, Plastics and Polymers **42** (1974) 239.

[4] G. I. Harris, CAV **11** (August 1978) 88.

[5] L. N. Phillips and D. J. Murphy, Royal Aircraft Establishment, Farnborough, Tech. Memorandum Mat. 322.

[6] Anon., Modern Plastics Intern. **7/10** (October 1977) 24.

[7] Etronax and N-Etronit-XY, products of Elektro-Isola, Vejly, Denmark.

6.2 POLY(P-VINYLPHENOL)

6.2.1 Structure and synthesis

Maruzen Oil Co., Osaka (Japan) and New York, offers poly(p-vinylphenol) under the name Resin M and brominated poly(p-vinylphenol) as Resin MB [1, 2]:

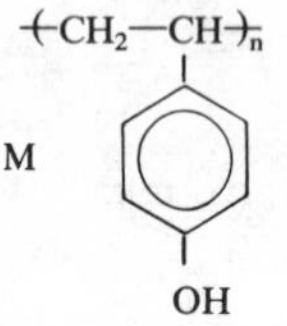

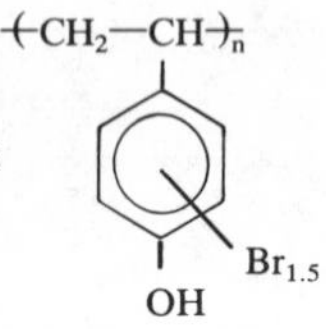

The resins are predominantly used as hardeners for epoxy resins; therefore, they are described here and not in chapter 2 dealing with saturated carbon chained polymers.

p-Vinylphenol is prepared from p-ethylphenol by dehydrogenation over a fixed-bed catalyst. Polymerization is carried out batchwise without initiators at temperatures between 100 and 150°C [3]. The bulk polymerized material contains relatively high levels of residual monomer and oligomers. At present, the resins are produced in Japan in a pilot plant with a capacity of 36 t/a; an expansion to 500 t/a was planned for 1981.

6.2.2   Properties and applications

The slightly orange colored powdery resins dissolve in alcohols, ketones, and esters but not in hydrocarbons. Resin MB is soluble also in chloroform and dichloromethane. Resin M shows good compatibility with phenolic, melamine, polyester, epoxy, urethane, alkyd, acrylate, and vinylacetate resins and with poly(vinyl alcohol). Other properties are listed in Table 6-2.

TABLE 6-2
Property values of Maruzen M and Maruzen MB [1].

Physical property	Physical unit	Property values of	
		M	MB
Density	g/cm^3	1.2	1.9
Weight average molar mass	g/mol	4000	8000
Water content	%	<3	<1
Monomer content	%	<1	<1
Oligomer content	%	<2	<2
Melt temperature	°C	160–200	190–220
Flammability temperature	°C	308	—
Bromine content	%	—	47–52
Hydroxyl equivalents	—	120	240

Both resins are amenable to polyaddition and condensation reactions via the hydroxylic group and to substitutions at the nucleus. The resins are crosslinked by epoxy resins. Typical properties of such thermosets are listed in Table 6-3.

The Maruzen Resins are used in the production of printed circuits. The epoxy-cured laminates contain 62–64% by weight glass and are copper-clad. The properties of such laminates are demonstrated in three examples. The formulation of MCL-2 consists of 63.9 pt Resin M, 100 pt Epikote 828 and, 1 pt BF$_3$/piperidine; MCL-4 consists of 51.5 pt Resin MB, 100 pt Epikote 1001, and 1.52 pt p-dimethylaminobenzaldehyde; and MCL-51 consists of 71.4 pt Resin MB, 32.6 pt Resin M, 100 pt Epikote 828, and 1.02 pt BF$_3$/piperidine.

TABLE 6-3
Properties of an epoxide cured Maruzen M resin.

Physical property	Physical unit	Property value
Density	g/cm^3	1.24
Tensile strength	MPa	42
Flexural strength	MPa	114
Impact strength (Izod)	N	13.9
Linear expansion coefficient ($\times 10^5$)		
between 30 and 185°C	K^{-1}	7.5
between 185 and 365°C	K^{-1}	16.6
Heat distortion temperature	°C	170–180
Volume resistivity	ohm $\times$ cm	8.1 $\times$ 10^{15}
Dielectric strength	kV/mm	31.3
Surface resistivity	ohm	5.5 $\times$ 10^{15}
Relative permittivity (60 Hz)	—	4.6
Dissipation factor (60 Hz)	—	0.0057
Arc resistance	s	89
Water absorption	%	0.4

Formation: 100 parts Araldit GY-260, 63.4 parts Maruzen M resin, 0.5 parts BF$_3$/piperidine as catalyst. Curing: 2 h at 175°C.

The laminates exhibit excellent thermal stability (Figure 6-6), improved flexural strength compared to commercial epoxy/glass laminates (Figure 6-7), and superior dimensional stability (Figure 6-8).

The laminates withstand soldering temperatures of 310–370°C for 1 min and 280–340°C for 10 min. The initial peel off strength is 17–19 N/cm. Other properties are: water absorption 0.07–0.14%, volume resistivity ca. 10^{16} ohm $\times$ cm, surface resistivity 10^{14} ohm, relative permittivity (1 MHz) 5, dissipation factor (1 MHz) 0.007–0.030, and arc resistance 140 s. Cured resins can be drilled and show no tendency to smear. Epoxy-cured resins are largely resistant to chemicals. The weight increase after 2 h boiling in solvents is 0.9% in xylene, 1.1% in trichloroethylene, 1.2% in dichloromethane, 0.8% in acetone, 1.3% in methanol, 0.8% in tetrahydrofuran; weight increase after 50 h of boiling in 10% NaOH is 2.6% and in 10% HCl 0.7%.

Other applications which have been recommended include insulating tapes, heat stable repair and lining materials, e.g., for auto-

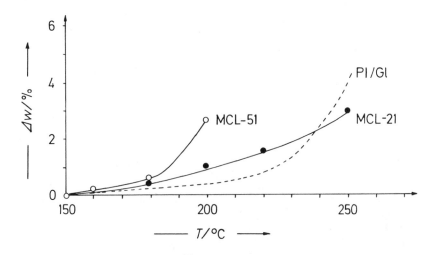

Figure 6-6 Weight loss of Maruzen MCL resins as compared to a glass fiber reinforced polyimide PI-Gl as function of temperature after 1000 h thermal aging [1].

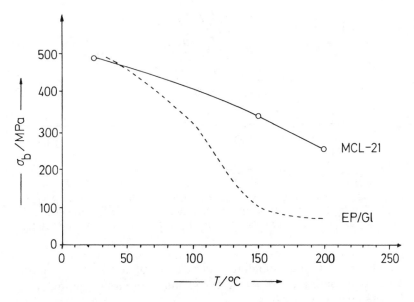

Figure 6-7 Temperature dependence of flexural strength of a Maruzen MCL resin and a glass fiber reinforced epoxide after heating for 24 h.

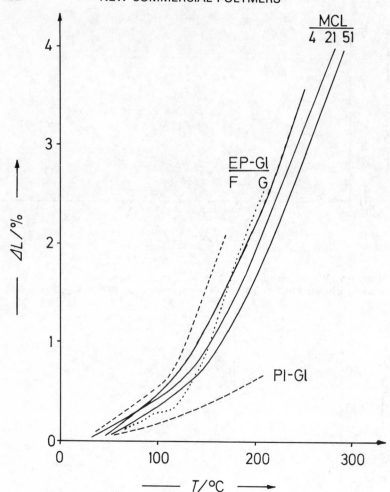

Figure 6-8 Dimensional stability, measured as expansion in z direction, for Maruzen MCL resins MCL-4, MCL-21, and MCL-51, a glass fiber reinforced polyimide PI-Gl, and two glass fiber reinforced epoxides EP-Gl (NEMA classes FR-4 and G-10 [1]).

mobile exhaust pipes, molding materials, metal adhesives, corrosion protective coatings, powder coatings, and photo-sensitive materials for off-set printing plates.

Resin M can be sulfonated and yields stable chelates with Cu(II) and Ni(II) in aqueous media. Nitration of Resin M leads to a material which is useful as a carrier in polypeptide syntheses. Resins M and MB react with aqueous formaldehyde under formation of methylol compounds which can be cured with resols at 100−150°C and find uses as molding materials, laminates, and adhesives.

Prices of Resin M and Resin MB were 6 US-$/lb and 5.20 US-$/lb, respectively, in 1977 [4].

References

[1] Maruzen Oil. Co., Technical Information, April 1976.
[2] H.-G. Elias, Polymer News **6** (1979) 26.
[3] Anon., Chem. Eng. **85**/11 (1978) 99.
[4] A. St. Wood, Mod. Plastics **54**/6 (1977) 52.

6.3 MODIFIED POLY(PHENYLENE OXIDES)

6.3.1 Introduction

Poly(oxy-2,6-dimethyl-1,4-phenylene) or poly(2,6-dimethyl-phenylene oxide) or PPO, the styrene-modified PPO types and some recent developments have been extensively described in the literature [1, 2, 3, 4].

At present, unmodified PPO reportedly is only produced in the USSR under the name Aryloxa and in Poland under the name Biapen [3]. Rockwell developed a proprietary polyether designated Rodoc which is prepared by oxidative coupling of a mixture of 2,6-dimethylphenol and 2-methyl-6-allylphenol. However, the polymer is not offered to the market [5].

Modified PPO types are produced by General Electric under the name ®Noryl; in Europe, they are marketed by General Electric and AKU as ®Arnox. Glass fiber reinforced ®Noryl grades are manufactured in the U.S. by LNP as Thermocomp and by Fiberfil-Dart. Recently introduced Noryl types supplementing the already broad product line include the non-reinforced flame-retardant grade N 190 and the 300 series where the code figure 300 indicates a heat deflection temperature of 300°F (149°C) which is the highest reached so far in the Noryl series [6]. Noryl GTX is a PPO/polyamide blend [11].

Mitsubishi Gas Chemical Co. produces PPO in a proprietary process which is claimed to be independent of General Electric's. In Japan, Borg-Warner and Mitsubishi Gas Chemical Co. founded a joint venture under the name Diamar. The new company announced the start-up of a plant with a capacity of 10 000 t/a modified PPO for 1983 [7].

Another US-Japanese joint venture, the Asahi-Dow, Ltd., developed a styrene grafted PPO which has been marketed under the name ®Xyron. Recently, Dow sold out of its joint venture. Asahi Chemical Industry Co. will integrate the venture's businesses into its own operations [11].

6.3.2 Poly(oxy-2,6-dimethyl-1,4-phenylene)-styrene-graft polymer

Poly(oxy-2,6-dimethyl-1,4-phenylene)-styrene-graft   polymer   or PPO-styrene-g-polymer is produced in a two-step process [8].

Oxidative coupling of 2,6-dimethylphenol yields PPO onto which styrene is then grafted. The technology is similar to the production of ABS. Asahi-Dow has developed a number of Xyron types: the three flame retardant grades 100V, 200V and 300V, two high-heat grades 410H and 500H, two glass fiber reinforced grades G 702H and H 0051, as well as the developmental products H0111 and H0121 with UL V-O rating.

Physical properties of Xyron [8] and the newer Noryl grades [9] are summarized in Table 6-4. They are equivalent to those of the modified PPO types [4, 6] but none of the Xyron grades reaches the good property balance of the new Noryl 300 series. The non-reinforced Noryl N 300 surpasses all non-reinforced Xyron grades with respect to stiffness, flexural strength, notched impact resistance, and heat deflection temperature. Of the glass fiber reinforced grades, Xyron G 702 H is equivalent to Noryl GFN 3 but Noryl has a higher heat deflection temperature. Uses for Xyron correspond to those of modified PPO types [4].

Prices of Noryl are between 0.97 US-$/lb for grade N 190 and about 1.75 US-$/lb for the flame retardant and reinforced grades of series 300 with high heat deflection temperatures. Prices of Xyron were expected to be in the range of 2.50−3.75 US-$/lb [10].

TABLE 6-4

Property values of PPO/styrene graft polymers and modified PPO's.

| Physical property | Physical unit | Property values of | | | | | | | |
| | | PPO/Styrene graft polymers (Xyron) | | | | | modified PPO's (®Noryl) | | |
		100V	300V	500V	G 702H	H 0111	N 190	N 300	GFN 3
Glass fiber content	%	0	0	0	20	20	0	0	30
Density	g/cm³	1.10	1.09	1.08	—	—	1.08	1.06	1.2
Melt index[1]	g/10 min	9	15	7	—	5	—	—	—
Tensile strength at yield	MPa	34.3	49.0	58.8	107.9	98.1	48.3	75.9	117.2
Tensile strength at break	MPa	—	—	—	—	—	—	—	—
Elongation at break	%	40	30	20	6	5	35	20	5
Modulus of elasticity	GPa	—	—	—	—	—	2.48	—	8.28
Flexural strength	MPa	58.8	78.5	93.2	152.0	138.3	56.5	104.1	137.9
Flexural modulus	GPa	2.06	2.16	2.35	5.20	4.90	2.24	2.41	7.59
Impact strength with notch	kJ/m	0.25	0.23	0.22	0.12	0.05	0.37	0.53	0.12
Hardness (Rockwell)		L 78	L 90	L 100	—	—	R 115	R 119	L 108
Heat distortion temperature (1.81 MPa)	°C	80	100	120	140	136	88	149	158
Continuous service temperature	°C						80	105	90
Volume resistivity	ohm × cm	10^{16}	10^{16}	10^{17}	—	—	10^{15}	—	10^{17}
Surface resistivity	ohm	10^{16}	10^{16}	10^{17}	—	—	10^{15}	—	10^{17}
Relative permittivity (1 MHz)		2.8	2.7	2.7	—	—	2.7	2.6	2.85
Dielectric strength	MV/m	29	29	29	43	—	24.8	20.0	—
Water absorption	%	0.10	0.10	0.10	0.06	—	0.07	0.06	0.06
Flammability (UL 94; 1.5 mm)	V-1	V-1	V-1	—	V-0	V-0	V-0	HB	—

[1] 250°C, 10 kg; grade 100 V at 200°C, 10 kg.

References

[1] A. S. Hay et al., Encycl. Polymer Sci. Technol. **10** (1969) 92.

[2] K.-U. Bühler, Spezialplaste, Akademie-Verlag, Berlin 1978.

[3] H.-G. Elias, Ullmanns Encyklopädie der technischen Chemie, Verlag Chemie, Weinheim, 4th ed., vol. **15** (1978) 421.

[4] U. Vogtländer, Kunststoffe **70** (1980) 645.

[5] J. A. Lincoln, Rockwell International, Rocketdyne Division, personal communication (1981).

[6] J. B. Titus, New Plastics, Plastec Report R48, Plastics Technical Evaluation Center, Picatinny Arsenal, Dover, N.J., 1977.

[7] Anon., Modern Plastics **58** (April 1981) 16.

[8] S. Izawa, Japan Plastics Age **16** (November 1978) 1.

[9] ®Noryl, General Electric Co., Technical Information.

[10] Anon., Modern Plast. Intern. **8** (February 1978) 8.

[11] J. Bussink, Kunststoffe **74**, 573 (1984).

6.4 POLY(2,6-DIBROMO-1,4-PHENYLENE OXIDE)

6.4.1 Structure and synthesis

Poly(2,6-dibromo-1,4-phenylene oxide) (I) can be prepared by decomposition of metal salts of the corresponding halogenated phenols or by oxidation of 2,4,6-tribromophenol in alkaline medium in the presence of oxidizing compounds such as PbO_2 or $K_3[Fe(CN)_6]$ [1]. Oxidative coupling of phenols as it is utilized in the PPO production process is not feasible with 2,6-dihalogen-phenols [2]. The bromination of dissolved poly(1,4-phenylene oxide) results in the formation of numerous brominated products [3]. Presumably, the synthesis of I starts from the commercially available 2,4,6-tribromo-phenol which is oxidized by one of the known oxidizing agents:

(6-2)

(I)

Poly(2,6-dibromo-phenylene oxide) is commercially manufactured by Velsicol Chemical Co. under the trade name ®Firemaster TSA.

6.4.2 Properties and applications

®Firemaster TSA is a white powder containing 63−65.5% bromine [4]. The bromine content calculated for I is 64.0%, the synthesized material contains 65.1% [1]. The softening range of the commercial product is between 200 and 230°C. It has a density of 2.07 g/cm^3 and it is insoluble or only slightly soluble at room temperature in water, ethanol, acetone, hexane, toluene, methylenechloride, and per-chloroethylene; the solubility in chloroform and tetrahydrofuran is 75 and 220 g/l solvent, respectively.

The commercial product has excellent thermal stability. A thermogravimetric analysis at a heating rate of 20°C/min shows weight losses of 25% at 495°C, 50% at 520°C, and 75% at 568°C. On decomposition hydrogen bromide is formed; therefore, the material must not be overheated.

The acute oral toxicity LD_{50} for rats was found to be greater than 21.5 g/kg and the acute dermal toxicity LD_{50} for rabbits was higher than 3.038 g/kg [4].

®Firemaster TSA is recommended as a flame retardant for glass fiber reinforced nylons, thermoplastic polyesters, and for other engineering plastics requiring high processing temperatures. Incorporation is achieved by melting the blended polymers. The molecular weight of I was found to be 3 150 g/mol [1]. Although the polymer has a relatively low molecular weight, it shows no tendency to migration and blooming in blends. The flow of polymer blends reportedly is improved by addition of Firemaster TSA.

References

[1] J. Cox, J. Appl. Polym. Sci. **9** (1965) 513.
[2] A. S. Hay, Polym. Eng. Sci. **16** (1976) 1.
[3] I. Cabasso, J. Jagur-Grodzinski, and D. Vofsi, J. Appl. Polym. Sci. **18** (1974) 1969.
[4] ®Firemaster TSA, Technical Information, Velsicol Chemical Corp.

6.5 HYDANTOIN-CONTAINING EPOXIDES

6.5.1 Structure and synthesis

Hydantoin derivatives are readily accessible from ketones, hydrocyanic acid, and carbon dioxide:

(6-3)

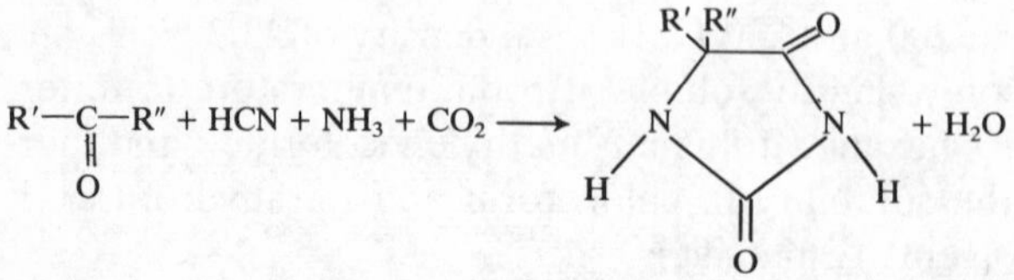

$$R'-\underset{\underset{O}{\|}}{C}-R'' + HCN + NH_3 + CO_2 \longrightarrow \quad + H_2O$$

The subsequent reaction to epoxides is carried out by several routes [1]. The products have been described in an early review [2]. In the meantime, the product range has been altered and supplemented [3, 4]. Hydantoin containing epoxides based on structure I are now commercially available from Ciba-Geigy under the designation ®Aracast (Table 6-5):

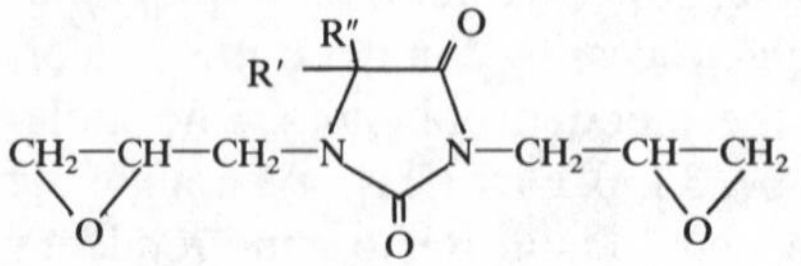

®Aracast XB 2826 is a mixture of XB 2793 and a conventional epoxy resin (®Araldit 6010).

TABLE 6-5
Structure of ®Aracast Hydantoin/Epoxide Resins.

| ®Aracast Type | Substituent | | Code |
	R_1	R_2	
XB 2793	Methyl	Methyl	I a
XU 238	Methyl	Ethyl	II a
XU 229	Ethyl	Amyl	III a
XU 231	——— Pentamethylene ———		IV a

6.5.2 Properties and applications

The properties of the uncured ®Aracast resins are summarized in Table 6-6. Hydantoin-epoxy resins can be cured by all conventional epoxy hardeners, e.g., with anhydrides, aromatic amines, and Novolak resins. Aliphatic amines as hardeners are recommended only for Aracast XU 229. The selection of hardener type and curing conditions depends on the intended end use, which varies considerably for the individual Aracast grade.

XB 2793 sees service in water-dispersible coating systems, preferably in formulations with acrylic resins. XU 238 shows good solubility in water and is used in water soluble coatings. Highly filled coating systems are achieved with XU 229 due to its low viscosity. XU 231 is designed for powder coatings. XB 2826 is a solvent free system to be used in chemically resistant coatings with good mechanical properties.

Numerous formulations and their property values have been published [2, 3, 4, 5]. The low priced XB 2793 which is produced from acetone has been the product most intensively investigated. The most obvious advantage of hydantoin resins is their low viscosity which in conventional epoxy resins can only be achieved by the addition of reactive thinners resulting in a certain loss of properties in the cured resins.

Curing of hydantoin resins gives the best results when stoichiometric 1:1 molar ratios of anhydride/epoxy are applied [3]. Useful hardeners are liquid anhydrides, such as hexahydrophthalic anhydride (HT) and isomerized methyltetrahydrophthalic anhydride (MT). Curing of hydantoin resins with HT results in improved mechanical properties and higher heat deflection temperatures compared to curing with MT [3].

In spite of a favorable balance of mechanical and thermal properties, the most obvious disadvantages of conventional epoxy resins are the unsatisfactory resistance to UV and the poor performance in arc and spark discharges. The non-aromatic hydantoin ring together with non-aromatic hardeners leads to epoxides with significantly improved resistance to UV. Compared to conventional epoxy resins, the hydantoin-epoxy resins have equivalent mechanical properties but higher heat deformation temperatures and improved dielectric strength. They withstand effects caused by arcs and tracking currents in transformer fluids, such as SF_6. The low coefficient of

TABLE 6-6
Property values of non-hardened ®Aracast resins.

Physical Property	Physical Unit	XB 2793	XU 238	Property values of XU 229	XU 231	XB 2826
Density	g/cm^3	0.76	—	0.70	—	0.74
Melt temperature	°C	25	—	—	90	—
Viscosity (25°C)	Pa × s	2.50	1.40	1.15	—	5.00
Epoxy equivalent/100 g		0.70	0.76	0.63	0.69	0.62
Flammability temperature	°C	149	110	179	>93	113
Color	—	light yellow	light yellow	Gardner 1	white	light yellow
Water solubility	—	dispensible	soluble	insoluble	insoluble	—

thermal expansion permits embedding of metal parts, e.g., copper or aluminum. Westinghouse uses hydantoin resins as embedding material for high-voltage connection in transformers [6].

References

[1] J. Habermaier, Angew. Makromol. Chem. **63** (1977) 63.
[2] H.-G. Elias, Neue polymere Werkstoffe 1969–1974, Carl Hanser Verlag, München 1975; New Commercial Polymers 1969–1975, Gordon and Breach, New York 1977.
[3] G. Buchi, SPE 37th Ann. Techn. Conf., Technical Papers **25** (1979) 803.
[4] ®Aracast, Technical Information Ciba-Geigy.
[5] E. H. Catsiff, H. B. Dee, and R. Seltzer, Modern Plast. **55** (July 1978) 54.
[6] R. Zucker, SPE 37th Ann. Conf., Technical Papers **25** (1979) 807.

6.6   PEROXIDE AND SULFUR CURABLE EPICHLOROHYDRIN ELASTOMERS

6.6.1   Structure and synthesis

For many years polymers of epichlorohydrin (CO) and copolymers with ethylene oxide (ECO) have been on the market as elastomers resistant to ozone, oil, and low temperatures. Curing is achieved by reacting the chlorine groups with N,N'-ethylene thiourea (ETU)/ metal oxides [1, 2].

Now, Nippon Zeon Co., Ltd., offers a copolymer of epichlorohydrin and allylglycidyl ether as well as a terpolymer (ETER) of epichlorohydrin, ethylene oxide, and allylglycidyl ether under the trade names ®Gechron 1100 and 3100, respectively [3]. The new ®Hydrin 400 supplied by B. F. Goodrich is also a terpolymer [4, 5].

The allylglycidyl ether $CH_2=CH-CH_2-O-CH_2\overbrace{}^{} CH_2$ is sus-

ceptible to polymerizations via the double bond and via the oxirane group. The copolymerization induced by ionic or coordination catalysts causing ring opening leads to amorphous co- and terpolymers with pendent allylic groups [6]:

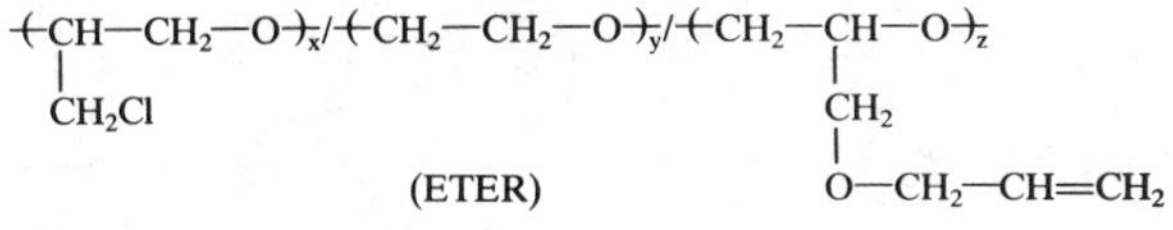

$$\{CH-CH_2-O\}_x/\{CH_2-CH_2-O\}_y/\{CH_2-CH-O\}_z$$

$$\begin{array}{ccc} | & & | \\ CH_2Cl & & CH_2 \\ & & | \\ \text{(ETER)} & & O-CH_2-CH=CH_2 \end{array}$$

6.6.2 Properties and applications

The older ECO elastomers require relatively long curing times and show a tendency to form deposits in the molding tools during processing [4]. The vulcanizates suffer a change in hardness during heat aging [3] and show poor resistance to H_2S and to the so-called "sour gas" in electronic fuel injection systems [7]. "Sour gas" is peroxide-containing partially oxidized fuels.

The new co- and terpolymers can be cured via the pendant allylic groups by peroxides or sulfur. In accordance with the CO- and ECO-types they are also curable with the common N,N'-ethylene-thiourea (ETU)/metal oxide systems. Table 6-7 lists compound formulations which have been suggested for curing by peroxides and co-activators (I, II) and by sulfur (III) [5]. The fast curing com-

TABLE 6-7
Formulations for the peroxide and sulfur vulcanization of Hydrin 400 [5].

Compounds	Composition (parts)		
	I	II	III
Hydrin 400	100	100	100
Stearic acid		0.8	
Nickel-dibutyldithiocarbamate	1.0	0.9	
Dibasic lead phosphate		2.0	
Potassium stearate	2.5		
Calcium oxide	2.5		
Dioctylphthalate	11.0	10.0	100
Carbon black N 326	20.0	30.0	20.0
Carbon black N 550	30.0	15.0	40.0
Tetraethyleneglyol dimethacrylate	2.5		
Trimethyloltrimethacrylate		2.5	
Dicumylperoxide	2.5		
2,5-Bis(t-butylperoxy)-2,5-dimethylhexane		1.1	
Methylniclat antioxidant		0.5	0.5
ZO-9 Plasticizer		0.3	
Zinc stearate			3.5
2-Mercaptobenzthiazol			1.0
Tetramethylthiuramsulfide			0.5
Sulfur			1.5
Isobutylniclat			1.0
Vulcanization			
Temperature (°C)	160	182	160
Time (s)	1200	51	900

pound II is suitable for injection molding processing.

The curing of ETER by peroxides and co-activators, e.g., polyfunctional methacrylates, proceeds faster than with CO. However, the vulcanizates exhibit lower mechanical and dynamic properties and poorer resistance to fuels when compared to vulcanizates obtained from cures with sulfur and accelerators. The vulcanizates prepared with sulfur or sulfur donors do not tolerate use temperatures above 120°C, whereas the upper continuous service temperature of peroxide-cured vulcanizates is about 150°C (Table 6-8).

ETER can also be vulcanized in blends with other elastomers. In synthetic 1,4-cis-polyisoprene, an addition of ETER improves heat stability, dynamic modulus, and the damping characteristics; in SBR and neoprene the conductivity and the resistances to heat, oil, and fuel are improved; and in blends with NBR ETER imparts improve-

TABLE 6-8
Property values of peroxide and sulfur vulcanized ®Hydrin 400 [5].

Physical property	Physical Unit	Property value of		
		I	II	III
Density	g/cm^3	1.39	1.38	1.39
Mooney Scorch (121°C; t_5)	—	50.0	57	52
	min	6.0	13	8.5
Modulus (100%)	MPa	2.62	0.89	2.00
Modulus (300%)	MPa	11.44	3.92	8.34
Tensile strength at break	MPa	12.68	12.12	13.64
Elongation at break	%	350	800	750
Hardness (Durometer A)	—	58	44	60
Compression set (method B)				
after 70 h at room temp.	%	5.2	—	—
after 70 h at 125°C	%	31.6	—	—
after 22 h at 100°C	%	—	58	84
Properties after oven aging (70 h at 150°C)				
Tensile strength at break	MPa	12.68	9.85	4.34
Elongation at break	%	250	330	160
Hardness (Durometer A)	—	71	60	62
Properties after 70 h in Fuel C at room temperature (ASTM)				
Tensile strength at break	MPa	6.27	4.34	11.85
Elongation at break	%	180	460	600
Hardness (Durometer A)	—	43	28	44
Volume increase	%	29.7	35	26

ments with regard to low temperature and weathering characteristics. ETER is also blended with CO and ECO.

Typical applications of ETER include sealings, packings, collars, diaphragms, tank caps, and tubings in service fields where an equally good resistance to heat, oil, hydrocarbon, and alcohol-based fuels as well as to chemicals is required. Blends of ETER and chlorinated poly(ethylenes) are used in cable jacketing. The adhesion between ETER and metals or fabrics is superior to those of ECO elastomers.

A detailed account on the new ETER polymers, a comparison of their properties with other elastomers, and their applications has been published [8].

References

[1] E. Scheer, Epichlorohydrin Elastomers, in R. O. Babbit, ed., The Vanderbilt Rubber Handbook, 1978, 275.

[2] W. Hofmann, Kautschuk-Technologie, Gentner Verlag, Stuttgart 1980, 173.

[3] ®Gechron 1100 and ®Gechron 3100, Nippon Zeon Co., Ltd., Tokyo, Technical Information.

[4] J. T. Oetzel and E. N. Scheer, Rubber Chem. Technol. **51** (1978) 860.

[5] ®Hydrin 400 Elastomer, B. F. Goodrich Chemical Group, Cleveland, Technical Information, February 1981.

[6] R. H. Yocum and E. B. Nyquist, Functional Monomers, vol. 2, M. Dekker Inc., New York 1974.

[7] E. N. Scheer, ®Hydrin Elastomer Review, Seminar, Nashville, September 26, 1979.

[8] W. Hofmann and C. Verschuet, Gummi, Asbest, Kunststoffe **33**/9 (1980) 590; **33**/10 (1980) 742; **34**/1 (1981) 24; **34**/3 (1981) 136.

6.7   POLY(ARYLETHERKETONES)

6.7.1   Structure and synthesis

The synthesis of poly(aryletherketones) of the general structure I by subjecting acid chlorides (II) to condensation under Friedel-Crafts reaction conditions was first achieved by DuPont [1] in 1962 and by ICI [2] in 1964.

In another synthesis, bis-4-fluorophenylketone (III) was condensed with the dipotassium salt of bis-4-hydroxy-phenyl-ketone

(IV) [3]. All reactions yielded only low molecular weight polymers because the initially formed crystalline oligomers precipitate from the solution and thus prevent higher conversions to high molecular weight products. However, using HF as a good solvent for I and BF_3 as catalyst high molecular weight I is obtained [4].

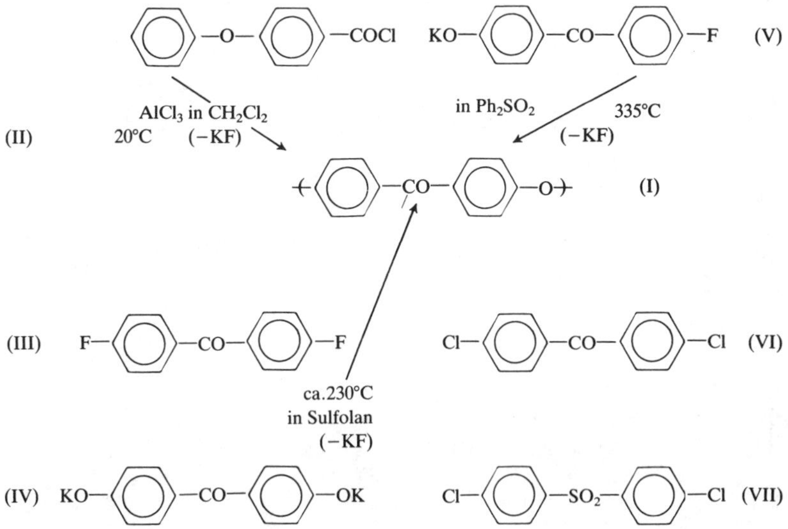

The last synthesis was the basis for the development of Stilan 1000 by Raychem. The polymer with the structure I has a melting point of 365°C and demonstrates high toughness, low flammability, and excellent resistance to chemicals [5]. Stilan 1000 had been designed to find applications as a high performance engineering plastic but the production was terminated in 1976 [6].

At ICI, arylsulfones have been found to be good solvents for the polycondensation of III and IV at 280−340°C and for the self-condensation of V at 335°C [7, 8]. The first reaction yields branched polymers with high gel content; the polymers with reduced viscosities R. V. (1 g polymer dissolved in 100 ml 98% sulfuric acid, measured at 23°C) between 102 and 195 ml/g are brittle and become flexible only at R. V. of above 195−219 ml/g. The second reaction yields flexible polymers already in the R. V. range of 103−187 ml/g.

The polycondensation of bis-4-chlorphenylketone VI with IV pre-

sents some difficulties whereas copolymers from III with VI and IV
are readily obtained at lower reaction temperatures. The reactivity
of III is also increased when III is partially replaced by bis-4-
chlorophenylsulfone (VII).

The poly(aryletherketones) prepared by ICI according to the
above mentioned synthesis routes and illustrated by the following
structural units all exhibit high melting temperatures T_m:

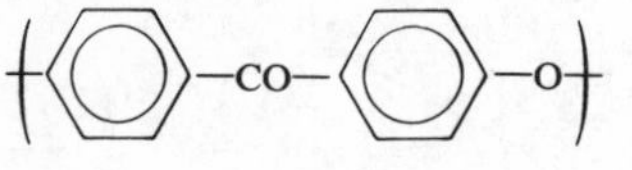

(VIII) T_g 154°C
T_m 367°C

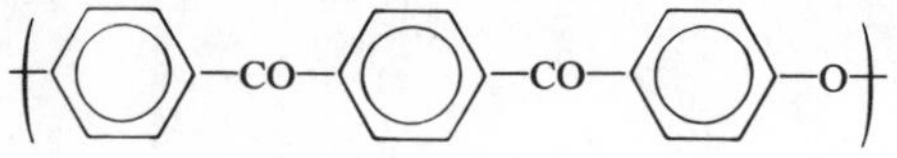

(IX) T_m 384°C

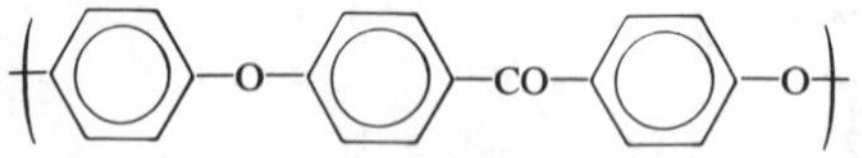

(X) T_g 144°C
T_m 335°C

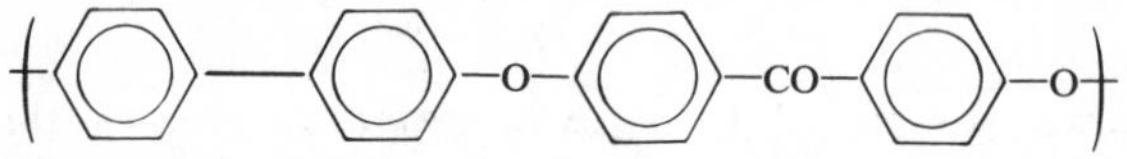

(XI) T_m 416°C

X-ray studies show almost identical patterns for the polymers
VIII, IX, and X; only polymer XI is slightly different. A 1:1 copolymer of VIII and XI with T_m = 345°C is crystalline, but a copolymer
of X and XI has a significantly reduced crystallinity and a broad
melting range. From the X-ray fiber diffraction pattern of IV, the
distance of two consecutive structural units was determined as 0.993
nm. The chain unit

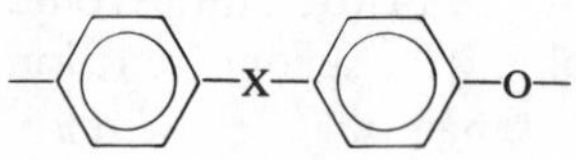

with X = O or CO is stretched and aligned in a plane. Ether and carbonyl groups are mutually interchangeable without loss of polymer crystallinity. The crystal structure is very similar to that of poly(p-phenyleneoxide) where the spacing of structural units is 0.972 nm.

In 1979, ICI marketed a so-called polyether-etherketone named PEEK [9, 12]. The polymer with the structural unit X has a T_g of 143°C and a T_m of 334°C. The older Stilan 1000 of Raychem presumably had the structural unit VIII.

PEEK is commercially available from ICI as a non-reinforced material, in two grades reinforced with 10% and 20% glass fiber [9], and since 1981 also in two grades with 20% and 30% carbon fiber. For a review see [14].

6.7.2 Processing of PEEK

Originally, PEEK was designed for extrusion wire coating. There exist also processing recommendations for injection molding, monofilaments, rotational molding, coatings, and laminates reinforced with glass or carbon fibers [9].

PEEK granulates must be pre-dried at 150°C for at least 3 h. For extrusion wire coating, an extruder with a high L/D ratio of 24:1 and screw compression zones similar to the processing of fluoropolymers is needed. Mass temperatures are between 360 and 380°C which require cylinder temperatures of 395 to 405°C. The coated wire should not be quenched in cold water which would lead to undesirable amorphous structures. Cooling with air and annealing at 200°C for 1 h give the best mechanical and chemical properties of the coating. PEEK residues are removed from the extruder by "rinsing" with high molecular PE at 240–280°C.

PEEK exhibits extremely high melt viscosities. At a shear rate of 1000 s^{-1} they are:

at 360°C 480 Pa × s (4900 poise)
at 380°C 400 Pa × s (4000 poise)
at 380°C 550 Pa × s (5500 poise) for PEEK with 10% glass fiber
at 380°C 800 Pa × s (8000 poise) for PEEK with 20% glass fiber
at 400°C 350 Pa × s (3500 poise).

Under appropriate precautions, processing temperatures of ca. 400°C cause no thermal degradation. Under exclusion of air, PEEK is thermally stable for more than 1 h at 400°C.

For screw injection molding, cylinder temperatures of 350–400°C and mold temperatures of 150–160°C are required. Annealing of the molded parts at 300°C for 2 min or at 200°C for 1 h results in a high degree of crystallinity.

6.7.3 Properties and applications

PEEK is a crystalline polymer with T_g = 143°C and T_m = 334°C. The maximum degree of crystallinity is obtained from the melt at 256°C and in molded parts by annealing at 185°C. The maximum level of crystallinity is 48%. The density of the amorphous polymer is 1.265 g/cm^3 and of the crystalline polymer 1.320 g/cm^3.

PEEK is an engineering plastic with high heat deflection temperatures. Its properties are significantly improved by reinforcement with carbon or glass fibers (Table 6-9 and 6-10). The upper service temperature limits of carbon fiber reinforced PEEK are 220–240°C.

The best mechanical properties are obtained by reinforcing PEEK with carbon fiber fabrics and rovings (Table 6-11). According to ICI

TABLE 6-9

Temperature dependence of the tensile strength at break and the flexural modulus of non-reinforced PEEK and carbon fiber reinforced PEEK.

Physical property	Physical unit	Property values for		
		PEEK non-rein-forced	PEEK with 20% C-Faser	PEEK with 30% C-Faser
Tensile strength at break at				
23°C	MPa	70	165	215
100°C	MPa	50	127	185
150°C	MPa	37	82	107
200°C	MPa		33	67
250°C	MPa		30	49
300°C	MPa		29	38
Flexural modulus at				
23°C	GPa	3.9	12.5	15.5
100°C	GPa	3.0	9.5	12.2
150°C	GPa	2.0	4.0	10.0
200°C	GPa		2.5	3.5
300°C	GPa			2.1

TABLE 6-10

Mechanical properties of laminates from PEEK with fabrics and rovings from carbon fibers (all measurements in fiber direction).

Physical property	Physical unit	Property value in fiber direction for		
		PEEK/ C-Fabric (50 wt. % or 41 Vol. %)	PEEK/ C-Rovings (60 wt. % or 52 Vol. %)	PEEK/ C-Rovings (70 wt. % or 62 Vol. %)
Density	g/cm^3	1.53	1.59	1.65
Tensile strength at break	MPa	360	1280	1600
Elongation at break	%	1.6	1.3	1.3
Modulus of elasticity at				
23°C	GPa	28	106	128
200°C	GPa	14	53	64
300°C	GPa	8	32	38

sources, PEEK laminates are the first thermoplastic composites which demonstrate higher strength and stiffness and at the same time a better resistance to chemicals than epoxy laminates [9].

PEEK is soluble only in concentrated sulfuric acid. The resistance to hot water is excellent; wires insulated with 0.25 mm thick PEEK coatings lose only 5% of the initial strength when immersed in 100°C hot water for 322 days. PEEK exhibits high stress corrosion resistance to ketones, halogenated hydrocarbons, alcohols, esters, and jet fuels. Embrittlement of PEEK on gamma irradiation is less than with poly(styrene) which is commonly used as an irradiation resistant polymer. Failures in PEEK insulations result only when a dose of about 1100 Mrad is applied from a ^{60}Co source. The resistance of PEEK to irradiation depends on the molecular weight; best results are obtained with polymers of melt viscosities of 400−500 Pa × s.

PEEK coated wires and cables are already being used in the nuclear industry. Blow molded containers can be used for the transport of spent nuclear fuel. Large volume parts are predicted to see service in aircrafts, and carbon fiber reinforced materials are intended to replace metals in jet engines [10, 11].

Reinforced PEEK has higher heat deflection temperatures and better mechanical properties than reinforced poly(phenylene-sulfide) and polysulfone types; however; it is also more expensive. In 1981, the price for non-reinforced PEEK was 33 US-$/lb.

TABLE 6-11

Property values of injection molded specimens from non-reinforced and reinforced PEEK [9].

Physical property	Physical Unit	Property values of				
		PEEK non-reinf.	PEEK with 10 % glass fiber	PEEK with 20 % glass fiber	PEEK with 20 % carbon fiber	PEEK with 30 % carbon fiber
Density	g/cm^3	1.30	1.37	1.44	1.40	1.44
Shrinkage	%	1.1	0.4	0.2		
Tensile strength at yield	MPa	91	149	135	165	215
Tensile strength at break	MPa	70				
Elongation at break	%	150	8.5	4.4	6	3
Modulus of elasticity (150°C)	GPa	1.1				
Flexural modulus	GPa	3.8	6.3	10.5	12.5	15.5
Flexural strength	MPa				260	248
Notched impact strength						
Charpy	kJ/m^2	54	6.5	7.6		
Izod	kJ/m^2	0.048	0.063			
Heat distortion temperature (1.81 MPa)	°C	135–160	209	286	300	300
Volume resistivity	Ohm × cm	(4–9) × 10^{16}				1.4 × 10^5
Surface resistivity	Ohm					1.2 × 10^2
Relative permittivity	—	3.2–3.4				
Water absorption (24 h)	%	0.14				
Limiting oxygen index	%	35				
Flammability (UL 94)						
1.5 mm	—	V-0				
3.2 mm	—	5V				

References

[1] W. H. Bonner, USP 3 065 205 (1962).
[2] I. Goodman, J. E. McIntyre, and W. Russell, Brit. P. 971 227 (1964).
[3] R. N. Johnson, A. G. Farnham, R. A. Clendinning, W. F. Hall, and C. N. Merriam, J. Polymer Sci. [A-1] **5** (1967) 2375.
[4] B. M. Marks, USP. 3 442 857 (1969).
[5] L. C. Glover, Jr., SPE 34th Annual Meeting **22** (1976) 148.
[6] L. C. Glover, Jr., Raychem Corp., personal communication, August 4, 1976.
[7] J. B. Rose, Brit. P. 1 414 421 (1975).
[8] T. E. Attwood, P. C. Dawson, J. L. Freeman, L. R. J. Hoy, J. B. Rose, and P. A. Staniland, ACS Polymer Preprints **20**/1 (1979) 191; Polymer **22** (1981) 1096.
[9] PEEK, Polyether-etherketone, ICI, Technical Information.
[10] Anon., Modern Plast. **58** (March 1981) 20.
[11] Anon., Plastics Technology **27** (March 1981) 33.
[12] T. C. Stening, C. P. Smith, and P. J. Kimber, Modern Plastics (November 1981) 86; Modern Plastics Intern. **12** (March 1982) 54.
[13] Modern Plast. Internat. **12** (July 1982) 52.
[14] R. B. Rigby, Polymer News **9**, 325 (1984).

7 Polyesters and polyanhydrides

7.1 INTRODUCTION TO THERMOPLASTIC POLYESTERS

In the recent years, the fast developing field of thermoplastic polyesters has been described in several publications [1−3]. Here, a brief summary of the commercially available polymer types and a review of the newest developments will be given.

The composition of thermoplastic polyesters derived from various structural units is summarized in Table 7-1. Poly(ethylene terephthalate) (PET) was the first engineering plastic of this class; it consists of alternating units of terephthalic acid (T) and ethylene glycol (EG).

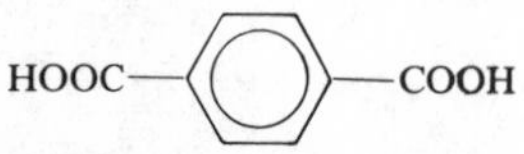

HOOC—⟨○⟩—COOH   (T)

HO—CH$_2$—CH$_2$—OH   (EG)

HO—CH$_2$—CH$_2$—CH$_2$—CH$_2$—OH   (BG)

The problems connected with the processing of the first PET types were the incentive to develop easier processable materials, such as poly(butylene terephthalate) (PBT) based on terephthalic acid (T) and 1,4-butylene glycol (BG) and a number of modified PET types and PET/PBT hybrids; see chapters 7.2 and 7.3.

Block copolymers composed of "soft" multibutyleneoxide terephthalate segments and "hard" butylene terephthalate segments are elastomers (®Hytrel); see chapter 4.1.

The reaction of terephthalic acid (T) and isophthalic acid (I) with ethylene glycol (EG) and/or 1,4-cyclohexanedimethylol (CHDM) leads to a number of amorphous and transparent copolyesters (see chapters 7.4 and 7.7.).

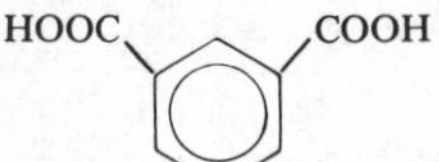

HOOC   COOH

⟨○⟩   (I)

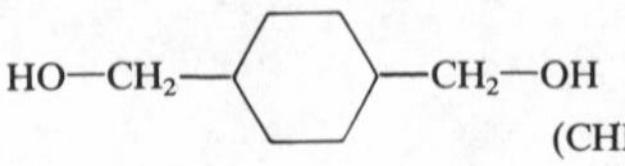

HO—CH$_2$—⟨○⟩—CH$_2$—OH   (CHDM)

TABLE 7-1

Composition of thermoplastic polyesters.

Polymer type or Trade name	Company	Dicarboxylic acid T	I	N	Hydroxy acid HB	Glycols EG	BG	CHDM	Diphenols DP	BP	DS	HC	Phosgene
PET	many	×				×							
PET	many	×				×							
PET/PBT	many	×				×	×						
Kodar PETG 6763	Eastman	×				×		×					
Kodar A-150	Eastman	×	×			×	×	×					
Ektar PCTG	Eastman	×						×[1]					
X7G	Eastman	×			×	×							
Hytrel	DuPont	×				×	×[2]						
Ekonol	Carborundum				×								
Ekkcel C 1000	Dart		×		×							×	
Ekkcel I 2000	Dart	×			×				×				
Q-Film	Teijin			×		×							
Polyarylate	some	×	×							×			
Polyester-carbonate	some	×								×	×		×
Merlon T	Mobay									×	×		×

1) With a second glycol.

2) Blockcopolymer, thermoplastic elastomer.

p-Hydroxybenzoic acid (HB) can be polymerized to a homopolymer; its polycondensation together with terephthalic acid (T) or isophthalic acid (I) and 4,4'-dihydroxydiphenylether (DP) gives copolymers with high heat deflection temperatures (see chapter 7.4).

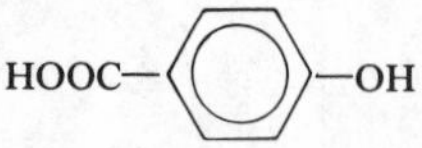

The polyesters with high heat deflection temperatures are also obtained when ethylene glycol (EG) is reacted with 2,6-naphthalic dicarboxylic acid (N) in place of the benzene dicarboxylic acids.

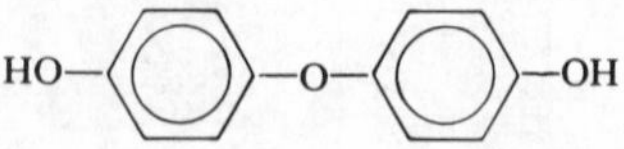

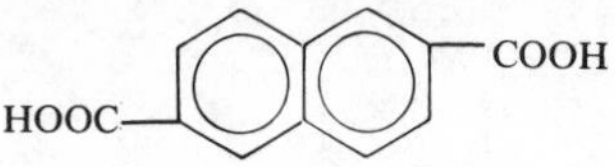

The polycondensation of benzene dicarboxylic acids (T and I) with diphenols, in particular with bisphenol-A (BP), yields aromatic polyesters which also exhibit high heat deflection temperatures (see chapter 7.6).

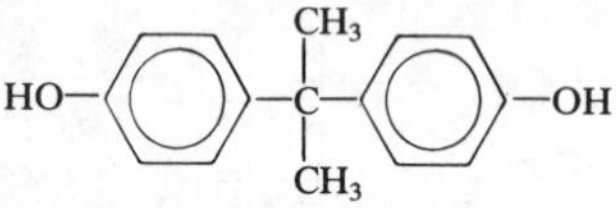

Structurally related to the polycarbonates on one hand and to polyarylates on the other are the polyestercarbonates obtained from bisphenol-A (BP), dicarboxylic acids, and phosgene $COCl_2$. Besides bisphenol-A (BP), 4,4'-dihydroxydiphenylsulfide (DS) is also used (see chapter 7.7).

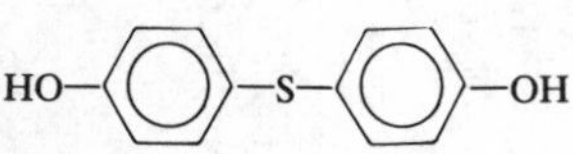

References

[1] J. R. Caldwell, W. J. Jackson, Jr., and T. F. Gray, Jr., in Encycl. Polym. Sci. Technol., suppl. vol. **1** (1976) 444.

[2] H.-G. Elias, Neue polymere Werkstoffe 1969–1974, Carl Hanser Verlag, München 1975; New Commercial Polymers 1969–1975, Gordon and Breach, New York 1977.

[3] M. Bakker, Encycl. Chem. Technol. **9** (1980) 118.

7.2   POLY(ETHYLENE TEREPHTHALATE)

7.2.1   Introduction

Poly(ethylene terephthalate) has been widely used since the 50's for high-strength fibers and films. First attempts in the 60's to modify PET for injection molding processing (AKU, Hoechst, DuPont) were not successful due to severe problems in processing and to the poor properties achieved. The first products had too high molecular weights and recrystallized too slowly causing prolonged cycle times; the molded parts exhibited poor notched impact resistance and dimensional stability.

In 1974–1976, DuPont developed a PET grade which could be extrusion blow molded to bottles for carbonated beverages (see chapter 7.5). Before the end of 1978, DuPont introduced glass reinforced PET types (®Rynite) containing a new nucleating agent which ensures high recrystallization rates. Presumably, ®Rynite also contains small amounts of ®Hytrel or nylon.

PET injection molding grades must have relatively low molecular weights because the crystallization rate decreases with increasing molecular weight. The intrinsic viscosities of the individual PET grades differ markedly:

PET for injection molding	$[\eta]$ = 45	ml/g
PET for fibers and films	$[\eta]$ = 60–65	ml/g
PET for bottles and extrusion coating	$[\eta]$ = 70–85	ml/g
PET for tire cord fiber	$[\eta]$ = 100	ml/g.

PET injection molding grades are now commercially available from several producers (Table 7-2). Furthermore, a number of PBT/PET blends have been marketed which contain 10–15% of PET or elasticizing components [1]. ®Ropet is a PET/acrylic polymer blend. The recent developments have been described [2].

TABLE 7-2
Poly(terephthalates) and Copolyesters for Injection Molding.

Company	Polymer Type	Trade Name	Non-reinforced Grades	Reinforced Grades Glass fiber content in %	Glass fiber/ mica content in %	Flame resistant Grades UL 94 V-0
Akzo	PET	Arnite AO4 102	×			
	PET	Arnite AO4 900	×			
	PET	Arnite AV2 360 S		33		×
	PET	Arnite AV2 370		36		
	PET	Arnite AV2 390		50		
	PET	Arnite AV4 340		20		
BASF	PET	Ultradur				
Hüls	PBT	Vestodur				
DuPont	PET	Rynite 530		30		
	PET	Rynite 545		45		
	PET	Rynite 555		55		
	PET	Rynite 935			35	
Eastman	Copolyester	Kodar PETG 6763	×			
Mobay	PET	Petlon 3530		30		
	PET	Petlon 3550		50		
	PET	Petlon 4530		30		×
Rohm & Haas	PET/Acrylic	Ropet 400		30		×
	PET/Acrylic	Ropet 500		30		

7.2.2 Properties and applications

Table 7-3 summarizes the properties of non-reinforced polyesters, PET (with small amounts of PBT) ®Arnite A04900, PBT Arnite T06200, and copolyester ®Kodar PETG 6763, glass reinforced polyesters PET ®Rynite with 30, 45, and 55% glass, PBT Arnite TV4261, and the PET/acrylic polymer blend ®Ropet 400, each containing 30% glass fibers [3−7].

PET possesses the highest heat deflection temperature and stiffness among the non-reinforced polyesters. PETG shows a greater notched impact resistance than PET and PBT; its heat deflection temperature is higher than that of PBT. PETG is amorphous and, in contrast to the partially crystalline PET and PBT, presents no problems in injection molding processing. However, the non-reinforced polyesters, including the copolyesters, do not offer enough stiffness for use as engineering plastics.

Glass fiber reinforced PET is superior to the corresponding PBT with regard to stiffness, strength, toughness, heat deflection temperature, flow, creep resistance, dimensional stability, abrasion resistance, and surface gloss. The mechanical properties and the heat deflection temperature increase with increasing glass content.

The advantages of glass reinforced PBT over PET, each containing 30% glass fibers, are the broader processing range, faster recrystallization, reduced cycle times, lower melting range, and higher dielectric strength and arc resistance; also it is easier to modify to reduce flammability.

All polyesters are susceptible to hydrolysis at the high processing temperatures and hence require rigorous drying prior to melt processing. The maximum moisture content of 0.005−0.02% is a commonly accepted limit. At an equivalent moisture content, PET hydrolyses more rapidly than PBT because of its higher processing temperatures.

Thin-walled objects manufactured from glass reinforced PET and PBT frequently show a tendency to pronounced warpage. This is overcome by partial replacement of glass fiber by mineral fillers, e.g., with mica in Rynite 935, but heat deflection temperature and notched impact resistance are hereby reduced. Another alternative is to use blends of PET with acrylic polymers, e.g., Ropet, which demonstrate high flowability and high stiffness at only a minor sacrifice in heat deflection temperature. Warpage is reduced [8] but

TABLE 7-3

Properties of non-reinforced and glass fiber reinforced poly(ethylene terephthalates) for injection molding.

Physical property	Physical unit	Non-reinforced polyesters			Glass-fiber reinforced polyesters				
		Arnite AO4 900	Arnite TO6 200	Kodar PETG 6763	Rynite 530	Rynite 545	Rynite 555	Arnite TV4 261	Ropet 400
Glass fiber content	%	0	0	0	30	45	55	30	30
Density	g/cm^3	1.38	1.30	1.27	1.56	1.69	1.80	1.52	1.53
Melt temperature	°C	255	223	—	254	254	255	223	260
Tensile strength at yield	MPa	81	52.0	48	—	—	—	134.4	—
Tensile strength at break	MPa	42	50.0	41	158	193	196	134.4	110.3
Elongation at yield	%	4	4	225	—	—	—	—	—
Elongation at break	%	70	200	—	2.7	2.1	1.6	3	2.7
Modulus of elasticity	GPa	2.80	2.60	—	9.9	14.6	—	9.0	—
Flexural strength	MPa	—	74.6	69	231	283	310	205	—
Flexural modulus	GPa	3.0	2.5	2.0	9.0	13.8	17.9	8.5	8.3
Notched impact strength	kJ/m	0.02	0.02	0.10	0.96	1.23	—	0.10	0.11
Hardness (Rockwell)	—	L107	L99	R105	R120	R120	—	L103	—
Linear expansion coefficient ($\times 10^5$)	K^{-1}	7.00	7.00	—	2.9	2.3	—	5.0	—
Vicat temperature	°C	235	210	82	—	—	—	215	—
Heat distortion temperature (1.81 MPa)	°C	70	55	63	224	226	229	205	218
Flammability	—	HB	HB	—	HB	HB	—	HB	—
Limiting oxygen index	%	25	23	—	20	20	—	20	—
Taber abrasion (CS 17)	mm/1000	10	18	—	5.7	6.8	—	45	—
Internal friction coefficient	—	0.24	0.24	—	0.28	0.17	—	0.22	—
Water absorption (24 h)	%	0.10	0.09	—	0.05	0.04	0.04	0.06	—

Property values

the notched impact resistance is lower compared to those of modified, glass reinforced PET types.

The modified and glass filled PET types, such as Rynite, are the engineering plastics with the greatest stiffness. They are used for parts demonstrating dimensional stability under load and at elevated temperatures. Reported applications for Rynite include down-sized electro-mechanical devices, miniaturized brake switch cases, spools, bobbins and relay bases, keyboards, precision parts for instrument gauges, deflectors in heat exchangers, and tape-driven window regulators in 1980 automobiles [5].

Competitive materials for PET are PBT and glass-reinforced grades of nylons, PC, PPO, and PSU, thermosets as well as zinc and aluminum, but PET is lower priced than most of these materials.

In 1980, the following prices were reported: PET Rynite 530 1.07 US-$/lb, Rynite 545 1.10 US-$/lb, PBT with 30% glass fiber 1.16 US-$/lb, and PBT with 40% glass fiber 1.26 US-$/lb.

In 1979, the consumption of PET injection molding material was still rather small; it reached 2000 t in West Europe and 5000 t in the U.S.; however, high annual increase rates of 15% have been predicted [2, 9]. U.S. production reached 46000 t in 1984.

References

[1] G. Forger, Plastics World (February 1980) 42.
[2] W. Fischer, J. Gehrke, and D. Rempel, Kunststoffe **70** (1980) 650.
[3] ®Arnite, Akzo Plastics, Technical Information.
[4] ®Kodar PETG 6763, Eastman Chemicals, Technical Information.
[5] ®Rynite, DuPont, Technical Information.
[6] ®Ropet, Rohm & Haas, Technical Information.
[7] ®Petlon, Mobay, Technical Information.
[8] N. L. Avery, K. E. Hansen, and W. J. Work, Paper presented at the SPE Nat. Techn. Conf., November 1979, Detroit, MI.
[9] Anon., Modern Plast. **56** (September 1979) 92.

7.3 POLY(BUTYLENE TEREPHTHALATE)

Poly(butylene terephthalate) PBT has been commercially available since the first half of the 70's and has enjoyed an exceptionally strong market development, in particular in the U.S. automotive industry. PBT has been reviewed in a number of publications [1-6]; therefore, only the more recent developments will be presented here.

The producers of PBT, the trade names, and the major product grades are summarized in Table 7-4. Many PBT types are copolymers. Usually, their compositions have not been disclosed.

The multitude of PBT producers induced some companies to withdraw from the over-crowded market. The period from 1977 saw the withdrawal of Allied (®Versel), DuPont, Diamond Shamrock, Eastman (only the reinforced grades), Goodyear, Hoechst (®Hostadur B), Hooker, and ICI.

Recently, development work on PBT has focussed on improvements of notched impact resistance, flame retardancy, anisotropic behavior of glass reinforced grades, and on attempts to raise the glass temperature.

High notched impact resistances are achieved through elasticization by blends or graft polymers. Warpage is reduced by partial substitution of glass fibers by glass microbeads, mica, or wollastonite. The effort to improve the troublesome processing of PET by addition of PBT has led to the market introduction of about 35 new PBT/PET hybrids.

Other suitable polymers for blending with PBT include polycarbonates, silicone-polycarbonate block copolymers, PTFE, polyamides, and thermoplastic elastomers based on styrene/dienes.

References

[1] J. R. Caldwell, W. J. Jackson, Jr., and T. F. Gray, Jr., in Encycl. Polym. Sci. Technol., Suppl. Vol. **1** (1976) 444.
[2] H.-G. Elias, Neue polymere Werkstoffe 1969–1974, Carl Hanser Verlag, München 1975; New Commercial Polymers 1969–1975, Gordon and Breach, New York 1977.
[3] F. Breitenfellner and J. Habermeier, Kunststoffe **66** (1976) 610.
[4] W. Fischer, J. Gehrke, and D. Rempel, Kunststoffe **70** (1980) 650.
[5] J. C. Weaver and T. F. Gray, Jr., SPE Techn. Paper **24** (1978) 292.
[6] Anon., Plastics World (February 1980) 42.

TABLE 7-4

Trade names and grades of poly(butyleneterephthalate) (January 1981).

Company	Trade Name	Without Fillers		with glass fibers (%)	With Fillers	
		without flame retardant	with Flame Retardant		with other fillers (%)	with flame retardant
Akzo	Arnite	X		20, 30, 35		
Amcel	Kelanex	X		15, 20, 30		
ATO	Orgater					
BASF	Ultradur	X				
Bayer	Pocan B	X		20, 30	Glass fibers/spheres 30	X
Celanese	Celanex	X	X	15, 18, 30		X
Ciba-Geigy	Crastine	X	X	20, 30	Glass fibers 20, 30	X
Dynamit-Nobel	PTMT			30		
Eastman	Tenite	X				
Fiberfil/Dart	Fiberfil					X
GAF	Gafite	X	X	15, 20, 26, 30	Glass fibers/spheres 26, 40	X
					Mica	X
GAF	Gaftuf	X		30		
General Electric	Valox	X	X	15, 30, 40	Mineral filled 10, 25	X
					Glass fiber/mineral 35, 45	
INP	Thermocomp			20, 30, 40	Carbon fiber 30	X
					PTFE 18, 20	
					Glass/PTFE 30/15	
					Silicone 2	
					Glass/PTFE/Silicate 30/13/2	
Mitsubishi Chemicals	PBT	X	X	15, 20, 30	Inorganic	X
Mitsubishi Rayon	Shinko-Lac	X	X	30		X
Mobay	TPE 2500	X	X			
Montedison	Pibiter	X		20, 30		X
Plastech Alloys	Plastalloy			20, 30, 40		X
SNIA	Snialen					

7.4 POLY(P-HYDROXYBENZOATES)

The older poly(p-hydroxybenzoates) have been extensively described in the literature [1−5]. Here, the developments since 1975 will be discussed.

®Ekonol is a homopolymer of p-hydroxybenzoic acid with the structural unit

$$-O-\!\!\!\bigcirc\!\!\!-CO-$$

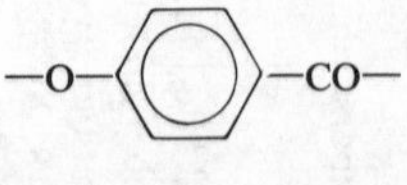

which is commercially available from Carborundum at 70 US-\$/lb. The polymer is used in compounds: Ekonol T 4000 and T 101 as blends of Ekonol with 60−70% PTFE, Ekonol E 4000 and D 6000 containing aluminum and bronze powders, and Ekonol D 5000 as an aqueous dispersion with PTFE. The flowability of the grades T 4000 and T 101 is too low for injection molding processing but allows compression molding and extrusion. Proven products are bearings, packings, O-rings, piston rings, and pump rotors. E 4000 and D 6000 are used by Metco, Inc., Westbury, for flame spray coatings. Glass fibers are impregnated with D 5000.

®Ekkzel C-1000 copolymer is based on p-hydroxybenzoic acid, isophthalic acid, and hydroquinone with the structural units

$$-O-\!\!\!\bigcirc\!\!\!-CO- \qquad \overset{-CO}{\underset{}{\bigcirc}}\!\!\overset{CO-}{} \qquad -O-\!\!\!\bigcirc\!\!\!-O-$$

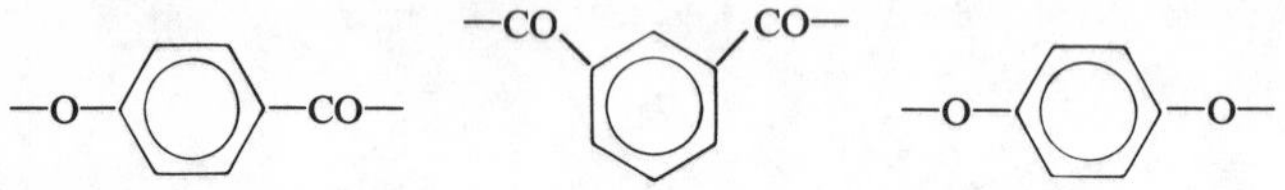

®Ekkzel I-2000 copolymer is based on p-hydroxybenzoic acid, terephthalic acid, and p,p′-dihydroxydiphenylether with the structural units

$$-O-\!\!\!\bigcirc\!\!\!-CO- \quad -OC-\!\!\!\bigcirc\!\!\!-CO- \quad -O-\!\!\!\bigcirc\!\!\!-\!\!\!\bigcirc\!\!\!-O-$$

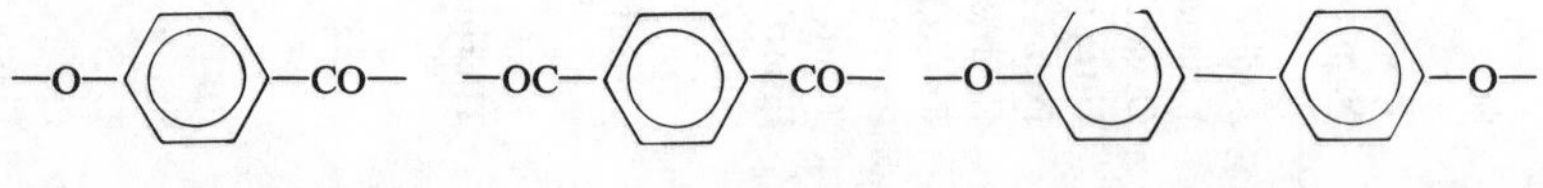

These copolymers and the reinforced grades CF-1006 (C-1000 with 30% glass fiber), CL-1150 (C-1000 with 25% PTFE), and IB 2006 (I-2000 with 30% glass beads) were produced by Carborundum until 1977. In 1978 the complete product line was sold to Dart which has been reported as planning a new pilot plant and introducing a new production process. The new products are said to have enhanced properties and lower prices. The prices last stated were 90 US-\$/lb for C-1000 and 30−40 US-\$/lb for I-2000. The copolymers exhibit a good balance of properties: long service life at elevated temperatures, good mechanical properties at temperatures above 280°C, low water absorption, irradiation resistance (10^8 J/kg), good electrical properties at elevated temperatures, and low flammability. The most obvious disadvantages are the brittleness of molded parts, the susceptibility to attack by strong alkali, acids, and hydraulic fluids, and the difficult processing which requires temperatures of almost 400°C in compression molding and the observance of a narrow temperature range between 370 and 385°C during injection molding. The molded parts suffer anisotropic shrinkage which can be avoided when the glass reinforced grades are used.

It remains to be seen if the poly(hydroxybenzoates) can stake a market share against the new and easier processable polyimides. Proven applications for the poly(hydroxybenzoates) include bearings, sealings, and piston rings. The Tribol Co., Santa Ana, California, supplies semi-finished goods. A major use is found in the manufacturing of printed circuits. Dart is considering the production of films and fibers.

A new polymer is the copolyester X7G developed by Eastman Chemical [6]. The synthesis is conducted in two steps: poly(ethylene terephthalate) with a molecular weight between 5000 and 80 000 g/mol is acidolyzed by p-acetobenzoic acid at 275°C. Segments containing terminal carboxyl and acetoxy groups are formed; in the final vacuum stage these condense generating acetic acid [7]:

(7-1)

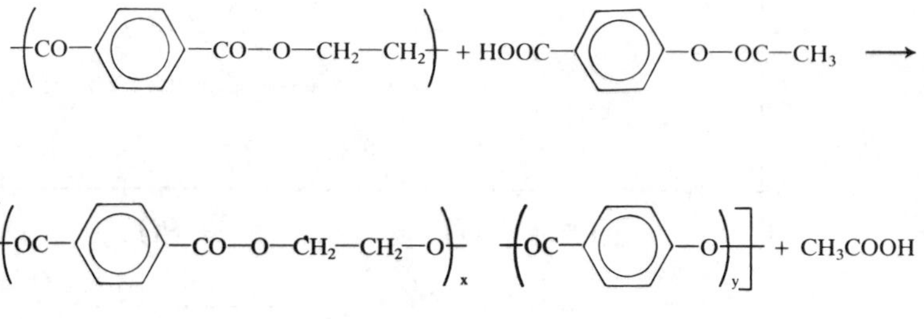

The amorphous copolyester is composed of about 60 mol-% p-oxybenzoyl and 40 mol-% terephthaloylethyleneglycol units. At this composition the melt viscosity reaches a minimum (Figure 7-1) and maximum values of flexural modulus, tensile strength, and notched impact resistance are obtained (Figure 7-2). Higher proportions of p-oxybenzoyl units result in polymers with higher heat deflection

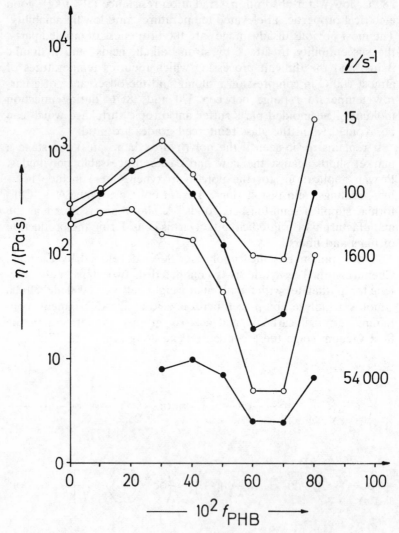

Figure 7-1 Melt viscosity at 275°C as function of the molar amount of p-hydroxy-benzoic acid units in copolyesters with terephthalic acid/ethylene glycol at various shear rates γ [7].

temperatures but the products tend to become insoluble and un-processable [7, 8].

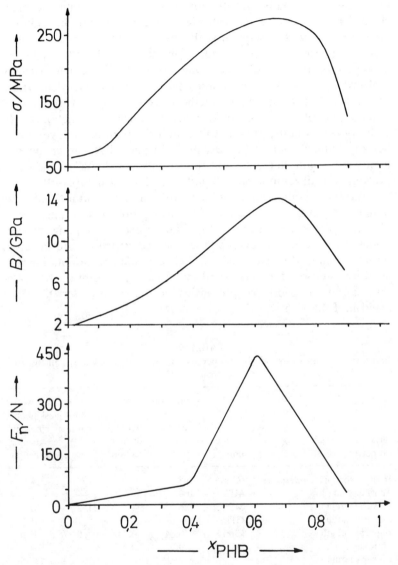

Figure 7-2 Tensile strengths at break, α, flexural moduli, B, and notched impact strengths F_n of various copolyesters from terephthalic acid and ethylene glycol with various mole fractions of p-hydroxy benzoic acid units [7].

The copolyester X7G displays a new physical structural principle. Mesomorphous structures are created through the alignment of long, flat, and relatively stiff molecular segments along the main chain. They cause a partial self-orientation of polymer chains in the direction of flow. Consequently, the melt viscosity decreases as such nematic phases are generated. Thermotropic liquid crystalline copolyesters are also obtained when terephthalic acid is partially replaced by 2,6-naphthalic dicarboxylic acid, p,p'-diphenylenedicarboxylic acid, or transcyclohexanedicarboxylic acid. Hydroquinone, other diphenols, or substituted hydroquinones can function as dihydroxy compounds, provided the linking group has an even number of chain elements. For example, bisphenol-A has an uneven number of chain elements, and this "break" or "flaw" in the chain prevents the development of liquid crystalline structures [9].

Contrary to compression molded parts, the injection molded parts obtained from the copolyester X7G display pronounced anisotropic physical properties (Table 7-5). The injection molding process forces the melt to orientate and the oriented structures are frozen in during cooling of the molded parts. The remaining orientation acts as fibrous self-reinforcement and increases the moduli of the material. Other non-isotropic properties of the copolyester X7G are listed in Table 7-6.

TABLE 7-5

Influence of anisotropy on the properties of injection molded specimens from polyester X7G.

Physical property	Physical unit	not reinforced		reinforced with 30% glass fibers	
		parallel	perpendicular	parallel	perpendicular
		to flow direction		to flow direction	
Heat distortion temperature	°C	65	55	79	74
Tensile strength	MPa	109	30	—	—
Flexural strength	MPa	126	34	152	76
Flexural modulus	GPa	14.0	1.76	14.2	4.4
Impact strength	kJ/m	0.84	0.24	0.27	0.16
Shrinkage	%	0.0	0.3	—	—
Linear thermal expansion coefficient ($\times 10^5$)	K^{-1}	0.0	4.5	—	—

TABLE 7-6
Other properties of polyester X7G.

Physical property	Physical unit	Property value
Density	g/cm^3	1.39
Heat distortion temperature		
0.45 MPa	°C	80
1.76 MPa	°C	65
Heat conductivity	W m^{-1} K^{-1}	0.22
Limiting oxygen index	%	30
Flammability UL 94		
thickness 0.318 cm	—	V-2
thickness 0.635 cm	—	V-1
Dissipation factor		
60 Hz	—	0.002
1 MHz	—	0.04
Relative permittivity		
60 Hz	—	3.8
0.1 MHz	—	3.7
Dielectric strength	V/25 μm	635
Arc resistance	s	103

Copolyesters with liquid crystalline structures can be processed to fibers which exhibit significantly higher tenacities and moduli after thermal treatment at temperatures below the melting point. In one example reported, tenacity rose from 44.1 to 144 km and tensile strength at break from 4.9 to 16 g/den [10]. Liquid crystalline polyesters are now also produced by Dart and by Celanese Engineering Resins.

References

[1] H.-G. Elias, Neue polymere Werkstoffe 1969−1974, Carl Hanser Verlag, München 1975; New Commercial Polymers 1969−1975, Gordon and Breach, New York 1977.
[2] K.-U. Bühler, Spezialplaste, Akademie-Verlag, Berlin 1978.
[3] J. Economy, R. S. Storm, V. I. Matkovich, S. G. Cottis, and B. E. Nowak, J. Polymer Sci.-Polymer Chem. Ed. **14** (1976) 2207.
[4] J. B. Titus, New Plastics, PLASTEC Report R48, Plastics Technical Evaluation Center, Picatinny Arsenal, Dover, N.J. 1977.
[5] R. S. Storm and S. G. Cottis, ACS Coatings and Plastics Preprints **34** (1974) 194.

[6] W. J. Jackson, Jr., H. F. Kuhfuss, and T. F. Gray, Jr., SPI Proc. Reinf. Plast. Comp. Inst., 30th Ann. Conf. (February 1975), Section 17 D, 1.

[7] H. W. Coover, Jr., T. F. Gray, Jr., and R. W. Seymour, Polimery **12** (1977) 393.

[8] R. W. Lenz and K. A. Feichtinger, ACS Polymer Preprints **20**/1 (1979) 114.

[9] H.-G. Elias, Polymer News **6** (1979) 51.

[10] DuPont, Belg, P. 828 935 (1975).

7.5   BARRIER POLYMERS BASED ON POLYESTERS

7.5.1   Structure and synthesis

Barrier polymers are polymers with extremely low permeabilities to gases, in particular to oxygen and carbon dioxide.

The first polymers with barrier properties were thermoplastics based on acrylonitrile. The syntheses, properties, and applications of nitrile resins as well as the physical principles of barrier materials have been extensively described [1–4]. After the FDA prohibited the use of nitrile resins for the manufacture of beverage bottles [5], Borg-Warner (®Cycopac 930), DuPont (®Vicabar), and Solvay (®Soltan S) terminated the production of their nitrile resins. Still on the market are the nitrile resins ®Barex of Vistron/Sohio, in Europe marketed by Lonza, and ®Cycle-Safe/®Lopac of Monsanto. Barex contains 90% of a copolymer (composed of 74% acrylonitrile and 26% methylacrylate) grafted onto 10% butadiene rubber. Cycle-Safe/Lopac is a copolymer of 70% acrylonitrile and 30% styrene. After the prohibition of nitrile resins, the polyester resins replaced them in the market. PET has the FDA approval for applications in contact with food. The FDA regulation 177.1630 is applicable for PET bottles.

1976 saw the first market introduction in the U.S. of stretch-blow molded, biaxially oriented beverage bottles. By 1980 use had extended to Europe and Australia. Beverage bottles require a special "bottle-grade" PET quality. The major producers of PET bottle-grades are: in the U.S., the companies Celanese, Eastman, Goodyear Tire & Rubber, Hoechst Fibers Industries, and Rohm & Haas; in Europe, the companies Akzo, Hoechst, ICI, Montedison, Rhône-Poulenc, and Snia-Viscosa.

The synthesis of PET bottle-grade is conducted according to known procedures from dimethylterephthalate or terephthalic acid and ethylene glycol with oxides of antimony, germanium, or titanium as catalysts. The use of GeO_2 leads to distinctly light colored products. The PET ®Cleartuf of Goodyear Tire & Rubber is produced without catalysts from terephthalic acid and ethylene glycol at 280°C. Terephthalic acid exerts an autocatalytic effect resulting in high reaction rates [6]. Celanese produces an amorphous PET copolymer. The incorporation of isophthalic acid, cyclohexanedimethanol, or neopentylglycol prevents the formation of crystalline regions [7]. Continuous melt polycondensations generally give polymers with superior thermal and thermo-oxidative resistances to polymers obtained from batch processes [8]. The melt polymerized pellets are post-polymerized in the solid phase at temperatures between 180 and 240°C. During this stage, purging with refined dry nitrogen and/or application of vacuum removes the volatile compounds. Only such procedures ensure the required low levels of acetaldehyde and oligomers. The level of carboxy groups is simultaneously reduced, thereby providing a PET with enhanced hydrolytic stability.

The acetaldehyde content is about 28 ppm after the melt polycondensation; it is minimized to 0.6 ppm during the solid phase post-polycondensation. Prolonged residence times at low temperatures are more favorable than short residence times at elevated temperatures. If the carrier gas contains oxygen as impurity or if one operates with a closed system at 200−240°C, acetaldehyde levels between 20 and 200 ppm are generated within a relatively short period of 1−3 h.

The cyclic trimeric terephthalate is the main oligomer formed during the polycondensation; linear oligomers, the bis(hydroxyethyl)terephthalate and the mono-hydroxyethylterephthalate are also found [6, 8]. The total oligomer content is 2.77% after melt polycondensation; the solid phase post-treatment reduces it to 1.36%. For this purpose, temperatures of 210−240°C and residence times between 3 and 6 h are required [8]. The reaction conditions must also take into account the thickness of the PET chips.

PET bottle-grades should have the lowest possible rate of crystallization. This is achieved with PET types having high molecular weights of about 25 000 (but not less than 23 000), increased levels of diethylene glycol, and the lowest possible degree of nucleation which is only obtained when minimum catalyst concentrations are used. Too high temperatures during the solid phase polycondensation result in polymers with high crystallinity; then the partially molten crystallites act as nucleating agents.

7.5.2   Properties, processing and applications

PET bottle-grades are commercially available with intrinsic viscosities, $[\eta]$, of 72–75, 85 and 98 ml/g. For example, Cleartuf has an intrinsic viscosity of 72 ml/g and a number average molecular weight of 24 000 g/mol [6]. The tendency goes to PET grades with increased molecular weights from which bottles can be manufactured exhibiting greater stiffness, reduced creep under internal pressure, enhance barrier characteristics, and lower weights [9].

Typical specifications of PET bottle-grade resins are listed in Table 7-7 [10]. In order to manufacture PET bottles with high strength, excellent transparency, and no impairment to taste, a number of parameters must be kept within narrow limits during processing.

TABLE 7-7
Specifications for bottle grade PET resins [10].

Physical property	Physical unit	Property values for		
		M 81	M 82	M 83
Intrinsic viscosity	mL/g	72 ± 2	80 ± 2	85 ± 2
Diethylene glycol	%	<1.0	<1.0	<1.0
Acetaldehyde	ppm	<2.0	<2.0	<2.0
Carboxylic end groups	mmol/kg	<20	<20	<20
Crystallinity	%	50	52	55
Density	g/cm³	1.4	—	—
Melt temperature	°C	260	260	260
Water content	%	<0.01	—	—

PET is supplied with a moisture content of less than 0.01%. It acts like a desiccant. Water absorption depends on the degree of PET crystallinity; values of 0.3 to 0.5% are reached after 20 to 40 days [8, 10]. Hydrolytic degradation of PET already starts at 150°C; the thermo-oxidative and the purely thermal degradations set in at only 200°C and 260°C, respectively. A mere 0.10% of water is sufficient to reduce the intrinsic viscosity by half at the melt temperature. Therefore, PET must be rigorously dried to moisture contents of 20–30 ppm. Drying is conducted at 150–175°C with hot and dry air (dew point $\leq$ −40°C), with a flow rate of 300 mm/s, at a residence time of 3 h at 175°C to 5 h at 150°C. A PET with an initial $[\eta]$ = 74 ml/g, containing 20 ppm moisture, reportedly suffered a degradation to $[\eta] = 72.5$ ml/g on melting; 50 ppm moisture caused a degradation to $[\eta] = 70$ ml/g and led to problems in processing [11]. Another benefit achieved by the rigorous drying is the reduction of the acetaldehyde content.

The manufacturing of PET bottles is conducted in two stages. First, PET is injection molded in water-cooled molds to produce clear, amorphous preforms. In the second stage, the preform is converted to the finished bottle by heating above T_g and blowing into the final mold. The stretch-blow technique gives a balanced biaxial orientation [10, 12, 13]. During processing, new acetaldehyde is generated. Starting with a resin containing 0.6 ppm acetaldehyde, levels of 5.8 ppm in the preform and 2.3 ppm in the bottle were detected [8]. The generation of acetaldehyde can be kept at a minimum when the preform is produced at the lowest possible melt temperature between 260 and $\leq 290°C$ at short residence times and little melt shearing. The tendency in developments goes to PET grades with reduced processing temperatures. Rhone-Poulenc has announced a PET which reportedly can be processed at temperatures which are $12-15°C$ lower than conventional PET grades [14]. Celanese developed a PET which can be processed with very little shear action [7].

Under the processing conditions mentioned above, the level of oligomers increases only slightly. Starting with a resin containing 1.36% oligomers, levels of 1.62% in the preform and 1.91% in the bottle were found [8]. Goodyear's Cleartuf contains only 0.062% which is extractable by methylene chloride [6]. For PET food packaging applications, the FDA has specified an upper limit of 0.05 mg/in^2, extractable with chloroform. Even at prolonged exposure over several days, only traces in the ppb range of oligomers, terephthalic acid, and ethylene glycol are extracted with hot water from PET bottles [6].

PET bottles with high transparencies are only obtained when the crystallization rate is reduced to a minimum. The crystallization rate decreases as molecular weight and diethylene glycol levels increase. The maximum rate is between 140 and 180°C [8, 10, 11]. At 170°C, PET is rendered cloudy within one minute and then turns to a white opaque appearance. At a temperature of 95° or 240°C, this phenomenon is observed only after hours [11]. Therefore, the molds for producing the preform should be cooled with coolants below 5°C, wall thickness should not exceed 4 mm, and the temperatures during the biaxial orientation must be kept between 90° and 120°C.

Bottles for carbonated beverages must comply with the following requirements: high strength in drop tests and under axial loads, satisfactory strength against internal pressures at 5 vol.-% or 11 g Co_2 per liter, maximum loss of 15% CO_2 over a period of 3 months, little creep or cold flow with a maximum of 3% volume change over a period of 3 months, high transparency, and approval for applications in contact with food.

In PET bottles, the mechanical and optical properties depend largely on the stretch-blow conditions and the degree of orientation hereby achieved [13, 15, 16]. Tensile strength, tensile modulus, and impact tensile strength increase as the stretch ratio increases (Table 7-8). Gas permeabilities are markedly reduced with increasing degree of orientation (see Table 7-9).

The barrier effect of PET towards CO_2 reaches only 10−20% compared to nitrile resins. Therefore, PET is primarily used in large bottle sizes where the greater wall thickness provides the necessary barrier characteristics. Furthermore, PET costs less than nitrile resins and has the additional advantage of greater toughness. In the 2-liter bottle size for carbonated beverages, PET controls a market share of 80−85% in the U.S. [17]. The most obvious advantages of PET bottles to glass bottles are the much lower weight (weight ratio of glass to PET is 13:1), far less fragility, and lower over-all energy consumption for raw materials, production, filling, transport, and recycling.

For the near future, good market opportunities are predicted for PET bottles in packaging of salad oils, mineral water, cider, wine, beer, liquors, foods, cosmetics, pharmaceuticals, and household chemicals. Competitive materials are, besides glass, the biaxially oriented transparent PVC [18] and PP bottles [19]. However, PP bottles show a high permeability to oxygen and are not suitable for packaging oxygen sensitive goods. An alternative exists in the co-extrusion of PP with barrier resins.

7.5.3 Recycling of PET bottles

The world-wide consumption of PET for containers could climb from about 200 000 t in 1980 to about 700 000 t in 1985 [20]. Anticipating a drastic increase in voluminous waste materials, some

TABLE 7-8

Properties of Nitrile and PET Resins with Barrier Properties (Company Literature of Lonza and Hoechst).

Physical property	Physical unit	Nitrile Resin Barex 210		PET Resin M 81	
		not oriented	oriented	not oriented	oriented
Density	g/cm^3	1.15		1.41	
Tensile strength at break	MPa	66		52	
Tensile strength at yield	MPa		138		83–200
Elongation at break	%	33	65	12	130
Modulus of elasticity	GPa	3.4			
Flexural modulus	MPa	99		67	125
Impact strength with notch	kJ/m	0.16		0.05	
Impact tensile strength	kN/m	301	831		
Average drop height (Staircase method)	m	2.18	>6.5		≥4
Heat distortion temperature	°C	74		>93	
Vicat temperature	°C	78			

TABLE 7-9
Gas Permeabilities of Barrier Resins [13].

Gas	Physical Unit	Nitrile Resins		PET		PP		PVC	
		Not oriented	oriented	Not oriented	oriented	Not oriented	oriented	Not oriented	oriented
Oxygen	$\dfrac{cm^3 \times mm}{m^2 \times d \times bar}$	0.3–0.4	0.15–0.25	4–4.5	1.8–3.5	80–90	35–45	6–7	3–3.5
Carbon	$\dfrac{cm^3 \times mm}{m^2 \times d \times bar}$	0.6–0.8	0.4–0.5	10–11	5–8	300–320	170–180	10–11	7.5–9
Water vapor (20°C: 85% RH)	$\dfrac{g \times mm}{m^2 \times d}$	0.5–0.8	0.3–0.5	3	0.9–2	0.10–0.15	0.04–0.07	0.8	0.6

countries have refrained from introducing PET bottles unless adequate recycling processes become available.

A typical 2-liter PET bottle consists of 63 g PET, 22 g HDPE base cup, 5 g paper label and glue, and 1 g aluminum used as screw cap. First, the plastic bottles are compressed to a density of at least 300 kg/m^3. The waste is ground or granulated and then the PET is separated by skimming, flotation, hydrocycloning or electric techniques [21].

Goodyear granulates whole bottles together with caps and separates paper and fine dust by air classification; the aluminum is recovered from an Eddy current separator. In a flotation stage using water, HDPE and EVAC (cap inner-linings) are separated from PET. At an annual plant capacity of 4500 t, the recycle costs, including purchase and transport of bottles, reportedly are only 0.40 US-$/kg, i.e., 0.035 US-$ per bottle [22].

The secondary products should not be reused in the manufacture of beverage bottles because of the higher acetaldehyde level. Suggested applications include injection molded articles, e.g., cases or hoops, and monofilaments for sewing threads, fabrics, fillings for pillows and sleeping bags, or floor tiles. Goodyear developed a composite material resembling wood which can be drilled, sawed, and nailed; another material presented was a corrugated sheet roofing [22].

Eastman dissolves PET waste in an excess of propylene glycol at 180°C. After the glycolysis the degraded product is esterified with maleic anhydride in the presence of manganese(II)acetate as catalyst. The obtained unsaturated polyester (UP) is dissolved in styrene and cured by conventional methods. The reaction time to prepare UP from PET is markedly shorter compared to that for conventional resins. The properties of glass fiber reinforced UP manufactured from PET are equivalent to the common UP laminates [23]. In the production of UP, phthalic and isophthalic acid could be replaced entirely by PET wastes.

References

[1] H.-G. Elias, Neue polymere Werkstoffe 1969−1974, C. Hanser Verlag, München 1975; New Commercial Polymers 1969−1975, Gordon and Breach, New York 1977; in Ullmanns Encykl. Techn. Chemie, vol. **15**, Verlag Chemie, Weinheim 1978.

[2] Barrier Polymers, in Encycl. Polym. Sci. Technol., suppl. vol. **1** (1976) 65.

[3] M. Salame and S. Steingieser, Barrier Polymers, Polym.-Plast. Technol. Eng. **8**(2) (1977) 155.

[4] Barrier Polymers, in Encycl. Chem. Technol., vol. **3** (1978) 480.

[5] Food and Drug Administration, final decision, in Federal Register, vol. **42** (September 23, 1977) 48528.

[6] E. N. Nowak, G. Cohn, D. D. Callander, and C. Bulman, SPE 38th ANTEC, New York (May 1980) 543.

[7] J. J. Baron, C. E. McChesny, and S. M. Sinker, 39th ANTEC, Boston (May 1981), SPE J. **27** (1981) 713; USP 4 182 841 (1980).

[8] G. Wick (Hoechst AG), Paper presented at the 1. Internat. Seminar on PET bottles, Antwerp, May 1980; Kunststoffe **70** (1980) 431.

[9] S. van Berkel (Akzo), Paper presented at the 1. Internat. Seminar on PET bottles, Antwerp, May 1980; Kunststoffe **70** (1980) 433.

[10] Hoechst Thermoplastic PET Resin, Technical Bulletin No. 2, American Hoechst Corp., Hoechst Fiber Industries.

[11] J. S. Schaul, SPE 38th ANTEC, New York (May 1980) 534.

[12] S. J. Szajna, Plastics Engineering, March 1981, 83.

[13] H. G. Fritz, Kunststoffe **71** (1981) 687.

[14] Anon., Modern Plast. Internat. **10** (March 1980) 22.

[15] C. Bonnenbat, G. Roullet, and A. J. deVries, 37th ANTEC, New Orleans 1979, 273.

[16] S. L. Kim, SPE ANTEC **27** (1981) 700.

[17] T. Adamski, Kunststoffe **70** (1980) 575.

[18] J. K. Presswood, SPE J. **27** (1981) 718.

[19] R. R. Shaffer, SPE J. **27** (1981) 722.

[20] A. D. Campion, Paper presented at the 2nd Internat. Seminar on PET bottles, Antwerp, May 1981.

[21] P. Klein, Paper presented at the 2nd Internat. Seminar on PET bottles, Antwerp, May 1981.

[22] J. S. Clements, SPE 38th ANTEC, New York (May 1980) 551; St. Stinson, Chem. & Engng. News (January 5, 1981) 30.

[23] R. Calendine, M. Palmer, and P. von Bramer, Modern Plast. Intern. **10** (June 1980) 34.

7.6 POLYARYLATES

7.6.1 Structure and synthesis

Polyarylates are polyesters of diphenols and aromatic dicarboxylic acids [1]. Copolyesters based on bisphenol-A, isophthalic acid and terephthalic acid were first prepared by V. Korshak [2] and A. Conix [3, 4].

At the end of 1975, Unitika, Ltd., marketed the first commercial polyarylate under the trade name ®U-Polymer [5, 6]. In the U.S. it is supplied by Union Carbide as ®Ardel and in Europe by Solvay under the name ®Arylef. Hooker Chemical holds a worldwide license from Gevaert on the patents of A. Conix. Pilot plant production started in 1978 and the product was designated ®Durel (formerly TP-4). Hooker Chemical was later acquired by Occidental Chemical; in 1984, Celanese bought Occidental's polyarylate business. Bayer introduced a polyarylate under the experimental code no. KL 1-9300 in 1978.

The U-Polymer is composed of terephthalic acid (I), isophthalic acid (II), and bisphenol-A (III) structural units in a molar ratio of 1:1:2. The composition of Durel is similar to the U-Polymer. The

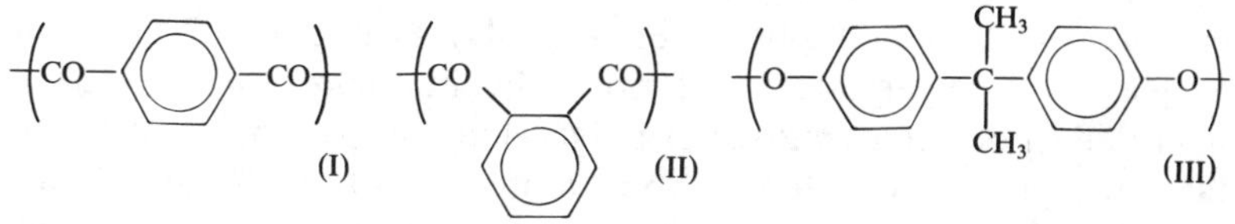

composition of Bayer's polyarylate has not been disclosed.

The polyarylates are presumably synthesized by interfacial polycondensation of the dicarboxylic acid chlorides, dissolved in chlorinated hydrocarbons, with an aqueous solution of the bisphenol-A alkali salts in presence of quaternary ammonium salts as catalysts [4].

Unitika and Solvay have marketed the basic grades U-100 and Arylef U-100, respectively; the modified grades U 1060, U 4015, U 8000, AX 1500, and AXN 1500 are also available [7, 8]. U 1060 contains 5% poly(ethylene) which imparts enhanced processability.

The AX series presumably contains small portions of nylon-6. Glass fiber reinforced grades are being developed. Union Carbide so far has marketed only the basic grade Ardel D-100. Besides the basic grade D-400 Hooker also offers the grades P 400, P 410, and P 430 as developmental products [9]. Bayer supplies a polyarylate with 30% glass fiber under the code KL 1-9301 [10, 11]. Polyarylates with 10, 20, 30, and 40% glass are commercially available from Fiberite in the USA under the designations RTP 2009 GF 10 to RTP 2009 GF 40.

7.6.2 Properties and applications

Polyarylates are engineering thermoplastics. They are amorphous polymers and combine toughness, very good flexural recovery, low notch sensitivity, high heat deflection temperature, high dimensional stability and retention of mechanical properties at elevated temperatures up to 140°C, good resistance to UV, inherent flame retardancy, low smoke toxicity, a very high autoignition temperature (545°C), and good electrical properties. Table 7-10 shows the properties of polyarylate basic grades compared to a glass reinforced grade (Bayer KL 1-9301).

The disadvantages of polyarylates are the inherent deep yellow color, sensitivity to stress cracking corrosion, the susceptibility to hydrolysis, and the relatively difficult processing.

The property spectrum of polyarylates resembles that of polycarbonates (PC) and polysulfones (PSU) (see Table 7-11). The toughness of polyarylates is inferior to PC but better than PSU. The heat deflection temperatures of polyarylates surpass those of PC by far; the upper service temperature limits of 140−150°C are similar to those of PSU. The electrical properties of polyarylates are equivalent to those of PC. Polyarylates, PC, and PSU are equally sensitive to stress cracking corrosion. The hydrolytic stability of PSU is greater than that of PC and polyarylates; however, modified polyarylate grades such as Arylef U 1060, U 4015, and in particular U 8000 show enhanced resistance to hydrolysis [8].

Polyarylates can be processed by injection molding and extrusion techniques. Polyarylates must be dried completely to less than 0.02% moisture prior to melt processing since the high melt temperatures of 320−370°C would cause severe hydrolysis. The melt viscosity of polyarylates is very high; small portions of crystalline polymers (PE, PA) improve the flowability but simultaneously reduce the heat deflection temperature slightly. It is recommended the mold temperatures be kept at 40−100°C. Reground waste can be used for many applications.

Polyarylates are suitable for a variety of outdoor applications, including solar energy collectors, tinted glazing, and safety glasses with high scratch resistance. Other uses are electronic and electrical hardware, flexible snap-fit applications such as electrical connectors, fasteners, hinges and springs, and thermally stable search light reflectors.

TABLE 7-10
Properties of Polyarylates.

Physical Property	Physical Unit	Ardel D-100 Arylef U-100	Durel D-400	Bayer KL 1-9300	Bayer KL 1-9301
Density	g/cm³	1.21	1.20	1.21	1.44
Glass temperature	°C	173	—	188	186
Tensile strength	MPa	66	69	62	108
Elongation at break	%	8	8	9	3.5
Elongation at yield	%	50	—	40	3.9
Modulus of elasticity	MPa	2000	2150	2100	6900
Flexural strength	MPa	81	103	62	66
Flexural modulus	MPa	2000	2206	2300	7800
Impact strength	kJ/m²	260–360	215*	no fracture	40
Impact strength with notch	J/m	220	—	250	8
Heat distortion temperature	°C	175	160	164	183
Dielectric strength	kV/mm	30	17	47	—
Relative permittivity (1 MHz)	—	2.62	2.96	3.2	—
Dissipation factor (1 MHz)	—	0.02	0.02	17.4	—
Volume resistivity	ohm × cm	3×10^{16}	2×10^{16}	10^{16}	—
Surface resistivity	ohm	$>2 \times 10^{17}$	—	10^{13}	—
Flammability (UL 94)	—	V-0	V-0	V-2	V-0
Limiting oxygen index	%	34	38	36	—

* after 3 months at 100°C; no fracture without aging.

TABLE 7-11

Mechanical properties of amorphous heat-resistant thermoplastics [11].

Physical property	Physical unit	Test method	Property Value			
			APE KL 1-9300	Polysulfone	Polyethersulfone	Polycarbonate
Impact strength	kJ/m²	DIN 53453	no break	no break	no break	no break
Impact strength with notch	kJ/m²	DIN 53453	22	5	4	50
ditto, Izod 3.2 mm	J/m	ISO-R 180	280	55	62	870
Flexural strength	MPa	DIN 53452	62	83	88	95
Flexural modulus	MPa	DIN 53457	2300	3200	3300	2200
Tensile strength at yield	MPa	DIN 53455	70	78	96	55
Elongation at yield	%	DIN 53455	9	5.4	6.7	6
Tensile strength at break	MPa	DIN 53455	62	55	69	65
Elongation at break	%	DIN 53455	56	76	26	110
Modulus of elasticity	MPa	DIN 53457	2100	2700	2800	2300
Hardness (Rockwell)	N/mm²	DIN 53456	110	130	150	110
Density	g/cm³	DIN 53479	1.2	1.2	1.4	1.2

In its fields of applications the polyarylates compete with PC and PSU. In services requiring high heat deflection temperatures where PC fails, polyarylates are used, benefited by their cost advantage over PSU.

Prices for polyarylate basic grades are 2.40 US-$/lb, for PC 1.23 US-$/lb and for polysulfones between 2.95 and 3.50 US-$/lb.

References

[1] G. Bier, Polymer **15** (1974) 527.
[2] V. Korshak, Dokl. Akad. Nauk SSSR **156** (1964) 880.
[3] A. Conix, Ind. Chim. Belg. **22** (1957) 1457.
[4] A. Conix (Gevaert NV), Brit. P. 901 605, USP 3 216 970.
[5] K. Hazman, Jap. Plastics **8**/3 (1974) 6; Jap. Plastics Age **12**/10 (October 1974).
[6] H. Sakara, SPE Ann. Techn. Papers **20** (1974) 459.
[7] ®Arylef, Solvay, Technical Information.
[8] A. M. Orban, Paper presented in Gothemburg, November 1980; Anon., Kunststoffe **70** (1980) 856.
[9] ®Durel, Hooker Chemical, Technical Information.
[10] KL 1-9300 and KL 1-9301, Bayer AG, Technical Information.
[11] D. Freitag and K. Reinking, Kunststoffe **71** (1981) 46.

7.7 POLY(CO-CARBONATES) AND POLY(ESTER-CO-CARBONATES)

7.7.1 Structure and synthesis

Polycarbonates of bisphenol-A (I) and phosgene (II) have been commercially available since 1958 by Bayer as ®Makrolon and since 1973 by General Electric as ®Lexan. They were later followed by some Japanese companies and also by Anic (1978) and ATO Chimie (1979). Dow Chemical is reported to be ready for the introduction of polycarbonate (PC).

The recent developments in PC have been reviewed in several publications [1, 2]. Development work focuses on improvements in notched impact resistance of thick-walled parts, increase of low temperature impact resistance, reduced flammability, higher heat deflection temperatures and enhanced melt stability. Impact and notched impact resistances are improved by PC/ABS blends; such products were introduced to the market in 1976 by Bayer as ®Bayblend and by Borg-Warner as ®Cycoloy [2]. Copolymers of I and II with halogenated bisphenols, in particular with tetrabromo-bisphenol A (III), show reduced flammability. There also exist bromine free PC types which meet the UL 94 VO classification.

The polycondensation of phosgene with 3,3′,5,5′-tetramethyl-4,4′-dihydroxydiphenyl-2,2-propane or tetrabromo-bisphenol A (IV) yields polycarbonates with higher heat deflection temperatures. Samples have become available for testing [2]. Improved notched impact resistance is also achieved with poly(co-carbonates) prepared from I, II and 4,4′-dihydroxydiphenyl-sulfide (V) which have been commercially available since 1980 from Mobay under the trade name ®Merlon T Resins [3, 4]. A polymer based on bisphenol-A with high heat deflection temperature and inherent flame-retardancy was announced by General Electric under the preliminary designation X-76 [5].

Structurally related to polycarbonates are the poly(ester-carbonates) prepared from I, II and VI. Mitsubishi Chemical Industries was reported planning a pilot plant in 1981 [6]. Allied Chemical [7] considers its products still as experimental. General Electric has recently introduced a "polyphthalate carbonate" under the trade name High Heat Lexan® PPC [8, 9].

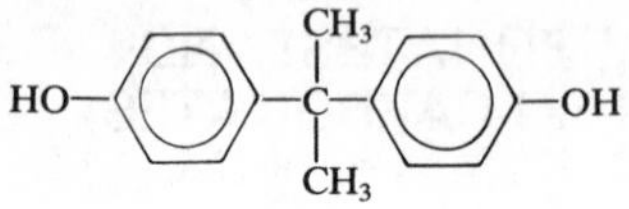

$$HO-\bigcirc-\underset{\underset{CH_3}{|}}{\overset{\overset{CH_3}{|}}{C}}-\bigcirc-OH \qquad\qquad I$$

$$COCl_2 \qquad\qquad II$$

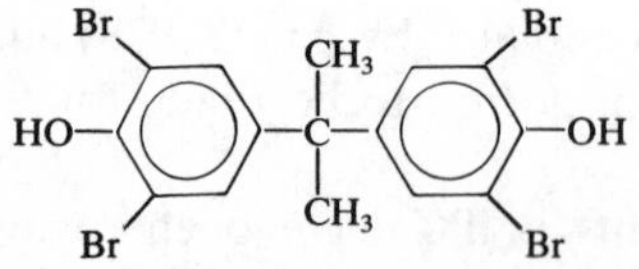

$$HO-\bigcirc-\underset{\underset{CH_3}{|}}{\overset{\overset{CH_3}{|}}{C}}-\bigcirc-OH \qquad\qquad III$$
(Br substituents on both rings)

$$HO-\bigcirc-\underset{\underset{CH_3}{|}}{\overset{\overset{CH_3}{|}}{C}}-\bigcirc-OH \qquad\qquad IV$$
(CH₃ substituents on both rings)

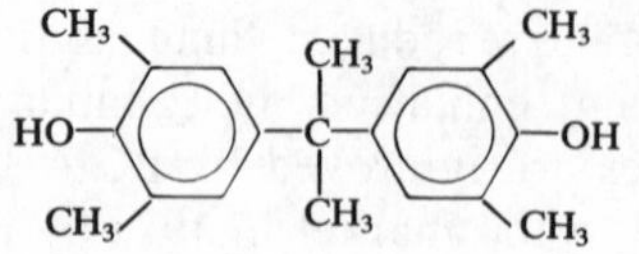

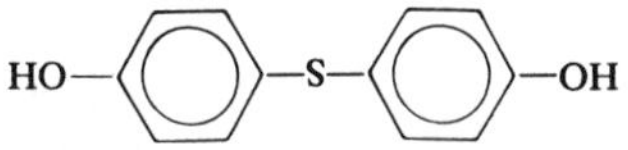

$$HO-\langle\bigcirc\rangle-S-\langle\bigcirc\rangle-OH \qquad V$$

$$HOOC-\langle\bigcirc\rangle-COOH \qquad VI$$

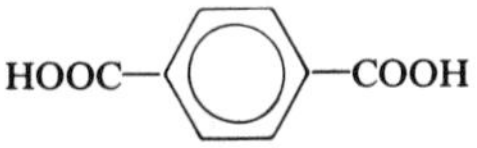

The preparation of PC has been extensively described [10]; thus only the syntheses of the recent poly(co-carbonates) and poly(ester-co-carbonates) will be dealt with here. Whereas PC is produced by a continuous interfacial polycondensation between phosgene dissolved in methylene chloride and the bisphenol-A alkali salt dissolved in water, the batch-wise production of specialty polycarbonates is conducted by transesterification of diphenyl carbonate with diphenols:

(7-2)

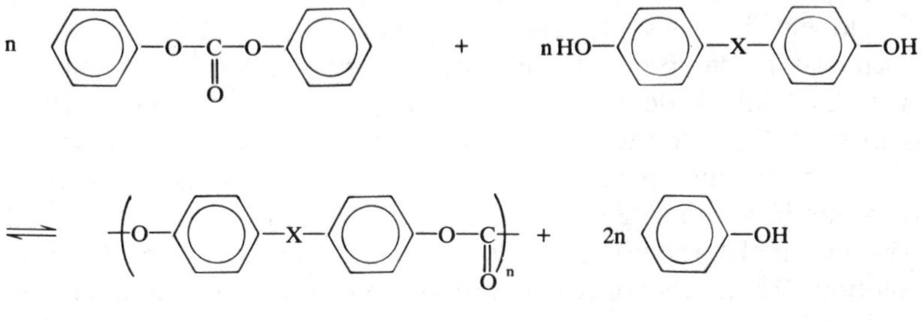

$$n\ \langle\bigcirc\rangle-O-\underset{\underset{O}{\|}}{C}-O-\langle\bigcirc\rangle \quad + \quad n\,HO-\langle\bigcirc\rangle-X-\langle\bigcirc\rangle-OH$$

$$\rightleftharpoons\ \left(O-\langle\bigcirc\rangle-X-\langle\bigcirc\rangle-O-\underset{\underset{O}{\|}}{C}\right)_n \quad + \quad 2n\ \langle\bigcirc\rangle-OH$$

Mobay studied a series of polycarbonates prepared from various diphenols where X represented an alkylene, ether, sulfide, or sulfone group [4]. Random copolymers were obtained with diphenols of equivalent reactivity. Block co- and terpolymers were prepared by mixing homo-prepolymers containing chlorocarbonate terminal groups, with subsequent condensation to high molecular weight products.

The first poly(ester-co-carbonate) (PEC) was synthesized by General Electric [11]. PEC is obtained when phosgene is added to a vigorously stirred solution of diphenols and free dicarboxylic acids:

(7-3)

$$HO-\langle\bigcirc\rangle-X-\langle\bigcirc\rangle-OH + R(COOH)_2 + COCl_2 \xrightarrow[25-30^{\circ}C]{Pyridine}$$

$$\left[\!\!\left[O-\langle\bigcirc\rangle-X-\langle\bigcirc\rangle-O-\overset{\overset{O}{\|}}{C}\right]_x\!\!\left[O-\langle\bigcirc\rangle-X-\langle\bigcirc\rangle-O-\overset{\overset{O}{\|}}{C}-R-\overset{\overset{O}{\|}}{C}\right]_y\right]_n$$

Reportedly, the x/y ratio in the PEC copolymers is easily regulated via the molar ratio of diphenol/dicarboxylic acid. Random and block copolymers were claimed with compositions depending on the sequence of monomer feeds.

Allied investigated the reaction between I, II, and VI in detail [12]. It was found that PEC prepared by the GE method [11] consisted of long bisphenol-A/PC segments and of short bisphenol-A/terephthalate segments. Thus, the properties of such copolymers should be equivalent to those of a pure bisphenol-A based PC. Furthermore, the products were found to contain impurities, such as free acid, anhydride, and terminal terephthalic acid groups, which above 250°C cause melt discoloration and melt viscosity reduction.

However, random copolymers without anhydride groups can be obtained in a two stage reaction [12]. First, 2 mols of I are reacted with one mol terephthaloyl chloride in pyridine/methylene chloride solution. Then, the obtained prepolymer containing terminal hydroxy groups is reacted with II. Technical useful PECs (Figure 7-3, the area below the curve) are formed only under certain reaction conditions. Molecular weight is controlled by p-tert-butylphenol. For injection molding processing, PEC's with reduced viscosities of about 70 ml/g are best. Parts compression molded at 320°C show only slight degradation and discoloration. Glass transition temperatures increase as the portion of terephthalic acid units increases (Figure 7-4); they also depend on the number average molecular weight according to

(7-4)

$$T_G = T_G^{\infty} - (A/\overline{M}_n)$$

with $T_G^\infty = 195°C$ and $A = 2.39 \times 10^5$ g °C mol^{-1}. The $\overline{M}_w/\overline{M}_n$ ratios of PEC with reduced viscosities of 170 ml/g and about 500 ml/g are ca. 2 and 2.5, respectively.

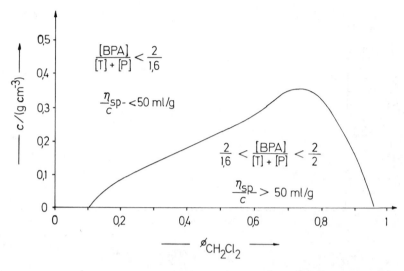

Figure 7-3 Influence of the concentration, c, of the reacting monomers as well as the volume fraction $\phi_{CH_2Cl_2}$ of methylene chloride in methylene chloride/pyridine mixtures on the reduced viscosities of the resulting polyestercarbonates [12]. [BPA], [T], [P] = molar concentrations of bisphenol A, terephthaloylchloride, and phosgene, respectively.

7.7.2 Properties and applications

Polycarbonates are engineering plastics and exhibit a good balance of properties: transparency, high stiffness, good dimensional stability, wide use temperature limits, high creep resistance, good electrical properties, and some flame retardancy. However, they have also obvious disadvantages as illustrated in Table 7-12 where properties of PC, PC/ABS, poly(co-carbonate), and poly(ester-co-carbonate) are compared.

The notch sensitivity of thick-walled PC parts especially at low temperatures can be reduced by blending with ABS but heat and weathering resistance must be partially sacrificed. Poly(co-carbonates) based on bisphenol-S (V) demonstrate further improvements: the notched impact resistance at low temperatures is superior

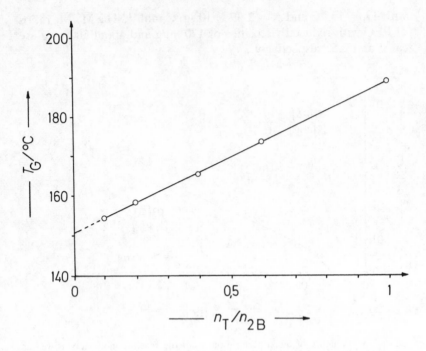

Figure 7-4   Glass transition temperature of polyestercarbonates of approximately the same molar mass as function of the ratio, n_T/n_{2B}, of terephthaloyl units to two times the bisphenol A units [12].

to that of PC or impact PC and nylon grades (Figure 7-5), with little or no sacrifice in heat deflection temperature (Table 7-12) or in transparency. ®Merlon poly(co-carbonates) are commercially available in three grades (T 4610, T 4530, and T 4340) which differ in comonomer ratio, molecular weight, and flowability. The impact resistance increases as the portion of bisphenol-S increases. The notorious notch sensitivity of PC is exhibited in notched impact tests when specimen wall thickness and notch apex radius are varied. In contrast to poly(co-carbonates), PC suffers a catastrophic loss in notched impact resistance at a specimen wall thickness of about 5 mm (see Figure 7-6). PEC is equally non-sensitive to notch effects as are the poly(co-carbonates) [13] and has the additional benefit of higher heat deflection temperatures compared to PC and the poly-(co-carbonates). On thermal aging, PEC shows a lower tendency to embrittlement than PC.

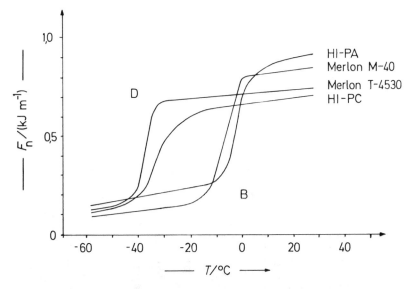

Figure 7-5 Notched impact strength, F_n, of two Merlon grades at various temperatures as compared to high impact polyamide (HI-PA) and high impact polycarbonate (HI-PC) [4]. D = deformational fracture, B = brittle fracture. Test: ASTM D 256 with 0.254 mm notch.

The highest heat deflection temperature for the polymers of this category is claimed by the new X 76 developed by the New Venture Group of General Electric [5]. The product is based on I; it is amorphous, transparent, and has a heat deflection temperature of 200°C. It displays good tensile strength but the impact resistance is slightly lower compared to PC; it is inherently flame retardant with UL 94 V-0 rating; it develops little smoke in combustion tests with low smoke toxicity. General Electric also develops a poly(ester-co-carbonate); properties, except the heat deflection temperature of 163°C [14], have not been disclosed.

PC based on IV is also thermally stable and has a Vicat temperature of 196°C; however, impact and notched impact resistances are markedly reduced compared to PC. Resistance to hydrolysis is higher but stress corrosion resistance is lower than in PC. Blends of PC based on IV with styrene polymers were commercially introduced in 1980 and are a supplement to the PC and PC/ABS product lines. The advantages of the blends over PC/ABS blends are the improved hydrolytic resistance, lower density, and higher heat de-

TABLE 7-12
Physical Properties of Polycarbonate, PC/ABS Blend, Polycocarbonate, and Polyester-co-carbonate.

Physical Property	Physical Unit	Polycarbonate Mobay Merlon M40	PC/ABS Mobay Bayblend MC-2500	Property values of	
				Polycocarbonate Mobay Merlon T 4530	Polyester-co-carbonate [2]
Density	g/cm³	1.199	1.120	1.20	—
Tensile strength at yield	MPa	67.7	58.6	61	66
Tensile strength at break	MPa	70.3	50.3	66	61
Elongation at yield	%	8	5	7	—
Elongation at break	%	120	15	115	62
Modulus of elasticity	GPa	2.3	2.62	2.5	2.10
Flexural strength	MPa	86.2	89.6	83	—
Flexural modulus	GPa	2.1	2.41	2.3	—
Impact strength with notch	kJ/m	0.12	0.56	0.87	*
Hardness (Rockwell)	—	M62	R117	M66	—
Vicat temperature	°C	—	—	—	—
Heat distortion temperature (1.81 MPa)	°C	136	116	126	—

* Impact strength with notch 0.43 kJ/m (0.1 inch thickness) and 0.32 kJ/m (0.55 inch thickness) [13].

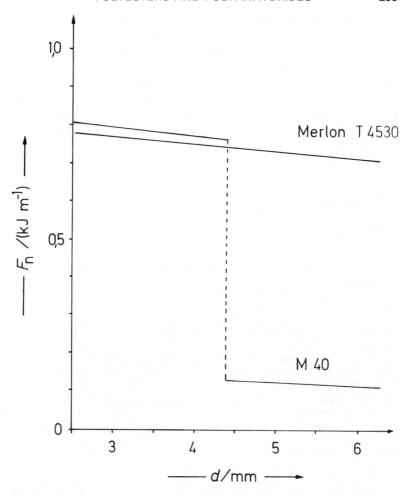

Figure 7-6 Dependence of notched impact strength of two Merlon grades on specimen thickness [4].

flection temperatures [2].

Recommended applications for the new products based on IV-PC are dishes for microwave ovens and head light reflectors [2]. The latter demonstrate high dimensional stability, low coefficient of thermal expansion, high heat deflection temperature, and good metal plating characteristics.

Primary applications of poly(co-carbonates) based on I and V will be lenses and glazings. They have greater refractive indices and Abbé coefficients than PC, poly(styrene), and poly(methylmeth-acrylate) [2]. Due to their high impact strength these products are better suited than PC, PMMA, or glass in applications where a high risk of fracture exists, e.g., in automobile lights and hazard warning flashers for construction or accident sites. Another recommended use is for the highly strained base plates of swivel office chairs. Such base plates are manufactured with decorative embossings which act as a multitude of notches.

References

[1] J. Bussink, Kunststoffe **66** (1976) 600.
[2] P. R. Müller, Kunststoffe **70** (1980) 636.
[3] ®Merlon T, Mobay, Technical Information.
[4] M. W. Witman, J. R. Thomas, S. Krishnan, and A. L. Baron, SPE Ann. Techn. Papers **26** (1980) 470.
[5] Anon., Modern Plast. **55** (November 1978) 14.
[6] Anon., Plastics Industry News **8** (1980) 114.
[7] L. Segal, Allied Chemical, personal communication, April 23, 1981.
[8] Anon., Modern Plastics Internat. (August 1984) 50.
[9] General Electric, company information.
[10] L. Bottenbruch, Encycl. Polym. Sci. Technol., vol. **10** (1969) 710.
[11] E. P. Goldberg, Belg. P. 570 530 and 570 531 (1958); USP 3 030 331 (1962) and USP 3 169 121 (1965).
[12] D. C. Prevorsek, B. T. Debona, and Y. Kesten, J. Polym. Sci.-Polym. Chem. Ed. **18** (1980) 75.
[13] D. C. Prevorsek, Y. Kesten, and B. Debona, ACS Polymer Preprints **20**/1 (1979) 187.
[14] Modern Plast. Intern. **9** (November 1979) 72.

7.8 POLY(GLUTARATES)

7.8.1 Structure and synthesis

Aliphatic polyesters of glycols and dicarboxylic acids, in particular with adipic or sebacid acid, are known as polymeric plasticizers for PVC. Their advantages to low molecular weight plasticizers are the low volatility and migration as well as the resistance to extraction by hydrocarbons. The disadvantages are the obstacles encountered in compounding with PVC, the lower plasticizing effect, and higher prices.

Considerable quantities of glutaric acid are obtained as a by-product in the production of adipic acid in the cyclohexane oxidation process [1]. Adipic acid yields cyclopentanone via a Dieckmann reaction which is then oxidized to glutaric acid:

(7-5)

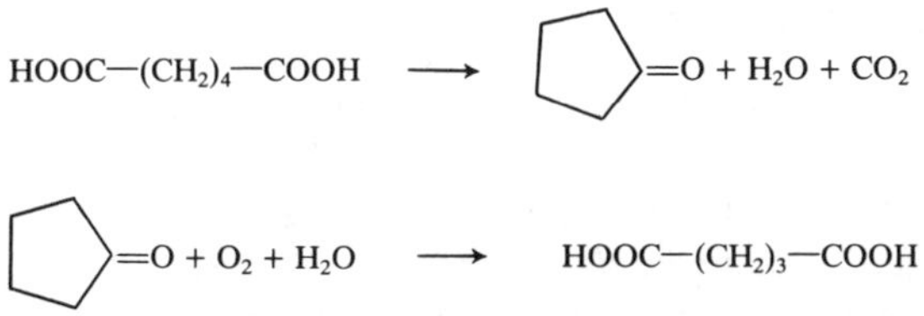

$$HOOC-(CH_2)_4-COOH \longrightarrow \bigcirc\!\!\!=O + H_2O + CO_2$$

$$\bigcirc\!\!\!=O + O_2 + H_2O \longrightarrow HOOC-(CH_2)_3-COOH$$

The C. P. Hall Co. has marketed viscous poly(glutarates) under the trade name ®Plasthall P-530, P-540, and P-550 with molecular weights between 1500 and 5000 g/mol [2, 3]. The poly(glutarates) are prepared by polycondensation of dimethyl glutarate with various glycols. The nature of the glycols has not been disclosed. Fatty acids or monofunctional alcohols are used to control the molecular weight [4].

7.8.2 Properties and applications

The eight poly(glutarate) grades supplied by C. P. Hall are highly viscous liquids with viscosities of 0.84−24 Pa × s at 25°C, saponification numbers of 333−559, acid numbers of 0.3−1.0, Gardner color values 4−9, volatiles (as weight losses at 155°C after 24 h) 1.4−5.4%, flash points 202−305°C, and freezing points −12 to −41°C [3].

Compared to poly(adipates), the poly(glutarates) impart to PVC superior low temperature properties and improved dry-blending characteristics. The medium-viscous Plasthall P-550 demonstrated a markedly reduced migration compared to poly(adipates) when heated at 70°C for 600 h [4]. The poly(glutarates) are recommended as plasticizers for PVC, adhesives, and synthetic rubbers.

References

[1] N. P. Chopey, Chem. Eng. **9** (1961) 70.

[2] Anon., Modern Plast. Intern. **9** (December 1979) 41.

[3] ®Plasthall, C. P. Hall Co., Technical Information.

[4] J. L. O'Brien, W. H. Whittington, and G. Chalfant, Plastics Compounding **4** (May/June 1981) 62.

7.9   POLY(AZELAIC ACID ANHYDRIDE)

7.9.1   Structure and synthesis

Aliphatic poly(carbonic acid anhydrides) have long been known but until recently had never found any practical use [1].

Emery Industries Inc. is now marketing a poly(azelaic acid anhydride) under the trade name ®Emery 9872 (PAPA) [2]. It is a linear polyanhydride with terminal carboxy groups:

$$HOOC(CH_2)_7COO\left[CO(CH_2)_7COO\right]H$$

Another producer is Anhydrides & Chemicals, Inc., which offers also poly(adipic acid anhydride) and poly(sebacic acid anhydride).

The synthesis of aliphatic poly(carbonic acid anhydrides) is conducted by heating the free acids with acetic anhydride or acetyl chloride under reflux [1].

7.9.2.   Properties and applications

Emery 9872 (PAPA) contains 36.6% anhydride groups and 6.5% carboxylic groups; it melts at 54°C and the melt viscosity at 98.9°C is 200 mPa × s. The material is soluble in toluene, xylene, methylethylketone, and methylisobutylketone. Solvents containing water cause hydrolysis.

PAPA is used as hardener in aromatic and cycloaliphatic epoxy resins. The recommended equivalent ratio of anhydride to epoxy is 0.8:1.0; the conventional liquid epoxy resins with an epoxy equivalent of 100 require about 70 phr PAPA. Liquid curing systems are obtained by mixing PAPA with Nadic-methylanhydride, dodecenyl succinic acid anhydride, or hexahydrophthalic acid anhydride (HHPA). A mixture of 35% PAPA and 65% HHPA forms an eutectic with a melting point of 10°C.

Epoxy resins cured with PAPA show improved flexibility and

toughness, resistance to thermal shocks (between −55 and 155°C), high dimensional stability under heat, and excellent electrical properties at elevated temperatures. To cured epoxy resins, PAPA in mixtures with hexahydrophthalic acids imparts improved elongation; in mixtures with Empol 1040 Trimeric Acid enhanced tensile strength is achieved.

The good solubility of PAPA in low-priced solvents makes it a candidate for the production of reactive laminates, prepregs, and adhesive tapes.

References

[1] R. Wegler, Poly(carbonsäureanhydride), in Houben-Weyl, Methoden der organischen Chemie, Band XIV/2, Makromolekulare Stoffe, Teil 2, Georg Thieme Verlag, Stuttgart 1963.
[2] ®Emery 9872 (PAPA), Emery Industries Inc., Technical Bulletin 104 A, November 1980.

8 Sulfur containing polymers

8.1 POLY(THIODIETHANOL)

8.1.1 Structure and synthesis

The American Cyanamid Co. produces pilot plant quantities of poly(thiodiethanol) under the name ®Cymax Polyester-Elastomer [1]. The starting material is thiodiethanol (I), also known as thioglycol or bis(2-hydroxyethyl)sulfide, which is obtained under anhydrous conditions from 2 moles ethylene oxide and 1 mol hydrogen sulfide (eq. 8-1). At elevated temperatures, I undergoes an acid catalyzed polycondensation to yield polythiodiethanol) (II) (eq. 8-2). The water is removed under vacuum.

(8-1)

$$H_2S + 2\ CH_2{-}CH_2 \longrightarrow HO{-}CH_2CH_2{-}S{-}CH_2CH_2{-}OH$$

(I)

(8-2)

$$n\ HO{-}CH_2CH_2{-}S{-}CH_2CH_2{-}OH \longrightarrow HO{\left(CH_2CH_2{-}S{-}CH_2CH_2O\right)}_n H + (2n - 1)\ H_2O$$

(II)

Due to its ordered structure one should expect polymer II to be crystalline. For example, poly(methylene sulfide) and poly(ethylene sulfide) are crystalline polymers with high melting points [2]. However, if small portions of other glycols are used together with I in the polycondensation, the crystallinity of II is largely suppressed. Moreover, another monomer is added for the purpose of introducing unsaturated pendant groups. ®Cymax is a quaterpolymer; the nature of the other monomers has not been disclosed.

8.1.2 Properties and applications

Cymax can be vulcanized with the conventional system of sulfur and zinc mercaptobenzothiazole. Improved aging characteristics are achieved by a new system of sulfur and polyamines without zinc.

TABLE 8-1
Properties of Vulcanizates of ®Cymax Poly(thiodiethanol).

Physical property	Physical Unit	Property value
Density	g/cm³	1.35
Tensile strength at break	MPa	11.7
ditto, after 70 h at 150°C	MPa	9.6
Elongation at break	%	250
ditto, after 70 h at 150°C	%	130
Hardness (Shore A)	—	70
ditto, after 70 h at 150°C	—	80
Tear strength (ASTM D 624 C)	kN/m	52.5
Compression set*		
after 22 h at 120°C	%	27
after 70 h at 150°C	%	49
Resilience (ASTM D 2632)	%	35
Abrasion resistance (ASTM D 1630)	—	115
Brittleness temperatures (ASTM D 2137, without break)	°C	−65
T_2	°C	−44
T_5	°C	−56
T_{10}	°C	−60
T_{100}	°C	−64
Property values after aging 70 h at 150°C in ASTM Oil No. 3		
Tensile strength at break	MPa	6.4
Elongation at break	%	150
Hardness (Shore A)	—	62
Volume increase	%	9
Property values after aging 70 h at 150°C in hydraulic fluid TL-4634		
Tensile strength at break	MPa	8.7
Elongation at break	%	168
Hardness (Shore A)	—	65
Volume increase	%	12

* postcured 16 h at 120°C.

The decahydropyrazine[2,3b]pyrazine (III) is particularly suitable.

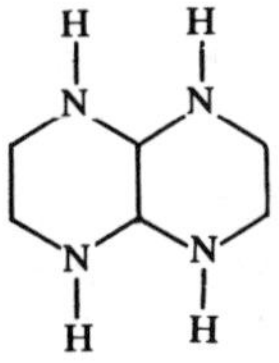

A typical compounding recipe consists of 100 pts by wt. Cymax, 50 pts HAF carbon black N 330, 1.0 pt processing aid, 0.5 pts resin oil, 0.7 pts sulfur, and 0.7 pts III. The vulcanization is carried out at 176°C for 30 min; in injection molding processing 200°C and curing times of 3 min or less are required. The properties of the vulcanizate prepared from the compound mentioned above are shown in Table 8-1.

The polymer has a glass temperature of −65°C and remains flexible at temperatures above 200°C. The mechanical properties and the aging characteristics are equivalent to those of NBR or sulfur-cured epichlorohydrin elastomers. Prolonged aging causes embrittlement in Cymax. Vulcanizates are resistant to fuels, hydraulic fluids, gasohol, and to media rich in H_2S. In a mixture of 20% methanol and 80% iso-octane/toluene (50/50) which is similar to a lead-free gasohol fuel, Cymax shows a volume swell of 49%; an epichlorohydrin elastomer under the same conditions swells 79% after 70 h. The similar (same ?) Polyether F-70 of American Cyanamid swells only little in methanol/gasoline mixtures; only fluorocarbon polymers performed better [3]. The use temperature limits of Cymax are −65° to 150°C [1]. Applications are foreseen in the automotive industry.

References

[1] T. M. Vial, Elastomerics **111**/10 (1979) 56.
[2] F. Lautenschlaeger and R. T. Woodham, in K. C. Frisch (ed.) Cyclic Monomers, High Polymers, vol. XXVI, Wiley-Interscience, New York 1972.
[3] I. A. Abu-Isa, Rubber Chem. Technol. **56** (1983) 135.

8.2   POLYSULFONES, POLYARYLENESULFONES, AND POLYETHERSULFONES

8.2.1   Structure and synthesis

The commercially available polysulfones, polyarylenesulfones, and polyethersulfones contain the following structural units:

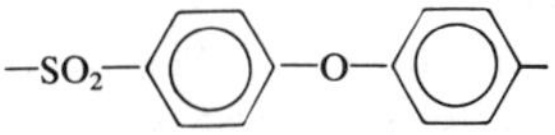

$-SO_2-\langle\bigcirc\rangle-O-\langle\bigcirc\rangle-$ Polyethersulfone 200 P (*ICI*)

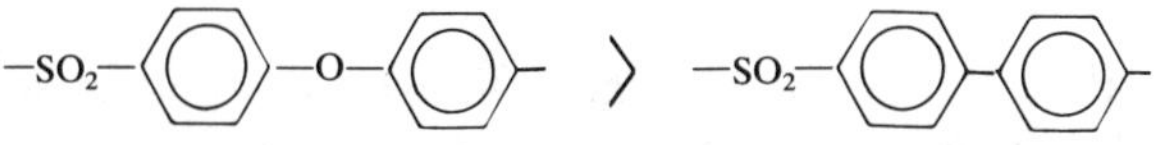

$-SO_2-\langle\bigcirc\rangle-O-\langle\bigcirc\rangle-$ ⟩ $-SO_2-\langle\bigcirc\rangle-\langle\bigcirc\rangle-$ Polyethersulfone 720 P (ICI)

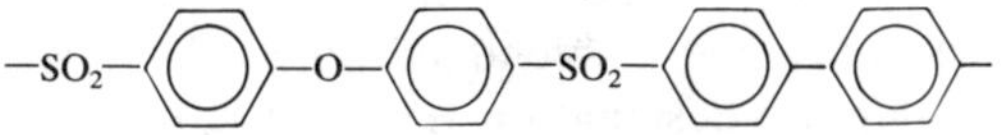

$-SO_2-\langle\bigcirc\rangle-O-\langle\bigcirc\rangle-SO_2-\langle\bigcirc\rangle-\langle\bigcirc\rangle-$ Polyarylsulfone ᴿ*Radel* (*Union Carbide*)

$-SO_2-\langle\bigcirc\rangle-O-$ ⟨ $-SO_2-\langle\bigcirc\rangle-\langle\bigcirc\rangle-$ Polyarylsulfone ᴿ*Astrel* (*Carborundum*)

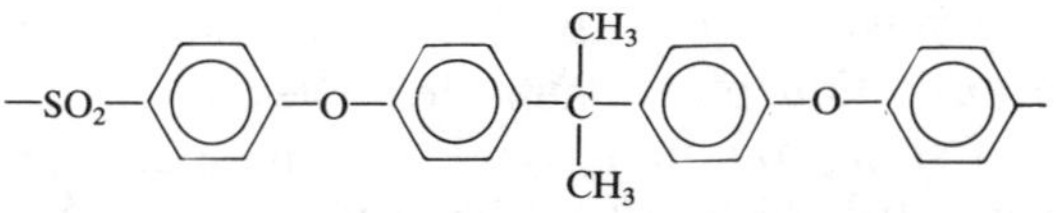

$-SO_2-\langle\bigcirc\rangle-O-\langle\bigcirc\rangle-\underset{\underset{CH_3}{|}}{\overset{\overset{CH_3}{|}}{C}}-\langle\bigcirc\rangle-O-\langle\bigcirc\rangle-$ Polysulfone ᴿ*Udel* (Union Carbide)

Strictly speaking, a differentiation among polysulfones, polyary-lenesulfones (generally polyphenylenesulfones), and polyethersulfones is not justified on grounds of chemical structures, since all commercial products likewise contain SO_2, ether, and phenylene groups. But apparently these designations are prevailing in the market [1].

The synthesis of these products and the developments up to about 1974 have been reported [2].

In 1976, Union Carbide Co. (UCC) introduced ®Radel. The 3 M Co. terminated the production of its ®Astrel after a lawsuit started in the U.S. between UCC and ICI over infringement on polysulfone patents. Later the court decided in favor of ICI. Carborundum became a licensee of ICI and aquired the production rights for Astrel from 3 M. In the U.S., ICI and UCC hold their own patent rights on polysulfones [3, 4].

®Arylon T of Uniroyal is a 50/50 blend of ®Udel and a propietary ABS. This ABS is based on alpha-methylstyrene and offers a higher heat deflection temperature. Recently, the U.S.S. Chemicals Div. of U.S. Steel acquired all rights for Arylon from Uniroyal. Also from Uniroyal, UCC acquired the rights for a polysulfone/ABS blend which is now supplied by UCC under the trade-name ®Mindel [5]. UCC has also marketed a polysulfone/SAN blend, trade-named ®Ucardel.

More recently, BASF introduced a polyethersulfone under the name ®Ultrason [18]. Five grades are supplied including two reinforced ones. Boeing, in cooperation with the Naval Air Command, announced the development of a so-called modified "NTS" polysulfone [19]. Ciba-Geigy has marketed a blend of polysulfone and polyfunctional epoxides under the designation BSL 914 [20].

Several reinforced polysulfones and polyethersulfones are commercially available. They include glass fiber, mineral, and carbon fiber reinforced grades, as well as compounds with glass fiber/PTFE or with PTFE alone. The commercially available grades are listed in Table 8-2.

Large scale production of polyarylenesulfones is conducted by two methods, i.e., polysulfonylation or polyether synthesis [6, 7, 8]. For the electrophilic substitution of aromatic hydrogen atoms by sulfonylium ions one can start from either an A-B type monomer (eq. 8-3) or carry out a polycondensation with the A-A and B-B type monomers (eq. 8-4). The reactions are performed at $100-250°C$ in solvents such as nitrobenzene, tetrachloroethylene, or sulfones with catalytic amounts of Lewis acids, e.g., with $FeCl_3$, $SbCl_3$, or $InCl_3$. A melt polycondensation has also been achieved [7].

(8-3)

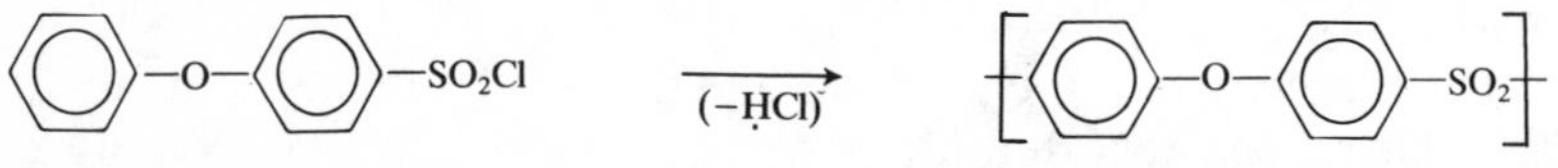

(8-4)

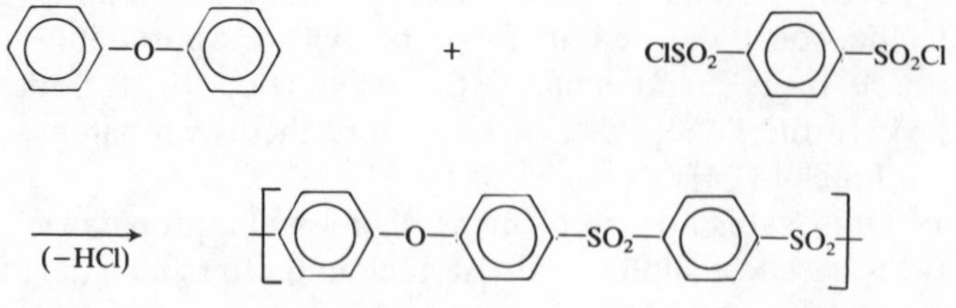

TABLE 8-2

Commercial polysulfones, polyether sulfones, polyarylene sulfones, and their composites.

Company	Trade name	not reinforced	reinforced — with glass fiber 15%	20%	30%	40%	with minerals	other
Carborundum	Astrel	360						
ICI	Victrex	100 P, 200 P 300 P, 600 P		420 P	403P			
Fiberite	Victrex RTP	720 P 900		903 1403	905 1405	907 1407		
INP	Thermocomp			GF-1004 JF-1004	GR-1006 GF-1006FR JF-1006	GF-1008 JF-1008	GX-4334 GX-4362	GC-1006 (30% carbon fiber) GL-4030 (15% PTFE) GFL-4022 and GFL-4036 (Glass/PTFE)
Thermofil				S-20 FG		S-40 FG		
Union Carbide	Udel	P 1700, P 1710 P 1720, P 1800 P 3500					P 6050 P 8000	
	Ucardel Radel	P 4174						
	Mindel A Mindel B Mindel M	M 5000, M 5010 A 600, A 650	B 315	B 322	B 330		B 390 M 800 M 825	
Uniroyal	Arylon T	Arylon T						
BASF	Ultrason							
Boeing/Naval Air Command	NTS							

Also in the nucleophilic substitution of aromatic halogen by phenoxy ions one has the choice between an A-B type polycondensation (eq. 8-5) and the A-A/B-B polycondensation (eq. 8-6).

The reactions are carried out at 130−250°C in solvents which dissolve both the reactants and the formed polymer. A mixture of chlorobenzene/dimethylsulfoxide is preferred.

(8-5)

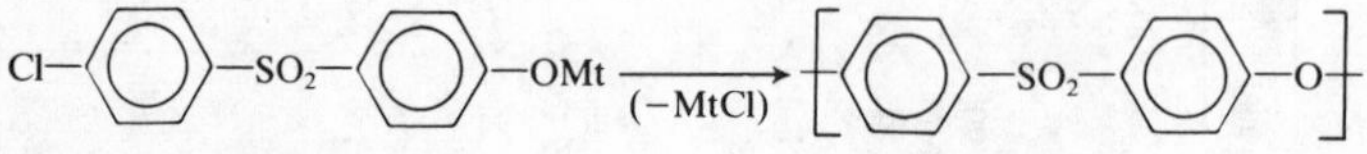

$$Cl-\hexagon-SO_2-\hexagon-OMt \xrightarrow[(-MtCl)]{} \left[\hexagon-SO_2-\hexagon-O\right]$$

(8-6)

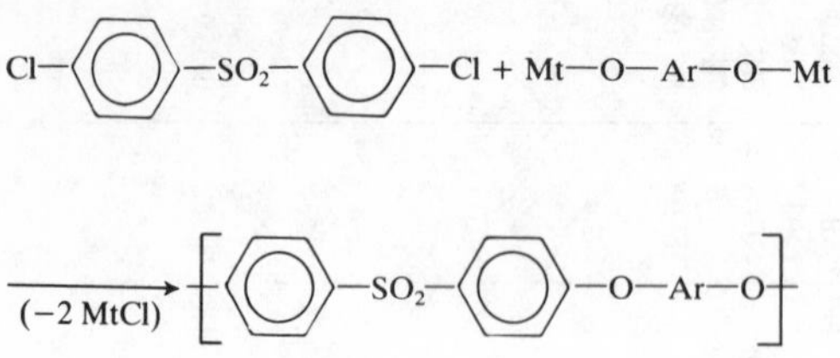

$$Cl-\hexagon-SO_2-\hexagon-Cl + Mt-O-Ar-O-Mt$$

$$\xrightarrow[(-2\,MtCl)]{} \left[\hexagon-SO_2-\hexagon-O-Ar-O\right]$$

Mt = metal (I)

The route according to eq. 8-3 yields almost exclusively p-substituted products, whereas the method following eq. 8-4 leads to about 80% para and 20% ortho substitution. Since polymer brittleness increases with increasing degree of ortho substitution [9], the route via eq. 8-3 appears to be the ideal procedure for the preparation of polyethersulfone 200 P. However, since the starting material is very expensive, reaction 8-5 is applied. Astrel 360 is prepared by reactions 8-3 and 8-4. For the preparation of Victrex 720 P copolymer, the reactions 8-5 and 8-6 are preferred. The route 8-6 is applied in the commercial production of Udel.

In nucleophilic polycondensations, polymers with reactive terminal groups are obtained. These terminal groups are capable of further condensation during processing; this results in an increase of molecular weights and hence in melt viscosity. In order to stabilize the melt viscosity, the remaining terminal phenolic groups are reacted with methyl chloride to give non-reactive methoxy groups.

Arylon polysulfone is prepared from dichlorodiphenylsulfone and a mixture of the potassium salts of bisphenol A and bis(4-hydroxyphenyl)-p-diisopropylbenzene in sulfolane/benzene [8].

Recently, copolymers composed of polysulfone blocks and poly-(dimethylsiloxane) [10] or polyamide 6 blocks [11] were developed, but the block copolymers have not yet been commercialized.

8.2.2 Properties and applications

All polyphenylenesulfones are amorphous and exhibit high creep resistance, stable electrical properties over a wide temperature and frequency range, transparency, and good thermal and hydrolytic resistance. Glass temperatures are between 192 and 290°C (Table 8-3). Structures and corresponding glass temperatures have been reported for 30 different polysulfones [8]. The heat deflection temperatures under a load of 1.81 MPa vary between 174 and 274°C. Victrex reputedly has a service life time of 20 years at 180°C [12] (see Fig. 8-1).

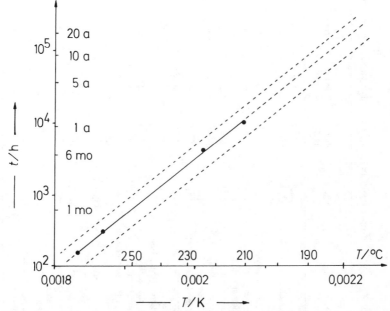

Figure 8-1 Life expectancy of polyethersulfone Victrex 300 P at various temperatures [12]. The life expectancy is defined as the time after which the tensile strength at room temperature has decreased to 50% of the initial value.

TABLE 8-3

Properties of polysulfones, polyarylene sulfones, polyether sulfones, and their composites.

Physical property	Physical Unit	Property values of							
		Victrex 200 P	Victrex 430 P[1]	Astrel 360	Udel P 1700	Radel	Mindel A 650	Arylon T	Mindel B 390[2]
Density	g/cm³	1.37	1.60	1.36	1.24	1.29	1.31	1.14	1.30
Glass transition temperature	°C	230	—	288	192	290	—	—	—
Melt index	g/10 min	—	—	—	6.5	10.0	—	—	—
Tensile strength at yield	MPa	84.1	140.3	89.6	70.3	71.7	51	51.7	—
Tensile strength at break	MPa	—	—	89.6	—	—	45	—	72.4
Elongation at yield	%	—	—	—	—	7	3.7	—	—
Elongation at break	%	40–80	3	13	50–100	60	30	80	3(–100
Modulus of elasticity	GPa	2.44	—	2.55	2.48	2.14	2.39	2.21	2.76
Flexural strength	MPa	129	190	118.6	106.2	86	92	75.9	114
Flexural modulus	GPa	2.57	5.79	2.72	2.69	2.28	2.52	2.07	3.31
Impact strength with notch	kJ/m	0.09	0.08	0.08	0.07	0.64	0.51	0.43	0.06
Hardness	—	M88	M98	M110	M69	—	—	R117	M74
Linear thermal expansion coefficient ($\times 10^5$)	K^{-1}	5.5	2.3	4.70	5.58	5.58	—	6.48	—
Vicat temperature	°C	226	226	—	—	—	—	—	—
Heat distortion temperature (1.81 MPa)	°C	210	216	274	174	204	150	149	169
Volume resistivity	Ohm × cm	1×10^{17}	1×10^{16}	3.2×10^{16}	5.0×10^{16}	8.9×10^{14}	3.6×10^{15}	1.5×10^{16}	—
Surface resistivity	Ohm	—	—	—	3.0×10^{16}	—	—	—	—
Dielectric strength	MV/m	16.0	16.0	13.8	14.6	14.6	17.0	16.9	18.1
Relative permittivity		3.5	3.5	3.9	3.2	3.4	3.1	3.1	3.3
Arc resistance	s	100	—	67	122	41	—	—	90
Flammability		V-0	V-0	—	V-0	V-0	—	—	V-0
Limiting oxygen index	%	34	—	—	30	38	—	—	—
Water absorption (24 h)	%	0.43	—	1.8	0.3	—	—	—	0.8

1) 30% Glass fiber.
2) Mineral filled.

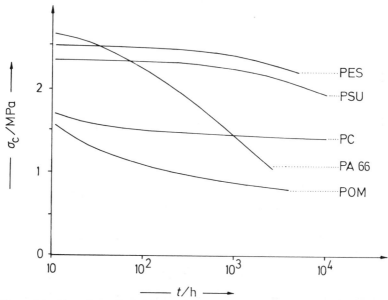

Figure 8-2 Time dependence of creep tensile strength for polyethersulfone (PES), polysulfone (PSU), polycarbonate (PC), polyamide 66, and poly(oxymethylene) POM [12b].

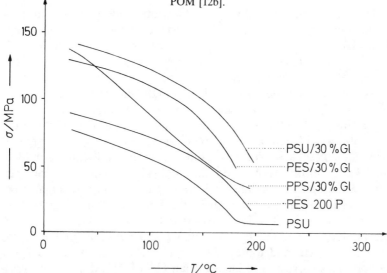

Figure 8-3 Temperature dependence of tensile strength for a polysulfone (PSU), a polyethersulfone (PES), and a poly(phenylenesulfide) (PPS) with or without 30% glass fibers [12b].

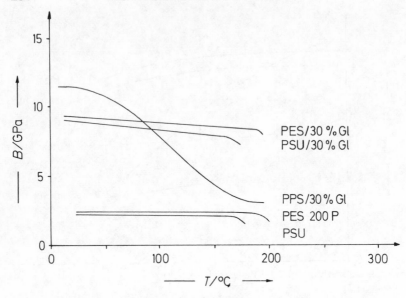

Figure 8-4 Temperature dependence of the flexural modulus for various polymers (see also figure 8-3).

The polyphenylenesulfones demonstrate good mechanical properties; however, notched impact resistance is only moderate except in Radel and the ABS modified types such as Arylon-T and Mindel-A. The creep resistance of all polyphenylenesulfones is excellent (Figure 8-2). At 150°C, creep resistance of unfilled grades, and with glass reinforced grades also tensile strength (Figure 8-3) and flexural modulus (Figure 8-4) are superior to other thermoplastic engineering resins, e.g., PPS. Here polyethersulfones perform better than polysulfones [12]. On aging at 260°C for 1500 h, the glass reinforced polyethersulfones are surpassed only by unfilled poly(amide-imides) and glass reinforced polyimides and PPS [13]. Polysulfones are resistant to the attack of diluted acids and bases. Many solvents, however, cause environmental stress-cracking. Udel is degraded in concentrated sulfuric acid; Astrel dissolves also in DMF, dimethylacetamide, m-cresol, DMSO, aniline, and pyridine, and Radel is soluble in DMF and dimethylacetamide. Radel shows a markedly better stress corrosion resistance than Udel [14]. Although polysulfones are resistant to oxidative attack the polymers are attacked by UV light.

Polysulfones can be processed on conventional injection molding machinery but Astrel, Victrex 720 P, and Radel require processing temperatures of 400–425°C. The ABS modified grades show enhanced processability. Thus, in a spiral flow test Arylon performed significantly better than modified PPO (140%), PC (230%), and polysulfone (350%).

Polysulfones [15] and polyethersulfones [15, 16] have found many applications. Proven examples are medical and household appliances which are sterilizable by hot air and steam, microwave cookware, corrosion-resistant piping, coffee makers, steam table pans, radomes, pump housings, bearing cages, power saw manifolds, hair dryer outlets, hot combs, and projector lamp grills. The many electric and electronic applications include automotive fuses, connectors, switch housings, coil bobbins, television components, capacitor films, battery cell frames, and semiconductor carriers [1]. Flexible printed circuits which can be used to 180°C are fabricated from polyethersulfone films. Integrated circuit sockets are molded from glass-filled material. Monsanto has marketed hollow fiber membranes, trade-named ®Prism, for the separation of hydrogen from CO/H_2 gas streams [17]. Udel and Victrex films serve as membranes in reverse osmosis processes; due to the high creep resistance at elevated temperatures and pressures as well as hydrophilic surface characteristics high separation efficiency is achieved.

Although injection molding is the primary processing technique, a number of polysulfones and polyethersulfone grades are suitable for extrusion, blow molding, vacuum or pressure forming, and sinter molding.

The following prices have been demanded for polysulfones: Udel (mineral filled) 1.80 $/lb, Udel (medical grade) 3.50 $/lb, Victrex 13.75 $/lb, Victrex (glass reinforced) 8.35 $/lb, Radel 15 $/lb, and ABS modified polysulfones 2 $/lb. The high-heat grades Victrex 720 P and Astrel were priced between 30 and 120 $/lb.

References

[1] Modern Plastics Encyclopedia 1980–1981, McGraw-Hill, New York 1980.
[2] H.-G. Elias, Neue polymere Werkstoffe 1969–1974, Carl Hanser Verlag, München 1975; New Commercial Polymers 1969–1975, Gordon and Breach, New York 1977.
[3] USP 4 008 203 (1977).

[4] USP 4 108 837 (1977).

[5] ®Mindel, Union Carbide Corp., Technical Information.

[6] J. B. Rose, Chimia (Aarau) **28** (1974) 561.

[7] V. J. Leslie, J. B. Rose, G. O. Rudkin, Jr., and J. Feltzin, Chem Tech **5** (1975) 426.

[8] K.-U. Bühler, Spezialplaste, Akademie-Verlag, Berlin 1978.

[9] I. K. Ahmed, J. Appl. Polymer Sci. **25** (1980) 821.

[10] L. M. Robeson, A. Noshay, M. Markus, and C. N. Merriam, Angew. Makromol. Chem. **29/30** (1973) 47.

[11] J. E. McGrath, L. M. Robeson, and M. Matzner, Polym. Sci. Technol. **4** (1974) 195.

[12] ®Victrex, ICI (England) and ICI Americas, Technical Information.

[13] J. E. Theberge and P. J. Cloud, Modern Plast. **55** (March 1978) 66.

[14] ®Radel, ®Udel, Union Carbide, Technical Information.

[15] A. A. Goenczy, Kunststoffe **69** (1979) 12.

[16] H. Domininghaus, Kunststoffe **69** (1979) 2.

[17] Anon., Chem. & Engng. News (November 19, 1979) 6; and (May 19, 1980) 57.

[18] Anon., Modern Plast. Intern. **12** (March 1982) 30; and **12** (April 1982) 64; ®Ultrason, BASF, Technical Information.

[19] Anon., Modern Plast. Intern. **11** (November 1981) 49.

[20] Anon., Modern Plast. Intern. **12** (March 1982) 64.

9 Polyamide Plastics

9.1 TRANSPARENT POLYAMIDES

9.1.1 Structure and synthesis

Numerous transparent polyamides have been developed over the last few years (Table 9-1), some of which have become commercially available. Transparent polyamides are based on a variety of different structural units (Table 9-2).

TABLE 9-1
Transparent polyamides.

Company	Trade name or grade
Allied Fibers and Plastics	Capron C-1000
BASF	Ultramid K 1297/2 (formerly K 4601)
Bayer	KL 1-2104
German Democratic Republic	FSTXA-7
Dynamit Nobel	Trogamid T
	Trogamid TG 35 (with 35% glass fiber)
Kay-Fries	Trogamid T and TG 35
LNP Corp.	Compounds from Trogamid
Fiberfil	Compounds from Trogamid
Ems Chemie	Grilamid TR 55
Union Carbide	Amidel TN 1540 and TN 1541
Mitsubishi Chemical	?
Hoechst	Hostamid LP 700
Chemische Werke Hüls	Vestamid X 4308
Phillips Petroleum	PACP 9/6
Unitika	CX 1004 and CX 1005
Upjohn	PA 7030 and 5050 (now Dow)

The first transparent polyamide was developed by W. R. Grace in 1960 [1, 2]. Since 1967, Dynamit Nobel as a licensee has marketed the product under the trade-name ®Trogamid T [3–8]. In the U.S., Trogamid T is supplied by Kay-Fries, Inc. Trogamid with 35% glass fiber and compounds with 25% PTFE are available in the U.S. from LNP Corp.; a flame-proofed grade with UL 94 V-0 rating is offered by Fiberfil.

TABLE 9-2
Monomer combinations used in the manufacturing of transparent polyamides

 I	$H_2N{-}CH_2{-}C(CH_3)_2{-}CH_2{-}CH(CH_3){-}CH_2{-}CH_2{-}NH_2$ II	$H_2N{-}CH_2{-}CH(CH_3){-}CH_2{-}C(CH_3)_2{-}CH_2{-}CH_2{-}NH_2$ III	Dynamit Nobel
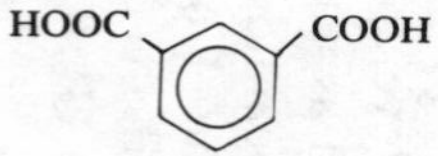 IV	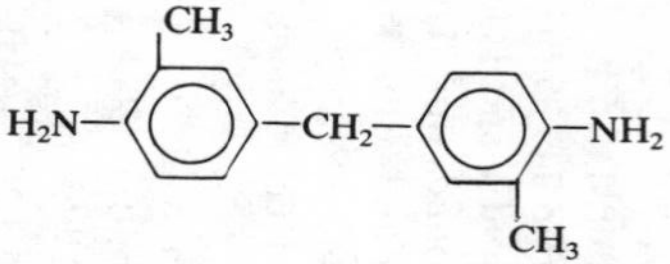 V	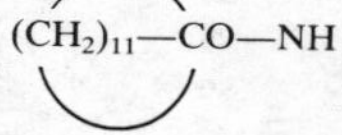 VI	Ems Chemie
$HOOC{-}(CH_2)_4{-}COOH$ IX	$H_2N{-}(CH_2)_6{-}NH_2$ X	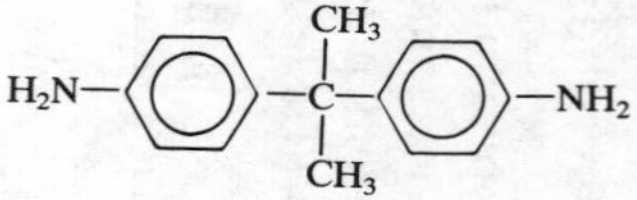 XI	BASF
 IV	$H_2N{-}(CH_2)_6{-}NH_2$ X		Bayer

TABLE 9-2

Monomer combinations used in the manufacturing of transparent polyamides.

$HOOC-(CH_2)_4-COOH$	$HOOC-(CH_2)_7-COOH$	$H_2N-C_6H_4-C(CH_3)_2-C_6H_4-NH_2$	Phillips Petroleum
IX	XII	XI	
$HOOC-$cyclohexane$-COOH$	$H_2N-CH_2-C(CH_3)_2-CH_2-CH(CH_3)-CH_2-CH_2-NH_2$	$H_2N-CH_2-CH(CH_3)-CH_2-C(CH_3)_2-CH_2-CH_2-NH_2$	DDR
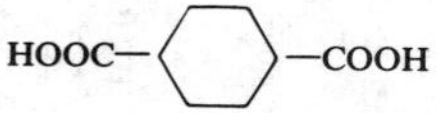 XIII	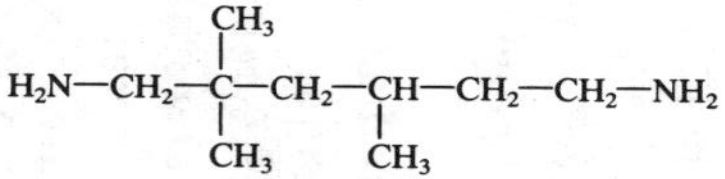 II	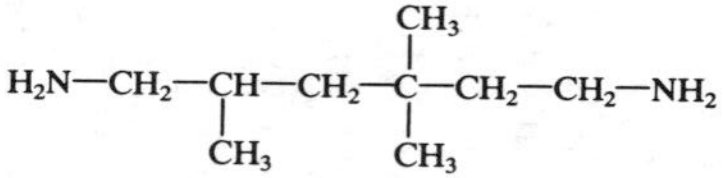 III	
$HOOC-C_6H_4-COOH$	norbornane$-CH_2-NH_2$ / H_2N-CH_2-	$(CH_2)_5-NH-CO$	Hoechst
 I	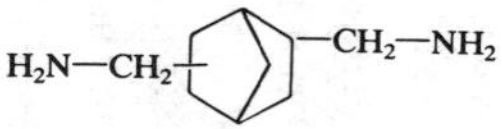 VII	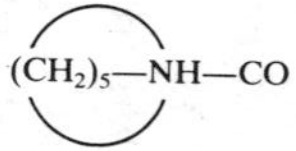 VIII	
$HOOC-(CH_2)_4-COOH$	$HOOC-(CH_2)_7-COOH$	$OCN-C_6H_4-CH_2-C_6H_4-NCO$	Upjohn (now Dow)
IX	XII	XIV	
$HOOC-C_6H_4-COOH$	$HOOC-(CH_2)_7-COOH$	$OCN-C_6H_4-CH_2-C_6H_4-NCO$	Upjohn (now Dow)
IV	XII	XIV	
		toluene diisocyanate (CH_3, NCO, OCN)	
		XV	

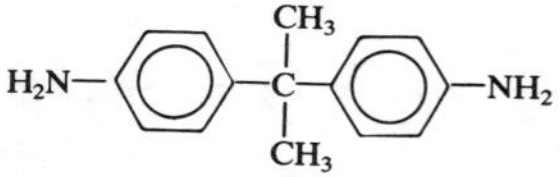

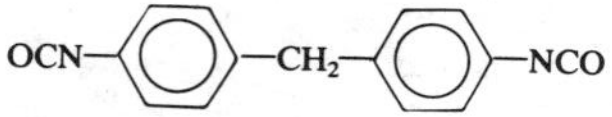

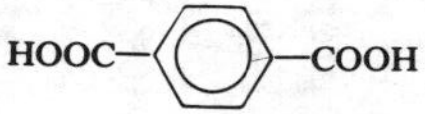

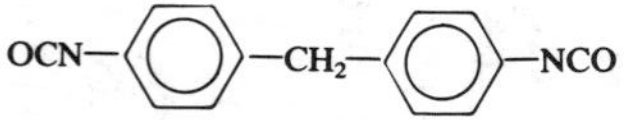

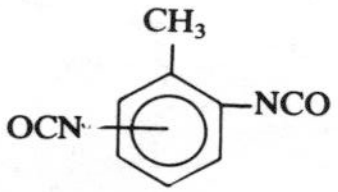

For the preparation of Trogamid T, the salt of the monomers I, II, and III (Table 9-2) is pre-condensated in a molar ratio of 2:1:1; the polycondensation of the pre-polymer is conducted in an extruder.

For the synthesis of ®Grilamid TR 55 the salts of IV and V are polycondensated with VI [8−12]. Grilamid TR 55 and the glass fiber reinforced grade TR 55 V 35H of Ems-Chemie AG (formerly Emser-Werke AG) were marketed in the U.S. under the name ®Amidel [13] by Union Carbide Corp. and in Japan by the Mitsubishi Chemical Corp. Amidel is no longer available after the contract between Ems-Chemie and Union Carbide was canceled in 1980 [34]. Grilamid TR55 V35 was withdrawn from the market because the advantages over PA 12 turned out to be only minor ones. In the U.S., Grilamid TR 55 is now supplied by Ems-American Grilon.

Hoechst developed the glass-clear polyamide ®Hostamid LP 700; it is prepared by polycondensation of I with a mixture of the isomers VII and VIII [7, 8, 14, 15]. Hoechst has never marketed its product [16].

BASF produced ®Ultramid K 1297/2 (formerly KR 4601) through polycondensation of IX, X, and XI [17]. The polymer has been withdrawn from the market [34].

®Durethan T40 (formerly experimental product KL 1-2104) Bayer's is a poly(hexamethylene-isophthalamide) prepared from IV and X [18−20]. The developmental product ®Zytel FE 3303 of DuPont is said to be identical in structure to Durethan T40 [34]. The polymer was reported in 1952 in the patent literature but it was not further developed because the required pure isophthalic acid was not available in commercial quantities at that time. The polymer is prepared from equimolar amounts of commercial grades IV and X in water without isolating the intermediate salt. A prepolymer is prepared under pressure at 210°C; then the water is distilled off and the polycondensation is carried out at 270°C until the desired molecular weight is reached [19].

Phillips Petroleum reported a glass-clear polyamide PACP 9/6 which is obtained by melt polycondensation of XI with a 60/40 molar mixture of XII and IX [21]. At present, the company has no intention of commercializing the product.

Poly(trimethylhexamethylene-cyclohexane-1,4-dicarbonamide), designated FSTXA-7, was developed in the German Democratic Republic. It has not been disclosed whether the polymer is produced on a technical scale. The synthesis is achieved by melt poly-condensation of trans-XIII with a mixture of II and III isomers [22].

The transparent products CX-1004 and CX-1005 of Unitika Ltd. are copolymers with nylon 6 [23, 24].

Whereas all the polymers mentioned above are prepared by poly-condensations of diamines and occasionally lactams with dicarbonic acids, Upjohn obtains transparent polyamides from the polyaddition reaction between dicarboxylic acids and diisocyanates [25−29]:

(9-1)

$$n\,HOOC-(CH_2)_x-COOH + n\,OCN-\langle\bigcirc\rangle-CH_2-\langle\bigcirc\rangle-NCO$$

$$\xrightarrow{-2\,n\,CO_2} \left[CO-(CH_2)_x-CO-NH-\langle\bigcirc\rangle-CH_2-\langle\bigcirc\rangle-NH\right]_n$$

The polyamide PA-7030, now called Isonamid® 7030 is prepared from the dicarboxylic acids XII and IX (molar ratio 70:30) and MDI (XIX). The polyamide PA 5050 is synthesized from the dicarboxylic acids IV and VII (molar ratio 50:50) and MDI (XIX) where 15% mol-% of MDI is replaced by 2,4-TDI (XV). By first reacting TDI with IV the formation of crystalline blocks of IV and XIV is prevented [28]. The reaction of MDI and azelaic acid, on the other hand, leads to a semicrystalline polymer [38].

Isonamid 7030 is available as a developmental product. [30]. Gel-free products without branching are only obtained when the poly-addition is carried out in solvents such as tetramethylenesulfone, diphenylsulfone or γ-butyrolactone at temperatures between 150 and 225°C. Within this range, the higher reaction temperatures are applied if the polymers tend to develop crystallinity. Metal alkoxides, phenoxides, lactamates, and preferably cyclic phosphorous oxides, are employed as catalysts [29]. Benzoic or stearic acid serve as chain stoppers to control molecular weights [27].

The most recent products in the family of transparent polyamides are ®Vestamid X 4308 of Chemische Werke Hüls [31] and ®Capron C-100 of Allied Fibers & Plastics Co. [32]. Vestamid X 4308 is based on PA 12, the nature of the comonomer has not been disclosed by Hüls. According to another source [34], Vestamid X 4308 is a co-polymer derived from laurolactam, isophorondiamine and iso-phthalic acid. In contrast to the transparent polyamides with amorphous structure, Capron C-100 is also transparent but is composed of nylon 6 with crystalline portions. Another nylon 6 based transparent polyamide was introduced by Ems-Chemie in September 1982 [34, 35].

9.1.2 Properties and applications

The transparent polyamides exhibit the typical properties of the conventional polyamides, i.e. high stiffness, tensile strength and toughness, good insulation and non-tracking characteristics, and good resistance to numerous solvating, swelling and saponifying media. Non-typical properties are the transparency, a yellowish color, little creep, and mechanical as well as electrical properties which are nearly independent of the conditioning state, i.e., the equilibrium moisture content (see Table 9-3). Recent developments resulted in minimizing the yellowish color and yielded products with nearly glass-clear appearance which is considered to be the major difference to polysulfones [34]. Due to their amorphous structure and contrary to PA 6 and PA 66, the transparent polyamides show a shear modulus curve which progresses nearly parallel to the x-axis and has a steep slope only at relatively high glass transition temperatures (Figure 9-1) [31]. The shear modulus and tensile modulus graphs illustrate that mechanical properties are retained over a broad temperature range.

With a view to applications at elevated temperatures, glass temperatures of amorphous polyamides should be at least 120°C. The commercial grades of amorphous polyamides have glass temperatures of 130−180°C and heat deflection temperatures of 108−160°C, whereas the heat deflection temperature of PA 6 and PA 66 is only 75°C. Due to the high glass transition temperatures, the amorphous polyamides exhibit significantly less creep at elevated temperatures (80°C) than PA 6 or PA 12 [12]. The glass transition temperature of poly(hexamethylene-isophthalamide) can be raised by co-condensation with cycloaliphatic amines (Table 9-4). Another possibility is the increased incorporation of aromatic groups as has been demonstrated in Upjohn's PA-5050 with the use of MDI/TDI. In some cases, the desired low melt viscosity is achieved through variation of comonomer molar ratio. For example, the PA 7030 of Upjohn with a 70:30 molar ratio of azelaic/adipic acid shows an eutectic at about 240°C (DSC) and hence favorable melt viscosity/temperature

TABLE 9-3

Properties of transparent polyamides as compared to polycarbonate and polysulfone.

Physical property	Physical Unit	Trogamid T	Grilamid TR55	KL-1-2104	Vestamid X 4308	PA 7030	PA 5050	Poly-carbonat[1]	Poly-sulfone[2]
					Property values of				
Density	g/cm^3	1.12	1.060	1.185	1.08	1.17	1.16	1.20	1.26
Tensile strength at yield	MPa	85	75	105	80	76	101	70	70
Elongation at yield	%	9.5	10	–	7	8.0	9.0	9.0	9.0
Modulus of elasticity	GPa	3.0	2.1	2.8	–	1.9	2.3	2.4	2.5
Tensile strength at break	MPa	60	60	65	70	83	82	59	59
Elongation at break	%	70	50–150	40	150	80–100	30	75	50
Flexural strength	MPa	91	105	155	70	105	117	93	106
Flexural modulus	GPa	2.62	1.75	3.20	2.40	2.11	2.37	2.34	2.69
Impact strength with notch									
Charpy at 20°C	kJ/m^2	10–15	6	4	5.4	–	–	–	–
–40°C	kJ/m^2	–	4	–	3.8	–	–	–	–
Izod at room temperature	kJ/m	0.07	0.060	–	–	1.09	0.23	0.8	0.07
Hardness DIN	MPa	125	95	180	155	–	–	–	–
Rockwell	–	M93	M89	–	–	M77	M76	R115	M69
Glass transition temperature	°C	140	160	130	–	140	180	149	190
Vicat temperature	°C	145	157	125	144	–	–	–	–
Heat distortion temperature (1.81 MPa)	°C	130	124	110	123	108–127	160	130	174
Limiting oxygen index	%	–	26	6	–	31.3	43.6	25	30
Water absorption (24 h)	%	0.41	3.3	6	5.5^3	0.4	0.34	0.14	0.3
Abrasion resistence (17/1000)	mg/1000	21	9	–	–	11	–	16	20
Relative permittivity	–	3.5	3.0	3.6	–	4.3	4.2	3.0	3.1
Volume resistivity	Ohm × cm	$>10^{14}$	3×10^{13}	$>10^{15}$	–	$>10^{15}$	$>10^{15}$	2.1×10^{16}	5×10^{16}
Surface resistivity	Ohm	$>10^{13}$	6×10^{12}	10^{14}	–	$>10^{16}$	$>10^{16}$	–	–
Dissipation factor (at 1 kHz)	–	0.03	0.012	0.020	–	0.025	0.017	0.002	0.001
Dielectric strength	kV/mm	25	50	28	–	–	–	–	–

1) Lexan (General Electric).
2) Udel (Union Carbide).
3) Value after saturation at 100°C.

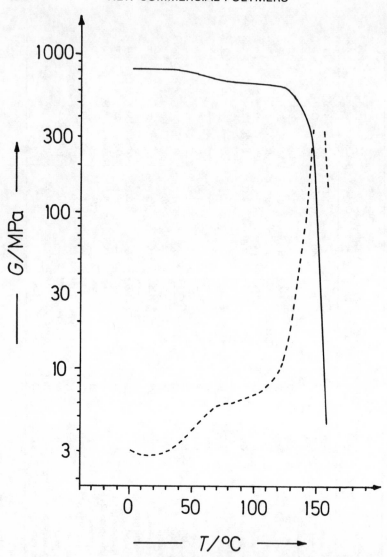

Figure 9-1 Temperature dependence of dynamic shear modulus of the amorphous polyamide X 4308. The broken line gives the logarithmic decrement of the mechanical damping [31].

characteristics [28] (Figure 9-2). At this composition, the degree of crystallinity and the crystallization rate are at a minimum, whereas the notched impact resistance reaches a maximum.

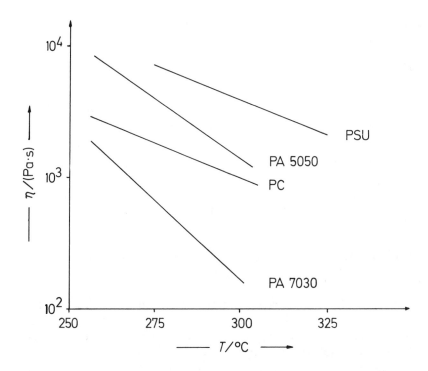

Figure 9-2 Temperature dependence of the melt viscosity of various engineering plastics [28]. PSU = polysulfone, PC = polycarbonate, PA 5050, and PA 7030 = transparent polyamides (Upjohn).

The tensile modulus of amorphous polyamides varies only slightly between −60°C and the glass temperature.

The loss in strength is small up to temperatures of 100−120°C, (Figure 9-3) [28]. The majority of transparent polyamides demonstrate good long-term characteristics, which is illustrated, for example, for Grilamid TR 55 by the creep curves at 23° and 80°C (Figure 9-4) [12]. The upper continuous service temperatures of Grilamid TR 55 are 80°C in air and in water and 90°C in oil; for short-time exposures, temperatures up to 140°C can be tolerated [12].

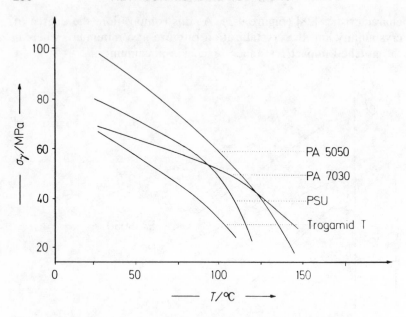

Figure 9-3 Temperature dependence of tensile strengths at yield for various engineering plastics [28].

The equilibrium water absorption is lower than in PA 6 and PA 66 and conditioning is not required. Stress-free parts made from amorphous polyamides are resistant to aqueous salt solutions and bases, aliphatic hydrocarbons, and esters, but they are attacked in general by strong mineral acids, the low-carbon fatty acids and alcohols, amines, phenols, methylene chloride, and strong aprotic solvents such as DMF. The polyamides PA 7030 and PA 5050 show a better resistance to the media mentioned thanks to their higher portion of aromatic units and to the small degree of crystallinity present in the polymers [26, 28]. For example, they do not dissolve, but only swell, in methylene chloride.

The surface of most of the transparent polyamides is rendered cloudy by water at temperatures above 70°C, this is caused by water absorption, e.g., of more than 5% at 90°C [18]. The use temperature of Trogamid T in water is limited to 80°C, whereas the transparency of Grilamid TR 55 is sustained for one year in boiling water [12]. Reportedly, it even survives exposure to boiling water for several years without becoming cloudy [34].

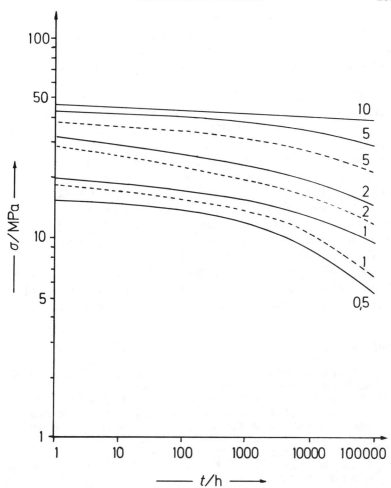

Figure 9-4 Time dependence of tensile strengths of Grilamide TR 55 at 50% relative humidity and temperatures of 23 (——) and 80°C (– – –) [11]. Numbers refer to the elongation in %.

Environmental stress cracking is observed in parts exposed to aromatic hydrocarbons, ethers, chlorinated hydrocarbons, and ketones [12, 19]; however, the resistance is still better than those of polycarbonates and polysulfones.

Injection molding processing requires cylinder temperatures from

260 to 300°C where stock temperatures should be at least 280 but less than 305°C. In order to avoid internal stress, the recommended mold temperatures for Durethan T 40 are 50−80°C and for Grilamid TR 55 90−110°C. The linear mold shrinkage is between 0.3 and 0.8%.

The transparent polyamides are an alternative to glass or other transparent polymers such as PMMA, PC, and polysulfones for many applications. Unlike PMMA, they possess high heat deflection temperatures and surface hardness. The advantages over PC are improved surface hardness and less tendency to stress cracking; advantageous to polysulfones are the better solvent resistance and, with some grades, higher tensile strength and ease of processability.

Applications include transparent covers, switch housings, clearvision screens, pressure and sight gauges, flow meters, filter and pump housings, gas lighter tanks, eyeglass frames, coffee filters, parts for milking machines, and tubing.

Amorphous polyamides form also the base of body exterior alloys for cars which are marketed as Bexloy® by DuPont [39]. The first application is for heck spoilers for the Pontiac Indy Fiero.

References

[1] R. Gabler, H. Müller, G. E. Ashby, E. R. Agouri, H. R. Meyer, and G. Kabes, Chimia (Aarau) **21** (1967) 65.

[2] US Patents 3 145 193, 3 150 113, 3 150 117, 3 198 771, and 3 294 758.

[3] G. Bier, in N. A. Platzer (ed.) Addition and Condensation Polymerization Processes (Adv. Chem. Ser. **91**), Amer. Chem. Soc., Washington, DC (1969) 612.

[4] H. Doffin, W. Pungs, and R. Gabler, Kunststoffe **56** (1966) 542.

[5] ®Trogamid T, Dynamit Nobel AG, Technical Information.

[6] H.-G. Elias, Neue polymere Werkstoffe 1969−1974, Carl Hanser Verlag, München 1975.

[7] H.-G. Elias, Polymer News **3**/5 (1977) 238.

[8] H.-G. Elias, Kunststoffe **70** (1980) 700.

[9] E. Schmidt and W. Griehl (Inventa A. G.), Ger. Offen. 2 642 244 (1976).

[10] ®Grilamid TR 55, Ems-Chemie A. G. (formerly Emser Werke A. G.), Technical Information 1979 and September 1980.

[11] S. Schaaf, manuscript and personal communication, December 14, 1977.

[12] S. Schaaf, Technische Rundschau (March 11, 1980) 33.

[13] ®Amidel, Union Carbide Corp., Technical Information 1978.

[14] E. Reske, L. Brinkmann, H. Fischer, and F. Röhrscheid (Hoechst A. G.), Ger. Offen. 2 156 723 (1971).

[15] H.-G. Elias, New Commercial Polymers 1969−1975, Gordon and Breach, New York 1977.

[16] J. Brandrup, personal communication (June 26, 1981).

[17] H. G. Peine, G. Falkenstein, H. Dörfel, P. Raff, and L. Schuster (BASF A. G.), Ger. Offen. 1 595 354 (1966) and USP 3 703 595 (1972).

[18] ®Durethan T 40 (formerly KL 1-2104), Bayer A. G., Technical Information 1980.

[19] W. Nielinger, B. Brassat, and D. Neuray, Angew. Makromol. Chem. **98** (1981) 225.

[20] R. Meyer et al. (Bayer A. G.), USP 4 205 159 (1980).

[21] H. W. Hill, Jr., R. W. Campbell, and R. S. Shue, 34th Ann. Meeting SPE **22** (1976) 372; Polymer Engng. Sci. **18** (1978) 36; R. W. Campbell (Phillips Petroleum Co.), USP 3 840 501 (1974) and 3 842 045 (1974); R. F. Kleinschmidt (Phillips Petroleum Co.), USP 4 028 476 (1977).

[22] H. Wolf, H. Hörhold and H. H. Seiffath, Faserforschung Textil-techn. **29/9** (1978) 590.

[23] Anon., Japan Chemical Week (October 16, 1980) 3.

[24] Anon., Plastics Industry News (November 1980) 102.

[25] J. T. Chapin, B. K. Onder and W. J. Farrissey, Jr., ACS Polymer Preprints **21/2** (1980) 130.

[26] B. K. Onder, P. S. Andrews, W. J. Farrissey, Jr., and J. N. Tilley, ACS Polymer Preprints **21**/2 (1980) 132.

[27] W. J. Farrissey, Jr., B. K. Onder and P. S. Andrews, Modern Plastics (March 1981) 82.

[28] B. K. Onder, W. J. Farrissey, Jr., J. T. Chapin and P. S. Andrews, SPE ANTEC, Boston, May 1981.

[29] B. K. Onder (Upjohn Co.), US Patents 4 061 622 (1977), 4 065 441 (1977), 4 087 481 (1978), 4 094 864 (1978) and 4 156 065 (1979).

[30] B. K. Onder, personal communication (December 15, 1980).

[31] ®Vestamid X 4308, Chemische Werke Hüls A. G., Technical Information, October 1981.

[32] Anon., Plastics Technol. **27** (December 27, 1981) 13.

[33] D. Michael, Kunststoffe **70** (1980) 634.

[34] W. Isler (Ems-Chemie A. G.), personal communication (October 1982).

[35] Ems-Chemie A. G., Technical Information 1982.

Supplement

[36] ®Vestamid X 4418, Chemische Werke Hüls A. G., Technical Information 1982. The material contains a lubricant, heat and light stabilizers.

[37] ®Zytel 330, DuPont, Technical Information 1982. The material has a light transmission value of 86% and haze value of 4%, tensile strength 96.5 MPa, elongation 150% and flexural modulus 27.6 GPa. Processing temperatures range from 277 to 321°C.

[38] K. Onder and A. T. Chen, Soc. Plast. Engineers, Annual Techn. Meeting (April 1984) 373.

[39] Anon., K-Plaste und Kautschuk-Ztg. **292** (7 February 1985) 3.

9.2 SUPER-TOUGH POLYAMIDES

9.2.1 Structure and synthesis

The notched impact strength of unreinforced polyamides such as PA 6, PA 66, and poly(hexamethylene-isophthalamide) PA 6,I ranges from 3 to 5 kJ/m^2, tested in dry, as-molded condition. These polyamides reach optimum toughness only through conditioning by absorption of water. Conditioning is time consuming and in addition has the disadvantage that tensile strength and tensile modulus decrease and electrical properties are effected.

New elastomer-modified materials require no conditioning to achieve good toughness. They are obtained by incorporating carboxy-groups containing ethylene copolymers or modified EPDM-, BR-, and SBR-graft copolymers [1, 2]. With BR graft copolymers, Charpy notched impact strengths of more than 50 kJ/m^2 are achieved.

Bayer A. G. has marketed developmental products based on elastomer-modified PA 6 and PA 6,I where are designated ®Durethan VP KL 1-2310/2 and KL 1-2311, respectively [3]. Super-tough modified PA 66 types are commercially supplied by DuPont under the trade names ®Zytel ST-801 and ®Zytel ST-811 [4, 5]. A flame-retardant grade ®Zytel ST-FR-80 is available as well. DuPont has not disclosed the nature of the modifier. DuPont is expanding its range of Zytel ST to include 14% and 43% glass-reinforced materials [8].

In 1982, Ems-Chemie A. G. introduced a series of modified PA 6 grades trade-named ®Grilon A28NX, A28NY, and A28NZ [6, 7], as well as the glass-reinforced grade ®Grilon PVN-3H and the polymer blend ®Grilon BT40. They cover a broad range of notched impact strengths. British Industrial Plastics, Ltd., has marketed two grades of acrylic-nylon 6 copolymers under the name ®Beetle Nylon AC 1 and EN 30 [9]. ICI developed an elastomer-modified nylon 66, trade-named ®Maranyl AD104 [8]. Unreinforced and glass-reinforced elastomer-modified nylon 66 grades are also supplied by the Italian company Lati SpA under the designations ®Latamid-6/6 PX15 and PX30 (unreinforced) and PX30 6/20 (20% glass) [8].

High notched impact strengths are also demonstrated by the so-called flexible polyamides (see chapter 9.6) and by thermoplastic polyamide elastomers (see chapter 4.5) which each show an area of property overlap with the super-tough polyamides.

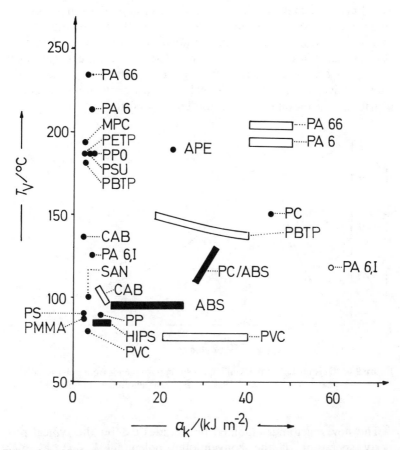

Figure 9-5 Vicat temperatures and notched impact strengths of various thermoplastics [2]. PA = polyamides, MPC = tetramethylbisphenol A polycarbonate, APE aromatic polyesters from terephthalic acid/isophthalic acid (1:1) and bisphenol A, PSU = polysulfone from bisphenol A and 4,4'-dichlorodiphenylsulfone, PA 6,I = polyamide from hexamethylenediamine and isophthalic acid, PET = poly(ethylene terephthalate), PPO = poly(phenylene oxide), PBT = poly(butylene terephthalate), PC = polycarbonate, CAB = cellulose acetobutyrate, PS = poly(styrene), HIPS = impact poly(styrene), PVC = poly(vinyl chloride). ●, ▬ = normal thermoplastics; ○, ▢ = rubber modified thermoplastics.

9.2.2   Properties and applications

Elastomer-modified polyamides exhibit the highest notched impact strengths among the unreinforced thermoplastics (Figure 9-5). Toughness reaches maximum with increasing elastomer content (Figure 9-6). As the elastomer portion increases, ball indentation hardness, flexural stress at conventional deflection, and Vicat temperature decrease. The particle size and the degree of modifier grafting exert an extensive effect on compound properties. First studies have shown that spherical modifier parts homogeneously distributed in the matrix yield the best properties [1].

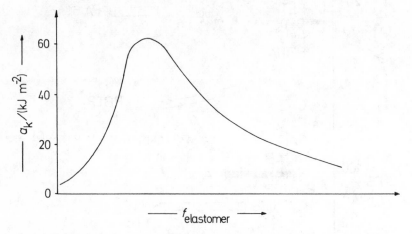

Figure 9-6   Dependence of notched impact strength of blends from poly(hexamethylene isophthalamide) with elastomers on the elastomer content [1].

The new elastomer-modified polyamides exhibit the typical property spectrum of the conventional polyamides, such as high strength and stiffness, low sensitivity to stress cracking and to attack of chemicals, good tracking characteristics, and so on. The distinguishing features are the considerably higher notched impact strengths and a reduced dependency of properties on the conditioning state (Figure 9-5). Although the new modified polyamides are claimed to require no conditioning to achieve good toughness, there is no doubt that notched impact strength can be further improved by conditioning. This is shown by comparing the room tem-

TABLE 9-4

Influence of diamines as substitutes for hexamethylene diamine on the glass transition temperature of poly(hexamethylene isophthalamide) [19].

Co-diamine Name	mol-%	$T_G/°C$
—	0	130
1,6-Diaminocyclohexane	15	148
4-Aminomethylcyclohexylamine	15	145
3-Aminomethyl-3,5,5-trimethylcyclohexylamine	15	149
Bis(4-aminocyclohexyl)methane	15	145

perature values of Izod (in J/m) and Charpy (in kJ/m^2) notched impact strengths for dry as molded and conditioned, i.e., equilibrated, specimens, (for the conditioned specimens, the values are given in brackets): Zytel ST 801 907 (1068 J/m), Maranyl AD104 225 (400 J/m), Grilon A 28NX 13 (40 kJ/m^2), Grilon A 28 NY 38 (110 kJ/m^2), Grilon A 28 NZ 42 kJ/m^2 (no break), Grilon A 28 NZ 1200 (1400 J/m) [6], Grilon PVN-3H 20 (35 kJ/m^2), Grilon BT 40 no break (no break), Beetle Nylon AC 1 11.3 (75 kJ/m^2), and Beetle EN 30 12 (50 kJ/m^2). Another characteristic feature of super-tough polyamides is the exceptional resistance to crack formation and crack propagation. The high Izod notched impact strengths obtained are nearly independent of notch radius, material orientation, wall thickness, or shape of the specimen.

With regard to properties, there is an area of overlap with flexible and elastomeric polyamides based on PA 11 and PA 12, e.g., ELY-1256 (new trade name is ®Grilamid ELY 60); see Table 9-5. In the order just mentioned, tensile strength, tensile modulus, and hardness decrease as expected. With respect to flexibility, some of the super-tough polyamides correlate with thermoplastic polyurethanes and with the flexible PA 11 and PA 12 types. The polyurethanes have the disadvantage of markedly higher densities, and the flexible polyamides usually are more difficult to process.

Some of the super-tough polyamides require no drying prior to melt processing. Stock temperatures should be in the range from 260 to 290°C but not to exceed 300°C. Mold temperatures of around 80°C are recommended. The super-tough polyamides were designed for injection molding processing; Grilon BT 40 can also be extruded.

TABLE 9-5

Comparison of properties of super-tough polyamides with standard polyamides and a thermoplastic polyamide elastomer.

Physical property	Physical Unit	Property value						
		KL 1 2310/2	KL 1 2311	Zytel[1] ST 801	Zytel ST 811	Grilon A28NZ	PA 6	PA 66
Density	g/cm³	1.08	1.08	1.04	1.01	1.07	1.15	—
Tensile strength at yield	MPa	60	70	—	—	45	80	83
Tensile strength at break	MPa	45	55	52	48	40	70	83
Elongation at yield	%	100	90	60	415	200	70	60
Flexural strength	MPa	70	105	68	—	65	82	120
Modulus of elasticity	GPa	2.10	2.10	—	—	1.05	2.90	3.40
Flexural modulus	GPa	2.13	—	1.69	0.48	1.2	3.20	2.83
Impact strength with notch (Charpy)								
at 23°C	kJ/m²	55—66	>50	—	—	40	4	4
at 0°C	kJ/m²	40	—	—	—	—	—	—
at −20°C	kJ/m²	22	—	—	—	20	—	—
at −40°C	kJ/m²	15	15	—	—	15	2	2
Impact strength with notch (Izod)								
at 23°C	J/m	1000	>1100	910	1174	1200	20—50	40
at 0°C	J/m	800	—	—	1068	1000	—	—
at −20°C	J/m	420	—	—	1014	600	—	—
at −40°C	J/m	130	—	160	694	250	—	33
Hardness (Sphere)	MPa	90	105	112	57	75	104	121
Hardness (Rockwell R)	—	—	120	—	—	—	—	—
Melting temperature	°C	217—221	185—190	255	215	155	217—221	240
Vicat temperature	°C	185—190	122	—	—	—	>200	—
Heat distortion temperature (at 1.81 MPa)	°C	—	103	71	53	52	54	90
Water absorption (at saturation)	%	<6	6	6.7	1.5[2]	6	8	8.5
Dielectric strength	kV/mm	>80	>80	—	—	40	>80	—
Surface resistivity	Ohm	10^16	>10^14	10^14	—	>10^15	10^14	10^15
Volume resistivity	Ohm × cm	10^15	>10^16	—	—	—	10^15	—
Relative permittivity (at 1 kHz)	—	3.4	3.6	3.2	—	3.2	3.7	3.9
Dissipation factors (at 1 kHz)	—	0.015	0.014	0.010	—	0.010	0.013	0.020

[1] Water content 0.2%.

[2] after 24 h.

Applications include sporting and recreational items, e.g., ski boots, blade support in skates, roller skate sole plates and trucks, surfboard mast joints, bike rims, and recently all-plastic bike wheels. In the roller skates, the usually used aluminum could be replaced by polyamide. Other applications are appliances, gearwheels, safety helmets, and wheel covers for 1982 model cars. Developing applications are in the automotive, motorcycle, and electro fields. Recently reported uses of reinforced super-tough polyamides were in the injection molding of a complicated one-piece motorcycle seat and a two-piece sun shield that will be fixed to the rear windows of European cars, and in the manufacture of racing-boat components [8].

References

[1] B. Brassat and W. Nielinger, Kunststoffe **71** (1981) 172.
[2] D. Neuray and K. H. Ott, Angew. Makromol. Chem. **98** (1981) 213.
[3] ®Durethan VP KL 1-2310/2 and KL 1-2311, Bayer A. G., Technical Information 1981.
[4] ®Zytel ST 801, ST 811, and ST FR 80 NC-10, DuPont, Technical Information 1980 and 1981.
[5] E. A. Flexman, Jr., Kunststoffe **69** (1979) 172.
[6] S. Schaaf and W. Isler (Ems Chemie A. G.), personal communications (October 1982).
[7] ®Grilon A28NX, ®Grilon A28NY, and ®Grilon A28NZ, Ems-Chemie A. G., Technical Information, June 1982; ®Grilon BT40, Ems-Chemie A. G., Technical Information, April 1982; ®Grilon PVN-3H, Ems-Chemie A. G., Technical Information, March 1982.
[8] A. Sternfield, Modern Plast. Intern. **12** (February 1982) 22.
[9] ®Beetle Nylon AC1 and EN30, British Industrial Plastics Ltd., Technical Information.

9.3 BLOCK COPOLYMERS OF POLY(DIOXAMIDE) AND POLYAMIDE 6

9.3.1 Structure and synthesis

The Sun Company (Philadelphia) produces pilot plant quantities of a new synthetic fiber designated "Fiber S" [1, 2]. It is composed of block copolymers which are prepared by reacting poly(dioxamides) and a spinnable polyamide, preferably polyamide 6, in the melt [3].

The synthesis of poly(dioxamides) is carried out by cyanoethylation of unsaturated nitriles (I) with diols (II), hydrogenation of the obtained dinitriles (III), and reaction of the formed diamines (IV) with dicarboxylic acids (V) to the intermediate salt which is subsequently polycondensated to VI:

(9-2)

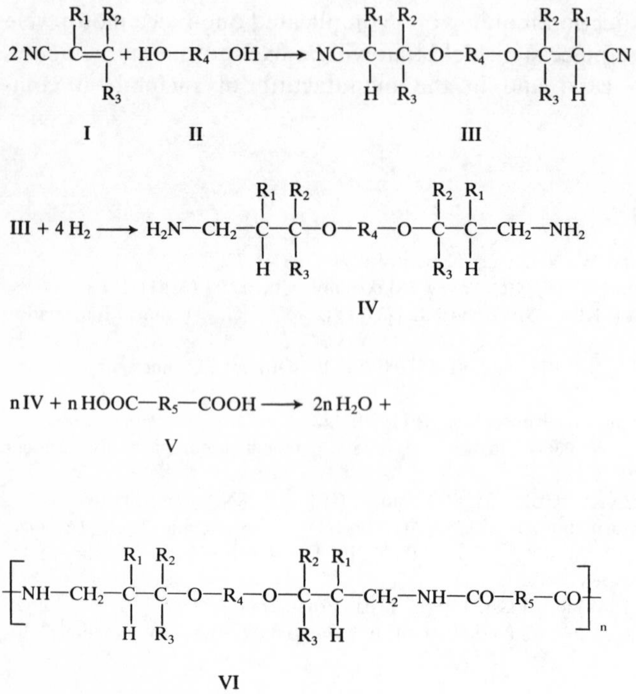

$$2\,NC-\underset{\underset{R_3}{|}}{\overset{\overset{R_1}{|}}{C}}=\underset{}{\overset{\overset{R_2}{|}}{C}} + HO-R_4-OH \longrightarrow NC-\underset{\underset{H}{|}}{\overset{\overset{R_1}{|}}{C}}-\underset{\underset{R_3}{|}}{\overset{\overset{R_2}{|}}{C}}-O-R_4-O-\underset{\underset{R_3}{|}}{\overset{\overset{R_2}{|}}{C}}-\underset{\underset{H}{|}}{\overset{\overset{R_1}{|}}{C}}-CN$$

I   II   III

$$III + 4\,H_2 \longrightarrow H_2N-CH_2-\underset{\underset{H}{|}}{\overset{\overset{R_1}{|}}{C}}-\underset{\underset{R_3}{|}}{\overset{\overset{R_2}{|}}{C}}-O-R_4-O-\underset{\underset{R_3}{|}}{\overset{\overset{R_2}{|}}{C}}-\underset{\underset{H}{|}}{\overset{\overset{R_1}{|}}{C}}-CH_2-NH_2$$

IV

$$n\,IV + n\,HOOC-R_5-COOH \longrightarrow 2n\,H_2O +$$

V

$$\left[NH-CH_2-\underset{\underset{H}{|}}{\overset{\overset{R_1}{|}}{C}}-\underset{\underset{R_3}{|}}{\overset{\overset{R_2}{|}}{C}}-O-R_4-O-\underset{\underset{R_3}{|}}{\overset{\overset{R_2}{|}}{C}}-\underset{\underset{H}{|}}{\overset{\overset{R_1}{|}}{C}}-CH_2-NH-CO-R_5-CO\right]_n$$

VI

R_1, R_2, and R_3 are hydrogen, alkyl, or isoalkyl groups, R_4 and R_5 are alkylene or isoalkylene groups, each with up to ten carbon atoms. The preferred reactants are evidently acrylonitrile, ethylene glycol, and adipic acid, which according to the above formulated reaction sequence yield poly(4,7-dioxadecamethylene-adipamide) VII. PA 6 and VII are melt reacted at temperatures from 282 to 305°C at residence times up to 3 h to give the block copolymer VIII which is spun directly from the melt:

$$\left[NH-(CH_2)_3-O-(CH_2)_2-O-(CH_2)_3-NH-CO-(CH_2)_4-CO \right]_x \Bigg/$$

$$\left[NH-(CH_2)_5-CO \right]_y \qquad \text{VIII}$$

The blocks x and y comprise from 10 to 130 units; the molecular weight of the block copolymers is between 5000 and 100 000 g/mol.

9.3.2 Properties

With increasing content of VII in VIII, the fiber absorbs increasingly more moisture. At 30% VII in VIII, moisture absorption is markedly higher than in polyamide 6 and reaches values that are equivalent to cotton (Table 9-6). On heat-aging, the copolymers suffer considerable loss of strength due to the ether groups present; the degradation is kept in limits by addition of suitable stabilizers. The receptivity for dyestuffs is superior to nylon 6.

TABLE 9-6
Comparison of block copolymer VIII with polyamide 6 and cotton.

Physical property	Physical unit	Property value		
		PA 6	VIII	Cotton
Inherent viscosity	mL/g	110	103	—
Melt temperature	°C	219	214–220	—
Tensile strength at break	g/tex	33.3	18.9	—
Elongation at break	%	45	65	—
Residual tensile strength after 2 h at 120°C	%	98	70[3]–108[4]	—
Water absorption				
at 65% RH	%	4.1	6.0	7.0
at 75% RH	%	4.5	8.6	9.5
at 85% RH	%	5.8	12.1	11.8
at 95% RH	%	7.6	15.5	14.5

[1] Block copolymer VIII with 30% VII, blended 30 min at 295°C, then spun. Draw ratio 3.7.
[2] In m-cresol.
[3] With stabilizer.
[4] Without stabilizer.

The S-Fiber has already been tested by several American and European fiber producers. The decision to start up commercial production with SNIA in Milan has not yet been made [2].

References

[1] Anon., Chemical Week (January 16, 1980) 33.
[2] W. J. Stout (Sun Co., Specialty Chemical Developments), personal communication (June 2, 1981).
[3] R. M. Thompson (Sun Ventures Inc.), USP 4 130 602 (1977).

9.4 POLYAMIDE-RIM-SYSTEMS

9.4.1 Structure and synthesis

Many companies are actively engaged in adapting the Reaction Injection Molding (RIM) process to other polymers besides the established PUR. RIM has several advantages over injection molding: less energy is consumed, faster machine cycles are possible, larger parts can be fabricated, clamping forces for tools are lower, and monomers cost less than polymers.

The feasibility of RIM processing has been studied with styrene-containing and styrene-free polyesters, epoxides, polyamides, poly(acrylates), poly(methacrylates), and poly(styrenes) [1−3]. The first commercial material was ®Asterite, introduced by ICI in 1981. It is composed of 28% methylmethacrylate, 70% silica, and small portions of pigments, dispersing agent, silanes, and curing agents [4, 5].

In 1981, Monsanto introduced a developmental "NBC-RIM" (Nylon-Block-Copolymer-RIM) [6, 7]. The block copolymer is prepared from poly(ethylene glycol), poly(propylene glycol) or poly(butadiene) containing hydroxy groups, and caprolactam:

$$H \left[O-R-O \right]_x \qquad \left[CO-(CH_2)_5-NH \right]_y H$$

By varying the polyether portion between 0 and 70%, the properties can correlate with those of a pure PA 6 or an elastomer. Preferably poly(propylene glycol) in portions of 20% is used. The system was commercialized in 1982; Monsanto has been marketing the unreinforced ®Nyrim 2000 and the ®Nyrim 2025 G containing 25% hammer-milled glass [8, 9, 10]. Monsanto supplies only the prepolymer and reactive catalyst formulations to be combined in the RIM process with caprolactam. It does not supply the latter which is a readily available commodity. One component contains the polyether and adipyl-bis-caprolactam as chain propagator. The catalyst is a caprolactam-magnesium bromide complex dissolved in caprolactam. It must be protected against moisture, oxygen, and CO_2.

The block copolymer is produced in RIM equipment which is specially designed to meet the nylon RIM requirements. By the end of 1982, the following companies were reported to have developed nylon RIM and/or RRIM machines: Admiral (L-10 Motoman), Afrós Cannon SRL, Cincinnati Milacron (HT-RRIM model), Battenfeld, Accuratio, Krauss-Maffei (NBC RIM-Star) and Elastogran (Puromat HT) [9, 11, 12].

Another route to polyamide-RIM-systems has been studied by Upjohn. Cyclic enamines were found to react readily with isocyanates at room temperature [13]. The reactivity can be controlled via ring size and/or functionality of the amines. The new systems have not been commercialized as yet. Still another nylon RIM system is being developed by Allied Fibers and Plastics but details have not been disclosed except that it is anionically polymerized with sodium catalysts [11].

Besides the far advanced and established PUR-RIM, nylon-RIM seems to be the most developed process. Other potential candidates are epoxy RIM and polyester RIM systems. Shell Chemical supplies the developmental epoxy RIM materials EPON DRH-504 and DRH-506 in pilot plant quantities [11]. Gloucester Engineering Co. introduced a developmental reinforced polyester RIM 3-component machine designed to impinge resin, catalyst/activator, and blowing agent in a 90:5:5 ratio [14]. The resin is a bisphenolmethyl-based molding compound supplied by Whitney & Co. A block copolymer resin formulation incorporating up to 20% butadiene is used to reduce brittleness and improve impact strength of both reinforced and unfilled materials.

An entirely different approach to new RIM systems utilizes the interpenetrating network (IPN) principle. It refers to a type of alloy consisting of intimate mixtures of two or more polymer networks held together by permanent entanglements. To prepare IPN's, the network preformed from the first component is swollen by the monomer of the second component. After polymerization and crosslinking of the second component, two mutually interpenetrating, but individually independent, networks exist [15]. The first commercial IPN is Shell Chemical's injection moldable ®Kraton GX-7500 which was introduced in April 1982. The styrene-butadiene block elastomer forms a highly synergistic IPN with an undisclosed modified polymer.

9.4.2 Properties and applications

In nylon RIM the polyol and nylon blocks are incompatible and form a two-phase system [16]. The sub-micron blocks of crystalline nylon 6 retain their identity and properties, and this accounts for the high melting point, rigidity, and toughness. The polyether phase is an elastomer with a T_g of $-60°C$; in small portions it acquires toughness and impact strength even at low temperatures. Properties of unfilled and reinforced PUR-RIM and nylon-RIM are compared in Table 9-7.

Nylon RIM moldings have excellent surface quality and can withstand automobile paint oven temperatures. For this purpose, the key property is sag resistance, which is said to be reliable at temperatures as high as $160-175°C$.

The claimed advantages of nylon RIM over PUR RIM are the better shelflife of the prepolymer mixture which only reacts catalytically at elevated temperatures and the viscosity of the components which is only 1/10 of the PUR components. Furthermore, with nylon RIM less metering and control equipment is needed; the properties depend on the preset composition of the components mixed outside of the machine, whereas PUR RIM requires the continuous control of a constant ratio for the components mixed in the machine. The reaction leading to nylon RIM is endothermic and requires temperatures of $130-140°C$; the reaction to PUR is strongly exothermic and allows no fabrication of very thick-walled parts. The cycle time for nylon RIM is $2-3$ min compared to PUR RIM with $2-4$ min. With nylon RIM, mold release agents are unnecessary. Nylon RIM scrap can be recycled for injection molding, though not for RRIM.

The disadvantages of nylon RIM compared to PUR RIM are the necessity of heated storage tanks and reactant stream lines, the higher reaction temperature, and the not yet fully-developed RIM

TABLE 9-7
Comparison of Nylon NBC (with 20% poly(propylene glycol)) with PUR-RIM.

Physical Property	Physical unit	Nylon NBC not reinforced	Nylon NBC Glass fiber reinforced	Property values PUR RIM 125 not reinforced	PUR RIM 125 Glass fiber reinforced	PUR RIM 200 not reinforced
Glass fiber content	%	0	25	0	16	0
Density	g/cm^3	1.1	—	1.0	—	1.0
Tensile strength	MPa	35.2	—	28.3–31.0	—	31.0
Elongation at break	%	270	—	100–120	—	50
Tear strength (C)	MPa	6.8	—	5.2–5.9	—	6.9
Flexural modulus						
at −29°C	GPa	1.72	—	1.52–1.80	—	2.62
at 22°C	GPa	0.79	1.90	0.86	1.46	1.38
at 70°C	GPa	0.43	—	0.26–0.31	—	0.41
Impact strength with notch						
Izod	kJ/m	0.69	—	0.27	—	0.11
Gardner, at Raumtemp.	in.-lb	160	140	80	16	—
at −29°C	in.-lb	160	—	<20	—	—
Shore hardness (D)	—	70	—	70	—	80
Shrinkage						
after 1 h at 121°C	mm	—	—	7.6–15.2	—	12.7
after 1 h at 163°C	mm	1.52	—	—	—	—
Linear thermal expansion coefficient ($\times 10^5$)	K^{-1}	126	52	108	61	—

machinery. Another drawback is that the catalyst formulation must be handled under a blanket of dry nitrogen. Also, as with nylon generally, moisture is a problem that can lead to dimensional instability. According to General Motors, a nonreinforced RIM nylon part expands 0.69% going from dry to 50% relative humidity. This would result in an unacceptable expansion in long fenders, with a loading of 25% milled glass fiber, expansion is cut to an acceptable 0.25% [17].

Nylon RIM is a strong candidate for exterior auto body components and large parts in appliance and business machine markets. Reinforced RIM nylon could be an attractive option for fenders, door panels, and fender extensions.

Monsanto sells the Nyrim prepolymer/catalyst components at 4 US-$/lb, yielding a formulated cost for a typical unreinforced material of 1.90 US-$/lb. A formulation containing 25% milled glass costs about 1.60 US-$/lb [9]. On a volume basis, reinforced RIM nylon costs 6.9 cts/cu. in. compared to 4.3 cts/cu. in. for reinforced RIM PUR and 4.5 cts/cu. in. for compression molded SMC [17].

A number of special RIM processing machines have been developed (see, e.g., [18, 19]). The properties and processing characteristics for nylon block copolymer RIM have been reviewed [20]. Recent papers discuss the application of RIM polyamides for automotive body panels and other parts [21, 22].

References

[1] R. S. Kubiak, Plastics Engineering **36** (March 1980) 55.
[2] G. Ferber, SPE ANTEC **25** (November 1979) 56.
[3] R. S. Kubiak and R. C. Harper, SPE ANTEC **25** (November 1979) 59.
[4] D. Sandiford, SPE **27** (May 1981) 374.
[5] ®Asterite, ICI Ltd., Technical Information 1981.
[6] A. St. Wood, Modern Plast. Intern. **11** (April 1981) 38.
[7] Nylon Block Copolymer (NBC-RIM), Monsanto Plastics & Resins Co., Technical Bulletin No. 6510, 1981.
[8] Anon., Chem. Engng. News (May 10, 1982) 42.
[9] Anon., Modern Plast. Intern. (June 1982) 15.
[10] D. Sandiford, Modern Plast. Intern. (July 1982) 28.
[11] J. Sneller, Modern Plast. Intern. (May 1982) 75.
[12] Anon., Modern Plast. Intern. **44** (October 1982) 78.
[13] L. M. Alberino and D. F. Regelman, Org. Coatings Plast. Chem. **44** (1981) 151.
[14] Anon., Modern Plast. Intern. (July 1982) 18.
[15] H.-G. Elias, Makromoleküle, 4th ed., Hüthig & Wepf Verlag, Basel-Heidel-

berg-New York 1981; Macromolecules, Plenum Press, New York, 2nd ed., 1984, vol. 2.

[16] Anon., Plastics Technol. **27** (April 1981) 29.
[17] A. St. Wood, Modern Plast. Intern. **12** (October 1982) 46.
[18] P. Wagner, Kunststoffe **73**/10 (1983) 588.
[19] Anon., Kunststoffe **73**/10 (1983) 595.
[20] C. R. Dupre, J. D. Gabbert and R. M. Hedrick, ACS Polym. Preprints **25**/2 (1984) 296.
[21] M. D. Skirha, ACS Polym. Preprints **25**/2 (1984) 298.
[22] Anon., Kunststoffe **74**/5 (1984) 300.

9.5   AROMATIC POLYETHERAMIDES

Two Japanese companies recently presented aromatic polyether-amides:

Hitachi, Ltd., produces an experimental aromatic polyamide through interfacial polycondensation of 2,2-bis[4-(4-amino-phenoxy)-phenyl]propane with a mixture of terephthalic and isophthalic acid chlorides [1]:

(9-3)

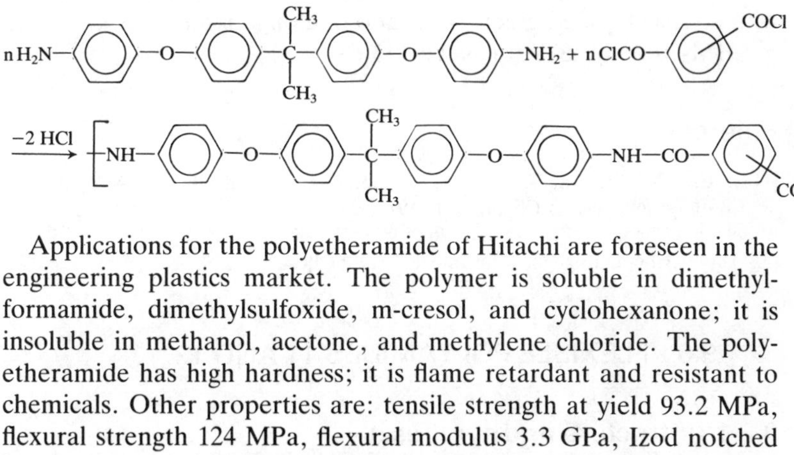

Applications for the polyetheramide of Hitachi are foreseen in the engineering plastics market. The polymer is soluble in dimethyl-formamide, dimethylsulfoxide, m-cresol, and cyclohexanone; it is insoluble in methanol, acetone, and methylene chloride. The poly-etheramide has high hardness; it is flame retardant and resistant to chemicals. Other properties are: tensile strength at yield 93.2 MPa, flexural strength 124 MPa, flexural modulus 3.3 GPa, Izod notched impact strength 0.12 kJ/m, and heat deflection temperature 198°C.

Teijin, Ltd., introduced another polyetheramide, under the de-signation HM-50, which is expected to compete with DuPont's ®Kevlar [2−4] (see also chapter 10). The polymer is obtained through polycondensation of terephthalic acid chloride I with a mixture of p-phenylenediamine II and 3,4'-diaminodiphenylether III in polar solvents [2]:

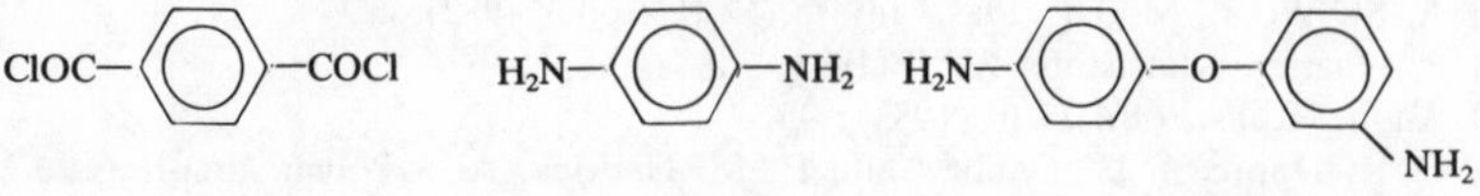

The molar ratio of the diamines applied in the polycondensation is given as 1:1 (see p. 294). After the reaction, it is neutralized by treatment with calcium hydroxide and the polymer is extruded into filaments [2]. The filaments are washed, dried, and drawn at 500°C [2].

The new polyetheramide HM-50 melts at 515°C, whereas Kevlar, i.e., poly(p-phenyleneterephthalamide), does not melt but decomposes at higher temperatures. HM-50 is said to be more resistant to sodium hydroxide, sulfuric acid, hydrochloric acid, nitrous acid, phosphoric acid, and acetic acid than Kevlar [2]. On the other side, HM-50 shows less stiffness than Kevlar [2]. Tenacity is reported to be 26 g/den, i.e., a breaking length of 234 km, compared to Kevlar with a tenacity of 22 g/den. The ratio of elongation to break is 4.2% for HM-50 and 3.8% for Kevlar [3].

References

[1] S. Era, M. Shitara, K. Nanumi, F. Shoji, and H. Kohkame, ACS Organic Coatings and Plastics Chem. **40**/1 (1979) 909.
[2] Anon., Chem. Week (October 28, 1981) 52.
[3] Anon., Chem. Engngn. (November 2, 1981) 19.

9.6   FLEXIBLE POLYAMIDES 11 AND 12

9.6.1   Structure and synthesis

With regard to properties, the flexible polyamides 11 and 12 can be classified between thermoplastic polyamide elastomers (see chapter 4.5) and elastomer-modified polyamides with high toughness. Of course, there is a certain overlap of properties among these three product types.

Improved flexibility is achieved by either incorporation of plasticizers or by modification of the basic PA 11 and PA 12. Chemische
Werke Hüls supplies the ®Vestamid grades L 1722, L 1723, L 1724,
L 2121, L 2122, L 2124, L 2224, and L 2128 which are PA 12 polymers containing plasticizers [1].

ATO Chimie has marketed the ®Rilsan Superflexible grades
BMN F15, BMN F25, BESN F15, and MBNO P40 based on PA 11
or PA 12 [2]. PA 12 based polymers with improved flexibility are
also supplied by Ems-Chemie AG. The grades ®Grilamid L25W40,
L25W20, and L25N150 are said to be equivalent to the Hüls products mentioned above [3]. The more recent ®Grilamid L25W40NZ
extrusion grade is a modified PA 12 [4, 5]. The nature of the modification has not been revealed.

9.6.2 Properties and applications

The flexibility of the toughened PA engineering plastics is inade-
quate for many applications, in particular at very low temperatures.
A crucial test for PA 11 and PA 12 is the notched impact test at
−40°C.

Flexibility can be improved with a trade-off in strength, flexural
and torsion moduli, hardness, heat deflection temperature, and
chemical resistance (Table 9-8). Compared to the conventional PA
11 and PA 12 grades, the elongation, impact and notched impact
strengths are improved thus enabling the flexible polyamides to
compete with thermoplastic elastomers of equivalent hardness.
However, the plasticizer containing flexible polyamides are not approved for direct contact with food.

The flexible polyamides can be processed by injection molding
and extrusion.

Applications include air pressure brake hoses [5], tubing [6],
rollers, gaskets, ski and hiking shoe solings, replacement of leather
in belts and bike saddles, mandrels in rubber tubing manufacture,
and cable jackets. Vestamid L 2124 is recommended for the production of very flexible monofilaments.

In air pressure brake conduits with reinforced hoses, PA 6 and
PA 66 can not be used because they are susceptible to environmental stress cracking when the non-ferrous metal couplings are
exposed to de-icing salt. The new flexible PA 12 meets the strict

TABLE 9-8

Properties of flexible polyamids 11 and 12. Property values for Vestamid measured immediately after injection molding; values for Grilamid after 14 days at 23°C/50% RH.

Physical property	Physical unit	Vestamid L2121	Vestamid L2128	Grilamid L25W40NZ	Rilsan BMNF25	Rilsan BMNF25	Rilsan BMNOP40
				Property value			
Density	g/cm^3	1.025	1.035	0.98	1.06	1.06	1.05
Water absorption (23°C/65% RH)	%	0.8	0.5	0.4	1	1	1.1
Tensile strength at yields	MPa	30	15	20	11	7	18
Elongation at yield	%	20	25	30	17	17	25
Tensile strength at break	MPa	45	24	30	38–42	30–31	47–53
Elongation at break	%	250	250	380	300–370	280–380	300–350
Modulus of elasticity	MPa	750	300	250	—	—	—
Flexural modulus	MPa	—	—	215	150	120	350
Flexural strength	MPa	19	7	14	—	—	—
Shore hardness (D)	—	72	56	57	52	43	63
Impact strength with notch							
Charpy at room temp.	kJ/m^2	n.b.	n.b.	n.b.	n.b.	n.b.	340
at –40°C	kJ/m^2	5	6	10	70	80	21
Melt temperature	°C	165–168	162–168	170–175	170–180	170–180	180–185
Vicat temperature (B)	°C	140	95	—	93–97	76–80	143
Heat distortion temperature (at 0.46 MPa)	°C	140	120	—	100–105	—	130–145
Volume resistivity	Ohm × cm	3×10^{11}	7×10^9	—	5×10^9	3×10^9	7×10^{10}
Surface resistivity	Ohm	5×10^{10}	1×10^{10}	—	—	—	—
Dissipation factor (1 kHz)	—	15×10^{-2}	55×10^{-2}	—	—	—	—
Relative permittivity (1 kHz)	—	10	27	—	—	—	—
Dielectric strength	kV/mm	55	45	—	—	—	—

n.b. = no break.

specifications SAE J844d and ISO TC 22/SC2/WG1 of the automotive industry [4, 5].

References

[1] ®Vestamid Polyamid 12, Chemische Werke Hüls, Technical Information, May 1981.
[2] ®Rilsan Superflexible, ATO Chimie, Technical Information, 1979.
[3] W. Isler (Ems-Chemie A. G.), personal communication, October 1982.
[4] ®Grilamid L25W40NZ, Ems-Chemie A. G., Technical Information, March and May 1982.
[5] S. Schaaf, Kunststoffe Plastics **12** (1981) 9.
[6] ®Rilsan tubings, ATO Chimie, Technical Information, March 1979.

9.7   POLYAMIDE 4,6

Poly(tetramethylene adipamide) was already known to W. H. Carothers about 50 years ago but the commercialization of this polyamide faced two problems: a low-cost source of 1,4-diaminobutane and the brown color of the polymer.

DSM (Geleen, The Netherlands) devised a new route to 1,4-diaminobutane. Addition of hydrogen cyanide to acrylonitrile leads to succinic nitrile which in turn is hydrogenated to 1,4-diaminobutane [1]

(9-4)

$$H_2C\!\!=\!\!CH\!-\!CN \xrightarrow{\;+\,HCN\;} NC\!-\!CH_2\!-\!CH_2\!-\!CN \xrightarrow{\;+\,H_2\;} H_2N\!-\!(CH_2)_4\!-\!NH_2$$

The development of polyamide 4,6 seems to be raw material driven: DSM has plenty of acrylonitrile (capacity: 150 000 t per year) and also hydrogen cyanide. Indeed, DSM is also looking into the use of 1,4-diaminobutane for other polyamides, e.g. polyamide 4,2 [2].

The polycondensation reaction follows the classical route: a salt is formed first from 1,4-diaminobutane and adipic acid. Polycondensation under medium pressure results in a prepolymer which is recovered as a solid. The prepolymer is then heated to about 250°C in an atmosphere of nitrogen and steam. The polycondensation process is clearly a solid state reaction since the melting temperature of polyamide 46 is at least 308°C [3] whereas the glass transition temperature is relatively low at 43°C [3].

Polyamide 46 has the amide group regularly spaced by 4 methylene groups each whereas in polyamide 6,6 the methylene groups occur in alternating arrays of 4 and 6. This feature explains why polyamide 4,6 has a higher crystallization rate and a greater crystallinity than polyamide 6,6.

Polyamide 46 is said to exhibit some outstanding physical characteristics [1]: tensile strengths are higher than those of existing polyamides, impact strength is twice as high as those of PA 6,6, PA 6, and PBTP, abrasion resistance is ahead of other engineering plastics.

Polyamide 46 will be marketed by DSM under the trade name Stanyl®. Commercialization is expected for 1989.

Polyamide 46 can also be spun into fibers which may find use as tire cords. The strength of polyamide 46 yarn is unchanged after heating under vulcanization conditions for an hour at 170°C. Flat-spotting does not seem to be a factor for Stanyl yarn.

References

[1] D. O'Sullivan, Chem. Engng. News (21 May 1984) 33.
[2] R. J. Gaymans, V. S. Vankatraman and J. Schuijer, J. Polym. Sci.-Polym. Chem. Ed. **22**, 1373 (1984).
[3] R. G. Beaman and F. B. Cramer, J. Polym. Sci. **21**, 223 (1956).

9.8 POLY(M-XYLYLENE ADIPAMIDE)

Poly(m-xylylene adipamide) was developed as fiber for tire cord in Japan many years ago [1]. It is now being marketed as engineering plastics by Solvay under the trade name Ixef® Polyarylamide (formerly also as Nyref or polyarylamide MXD 6) [2–4].

Ixef exhibits glass transition temperatures between 85 and 100°C and melting temperatures of 235–240°C. Its LOI value is 27.5. The modulus of elasticity in the dry state is 13.4 GPa, after saturation with water (1.7%) at 65% RH 11.3 GPa. The tensile strength at break is 177 MPa, the elongation 2.7%. The polymer is very impact resistant: Izod impact strength is 1100 J/m, notched impact strength 159 J/m. Grades with 30 and 50% glass fibers are also available.

References

[1] H.-G. Elias, Makromoleküle, Hüthig and Wepf, Basle 1971; H.-G. Elias, Macro-molecules, Plenum, New York 1977.
[2] Deutsche Solvay, personal communication (6 September 1984).
[3] Anon., Modern Plastics International **14**/3 (1984) 40.
[4] Anon., Kunststoffe **74**/10 (1984) 565.

10 Aramide fibers

10.1 INTRODUCTION

The US Federal Trade Commission has defined an aramide fiber as "a manufactured fiber in which the fiber-forming substance is a long-chain synthetic polyamide in which at least 85% of the amide linkages are attached directly to the aromatic rings". ISO is considering also those polymers as aramides in which up to 50% of the amide groups are replaced by imide groups. Polyamide-hydrazides and aromatic polyamides with heterocyclic groups other than imide structures are not covered by these definitions; however, the properties of these polymers are comparable to aramides.

Aramide fibers acquired a significant market in the 70's and many new fields of application have been exploited. Aramide fibers have been extensively described in several books and review articles [1–8]. Since 1974, a number of new developments and changes in product patterns have resulted.

10.2 STRUCTURE AND SYNTHESIS

Table 10-1 lists structures, producers, and trade names of commercially available aramide fibers together with developmental products which are or were on the market.

The preferred technical aramides production procedure is the polycondensation of aromatic acid dichlorides with aromatic diamines in solution at low temperatures. Solvents are dimethylacetamide (DMAC), N-methylpyrrolidone (NMP), hexamethylphosphoric triamide (HPT), and tetramethylurea (TMU). Dimethylformamide and dimethylsulfoxide are not useful because of their rapid reaction with acid chlorides. Polymers with higher molecular weights and enhanced solubility are obtained when the solvents contain inorganic salts, such as LiCl or $CaCl_2$.

The first aramide fiber was poly(m-phenylene isophthalamide), MPD-I, with structure I (Table 10-1). ®Nomex is prepared by polycondensation of isophthaloyl chloride with m-phenylenediamine in

TABLE 10-1
Structures, producers, and trade names of aramide fibers

	Structure	Producer	Trade name
I	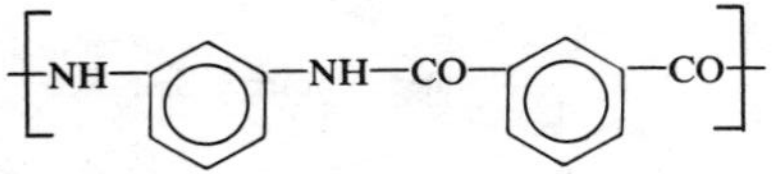 MPD-I	DuPont DuPont Teijin USSR DuPont Monsanto } Firesafe	Nomex Nomex III Conex Phenylon HT-4 Durette
II	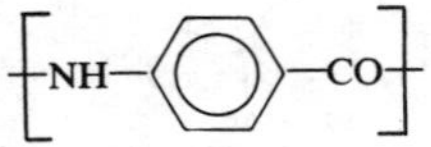 PPB	DuPont USSR	Fiber B Terlon
III	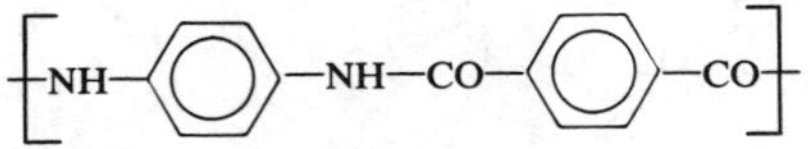 PPD-T	DuPont Enka USSR	Kevlar Arenka Vniivlon
IV	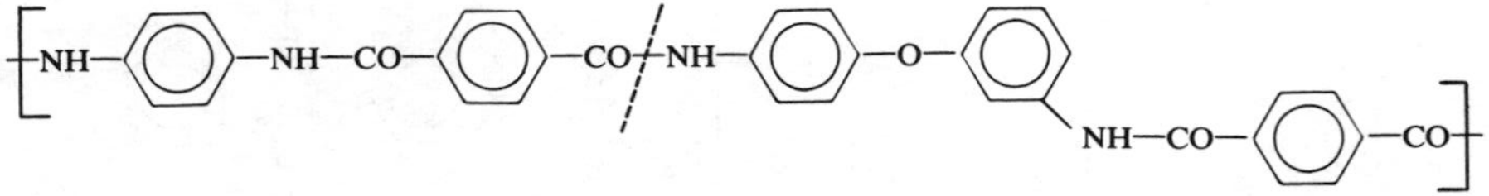	Teijin	HM-50
V	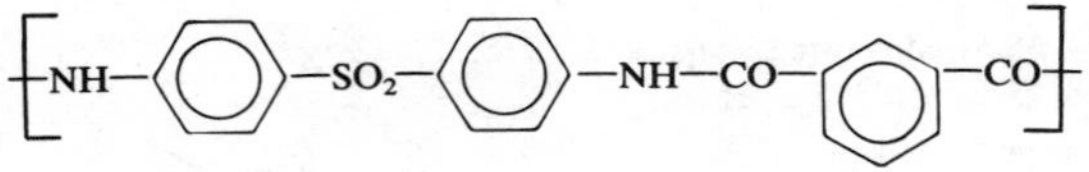	USSR	Sulfon I
VI	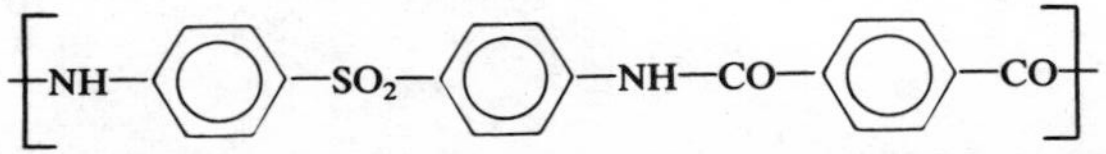	USSR	Sulfon T

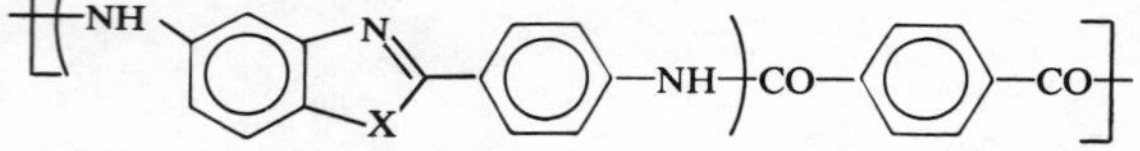

USSR — SVM

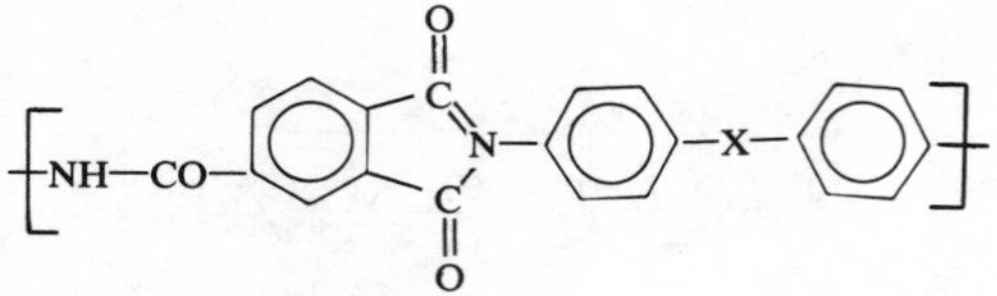

Rhone-Poulenc — Kermel (X = CH$_2$ oder O)

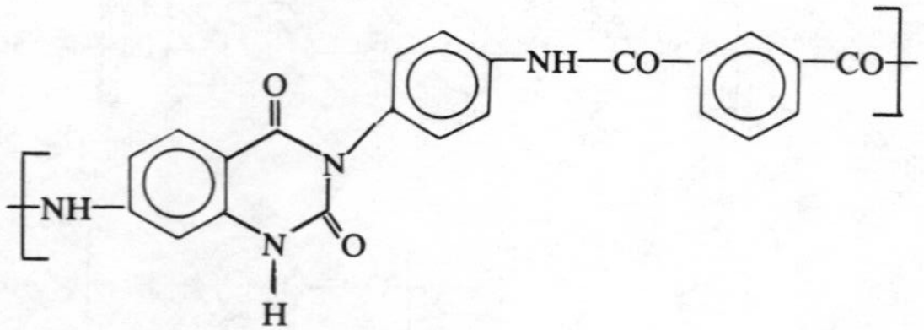

Bayer — AFT-2000

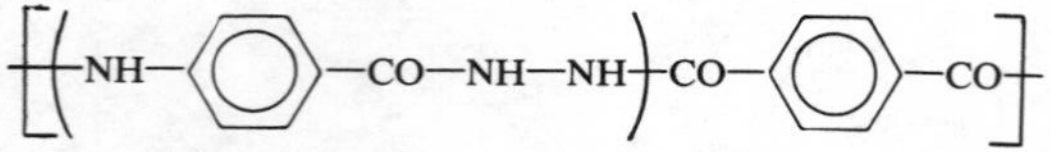

Monsanto — X-500-Series

Goodyear — Flexten

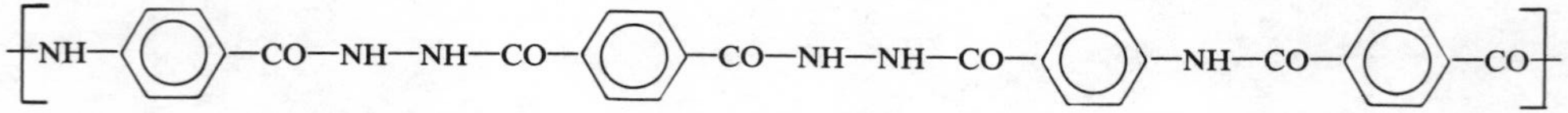

X = —O—, —S—, —NH—; also copolymers with p-phenylenediamine

non-ordered structure

ordered structure

DMAC [9] which acts as a scavenger for hydrogen chloride. After polymerization, it is neutralized with $Ca(OH)_2$; the formed $CaCl_2$ increases the solvating power of DMAC and the solution is then dry-spun. ®Nomex T 430 and T 450 are available as staple fibers of 1.6, 2.2, and 12 dtex, cable of 12 dtex, and continuous filament yarns of 100−1200 dtex, as well as electrical insulating paper. New are the ®Nomex fiber T 455 which is also named ®Nomex III and the developmental ®Nomex T 456 [10]. Nomex III offers improved flame protection and less shrinkage under the influence of flames compared to the earlier Nomex types. This is probably achieved by incorporation of tetrakis(hydroxymethyl)phosphonium chloride and subsequent cross-linking by means of melamine-formaldehyde resins [5, 11]. It is possible that the new Nomex types are identical with the fiber designated HT-4 by DuPont. HT-4 is an experimental Nomex-type fiber with high LOI values and high dimensional stability in a flame and reportedly has a higher heat resistance than Nomex [8]. The DuPont capacity for Nomex was 9100 t/a in 1975 [5]. The total capacity for aramide fibers including ®Kevlar was expanded to ca. 30 000 t/a in 1982 [12].

The production of "Fenilon" or "Phenylon", which also has structure I, is estimated to be less than 4500 t/a [5]. It has been employed as a molding material [6] as well as a fiber. Teijin also produces MPD-I; however, a two-stage interfacial polycondensation is presumably applied [13]. The ®Durette fibers developed by Monsanto, based on MPD-I, are treated with hot chlorine, sulfur chlorides, or sulfuroxychlorides and exhibit improved flame resistance. The fibers contain 9−10% chlorine. Monsanto has cancelled its production of Durette and sold the product line to Firesafe Products Inc., St. Louis.

Poly(p-benzamide), PPB, formerly designated Fiber B by DuPont, has structure II. It can be obtained by polycondensation of p-aminobenzoylchloride hydrochloride [14] or 4(4'-aminobenzamido)benzoyl chloride hydrochloride [15] in DMAC or TMU. DuPont stopped production of PPB, probably because of high monomer costs, low storage stability of the spinning dope, and the difficulties encountered in redissolving the polymer once precipitated [8]. PPB is still produced in the Soviet Union on pilot plant scale under the designation "Terlon" [5, 16, 17].

The technically and economically most interesting aramide fiber is poly(p-phenyleneterephthalamide), PPD-T, with structure III, which was first introduced by DuPont under the name ®Kevlar. The production of Kevlar starts with mixing a 10−30°C cold solution of p-phenylenediamine in HPT with 85−120°C hot terephthaloyl chloride [18]. The polycondensation is carried out in two stages, at below 95°C and at 40−95°C, respectively, under cooling and in the second stage under high shear. Within seconds, a polymer solution with 6−12% solids is formed. Water is added and the precipitated polymer is filtered and dried. The polymer is redissolved in sulfuric acid and is dry jet-wet-spun into an aqueous sulfuric acid coagulation bath. Use of mixed solvents in the preparation of PPD-T leads to synergistic effects. The highest molecular weights are obtained in a 1:2 weight ratio of HPT to NMP and at about 1:1.4 weight ratio of DMAC to HPT [19]. Solutions of PPD-T with above 6−7% solids can yield anisotropic solutions with the characteristics of liquid crystals. The viscosity of spinning solutions may be decreased by increasing the polymer solids concentration above a certain critical level by changing the solution from the isotropic to the anisotropic state [20]. Fibers spun from anisotropic solutions demonstrate high strength even prior to hot drawing due to the high degree of pre-orientation of the polymer in solution [18]. The production of high strength aramide fibers might be simplified considerably if one could succeed in melt spinning low molecular weight polymers with liquid crystal characteristics to be followed by a solid state polycondensation [17].

DuPont offers PPD-T as tire cord fiber, as ®Kevlar 29, and as high-modulus fiber ®Kevlar 49. The high modulus properties of Kevlar 49 are achieved by hot drawing the fiber in an inert atmosphere at temperatures between 250° and 550°C. Kevlar 29 filaments of 220−16 700 dtex as well as staple, felt, and pulp are available; Kevlar 49 is offered as filament yarn with 220−2350 dtex and as roving for filament winding with 5070 and 7890 dtex.

The Russian fiber "Vniivlon" is composed of PPD-T and heterocyclic chain units [5, 17]. Akzo/Enka introduced two PPD-T fibers under the designations ®Arenka and ®Arenka HM in 1976 [21, 22]. The company has decided to start up a plant with a capacity of 5000 t/a in 1985. DuPont sued Enka in England and in France for patent infringement on spinning of PPD-T from anisotropic solutions containing sulfuric acid [12].

Teijin's high modulus fiber HM-50 with structure IV is produced by polycondensation of terephthaloyl chloride with a 50/50-molar mixture of p-phenylenediamine and 3,4'-diaminodiphenylether in NMP solvent containing $CaCl_2$. Dry-jet-wet-spinning of the isotropic solution with 6% solids followed by hot drawing at temperatures of 460–500°C yields the high modulus fiber [23] (see also chapter 9.5).

A higher flexibility of aramides is achieved by incorporating into the backbone flexibilizing moieties such as -S-, $-SO_2-$, -O-, -CO-, or $-CH_2-$. The best heat resistances are demonstrated by products obtained from the polycondensation of 4,4'-diaminodiphenylsulfone with isophthalic acid or terephthalic acid. The polymers are produced in pilot quantities in the Soviet Union under the designations "Sulfon I" (V) and "Sulfon T" (VI), respectively [24].

Aramides with heterocyclic chain units are produced in the Soviet Union from a benzimidazolediamine and terephthaloyl chloride [5, 17, 25]:

(10-1)

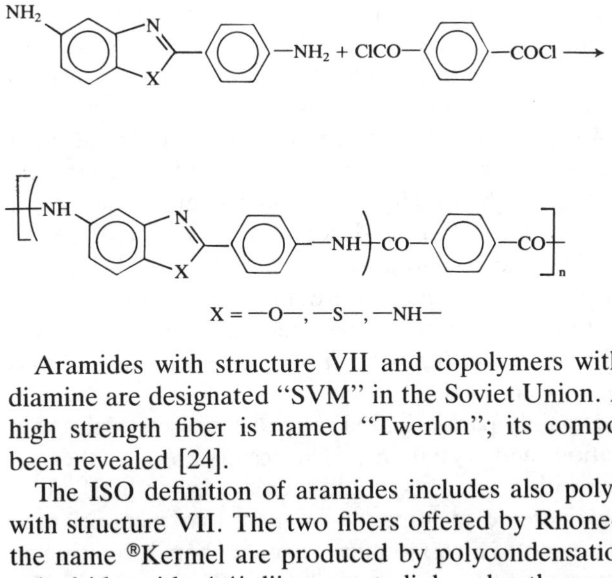

$$X = -O-, -S-, -NH-$$

Aramides with structure VII and copolymers with p-phenylenediamine are designated "SVM" in the Soviet Union. Another Soviet high strength fiber is named "Twerlon"; its composition has not been revealed [24].

The ISO definition of aramides includes also poly(amide-imides) with structure VII. The two fibers offered by Rhone-Poulenc under the name ®Kermel are produced by polycondensation of trimellitic anhydride with 4,4'-diisocyanatodiphenylmethane or 4,4'-diisocyanatodiphenylether in DMAC or NMP at 160–200°C with subsequent spinning from solution [8, 26]. Bayer developed an aramide type fiber designated ATF-2000 (IX) with quinazolinedione structure. It is prepared from 3-(p-aminophenyl)-7-amino-2,4(1H,3H)-

quinazolinedione and isophthaloyl chloride in DMAC [8, 27] (see also chapter 12.1).

The poly(amide-hydrazides) developed by Monsanto under the designation X-500 are a series of fibers with disordered (X) and ordered (XI) structures. The rigid and semi-rigid aramides are based on poly(p-benzamide) and poly(terephthalamide) with p-amino-benzhydrazide [1, 3, 8]. Up to 15 mol-% of meta structures can be incorporated in order to enhance solubility and processability. Goodyear uses PABH-T as fiber cord under the designation "Flexten" which is produced from p-aminobenzhydrazide and terephthaloyl chloride.

10.3 PROPERTIES

In view of practical applications and based on property patterns aramide fibers are categorized in two main groups. The first includes heat and flame resistant fibers and the second type in addition exhibits high strength and high modulus (Table 10-2).

The dominant properties of aramides are the high glass transition and decomposition temperatures. Glass transitions range from 250°C to above 400°C. Weight losses in an inert atmosphere begin at about 425°C, although some of the para-linked rod-like polymers do not lose substantial weight to 550°C. Aramide fibers do not melt. Upon burning, the fibers start to decompose at temperatures of 400°C to above 500°C thereby producing a thick char which acts as a thermal barrier that leaves the texture of a fabric intact for a prolonged period of time. The earlier aramide fibers showed a pronounced heat shrinkage which has been overcome with later developed types, e.g., ®Nomex III.

Completely or mainly para-oriented aramides possess high chain packing, orientation, and crystallinity. The results are high thermal stability together with a three to sixfold increase in modulus and a two to threefold increase in strength compared to meta-oriented aramides. The strength of high modulus aramide fibers is superior to all other fibers prepared from organic polymers. The specific modulus, i.e., modulus divided by density, of such aramides is by far higher than that for steel or glass but does not reach the values for high modulus graphite fibers (Table 10-3).

Aramide fibers demonstrate high tensile strength at elevated tem-

TABLE 10-2
Properties of aramide fibers [8, 17, 23, 24].

Fiber Trade name	Density in g/cm³	Tensile strength in cN/dtex	Elongation at break in %	Tensile modulus in cN/dtex	Water absorption in % at 21°C/65% relative humidity	Oxygen index LOI value
Nomex	1.38	4.4–5.3	15–30	75–145	4.5–5	27–28
Durette	1.41–1.43	2.3–2.9	12–17		4.1	35–37
Conex	1.37	4.4	35–50	76–119		26–30
Phenylon	1.37	4.4	20–25	88–118		25–38
Terlon	1.4	7.8–9.8	1–1.5	595		
Kevlar	1.44	18–21	4–5	300–400	1.5	31
Kevlar 29	1.44	19	4	408		
Kevlar 49	1.45	19	2.1	900–1000		
Arenka	1.44	18.8	3.7	424	6	29
Arenka HM	1.45	18.8	2.3	862	3.5	29
Twerlon	1.4	21.4	4.4	569		
HM-50	1.40	26.5	5.5	495		
Sulfon-T	1.45–1.46		16–18	59		28–30
AFT-2000	1.34		15–20	50–90	9–12	36–40
Kermel 1	1.34	2.5–3.0	10–25	35–40	3.1	
Kermel 2	1.39	4.0–6.0	10–20	45–75		
X-500	1.45–1.47	12–14	3.2	700–800		

TABLE 10-3
Comparison of Aramide high modulus fibers with other fibers (updated from [3, 28]).

Fiber	Density in g/cm^3	Tensile Strength in MPa	Modulus of Elasticity in GPa	Specific Modulus in N·cm·g^{-1}
Kevlar 49	1.45	2760	138	951
Arenka HM	1.45	2700	125	862
X-500 PABH-T (G)	1.47	2100	95	646
Polyester	1.39	620	10	72
Polyamide 6	1.14	600	4.5	39
Graphite A-S[1]	1.75	2820	208	1189
Graphite HT-S[1]	1.77	2820	234–253	1322–1429
Graphite HM-S[1]	1.91	2340	345–375	1806–1963
Graphite UHMS[1]	1.96	1860	485–517	2474–2638
E-glass	2.54	3440	72.5	285
S-glass	2.49	3448	86.2	346
Boron with tungsten core	2.65	3410	386	1457
Aluminum alloy	2.68	520	70	261
Beryllium	1.83	350	320	1749
Steel, drawn	7.75	2600	210	270
Whiskers from				
Si			159	
C	1.9	20000	1034	5442
Fe	7.8	13000	200	256
Quartz			76	
Zirconium			427	

[1] A = high strain/low modulus; HT = high tensile; HM = high modulus; UHM = ultrahigh modulus; S = surface treated.

peratures. The continuous service temperatures of Kevlar 29 and 49 are reportedly 150–180°C (see figure 10-1) [36]. The ®Arenka fibers retain 90% of their mechanical properties after 48 h at 200°C [22], whereas Nylon-6,6 fibers lose strength almost completely at 205°C. For continuous use at elevated temperatures, the meta-type polymers with high elongations at break are better suited than the para-type high modulus aramides. The latter have rather low elongations and they loss elongation rapidly after being heated, due to thermo-oxidative attack, thereby becoming too brittle to be especially useful [5]. Nomex in the form of paper has been reported to have a useful lifetime of only 40 h at 300°C but a useful lifetime of about 1400 h at 250°C. At 177°C, the Nomex fiber retains 80% of its strength after exposure to air for several thousand hours.

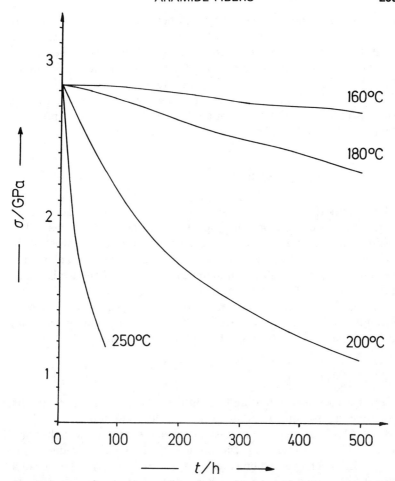

Figure 10-1 Tensile strength vs. time of Kevlar 29 yarn at different temperatures [36]. All measurements at room temperature.

Aramide fibers show better hydrolytic stability than polyester fibers, but they are not as acid resistant, except at elevated temperatures. Compared to nylon-6,6, the aramides have better acid resistance and similar hydrolytic stability. Aramides are not attacked by most of the organic solvents nor by oils, fats, or hydraulic fluids. Their excellent X-ray and gamma radiation resistance is a characteristic property of the aramide fibers. A disadvantage of aramide fibers is that they are exceedingly difficult to dye. Also, they are very susceptible to degradation by ultraviolet light. Upon UV irradiation, the decrease in strength is the more pronounced the thinner the fabric is. The aliphatic polyamides show a similar susceptibility to UV whereas polyesters perform much better [21]. To date, photo-quenchers and UV screening agents, which have been highly successful in the stabilization of aliphatic polyamides, have not been found for aramides [5].

10.4 APPLICATIONS

The remarkable economical success of the aramide fibers is attributed to the unique property spectrum which recently has led to many new and, in some cases, spectacular applications. Most of the information refers to the DuPont aramides, Nomex and Kevlar, because aramides of other companies have not yet been available on the market in sufficient quantities.

Aramide fibers are relatively expensive. Kevlar tire cord costs around 9 $/lb; the prices of Kevlar 29 and 49 vary widely according to yarn fineness in the range from 5 $/lb to 40 $/lb [12].

10.4.1   Tire cord

Kevlar is used as reinforcing fiber in tires for passenger cars, trucks, tractors, cross-country and pleasure vehicles, and motor cycles [29–31]. Kevlar is used in tire carcasses and as the belt in bias-belted and radial-belted tires where it can replace steel. A weight reduction of 0.5–1 kg is achieved when steel cord is replaced by Kevlar in radial tires of passenger cars or, in other words, 1 kg of aramide fiber replaces 5 kg of steel cord while maintaining the same strength. In truck tires tread wear and rolling resistance can be reduced by 10–15% and 8–9%, respectively [31]. Aramide cord is more flexible than steel cord, resulting in more comfortable driving and lessening of the noise level. Tires with aramide are superior to those with steel or rayon under conditions of high lateral guiding forces at steep angles [30]. The advantages of aramide fibers compared to glass fibers are better wear resistance, enhanced compatibility with the matrix, and trouble-free processability on conventional tire building machinery. In the U.S., the share of belts in bias-belted passenger cars is around 70% steel, 20% glass fiber, and 10% Kevlar [31].

10.4.2   Reinforcing material in composites

The use of aramides as reinforcing material in composites has been described in several books and articles [28, 31−38]. Figure 10-2 illustrates the position of Kevlar 49 in relation to other fibers in composites [28]. Epoxy resins are the preferred matrix materials. Table 10-4 lists the properties of unidirectional reinforced EP resins with fiber contents of 60 vol.-% [28, 35, 37].

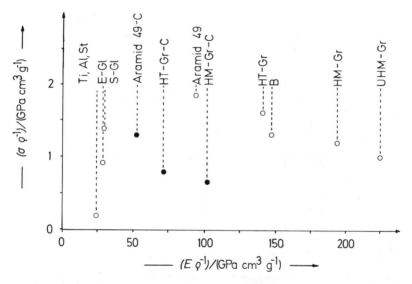

Figure 10-2   Relation between specific tensile strength and specific tensile modulus of fibers (○) and composites (●) [28]. σ = tensile strength, ρ = density, E = tensile modulus, Gr = graphite, Ti = titanium, Al = aluminum, St = steel, Gl = glass, B = boron fiber, HT = high tensile, HM = high modulus, UHM = ultra high modulus.

The use of aramides leads to significant savings in laminate density. The laminates exhibit high tensile and impact strength, good vibration damping, excellent fatigue life, and good electrical properties. Less favorable compared to laminates with graphite is the performance in compression or bending. Unidirectional composites reinforced with Kevlar 49 have a linear stress/strain curve to failure

TABLE 10-4

Properties of fiber reinforced epoxy resins [28, 35, 37].

Composite properties	Physical unit	Property values after reinforcement with						
		Graphite fibers[1]					Kevlar 49	Glass fibers
		A-S	HM-S	HT-S	T 300	GY-70		
Density	g/cm³	1.54	1.63	1.55		1.69	1.38	1.82–2.08
Tensile strength								
longitudinal	MPa	1517	1207	1482	1448	586	1517	1100–1241
transverse	MPa	62	86	90	45	41	28	35–41
Tensile modulus								
longitudinal	GPa	138	221	172	138	276	83	43
transverse	GPa	9.0	13.8	9.0	10.3	8.3	5.5	
Compression strength								
longitudinal	MPa	1172	621	1069	1448	517	276	586–758
transverse	GPa	110	172	165	138	262	76	41
Flexural strength	MPa	1551	1034	1689	1793	931	621	1172
Flexural modulus	GPa	117	193	159	138	262	76	41
Interlaminar shear strength, short beam	MPa	97	72	114	97	52	48–69	72–83
Prepreg cost	US $/lb	45	75	65	45	60	45	6.2

[1] A-S, HM-S and HTS supplied by Hercules; T 300 HT-type supplied by Union Carbide/Toray; GY-70 UHM-type supplied by Celanese/Toho-Belson.

when tested in tension. However, when tested in compression or bending, composites reinforced with Kevlar show metal-like ductility at high strain (Figure 10-3). This property of being elastic at low strain and ductile at high strain is unique when compared to graphite and all inorganic reinforcing fibers.

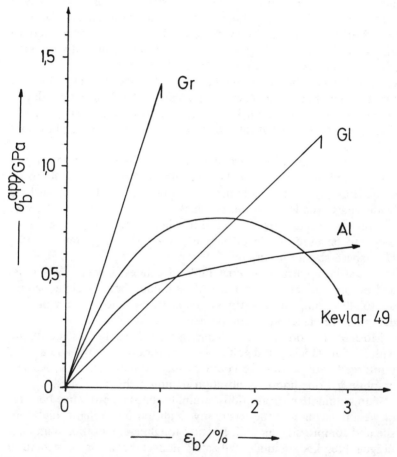

Figure 10-3 Apparent flexural stress vs. flexural elongation of graphite (Gr), glass (Gl), aluminum (Al), and Kevlar 49 fibers in epoxy resins [36].

Aramide fibers can be combined with graphite, boron, or glass fibers in hybrid composites which allow the design engineer to achieve a balance of properties unavailable from single reinforcing fibers [36–38]. Addition of glass fibers improves the compression strength and addition of graphite fibers improves the compression, impact, and flexural yield strengths in composites with aramide fibers.

Composites with Kevlar 49 are used in the aircraft and aerospace industries; however, structural and semi-structural reinforced plastic applications with aramide fibers are also increasing in the civil sector. Weight savings up to 30% are achieved when compared to similar parts made with glass. In newly designed aircrafts (Boeing 757, 767, and Airbus), aramide/graphite hybrids replace the conventional glass fiber reinforced structural parts subjected to high stresses [38]. For this particular application, DuPont is now developing a second generation Kevlar 49 fiber with improved compression strength; its market introduction is expected to take place in a few years.

In aircraft applications, aramide reinforced parts are interior and exterior finishings and trim, honeycomb elements, doors, flooring, seats, fuselage skins, radomes, helicopter rotor blades, turbine blade rings, rudders, elevators, flaps, ailerons, vertical tails, etc. Pressure vessels are manufactured by filament winding and are proposed to be used as rocket engine parts and storage tanks in the U.S. space shuttle program.

In marine applications, boat hulls of canoes, kayaks, sail-boats, and powerboats are fabricated with aramide fibers. Weight savings up to 40%, improved stiffness, impact resistance, and vibration damping are achieved in comparison to glass fibers.

Most of the formula-one racing cars have car bodies reinforced with Kevlar. Hockey and golf clubs, tennis rackets, bows and arrows reinforced with Kevlar are commercially available. In skis, Kevlar/graphite hybrids improve vibration damping characteristics.

In automotive applications, aramides become attractive because of weight saving and fuel economy, high tensile strength, high tensile and compressive moduli, fracture toughness, vibration damping, fatigue life, and unique "fail-safe" impact behavior in structural parts. Potential applications are brackets and door hinges, bumper and anti-intrusion door beams, radiator or transmission supports, and leaf springs. An experimental driveshaft of epoxy resin reinforced with 70% aramide fiber was demonstrated by DuPont in a 1982 experimental car. Weight savings are more than 50% compared to a steel driveshaft. The most spectacular examples of weight savings achieved with aramide fibers are the ultra-light airplanes, Gossamer Albatros and Solar Challenger, where the fibers were used as reinforcement and in stabilizing span wires. The Albatros weights only 32 kg at a wing span larger than that of a DC 9.

In friction products, Kevlar 29 is preferred in clutch and brake friction linings, packings, and gaskets where it can replace asbestos [35]. Kevlar reinforced clutch faces offer weight savings compared to asbestos reinforced clutches and double the values of wear resistance, bursting, and tensile strengths. However, aramide fibers are more difficult to mix with fillers and resins than asbestos fibers. The phenolic resins used in friction products can give relatively low adhesion to chopped Kevlar fibers. A new packing cord of aramide fibers impregnated with a PTFE dispersion is offered by DuPont under the designation Kevlar IT.

10.4.3 Protective clothing and body armor

Heat and flame protective clothing for the military and civil sector are manufactured from the aramide fibers Nomex and Nomex III [39, 40]. Fabrics do not melt and show practically no flame propagation and smoke development. They have good heat insulating capacity, high tenacities and wear resistance, a long service life, as well as good washability and wear comfort. Applications are auto racing drivers' suits, pants, shirts, and coats, smocks for workers in laboratories, foundries and chemical plants, welders' clothing, flight and astronaut suits, and protective clothing for fire fighters, refinery personnel, and tank truck drivers. Other applications include home ironing board covers, press cloths for industrial presses, threads for high-speed sewing, and paper machine felt for high-speed paper machines.

Comparative tests with fabrics from Nomex III (227 g/m^2) and flame-retardant cotton (340 g/m^2) on a dummy showed 23.9% second degree and 4.1% third degree burns after 21 s when the cotton clothing was exposed to flames, whereas with Nomex III only 12.6% second degree burns, based on total body surface area, were observed [40]. Fabrics made of Nomex III are resistant against 95% sulfuric acid at 82°C for a prolonged period of time, whereas a cotton fabric is destroyed in less than 15 s. Nomex III textiles have weight savings of up to 40% compared to flame retardant cotton.

Other applications for Nomex are chemically resistant filter cloth and bags for hot stack gases at continuous service temperatures of up to 200°C and with short-time peaks of 240°C.

Kevlar 29 is preferably used for ballistic protective systems [41, 42]. Applications include bullet and splinter proof clothing, helmets, and armor for cars, planes, and helicopters. Kevlar offers the same bullet protection as nylon at only half the weight. Considerable weight savings while maintaining full protection can be achieved by combining aramides with steel or ceramics in the form of claddings, composites, and sandwich elements.

Cut and stitch proof gloves of Kevlar 29 are commercially available [43]. Woven fabrics of Kevlar demonstrate 100% higher cut and stitch resistance than leather; the insulating property is far better than with asbestos.

10.4.4   Ropes, cables, hoses, belts

Kevlar 29 and 49 are used in the production of ropes and cables [44, 45]. Applications are antenna holding devices, hoist cables for helicopters and cranes, etc., alpinist's ropes, riggings, anchor chains, floating oil barriers, tethering cables, tensioning ropes and cables in oil drilling, and cables for captive weather balloons. In continuous service, aramide cables are superior to steel-wire cables. Replacement steel by Kevlar cables reportedly reduces the costs of a floating platform in off-shore oil drilling by about 3 million US $ [45].

In electro-mechanical cables, aramides are used in armored deepsea telephone cables, in overhead transmission lines, and in sonar cables.

Aramide reinforced materials see service in many other technical fields [31]. In floating oil loading hoses, the use of Kevlar permits savings in buoyant bodies. In automotive engineering, hoses reinforced or armored with aramides are used in radiators, heaters, air conditioning, hydraulic fluids, and fuel lines. Aramide fibers replace glass, polyester, nylon, rayon, and steel fibers in V-belts, transmission and conveyor belts, high-pressure hoses, particularly in heavyduty conveying and transporting equipment in quarries, and in coal and ore mining operations. A new nonskid snow chain consists of a rubber network with a Kevlar core; it has half the weight of a metal chain and reportedly allows speeds of up to 60 km/h on ice, slush, and snow.

10.4.5 Coated fabrics

Relatively new is the use of Kevlar in coated fabrics [31, 46]. Proven applications include air-inflated structures, such as balloons, dinghies, life rafts, and aircraft evacuation slides. An inflatable boat with Kevlar fabric for 10 people with a total load capacity of 2000 kg weighs only 90 kg. In coal mining operations, bellow-framed fresh-air ducts of Kevlar coated fabric are used.

A well-known example for Kevlar coated fabrics is the retractable roof over the Olympic stadium of Montreal. It consists of $18\,000$ m^2 of Kevlar and PVC composite fabric trade-named ®Tolvar and is suspended by Kevlar tension ropes. The Tolvar fabric weighs 51 tons, a similar roof of woven glass or polyester fabric would weigh far more than 100 tons.

Aramide fabrics can be coated with most of the common polymers. In the building industry, coatings with PVC, neoprene, polyurethane, and PTFE are mainly applied. Aramide fabrics must be protected against UV irradiation by impregnation or sheathing. Proven impregnations are polyurethanes and sheathings with polyester or PVF films. Good market opportunities exist for PTFE coated aramide fabrics in inflatable structures for above ground, super structure, and foundation construction work. In exterior applications, the PTFE must contain pigments because the transparent PTFE has no UV protective effect on aramide fibers.

Still under development are applications of aramide coated fabrics in parachutes, parabrakes, sails for racing yachts, and blimps.

References

[1] B. Black and J. Preston (eds.), High-Modulus Wholly Aromatic Fibers, M. Dekker, New York 1973.

[2] G. B. Carter and V. T. J. Schenk, Ultra-high Modulus Organic Fibres, in I. M. Ward (ed.) Structure and Properties of Oriented Polymers, Halsted Press, New York 1975.

[3] H.-G. Elias, Neue polymere Werkstoffe 1969−1974, C. Hanser, München 1975; New commercial polymers 1969−1975, Gordon and Breach, New York 1977.

[4] L. B. Sokolov, Heat Resistant Aromatic Polyamides, Khimiya, Moskau 1975.

[5] J. Preston, Aramid Fibers, in Kirk-Othmer, Encycl. Chem. Technol., vol. 3 (1978).

[6] K. U. Bühler, Spezialplaste, Akademie-Verlag, Berlin 1978.

[7] P. E. Cassidy, Thermally Stable Polymers, M. Dekker, New York 1980.

[8] B. v. Falkai, Synthesefasern, Verlag Chemie, Weinheim 1981.

[9] W. Sweeny (DuPont), US-P. 3287324 (1966).

[10] Flamm- und Hitzeschutzkleidung aus ®Nomex III, Techn. Information DuPont, Geneva 1980 (= flame and heat protective clothings).

[11] B. R. Baird (DuPont), US-P. 3519355 (1976).

[12] P. L. Layman, Chem. Engng. News, February 8, 1982, 23.

[13] T. Ono, Japan Text. News **243** (Febr. 1975) 71; Teijin, DOS 1908297 (1969) and DOS 2325139 (1973).

[14] S. L. Kwolek (DuPont), US-P. 3600350 (1971), US-P. 3819587.

[15] J. Pikl (DuPont), US-P. 3541056 (1970).

[16] Khim. Volokna **6** (1972) 20.

[17] G. A. Butnitskii, Khim. Volokna (March/April 1981) 11; translated from the Russian: Fibre Chemistry (Plenum Press) **13** (March/April 1981) 81.

[18] H. Blades (DuPont), US-P. 3767756 (1973), US-P. 3869429 (1975); J. A. Fitzgerald and K. K. Likhyani (DuPont), US-P. 3850888 (1974), DAS 2219703 (1972); T. I. Bair and P. W. Morgan, US-P. 3673143 and 3817941.

[19] A. Federov et al., Vysokomol. Soedin. [Ser. B] **51**/1 (1973) 74, cited in [5].

[20] S. W. Kwolek, P. W. Morgan, J. R. Schaefgen, and L. W. Gulrich, ACS Polym., Preprints **17**/1 (1976) 53.

[21] K. H. Hillermeier, Kunststoffe **66** (1976) 802.

[22] ®Arenka, a high-tensile, high-modulus aramide, Technical Inform. Enka bv, 1981.

[23] K. Matsuda, ACS Polym. Preprints **20**/1 (1979) 122.

[24] T. Swarski, Lenzinger Berichte **45** (May 1978) 28.

[25] G. I. Kudryavtsev et al., Khim. Volokna **6** (1974) 70.

[26] Soc. Rhodiaceta, DOS 1928435 (1969).

[27] B. v. Falkai, Lenzinger Berichte **36** (1974) 48.

[28] J. Delmonte, Technology of Carbon and Graphite Fiber Composites, Van Nostrand-Reinhold Co., New York 1981.

[29] R. A. Rohlfing, Rubber World **171** (1974) 94.

[30] J. W. Rothuizen, Gummi, Asbest, Kunststoffe **30**/6 (1977) 364.

[31] Symposium DuPont on ®Kevlar, Frankfurt, June 19, 1980; Kautschuk, Gummi, Kunststoffe **34**/1 (1981) 52.

[32] Kohlenstoff- und aramidfaserverstärkte Kunststoffe, VDI, Düsseldorf 1977.

[33] B. Yates, Fabrication Techniques of Advanced Reinforced Plastics, IPC Science and Technology Press Ltd., Guilford 1980.

[34] T. R. Shires and W. A. Willard, Eds., Advanced Composites, NBS Special Publication No. 563, National Bureau of Standards, Washington 1980.

[35] J. H. Sinclair and C. C. Chamis, 34th Ann. Techn. Conf., SPI, RP/C-Institute, Section 22 A, New Orleans, February 1979.

[36] J. C. Norman, ®Kevlar Fiber Developments in Reinforced Plastics and Friction Materials, Technical Information, DuPont, Wilmington, 1979.

[37] D. C. Vanthier, ®Kevlar 49 Aramidfaser, Verstärkung von Kunststoffen, Technical Information, DuPont, Geneva 1981 (reinforcement of plastics).

[38] a) P. G. Riewald and C. Zweben, ®Kevlar 49 Hybrid Composites for Commercial and Aerospace Applications, 30th Ann. Conf. of the SPI Reinforced Plastics/Composites Institute, Washington, February 1975; b) St. Wood, Modern Plastics Intern. **11** (November 1981) 46; c) D. Brownbill, Modern Plastics Internat. **12** (March 1982) 30.

[39] Flamm- und Hitzeschutzkleidung aus ®Nomex für Einsatz Verteidigungs-zwecke, Technical Information, DuPont, Geneva 1980.

[40] ®Nomex III, Technical Information, DuPont, Geneva 1981.

[41] Preliminary Information on ®Kevlar Aramide Fibers in Personnel Protection, Technical Information, DuPont, August 1975.

[42] Leichte Panzersysteme mit ®Kevlar, Technical Information, DuPont, Geneva 1982 (body armor with Kevlar).

[43] Handschutz mit ®Kevlar 29 Fasern, Technical Information, DuPont, Geneva 1982 (hand protection).

[44] Vorläufige Bewertung von ®Kevlar 29 und ®Kevlar 49 bei der Verwendung in Kabeln und Seilen, Technical Information, DuPont, Geneva 1978 (preliminary rating of Kevlar for the use in cables and ropes).

[45] N. O'Hear, Manuskript: Seile und Kabel aus ®Kevlar, DuPont, Geneva 1980 (manuscript: ropes and cables).

[46] Eigenschaften von beschichteten Geweben mit besonderer Berücksichtigung von Schalentragwerken, Technical Information, DuPont, Geneva 1976 (properties of coated fabrics especially for monocoque structures).

11 Polyimides

11.1 INTRODUCTION

Polyimides comprise the largest and most important group of thermally stable polymers. Their production climbed from 300 t in 1971 to more than 3000 t in 1980. Thereof 40% are true polyimides, the remaining being modified polyimides containing other structural units beside the imide group.

All polyimides are characterized by the group   -N⟨ CO- / CO-

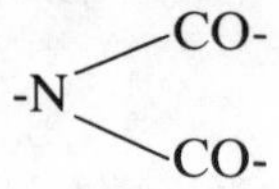

and also include the poly(amide-imides), poly(ester-imides), poly-(ether-imides) and poly(heterocyclic imides).

Various routes exist for the synthesis of polyimides. The first approach is to form the imide group simultaneously with the formation of the polymer. Another general method is a two-step condensation: the first step is the preparation of a soluble processable intermediate polymer; the second step involves the ring closure resulting in an in situ formation of imide groups. Another approach starts from preimidized monomers or prepolymers which are then converted into polyimides by addition polymerization or polycondensation reactions.

Because of the comprehensive books and review aricles available on synthesis and properties of the older polyimides [1−8], the following sections are devoted to the discussion of newer developments (see also [76, 78]).

Certain trends are seen to supply curable polyimides with enhanced processability. To accomplish this one can start from oligomers capped with reactive functional groups or even directly from the monomers as in the PMR process (*Polymerization of Monomeric Reactants*). Furthermore, the recently introduced thermoplastic and soluble polyimides also comply with the processor's request to gain improved practicability.

11.2 POLYIMIDES WITH *IN SITU* FORMED IMIDE GROUPS

11.2.1 Polyimides from polycondensation of tetracarboxylic acids and diamines

The classic method of polyimide synthesis is the two-step reaction of pyromellitic acid dianhydride with an aromatic diamine. The first step is carried out in polar solvents and yields an intermediate poly(amic acid). Then the second step takes place at elevated temperatures of up to 300°C to complete the dehydration ring closure, resulting in an insoluble polyimide with the structural unit I (Table 11-1).

The principal drawbacks of this route are the limited shelflife of the poly(amic acid), its sensitivity to hydrolysis, in particular at high solids concentrations and at elevated temperatures, and the problem of removing the off-gas water and residual solvent. These by-products cause voids in finished products unless completely removed during the final cure.

Enhanced hydrolytic stability of the poly(amic acid) intermediate is achieved by applying amine-substituted o-dicarbonic acids in which one carboxylic group is blocked through esterification:

(11-1)

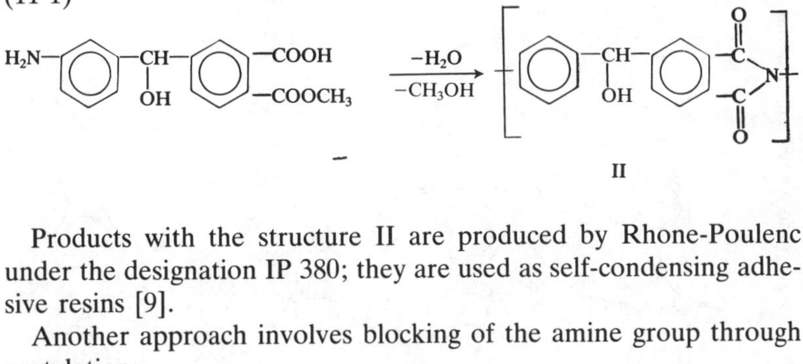

II

Products with the structure II are produced by Rhone-Poulenc under the designation IP 380; they are used as self-condensing adhesive resins [9].

Another approach involves blocking of the amine group through acetylation:

(11-2)

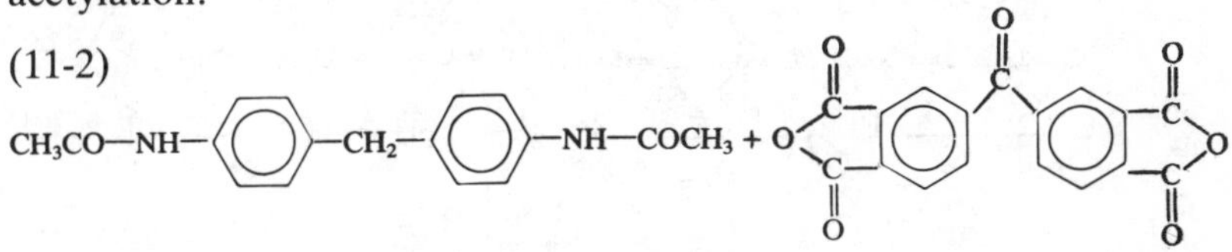

TABLE 11-1 Structures, trade names and manufactures of commercial polyimides

I

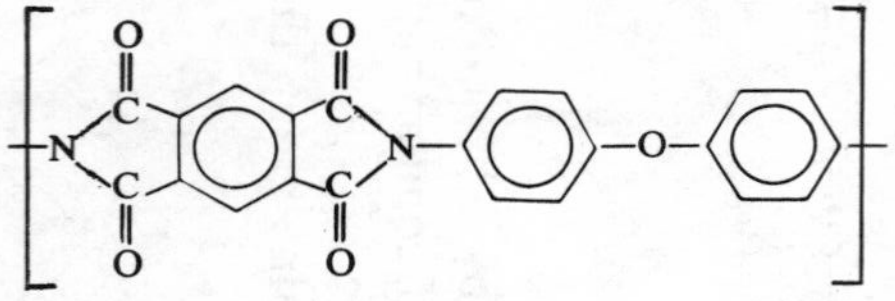

DuPont: **Kapton, Pyralin**
Pyre ML
Vespel
Bayer: Produkt 2225
UdSSR: Parimid

II

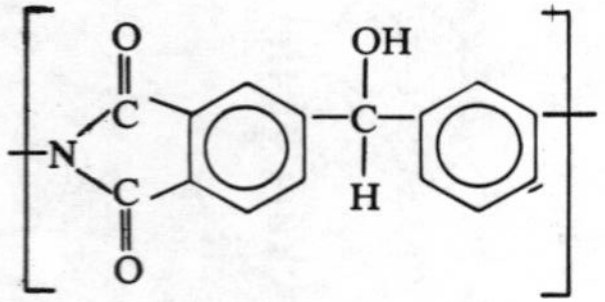

Rhone-Poulenc: IP 380

III

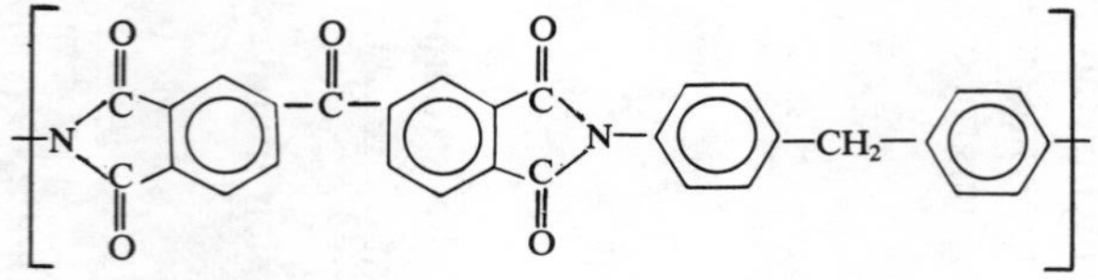

ICI: QX 13
Yorkshire Chem.: PI 212
Upjohn: Polyimid 2080

IV

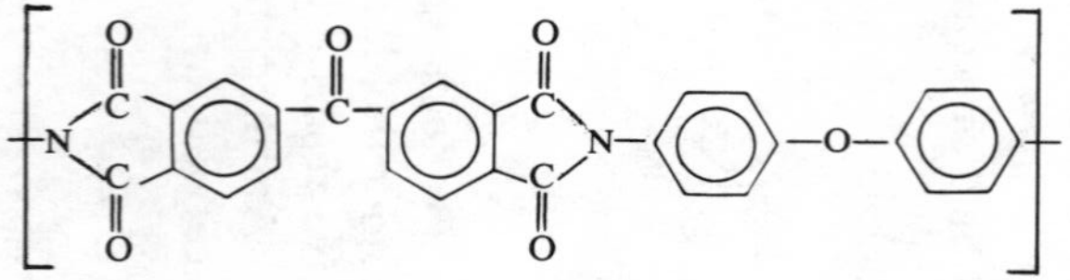

Monsanto: Skybond

Aramid P1 DFO — USSR — V

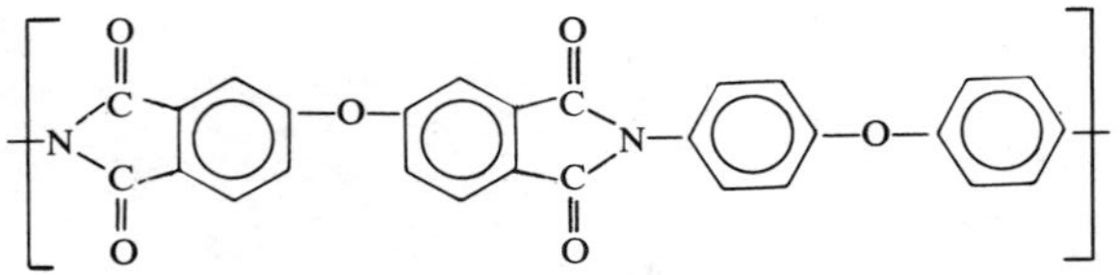

LARC-TPI — NASA — VI

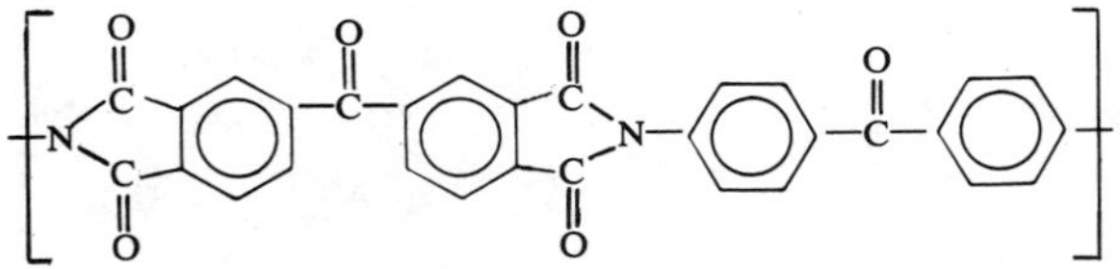

XU 213 / XU 218 — Ciba-Geigy — VII

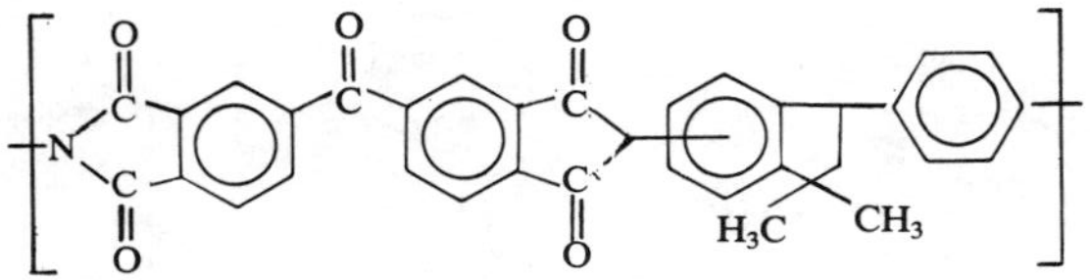

NR 150 B2 / NR 150 A2 / NR 056 X — DuPont — VIII

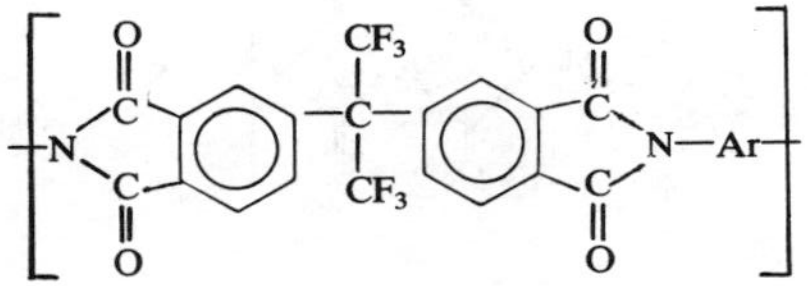

IX   TRW

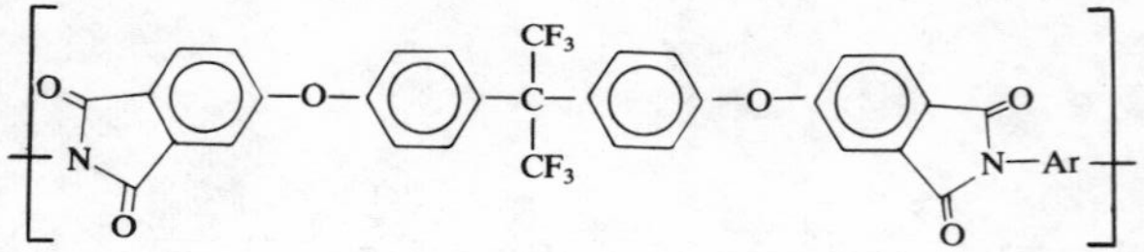

X   TRW

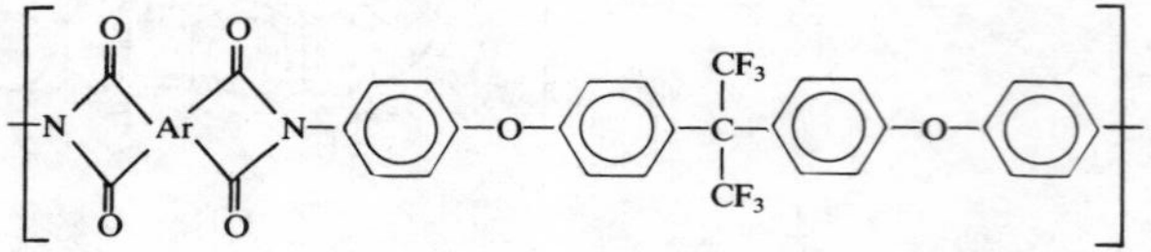

XI   Gulf Specialty Chemicals Thermid M

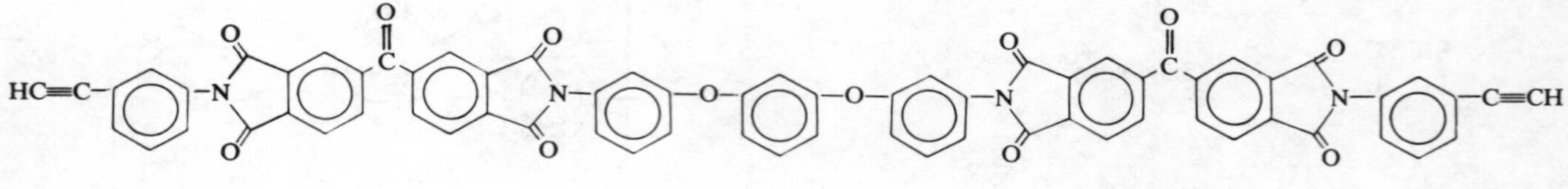

XII   Gulf Specialty Chemicals (if R=H: Thermid LR)

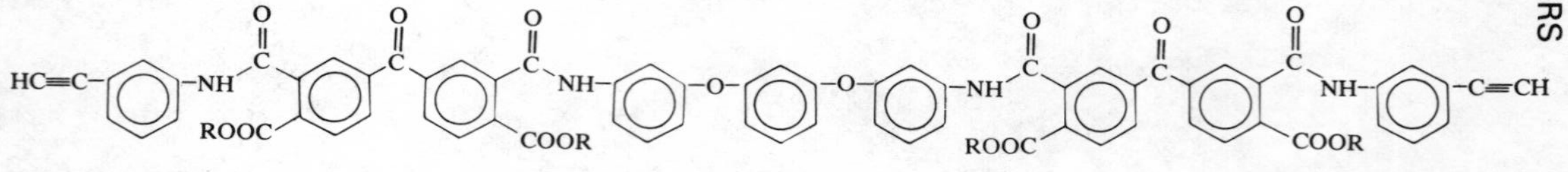

(if R = C₂H₅: Thermid AL 600)

XIII   ditto

TRW
Ciba-Geigy:
P 13 N

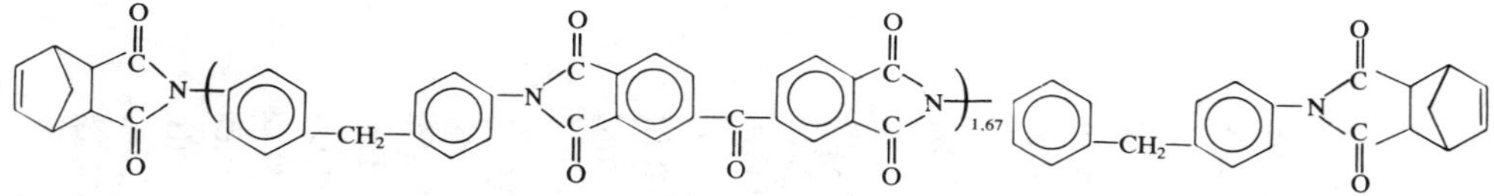

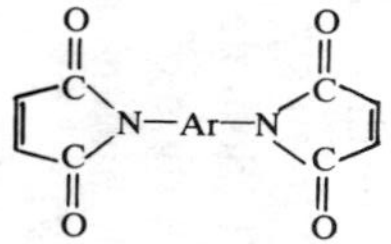

Rhone-Poulenc
Kerimid 353

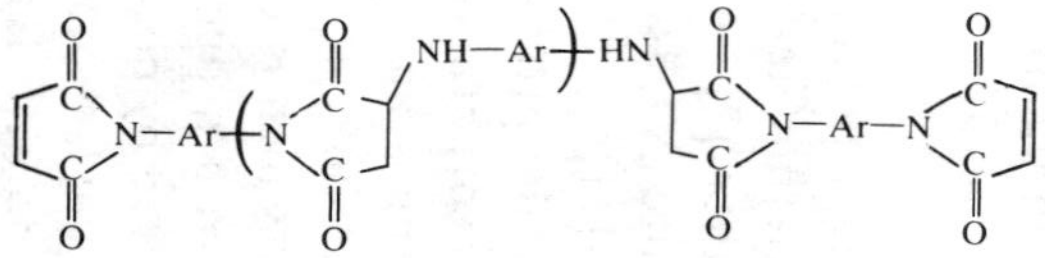

Rhone-Poulenc:
Kerimid 601

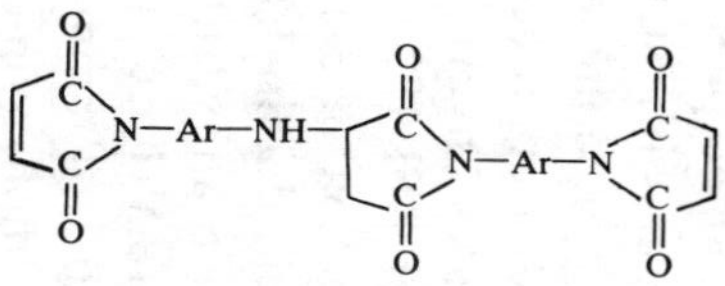

Rhone-Poulenc

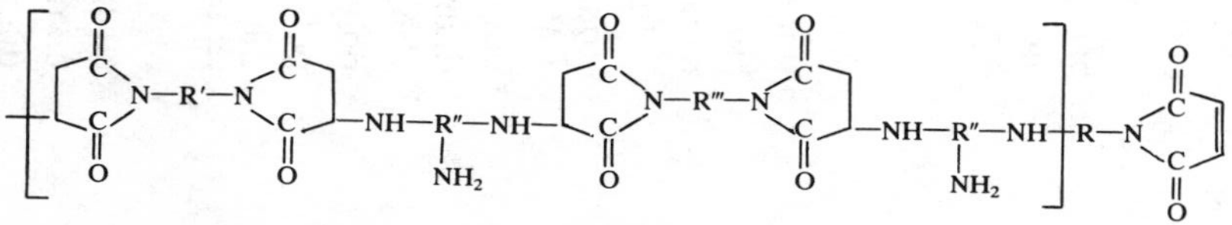

TRW:
FPI

XIV

XV

XVI

XVII

XVIII

This route was followed by ICI for the production of polyimide "QX 13". Later, the polymer was supplied by Yorkshire Chemicals but its production was eventually terminated.

The majority of polyimides contain flexibilizing moieties in the backbone in order to gain enhanced flexibility, solubility and, above all, improved processability, e.g. through thermoplastic processing.

Monsanto produces a polyimide with structure IV from 3,3',4,4'-benzophenone tetracarboxylic acid dianhydride (BTDA) and 4,4'-diaminodiphenylether.

BTDA has been commercially available from Gulf Oil Chemicals since 1975; depending on purity, it costs from 4.50 to 5.10 US-$/lb [10]. A number of new polyimides has been developed since BTDA became available.

In the Soviet Union, a polyimide with structure V is prepared from 3,3',4,4'-diphenylether tetracarboxylic acid dianhydride and 4,4'-diaminodiphenylether [2]. A linear thermoplastic processable polyimide with structure VI is obtained from BTDA and 3,3'-diaminobenzophenone [11].

Another approach to the production of soluble and thermoplastic polyimides involves incorporation of cycloaliphatic groups. The polymer XPI 182 of American Cyanamid presumbably is based on the reaction of 3,4-dicarboxy-1,2,3,4-tetrahydro-1-naphthalene-succinic acid dianhydride (which is obtained from styrene and 2 moles of maleic anhydride) with a diamine [7]. Soluble polyimides with structure VII are prepared by polycondensation of a mixture of 5- or 6-amino-1-(4'-aminophenyl)-1,3,3-trimethylindane and up to 25% of diaminodiphenylmethane with BTDA [12].

The introduction of flexibilizing moieties into the backbone generally results in a partial tradeoff in thermal stability. Improved polyimides with both enhanced flexibility and thermo-oxidative stability are obtained through incorporation of hexafluoroisopropyl groups. The first products of this kind were DuPont's NR 150 series [13, 14]. The preparation of polyimides with structure VIII is conducted by polycondensation of 2,2-bis(3',4'-dicarboxyphenyl)-hexa-fluoropropane with various diamines [15, 16]. For NR 150 B2, a mixture of 95 mol-% p- and 5 mol-% m-phenylenediamine is used. The product is supplied as solution dissolved in a mixture of N-methylpyrrolidone and ethanol. The amine component in the preparation of NR 150 A2 is diaminodiphenylether, and for NR 056 X a mixture of 75 mol-% diaminobenzene and 25 mol-% diaminodiphenyl dissolved in diglyme is employed. The most recent commercial polyimides of this kind with structures IX and X have been developed by TRW [17].

Recently, Toray Ind. introduced a polyimide named "TI" for injection molding. Toray has only disclosed that its polymer is produced from "polyhydric amine and polyhydric carboxylic acid" [18]. Another Japanese polyimide supplied by Dainippon Ink and Chemical Inc. reportedly is thermally stable at temperatures of up to 800°C [19]. It has been revealed that a new tetraacid is applied which is said to be available from piperylene and phthalic acid.

No details with regard to structure have been disclosed for a number of commercial polyimides, e.g. for ®Envex (Rogers), ®Feurlon (Bemol), and ®Meldin (Dixon). The polyimide Poly-X of Raychem reputedly is an aliphatic-aromatic PI [12].

In 1979, Rohm & Haas introduced thermoplastic "modified poly-aliphatic" imides under the designation ®Kamax [20, 21]. The starting materials are said to be the same as those used in the preparation of acrylic polymers. Kamax was not developed in competition to aromatic polyimide types; temperature resistance is lower, but also costs are considerably lower.

11.2.2 Polyimides from polyaddition of carboxylic acid dianhydrides and isocyanates

One approach to circumvent the cumbersome dehydration ring closure is the reaction of tetraacid dianhydrides with diisocyanates:

(11-3)

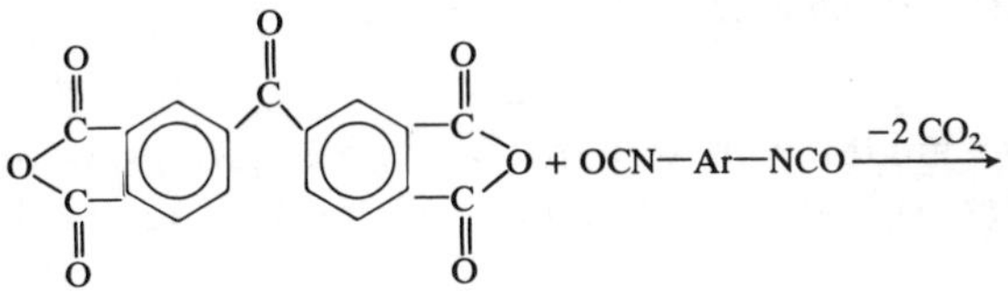

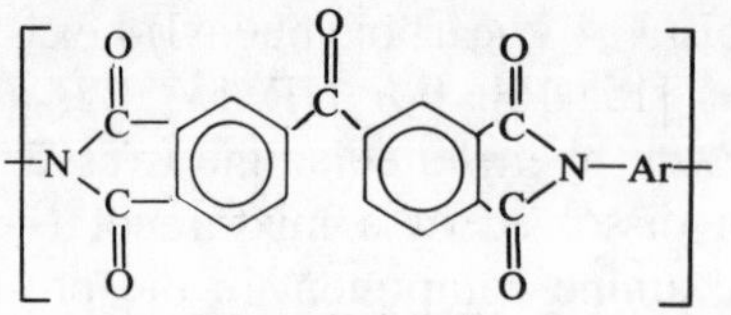

The process was developed by the Upjohn Co. The diisocyanates preferably used are MDI and 4,4'-methylene-bis-phenylisocyanate. The reaction is carried out in aprotic solvents at temperatures of below 100°C. Strong alkaline catalysts, e.g. alkoxides, alkali lactams or tertiary amines, and traces of water are required for this reaction [22]. N-carboxyanhydride is formed as an unstable intermediate which, with subsequent loss of CO_2, yields the imide [23]:

(11-4)

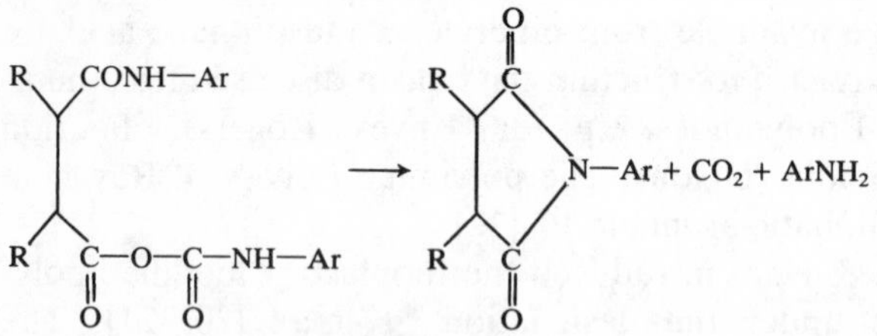

Polyimides obtained from this reaction are thermoplastic and show good solubility. Films can be cast from the solution. Filaments are obtained directly from solution by wet or dry spinning techniques [24]. The polyimide 2080 of Upjohn is fully imidized and is illustrated by structure III.

11.3   POLYIMIDES WITH PREFORMED IMIDE GROUPS

11.3.1   Oligomeric imides with terminal acetylene groups

The first oligomeric imides with terminal acetylene groups were developed by Hughes Aircraft Co. between 1969 and 1974 [25–27].

Hughes Aircraft terminated the production of its resins HR-600, HR-650 and HR-700 [1]; the Gulf Oil Chemical Co. acquired an exclusive license under which pilot plant quantities of the resins ®Thermid MC-600 (XI), LR-600 (XII), and AL-600 (XIII) have been prepared since 1977 [28]. Only Thermid MC-600 is a fully imidized product; it is identical in structure with the older HR-600.

Oligomeric imides are prepared by first reacting 10% solutions of BTDA and 1,3-bis(3-aminophenoxy)benzene dissolved in aprotic solvents, e.g. NMP, at room temperature. After 1/2-1 h a solution of the ethynylated amine (3-amino-phenyl-acetylene) is added, and water is removed over 4-6 h by azeotropic distillation with benzene at 150°C [27].

Thermid MC-600 is supplied as powder, LR-600 as 50% solution in NMP, and AL-600 as solution in ethanol with 75% solids. The acetylene groups undergo cyclization at temperatures above 200°C without need of catalysts, however, postcuring is required, e.g. for 16 h at 315°C or 4 h at 371°C [28]. The cyclization can be accomplished also at room temperature. For this purpose, graphite fibers are impregnated with a solution of Thermid MC-600, the solvent is removed, and the system is then subjected to a cathodic polymerization. This process is still at the laboratory stage [29].

11.3.2 Oligomeric imides with terminal norbornene groups

Oligomeric imides with terminal norbornene groups were developed by TRW at the end of the '60s [30, 31]. An oligomeric amic acid is prepared by condensation of 4,4'-diaminodiphenylmethane with BTDA in a molar ratio of 2.67:1.67 in DMF solution. Chain termination occurs through 2 moles of ®Nadic (eq. 11-5). Nadic is Allied Chemical's trade-name for 4-endo-methylene-tetrahydronaphthalic acid anhydride. It is produced by Diels-Alder reaction from cyclopentadiene and maleic anhydride. Endo-cis-bicyclo[2.2.1]-heptene-2,3-dicarboxylic anhydride, what is the correct nomenclature for the above product, is also commercially available from Hitachi Chemicals under the designation Hitachi HIMIC Anhydride.

The oligomeric amic acid is isolated by spray drying and is then

(11-5)

cyclized at about 200°C to yield the imide with structure XIV. The oligomer XIV was named P13N, with P standing for polyimide, 13 for a molecular weight of about 1300, and N for Nadic. Curing is carried out by compression molding at temperatures of above 260°C. Hereby, a reversion of the Diels-Alder reaction takes place. Cyclopentadiene, the oligomeric bismaleide and the vinyl group of Nadimic copolymerize with formation of a crosslinked polyimide. Nadimic is the trivial name for endo-5-norbornene-2,3-dicarboxyimide or endo-bicyclo-[2.2.1]-hept-5-ene-2,3-dicarboxy-imide.

On curing, small amounts of cyclopentadiene are formed which do not react and remain in the resin. The resin becomes brittle and the thermo-oxidative stability is less than expected [16, 31]. P13N was produced by Ciba-Geigy until the end of the '70s. The brittleness and problems in processing led to the further work at NASA Lewis Research Center. This effort culminated in the concept of using polymerizable monomer reactants PMR (see chapter 11.4).

11.3.3 Oligomeric polyimides with terminal bismaleimide groups

Bismaleimides are obtained from maleic acid and appropriate diamines:

(11-6)

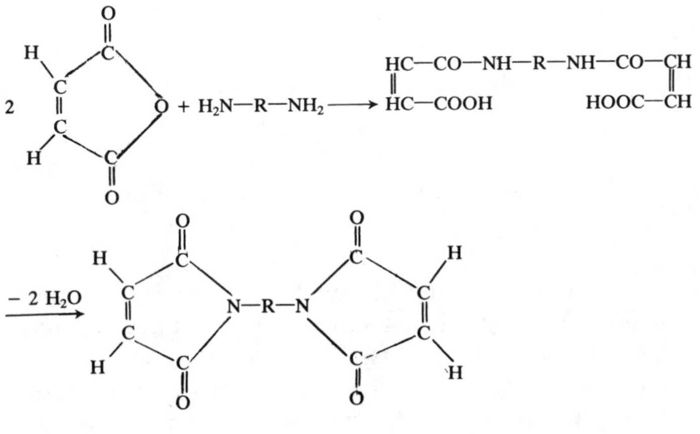

The preferred amino component is 4,4′-diaminodiphenylmethane.

The curing scheme involves a number of addition reactions [7]. Diamines do crosslink but yield no usable products. However, if such amines are reacted in non-stoichiometric amounts at temperatures from 200 to 280°C, addition of the amine group with the double bond occurs simultaneously with the polymerization of the double bonds. Polymers with good properties are thus obtained. Crosslinked polyimides with enhanced thermal stability can be prepared when preformed imide structures are incorporated in the bismaleimides.

This type of polyimides is also classified as poly(amino-bismaleimides) or poly(aspartimides). They were developed by Rhone-Poulenc [32, 34]. Oligomeric polyimides with structures XV and XVI are commercially available under the trade-names ®Kerimid 353 and ®Kerimid 601, respectively. Kerimid 353 is a eutectic ternary system [35]. Kerimid 601 is the reaction product of methylene dianiline and excess methylene-dianiline bismaleimide, sold in the U.S. by Rhodia Corp. Glass or graphite fiber reinforced and PTFE filled resins are commercially supplied as ®Kinel, and a metal adhesive is sold under the trade-name ®Nolimid A-380.

A new oligomeric polyimide with structure XVII demonstrates improved thermo-oxidative stability [35]. It is prepared from bismaleimide by chain extending with aromatic amino-maleimide:

(11-7)

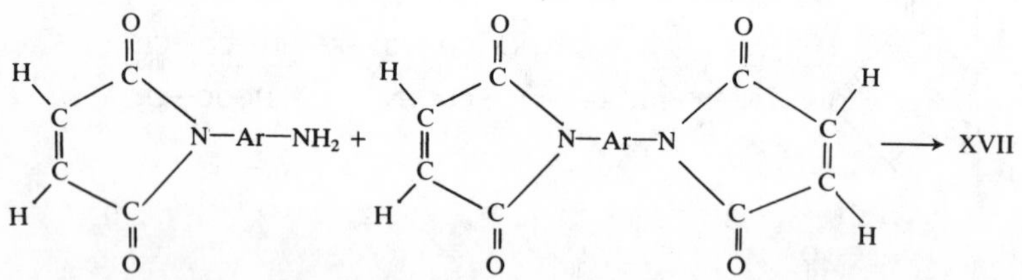

All these products are soluble and meltable. For example, molten Kerimid 353 has a shelf life that is sufficiently long to permit processing without the need of solvents.

Similar products are supplied by General Electric under the tradename ®Gemon [7]. An improved bismaleimide resin with a glass transition temperature T_g greater than 370°C is available from the U.S. Polymeric Division of Hitco Inc. under the designation V-378

(11-8)

A [36]; it is aimed at high temperature applications.

Many poly(amino-bismaleimides) require curing and post-curing temperatures of more than 280°C which renders many applications troublesome or even impossible, particularly in composites with aluminum. TRW has developed polyimides that can be processed at temperatures of ≤200°C [17, 37]. By a novel Diels-Alder reaction bis(furanyl)imides are reacted with bismaleimides. The intermediate △4-endooxy-tetrahydronaphthalimide forms aromatic structures at elevated temperatures.

These polyimides are prepared from bisfuranylimides and bis-maleimides as eq. (11-8) on page 323 shows.

Poly(amino-bismaleimides) can also be cured with epoxy resins at room temperature or slightly above [38]. Of several systems developed by TRW, the best performance is shown by a polyimide prepared from a bismaleimide with ®Jeffamine ED-900 (A), ®Jeffamine AP-22 (B) and bis(4-maleimidephenyl)methane (C) at 80°C.

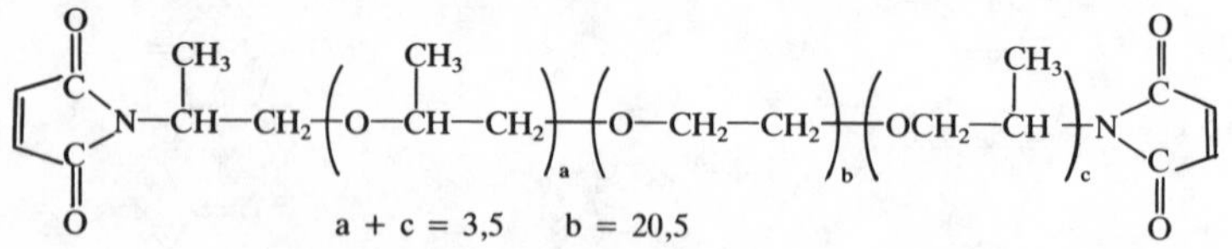

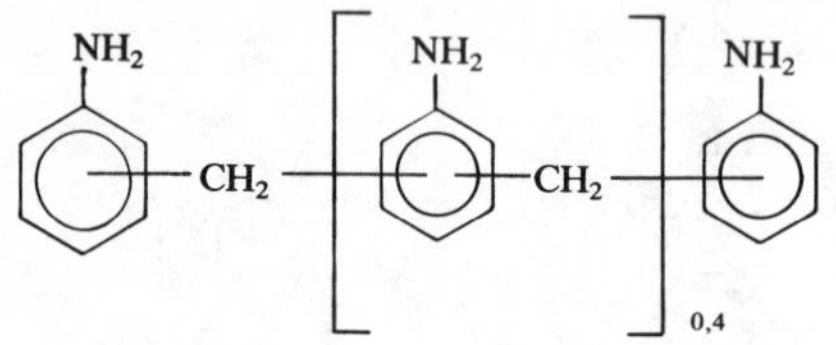

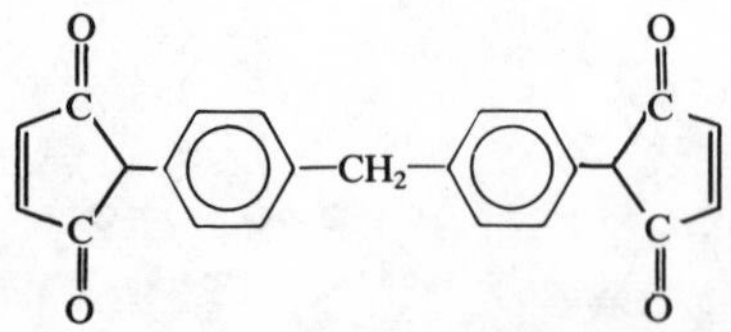

®Jeffamine ED-900 and ®Jeffamine AP-22 are commercially manufactured by Texaco Chemical Co. [39]. The resulting oligo-

meric polyimide with structure XVIII is liquid and is cured at room
temperature with epoxy resin DER 736 of Dow Chemical.

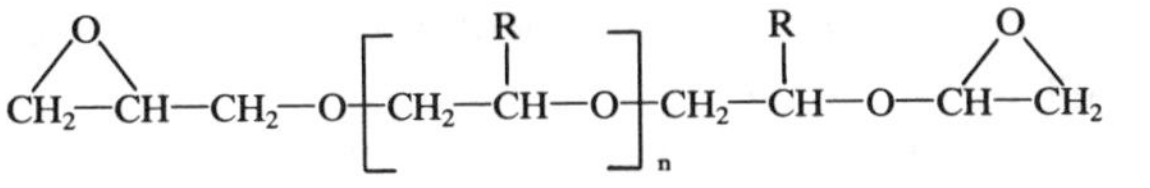

Curing and crosslinking of XVIII takes place via terminal maleimide groups as well as by reaction of the internal amino-groups with
the polyfunctional epoxy resin. The resulting polyimides are flexible
to −54°C and are targeted to replace polysulfides in fuel tank
sealings.

Epoximides also include the ®Novimide Resins of Isochem Resins
with the designations 700, 700/6 and 700/55 [40]. They are novolak-
type epoxides coupled with carboxylimide anhydrides. Curing with
catalysts is carried out at 95−100°C for 2 h with subsequent post-
curing at 120−215°C for 2−8 h.

Another epoxidimide was developed by Ferro Corp. and is described as N,N′-diglycidyl-benzophenone tetracarboxylic acid diimide:

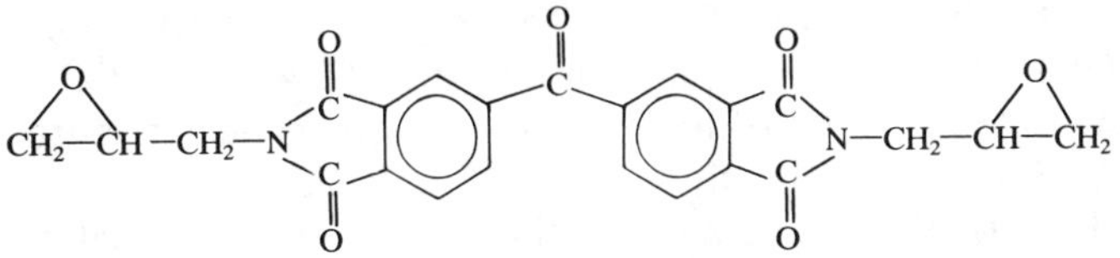

11.4 PMR (*IN SITU* POLYMERIZATION OF MONOMER REACTANTS)—POLYIMIDES

The concept of the PMR method is to start from a solution containing the mixed monomers. The mixture comprises the semi-ester
of a tetracarboxylic acid, a diamine and a chain terminator, usually
dissolved in methanol or ethanol. Suitable chain terminators are p-
aminostyrene, m-amino-phenylacetylene and Nadic-ester (NE).
Nadic-ester is the semi-ester of 4-endo-methylenetetrahydrophthalic
acid and is the preferred chain terminator in commercial PMR
systems.

The monomer mix does not react at room temperature and thus

provides excellent shelf-life. Fibers intended for reinforcement are impregnated with the solution. After removal of the alcoholic solvent, the temperature is slowly raised to 316°C (600°F). The monomer mix begins to melt under 100°C; in situ imidizing starts at 140°C, while the chain is simultaneously formed via the poly(amic acid). At approximately 200°C, the synthesis of the poly(amic acid) is completed. The resulting norbornyl end-capped prepolymer has a melting range from 175 to 250°C. At higher temperatures of up to 340°C, crosslinking through addition reaction of the double bonds is conducted under pressure in an autoclave or press [31, 41]. The idealized reaction scheme for polyimide PMR 15 is illustrated by eq. (11-9) on p. 327.

The PMR process developed by NASA [31, 42] has significant advantages over other polyimide syntheses. The preparation of prepolymers with subsequent separation is circumvented. The complete removal of the alcoholic solvents is accomplished more rapidly than with the commonly used high-boiling aprotic solvents, hence, the chances for residual solvents remaining in the resin are minimized.

PMR 15 has been termed the "first generation PMR"; it is prepared from the semi-ester of 3,3',4,4'-benzophenone tetracarbonic acid (BTDE), 4,4'-diaminodiphenylmethane (MDA) and Nadicester (NE) in a molar ratio of n: (n + 1): 2. Prepregs manufactured from PMR 15 are commercially available (Table 11-2).

Polyimide PMR II is prepared from the di-ester of 4,4'-(hexafluoropropylidene)-bis-phthalic acid (HFDE), p-phenylenediamine (PPDA) and NE. It is superior to PMR 15 with regard to thermooxidative stability and mechanical properties at 316°C. DuPont has developed a similar product from HFDE/PPDA/NE in a molar ratio of 4:5:2. It is supplied by U.S. Polymeric under the designation PMR-15II [41].

A further advancement is the solvent-free PMR LARC-160 developed by the NASA Langley Research Center [43]. It is prepared from BTDE, NE and ®Jeffamine AP-22 in a 0.335:0.610:0.539 molar ratio. The viscous liquid as well as prepregs made thereof are commercially available. LARC-160 is being used as matrix by Boeing, Lockheed, and Rockwell [16, 44].

The PMR technique is also used by other companies. Ferro Corp., Fiberite, Hexcel, Narmco Materials (Celanese) and Riggs Engineering supply prepregs and semi-finished articles [3]. It is not

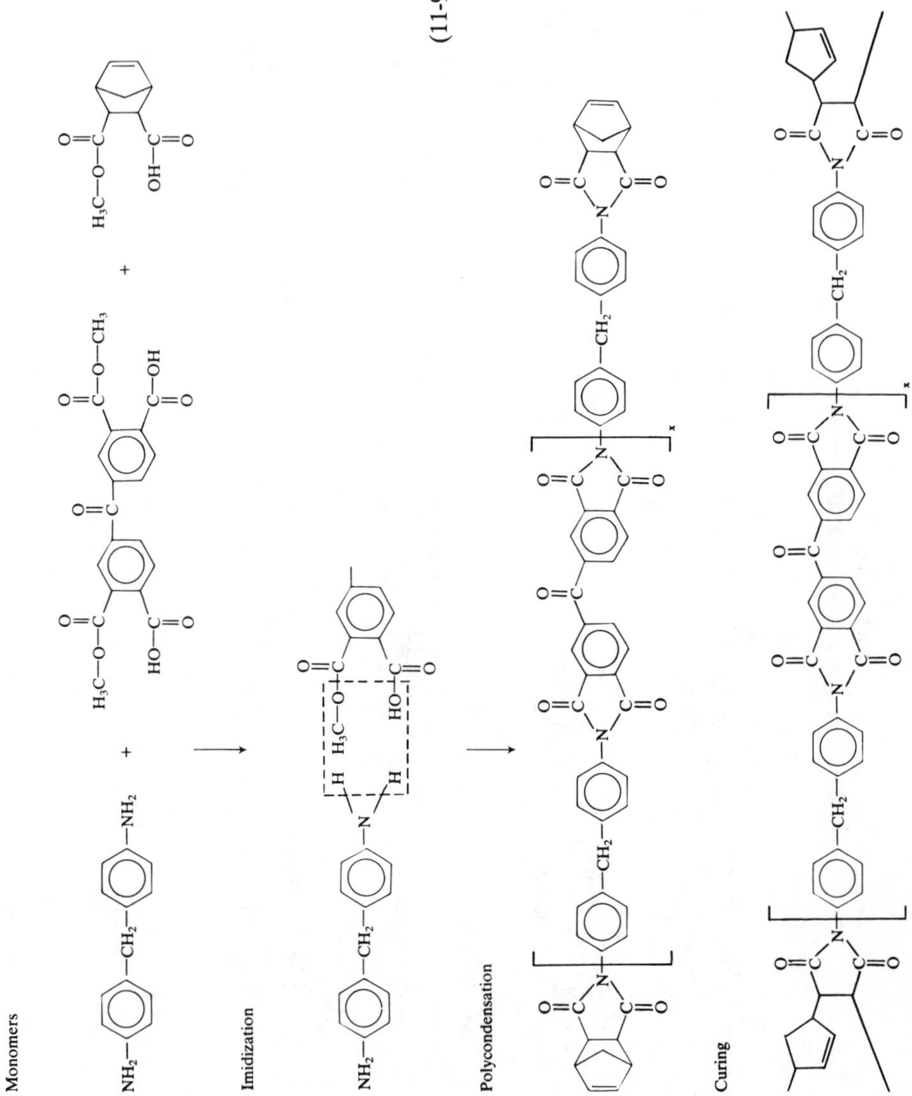

(11-9)

Monomers

Imidization

Polycondensation

Curing

TABLE 11-2
Monomer combinations for PMR Systems.

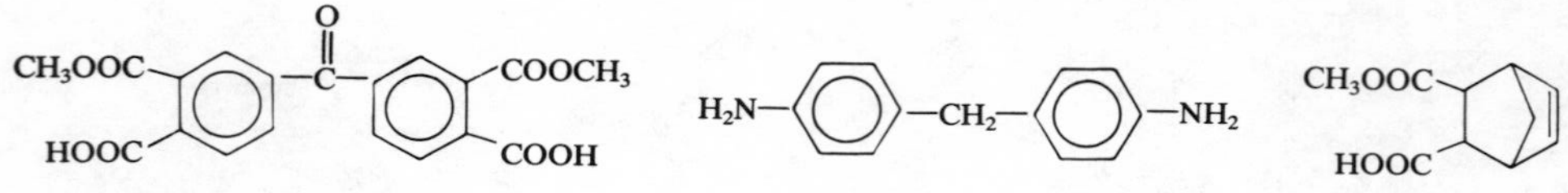

NASA
PMR 15

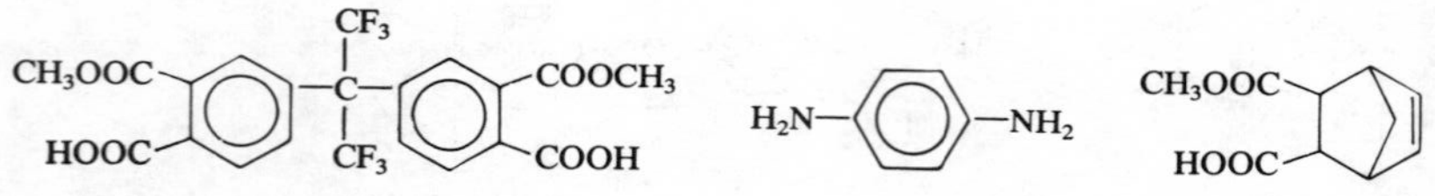

NASA
PMR II

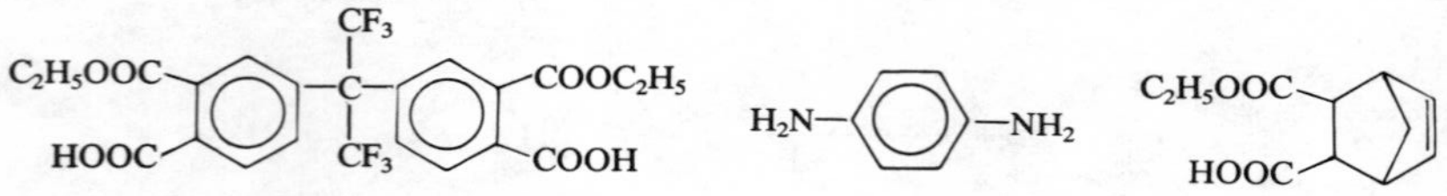

DuPont
PMR-15-II

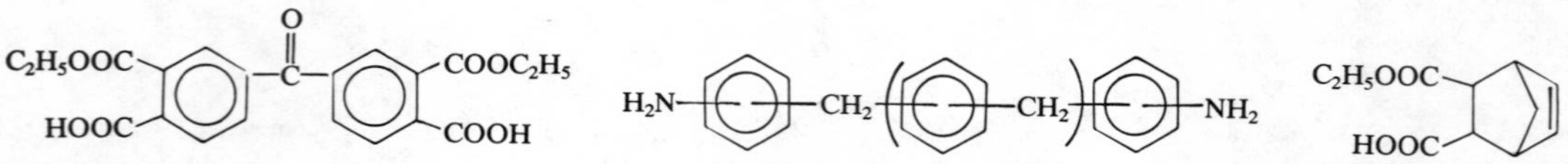

NASA
LARC-160

CH$_3$OOC
HOOC

CH$_3$OOC
HOOC

C$_2$H$_5$OOC
HOOC

C$_2$H$_5$OOC
HOOC

H$_2$N— —CH$_2$— —NH$_2$

H$_2$N— —NH$_2$

H$_2$N— —NH$_2$

H$_2$N— —CH$_2$—(—CH$_2$)— —NH$_2$

COOCH$_3$
COOH
CH$_3$OOC
HOOC

COOCH$_3$
COOH
CF$_3$—C—CF$_3$
CH$_3$OOC
HOOC

COOC$_2$H$_5$
COOH
CF$_3$—C—CF$_3$
C$_2$H$_5$OOC
HOOC

COOC$_2$H$_5$
COOH
C$_2$H$_5$OOC
HOOC

known in detail whether these companies also produce the resins or only deal with processing and manufacturing.

11.5 POLYAMIDE-IMIDES

Polyamide-imides have been commercially available since 1964. Their synthesis, properties and applications have been extensively described [1, 2, 7, 8].

The presently available poly(amide-imides) can be classified into the following product groups:

Molding materials:	Torlon 4000, 5000, 6000 and 7000 series, filled and unfilled grades, for compression and injection molding, and extrusion, supplied by Amoco Chem. Corp.
Wire and baking enamels:	AI-10, AI-830, AI-1130 L and AI-Lite, Amoco; AI-600, General Electric; Tritherm, P. D. George Co.; Rhodeftal, Rhone-Poulenc; XWE-960 A, Schenectady Chemicals; Fenogrant, formerly Schramm; HI 200 and HI 400, Hitachi;
Films:	Hitachi; Amanim, Westinghouse;
Fibers:	Kermel filaments 201, 203 and 223; Kermel 234 stapel fiber, all Rhone-Poulenc;

Torlon is prepared by phosgenation of trimellitic acid anhydride, the resulting acid chloride is reacted with 4,4′-diaminodiphenyl-methane in N-methylpyrrolidone at room temperature. The inter-mediate poly(amic acid) undergoes the usual dehydration ring closure resulting in a polyamide-imide.

For the production of wire and baking enamels one prefers the direct reaction of trimellitic acid anhydride and MDI in aprotic solvents. The diisocyanate reacts with both carboxylic acid and anhydride with subsequent loss of CO_2 to give an amide and imide, respectively. No second curing step is needed for imide formation.

(11-10)

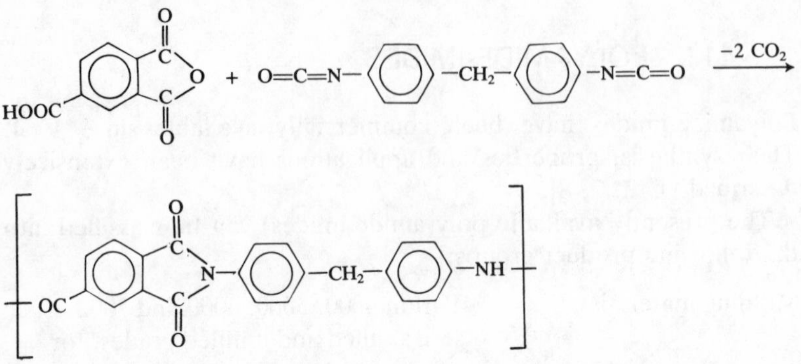

By the partial replacement of the triacid with an aromatic diacid or tetraacid dianhydride, the amide or imide content respectively can be increased. The thermal stability increases with increasing imide content.

Lacquers are commercially supplied as solutions in NMP or NMP/xylene. The same synthesis is also used in the production of ®Kermel fibers which are spun directly from the solution.

11.6 POLYESTERIMIDES

Polyesterimides have been available since 1966. Their manufacture, properties and applications have been described in several reviews [1, 2, 4, 5, 7].

Polyesterimides are mainly used for temperature resistant wire lacquers. The following products are commercially available: Imidex (General Electric), Teritherm (P. D. George Co.), Isomid (Schenectady Chemicals), Terabec and Allobec (Dr. Beck and Co., a subsidiary of BASF), E 3535 (Dr. Kurt Herberts, a subsidiary of Hoechst), and Cellatherm (Reichhold Chemie). Enamal Omega (Westinghouse Electric Co.) is a polyester-imideamide.

Polyesterimides are synthesized by reaction of trimellitic acid anhydride with diphenol esters or the ethylene glycol ester of terephthalic acid:

(11-11)

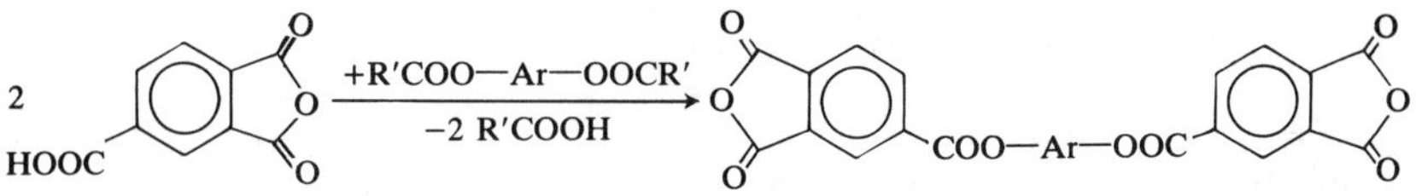

(11-12)

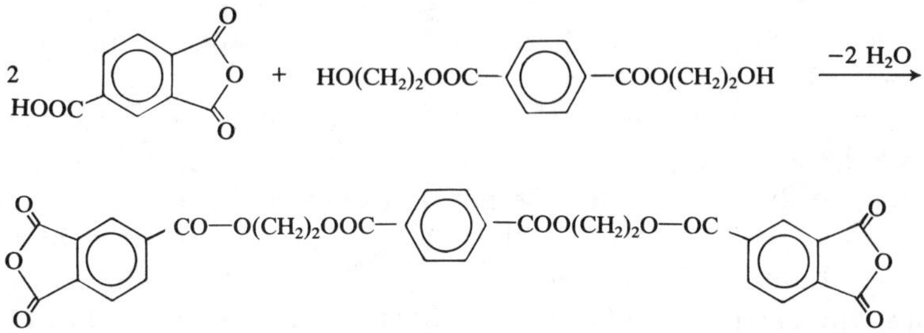

The intermediate products with terminal anhydride groups are then reacted with diamines, e.g. with 4,4'-diaminodiphenylmethane, in aprotic solvents, and the resulting poly(amicester acid) is imidized in the usual manner. Solutions for wire enamels contain polymers with fully imidized structure.

For the majority of commerical products the starting materials most frequently used include trimellitic acid anhydride, pyromellitic dianhydride and benzophenone tetraacid dianhydride, as well as diaminodiphenylmethane, p-phenylenediamine, p-amino-benzoic acid, aminoethanol or amino-acetic acid for imide formation. The monomers for ester formation include terephthalic acid, itaconic acid, phenylindane diacid, benzophenone diacid, ethylene glycol, glycerine, and tris(2-hydroxyethyl)isocyanurate [2]. When tri- and tetrafunctional compounds are used, crosslinking is achieved during the baking process.

When excess diamine is used, the polycondensation results in the formation of poly(ester-amide-imides). The products of Westinghouse are a mix of polyester, poly(ester-amide-imide), and a phenol-blocked isocyanate. The particular poly(ester-amide-imide) is prepared from trimellitic acid anhydride, dimethylterephthalate, tall oil acids, ethylene glycol, and diamino-diphenylmethane.

11.7 POLYETHERIMIDES

11.7.1 Structure and synthesis

Over ten years of development preceded the introduction of
®Ultem Resin by General Electric in February 1982 [45, 45a]. Commercially available are the non-reinforced grade Ultem 1000 and the
glass reinforced grades Ultem 2100 (10% glass), Ultem 2200 (20%
glass) and Ultem 2300 (30% glass). Mineral filled grades are under
development. Ultem is supplied in Europe by General Electric in
Bergenop-Zoom, Netherlands, and in Japan by Nippon Engineering
Polymers, Tokyo.

General Electric has published only a few details on the structure
and synthesis of its polyetherimide [45b]. However, its numerous
patents [46−49] indicate that the new polyetherimides are prepared
by reaction of aromatic bis(ether-phthalic acids) or anhydrides
thereof with aromatic diamines.

The general structure of bis(ether-phthalic acid anhydrides) I or II
is

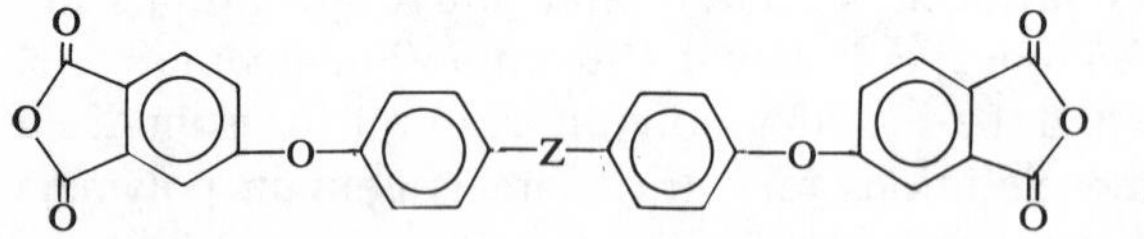

I

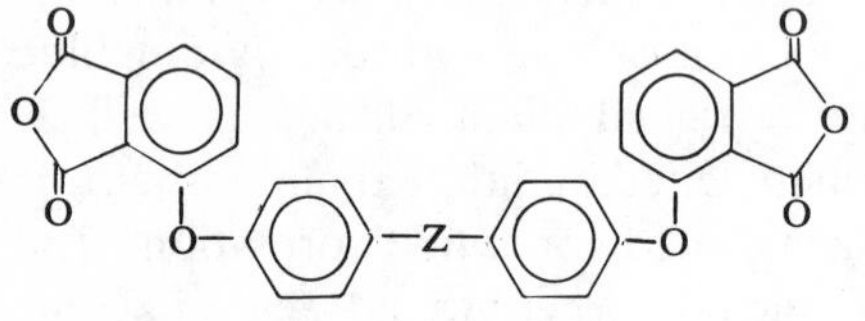

II

Z stands for the groups C_nH_{2n}, $C(CH_3)_2$, CO, SO_2, O or S. Presumably, Ultem has the following composition [45a]:

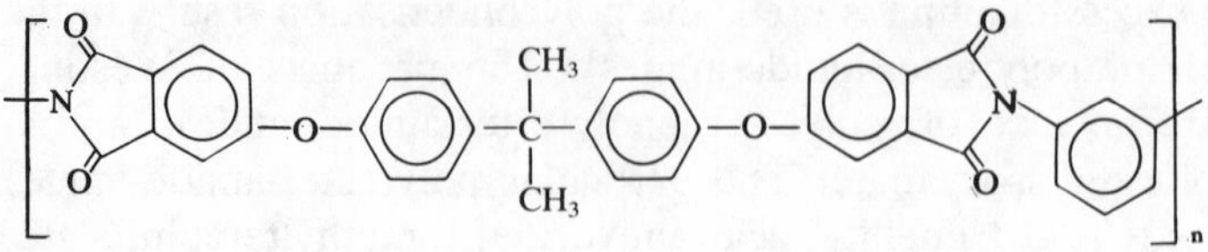

Polyetherimides I or II with $Z = (CH_3)_2C$ are prepared from the corresponding N-phenylnitrophthalimides, e.g. for I from N-phenyl-4-nitrophthalimide (III), by reaction with the disodium salt of bisphenol A (IV), (eq. 11–13, reaction 1). In this unconventional reaction, the nucleophilic attack of the bisphenol di-anion induces elimination of NO_2^- as the leaving group. Hydrolysis with aqueous NaOH at 160–175°C causes cleavage of the phthalimide groups in V, aniline is removed by steam destillation, and acidification results in the formation of 2,2-bis-4-(3,4-dicarboxy-phenoxy)phenyl-propane (VI) (reaction 2). Polycondensation of (VI) or its anhydride (VII) with aromatic diamines gives polyetherimides (VIII). The amine preferred for this reaction is presumably 4,4'-diaminodiphenylmethane (reaction 3); other diamines frequently cited in the patent literature are m-phenylenediamine and 4,4'-diaminodiphenylether.

The polycondensation is carried out in the usual manner in solution, e.g., with NMP, o-dichlorobenzene, m-cresol, toluene, etc., whereby the intermediate poly(ether amic acid) forms more or less imidized polymers depending on reaction conditions. These solutions can be used for coatings, wire enamels, film casting, and as precursor for carbon fiber reinforced matrices [46].

The production of the thermoplastic polyetherimides is presumably conducted by continuous melt polycondensation of anhydrides (VII) and diamines by means of an extruder at temperatures from 200 to 290°C [47]. The extruder is equipped with a degassing zone to remove the liberated water. Molecular weight can be controlled by addition of phthalic anhydride [48]. The polycondensation rate is accelerated by catalytic amounts of NaCl or $Fe_2(SO_4)_3$ [49].

(11-13) (*see next page*).

11.7.2   Properties and applications

®Ultem is an amorphous, transparent thermoplastic polyetherimide of yellowish color. It combines high heat resistance, exceptional strength and modulus, inherent flame resistance with low smoke evolution, stable dielectric constant and dissipation factor over a wide range of temperatures and frequencies, broad chemical resistance, and outstanding processability on conventional molding equipment (Table 11-3). The high heat deflection temperature re-

$$2 \quad \text{III} \quad + \quad \text{Na}-\text{O} \cdots \text{O}-\text{Na} \quad \text{IV}$$

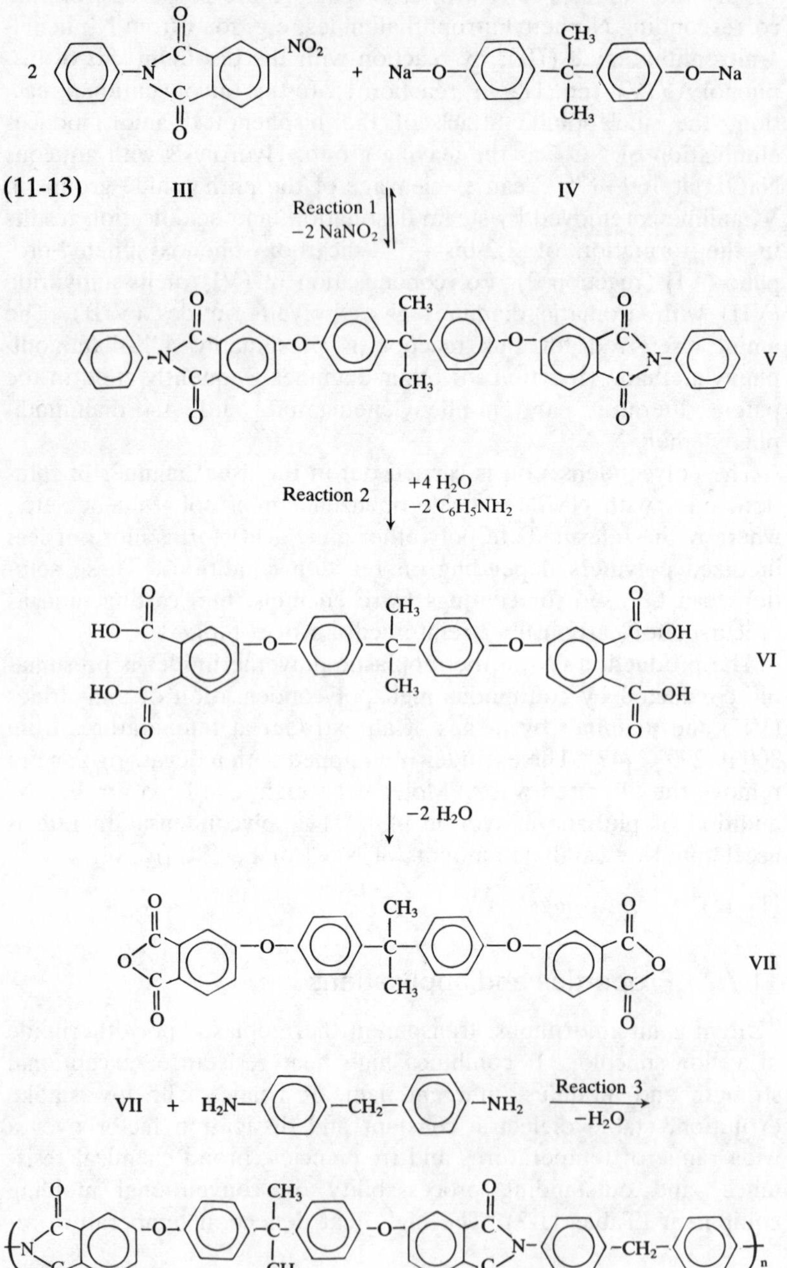

(11-13)   III

$$\text{Reaction 1} \quad -2\,\text{NaNO}_2$$

V

$$\text{Reaction 2} \quad \begin{array}{c} +4\,\text{H}_2\text{O} \\ -2\,\text{C}_6\text{H}_5\text{NH}_2 \end{array}$$

VI

$$-2\,\text{H}_2\text{O}$$

VII

$$\text{VII} \quad + \quad \text{H}_2\text{N}-\text{CH}_2- \text{NH}_2 \quad \xrightarrow[-\text{H}_2\text{O}]{\text{Reaction 3}}$$

VIII

sults from the imide structure, and the good processability is accomplished by the flexibilizing ether bridges. Ultem has a continous use temperature rating of 170°C. The high flame resistance with UL 94 V-0 rating at 0.76 mm is achieved without the use of additives. The limiting oxygen index of 47 is the highest of any commonly used engineering thermoplastics. Ultem exhibits extremely low levels of smoke generation and is superior in this regard to polyethersulfones, polycarbonates and polysulfones [45, 50].

Ultem is resistant to mineral acids, mineral salt solutions, and dilute bases. More than 90% of initial tensile strength is retained by Ultem 1000 after 10000 h of immersion in water at 100°C. Ultem is unaffected by most hydrocarbons although the polymer is soluble in halogenated hydrocarbons. Ultem is inherently resistant to UV radiation without the addition of stabilizers.

Ultem resins should be dried to less than 0.05% moisture prior to processing. Ultem is processed by injection molding, extrusion, blow molding and structural foam molding techniques. Because of the good thermal stability, melt temperatures from 340 to 425°C are common, 360°C being typical for most applications. Mold temperatures should range from 65 to 175°C. For ease of subsequent start up of processing machines, Ultem resin should be purged by HDPE or PC prior to shutdown. Regrind levels of 50% can be used with no measurable changes in heat deflection temperature, tensile strength, elongation, or impact strength. At high shear rates, the resin shows a low viscosity and allows the injection molding of sections as thin as 0.25 mm. Since Ultem resins display notch sensitivity, stress concentrators such as sharp corners should be minimized in the design of molded parts.

Ultem has been designed to close the gap in engineering resins between polysulfones and polycarbonates on the one hand and the commonly more expensive polyamide-imides and fluoropolymers on the other hand. Potential applications are in the electronic, electric, automotive, communications, aircraft and space industries. In some fields, Ultem can replace metals such as aluminum, die-casted zinc, steel, brass or bronze. Ultem has already been used in microwave oven parts, heat exchangers, automotive fuel systems, cable jacketing, and in thermally resistant connectors for electric and electronic appliances.

Prices for Ultem range from 4.25 US-$/lb for Ultem 1000 to 3.40 US-$/lb for Ultem 2300.

TABLE 11-3
Properties of ®Ultem Polyether-imide.

MECHANICAL	ASTM TEST	UNITS	ULTEM 1000	ULTEM 2100a)	ULTEM 2200b)	ULTEM 2300c)
Tensile strength, yield	D638	N/mm²	105	120	140	160
Tensile modulus, 1% secant	D638	N/mm²	3000	4500	6900	9000
Tensile elongation, yield	D638	%	7–8	5	—	—
Tensile elongation, ultimate	D638	%	60	6	3	3
Flexural strength	D790	N/mm²	145	200	210	230
Flexural modulus, tangent	D790	N/mm²	3300	4500	6200	8300
Compressive strength	D695	N/mm²	140	160	170	160
Compressive modulus	D695	N/mm²	2900	3100	3500	3800
Gardner impact		N-m	36	—	—	—
Izod impact	D256					
notched (3.2 mm)		J/m	50	60	90	100
unnotched (3.2 mm)		J/m	1300	480	480	430
Shear strength, ultimate	—	N/mm²	100	90	95	100
Rockwell hardness	D785	—	M109	M114	M118	M125
Taber abrasion (CS 17, 1 kg)	D1044	mg wt. loss/1000 cycles	10	—	—	—
THERMAL						
Deflection temperature, unannealed	D648					
@ 1.82 N/mm² (6.4 mm)		°C	200	207	209	210
@ 0.45 N/mm² (6.4 mm)		°C	210	210	210	212
Vicat softening point, method B	D1525	°C	219	223	226	228
Continuous service temperature index (UL Bulletin 746B)		°C	170	170*	170*	170*
Coefficient of thermal expansion (−18 to 150°C), mold direction	D696	m/m·°C	6.2×10^{-5}	3.2×10^{-5}	2.5×10^{-5}	2.0×10^{-5}
Thermal conductivity	C177	W/m·°C	0.22	—	—	—

	Test	Units				
FLAMMABILITY						
Oxygen index (1.5 mm)	D2863	%	47	—	50	—
Vertical burn (UL Bulletin 94)	—	—	V-0 @ 0.76 mm, 5V @ 1.9 mm	V-0 @ 1.6 mm	V-0 @ 1.6 mm	V-0 @ 1.6 mm
NBS smoke, flaming mode (1.5 mm)	E662					
D_s @ 4 min		—	0.7	—	1.3	—
D_{MAX} @ 20 min		—	30	—	27	—
ELECTRICAL						
Dielectric strength (1.6 mm)	D149					
in oil		kV/mm	24	—	—	—
in air		kV/mm	33	—	—	30
Dielectric constant @ 1 kHz, 50% RH	D150	—	3.15	3.5	3.5	—
Dissipation factor	D150					
@ 1 kHz, 50% RH, 23°C		—	0.0013	0.0014	0.0015	0.0015
@ 2450 MHz, 50% RH, 23°C		—	0.0025	—	—	—
Volume resistivity (1.6 mm)	D257	ohm-m	6.7×10^{15}	1.0×10^{15}	7.0×10^{14}	3.0×10^{14}
Arc resistance	D495	seconds	128	—	—	85
OTHER						
Specific gravity	D792	—	1.27	1.34	1.42	1.51
Mold shrinkage	—	%	0.5–0.7	0.4	0.2–0.3	0.2
Water absorption	D570					
@ 24 hours, 23°C		%	0.25	0.28	0.26	0.18
@ equilibrium, 23°C		%	1.25	1.0	1.0	0.9

a–c) Reinforced with 10, 20, or 30% glass fibers, resp.

* Provisional rating.

Final rating may be higher.

11.8 POLYIMIDES WITH HETEROCYCLIC BACKBONE UNITS

11.8.1 Poly(imide-co-isoindolo-quinazolinedione)

11.8.1.1 Structure and synthesis

Hitachi Chemical Co. commercially manufactures a prepolymer for the preparation of poly(imide-co-isoindolo-quinazolinedione) under the designation ®PIQ [51].

The synthesis of the prepolymer is accomplished by reacting in solution 2 moles of an aromatic dianhydride with one mole of an aromatic diamine and one mole of a novel aromatic diaminodicarbonic amide of the general structure

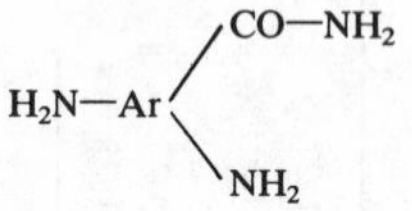

$$H_2N-Ar \begin{array}{c} {}^{CO-NH_2} \\[1em] {}_{NH_2} \end{array}$$

The reaction scheme is similar to the known [2] synthesis of poly-(isoindoloquinazolinedione) with the exception that dicarbonic diamides are replaced by mono amides and that the imide forming components are used in addition. Hitachi has not disclosed the nature of the employed aromatic components. The general reaction scheme for the synthesis of poly(imide-co-isoindoloquinazoline-dione) involves several steps (see eq. 11-14).

The commercial PIQ prepolymer is a poly(amic acid) dissolved in N-methylpyrrolidone with 14.5% solids and viscosity 1.6 Pa × s [52].

11.8.1.2 Properties and applications

®PIQ is used exclusively in the manufacture of LSI (Large Scale Integrated) circuits. In LSI circuits, increasingly higher packing densities or integration degrees are required. The PMP (Planar Metallization with Polymer) technique uses polymers as noncon-

(11-14)

ducting layers to replace the usually vapor deposited SiO_2. To this end, the prepolymer solution is dripped onto a metallized silicon wafer [53] with epitaxially grown MOS-(SiO_2) structure and is evenly spread on the surface through rotation [51, 52]. The solvent is removed under vacuum at 80−100°C. Imidization is achieved by heating at 205°C in an air circulating oven for 1−2 h and subsequent heating at 350°C for 30 min. The degree of cyclization can be followed by determining the permeability of water vapor through films. A minimum water vapor permeability was found for films made from PIQ which were cyclized at 350°C in either N_2 or O_2.

In the subsequent masking operation, a negative photo resist mask is used with hydrazine sulfate as etchant. The most obvious advantage of the PMP technique is the possibility of creating a completely planar surface which does not show the usual oxide steps on the chip surface.

PIQ serves equally well as nonconductive interlayer and as protective top coating. The layers made from PIQ must demonstrate excellent thermal stability and insulating property. PIQ films on silicon wafers survive heating in air at 450°C for 5 h. Under the same conditions, the common polyimides experience a 15−20% loss of their initial film thickness. No changes are observed in the infrared spectrum of PIQ films on heating at 450°C under N_2 for 5 h. The activation energy for the thermal decomposition of PIQ film under N_2 was found to be 289 kJ/mole [51]. However, the high thermal stability of PIQ is only ensured when the synthesis of PIQ is conducted under absolutely anhydrous conditions. Traces of water in the starting materials cause hydrolysis of dianhydrides to the free tetra acids which do not take part in the polycondensation and thus remain in the prepolymer. The dianhydrides are formed anew and evaporate during the thermal cyclisation. Also, the starting materials must be nearly free of sodium ions. Only PIQ containing less than 3 ppm sodium ions exhibits no influence on the characteristics of npn or pnp transistors. Hitachi can now supply PIQ with less than 0.5 ppm sodium. Measurements of electron, ion (Na) and surface conductivity, carried out by Texas Instruments at temperatures of up to 300°C, confirmed the applicability of PIQ for semiconductor systems [54].

11.8.2 Bismaleimide-triazine resins

11.8.2.1 Structure and synthesis

The BT-Resins of the Mitsubishi Gas Chemical Co., Inc., contain maleimide and cyanate groups. Starting materials are bismaleimide and the dicyanate which is obtained from the reaction of bisphenol A and cyanuric chloride. Heat curing results in the formation of a thermoset which incorporates triazine and triazine-imidazole ring structures [55]

(11-15)

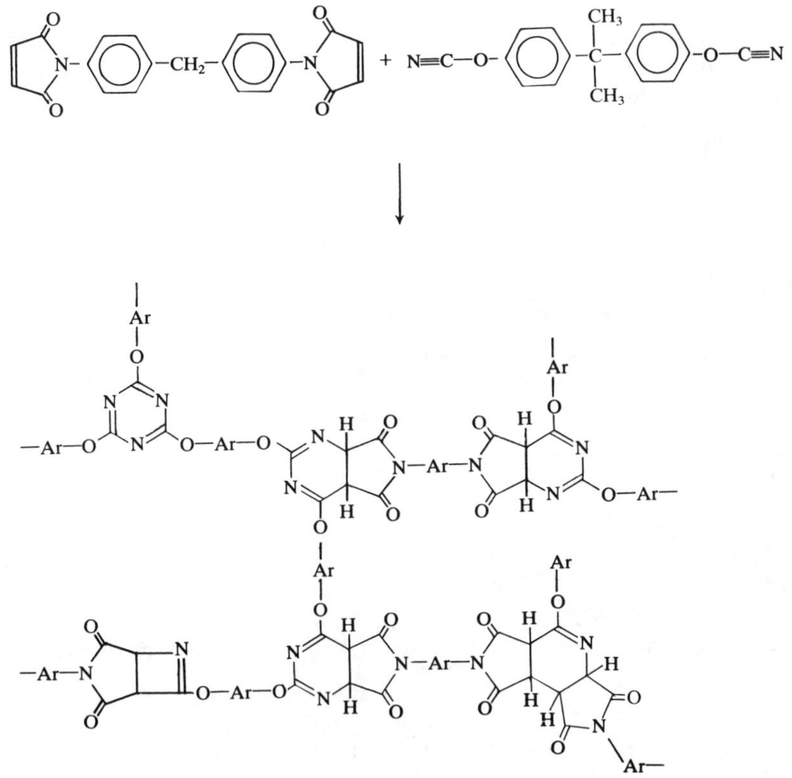

Triazine rings are formed through trimerization of cyanate groups:

(11-16)

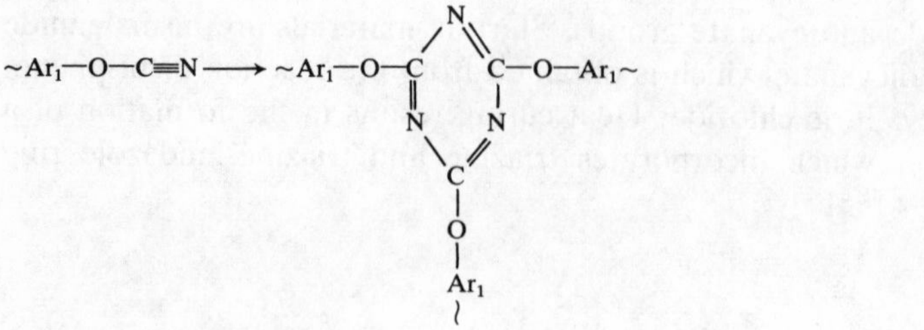

~Ar_1—O—C≡N ⟶ ~Ar_1—O—C ... C—O—Ar_1~   s-Triazine

Thus, the BT-Resins show properties similar to the triazine resins (see chapter 12.7). By reaction of cyanate with maleimide groups triazine-imidazole rings are formed (eq. 11-17).

(11-17)

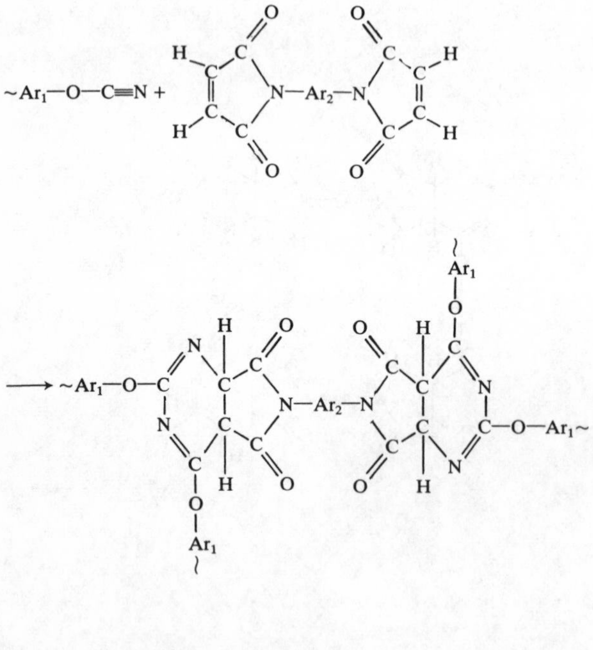

Triazine Imidazole

Furthermore, BT-Resins can react with epoxides under formation of oxazole rings (eq. 11-18). Cyanates react in the usual manner with hydroxy, carboxy and secondary amino groups (eq. 11-19 to 11-21). Maleimide groups can undergo self-polymerization (eq. 11-22), and the cyanate groups can react with dicyclopentadiene (eq. 11-23).

This variety of reactions can be utilized in the modification of a great number of polymers. The following polymers have been mentioned: epoxy, acrylic, silicone, silicone-epoxy, alkyd, polyester, polyurethane, phenolic, melamine, urea, xylene, imide, polybutadiene, diallyl phthalate, polybutene, cyclopentadiene, and polyvinyl butyral resins.

(11-18)

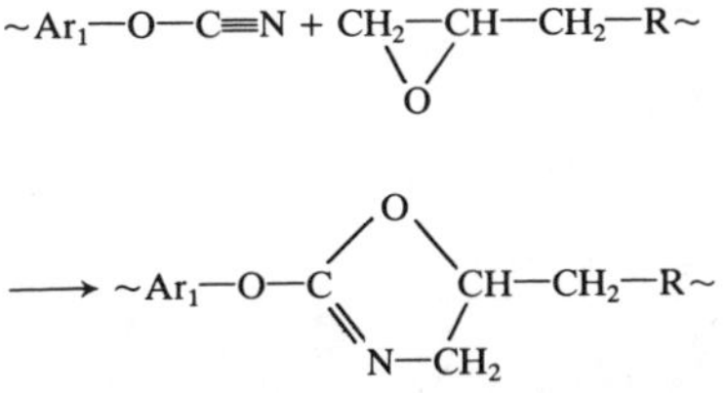

$$\sim Ar_1-O-C\equiv N + CH_2-CH-CH_2-R\sim$$

Oxazole ring or

Iso-oxazole ring

$$\longrightarrow \sim Ar_1-O-C \underset{N-CH_2}{\overset{O}{\diagup\diagdown}} CH-CH_2-R\sim$$

(11-19)

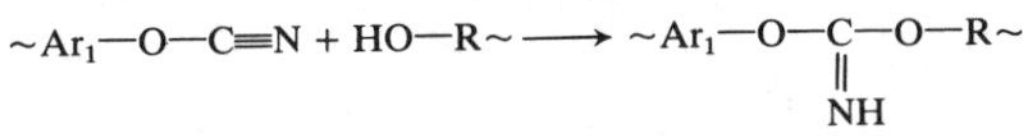

$$\sim Ar_1-O-C\equiv N + HO-R\sim \longrightarrow \sim Ar_1-O-\underset{NH}{\overset{\parallel}{C}}-O-R\sim$$ Alkylarylimide carbonate

(11-20)

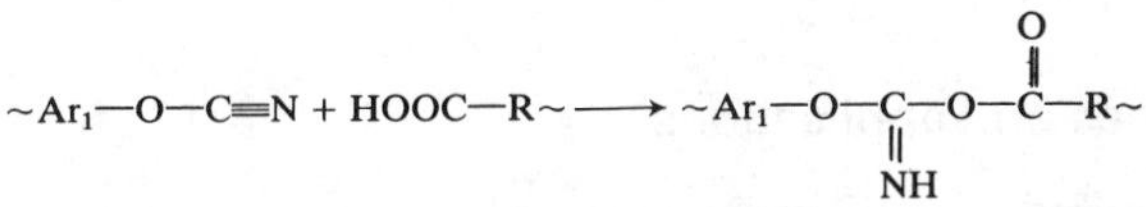

$$\sim Ar_1-O-C\equiv N + HOOC-R\sim \longrightarrow \sim Ar_1-O-\underset{NH}{\overset{\parallel}{C}}-O-\overset{O}{\overset{\parallel}{C}}-R\sim$$

(11-21)

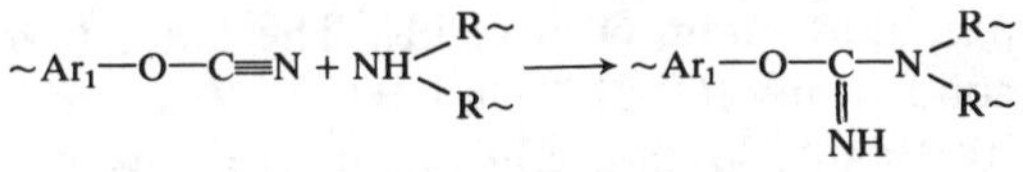

$$\sim Ar_1-O-C\equiv N + NH{\overset{R\sim}{\diagdown}}_{R\sim} \longrightarrow \sim Ar_1-O-\underset{NH}{\overset{\parallel}{C}}-N{\overset{R\sim}{\diagdown}}_{R\sim}$$ O-Arylisourea

(11-22)

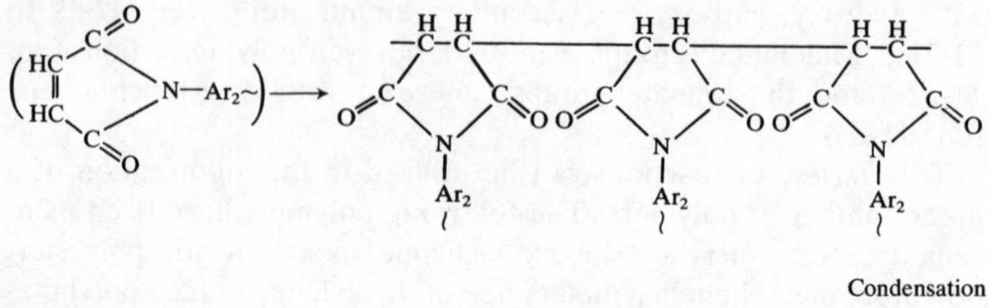

Condensation

(11-23)

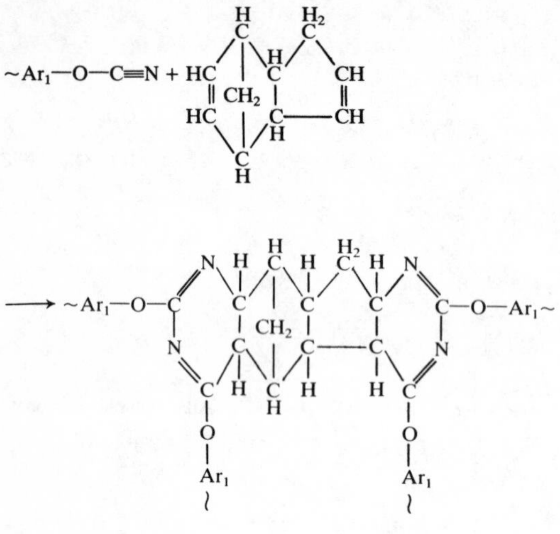

11.8.2.2 Properties and applications

BT-Resins are supplied as solid and liquid resin as well as solutions (Table 11-4).

The BT 2060 and 2067 grades are supplied as solutions in methyl-ethyl ketone with 70% solids; grades BT 2110 and 2117 are solutions in a MEK/DMF mix and contain 60% solids. The latter two are prepared from the solid resins BT 2170 and BT 2177, respectively. The BT-Resins 2060 and 2067 are equivalent to the Bayer

TABLE 11-4
Properties of BT Resins.

Grade	Physical state	Color	Density in g/cm³	Melting temperature in °C	Glass transition temperature[1] in °C	Continuous service temperature[2] in °C
BT 2100	solid	brown	1.24	50–60	260	160
BT 2170	solid	brown	1.24	60–70	260	180
BT 2177	solid	brown	1.24	60–70	250	180
BT 2300	solid	dk. brown	1.25	50–60	270	180
BT 2400	solid	dk. brown	1.28	50–60	290	200
BT 2600	solid	dk. brown	1.31	50–60	310	200
BT 2670	solid	dk. brown	1.31	60–70	310	230
BT 3103	liquid	brown	1.24	—	230	150–170

[1] cured resin.
[2] for 20000 h service of cured resin.

A. G. resins KL 3-4000 and KL 3-4005, respectively.

The molecular weights of BT-Resins range from 290 to 2100, depending on the grade. Liquid BT-Resins can be diluted with reactive thinners, such as glycidyl ethers, acrylic esters, diallyl phthalate, styrene and methyltetrahydrophthalic acid anhydride. The elasticization of BT-Resins is accomplished by addition of flexibilized epoxy resins, diene elastomers, or polyvinyl butyral.

Curing of BT-Resins requires temperatures in the range from 170 to 250°C. Prolonged postcuring at 200−250°C increases the glass transition temperature of the cured resins. Curing is accelerated by catalysts, such as metal salts of organic acids or chelates, per se or in combination with tertiary amines, imidazoles or peroxides. For example, a combination of zinc octoate and triethylene diamine or N,N-dimethylbenzyl-amine is recommended.

BT-Resins are predominantly used for the manufacture of pre-pregs and laminates. The properties of laminates made from BT-Resins are comparable to those made from Triazine A or polyimide (Table 11-5). Other applications include impregnations, compression molding and powder coating materials, lacquers, adhesives, and casting resins.

A casting resin of BT 2600 shows no weight loss on heating in air to 350°C, whereas BT 2170 loses ca. 5% of its weight. A compression molding compound of 22 pt BT 2170, 6 pt cresol-novolak-epoxy resin, 39 pt silica, and 33 pt glass staple fiber exhibits the following properties: water absorption 0.11%, heat deflection temperature 290°C, flexural strength 100 MPa, flexural modulus 12 GPa, tensile strength 42 MPa; at 200°C the same mechanical properties have values of 80 MPa, 10 GPa and 35 MPa, respectively.

In copper-clad laminates tested at temperatures of up to 250°C, BT-Resin 2600 demonstrates a lower relative permittivity and lower dissipation factors than epoxy, Triazine-A and polyimide resins. Also with BT 2600, the electric properties are less dependent on frequencies than with the other resins.

11.9   UNFILLED POLYIMIDES

Unfilled polyimides are commercially available as thermoplastic and thermoset type polyimides. All are of yellow color. Polyimides combine good mechanical properties with excellent thermal stability, for

TABLE 11-5

Comparison of properties of laminates from BT resin with those of other resins.

Physical Property	Physical Unit	Epoxide	Xylok	Property value of Triazin A	BT 2600	Polyimid
Resin	%	60	52	50	55	50
Viscosity	—	low	high	very low	very low	very high
Gel time at 168°C	s	100	255	70	100	—
Resin flow	%	30	25	20	25	35
Glass transition temperature	°C	110	225	255	310	290
Heat resistivity[1]	—	5	3	2	1	1
Smoldering resistivity						
Temperature	°C	—	—	350	400	—
Time	s	—	—	20	20	—
Oxidation resistivity[1]	—	5	2	3	1	1
Chemical resistivity[1]	—	5	2	2	2	2
Flammability (UL)	—	94 HB	94 HB	94 HB	94 V-1	94 V-0
Relative permittivity (1 MHz)	—	4.8	4.5	4.2	4.15	4.5
Dissipation factor (1 MHz)	—	0.0200	0.0150	0.0026	0.0020	0.0100
Surface resistivity	Ohm	5×10^{13}	10^{14}	5×10^{14}	10^{14}	10^{14}
Volume resistivity	Ohm × cm	10^{14}	10^{15}	10^{15}	10^{15}	10^{15}
Interlaminar shear strength[1]	—	4	4	4	4	5
Drillability[1]	—	4	4	2	1	2
Impregnability[1]	—	4	4	3	3	5

[1] Relative scale from 1 (excellent) to 5 (fair).

some grades up to 300°C (Table 11-5). In an inert atmosphere, they survive short term exposure to temperatures of up to 500°C; in air, slow oxidative degradation occurs. Unfilled molded parts have low coefficients of thermal expansion. Compression creep is very small even at high temperatures, however, notched impact strength is only moderate. Polyimides are unaffected by exposure to dilute acids, aromatic and aliphatic hydrocarbons, esters, ethers, alcohols, hydraulic fluids, and kerosene. Polyimides are attacked by dilute alkali and concentrated inorganic acids. There are no significant differences in properties of thermoset, thermoplastic, in situ formed or preformed polyimides, although some peculiarities exist, of course, which can be attributed to different structures. For example, aliphatic polyimides such as ®Kamax exhibit a relatively low heat deflection temperature when compared to aromatic polyimides. Their advantages, however, are improved notched impact strength, enhanced processability, and lower price.

Unfilled polyimides are commercially available as molding compounds and solutions, fibers, films, stock rod and sheet as well as parts molded to customer's order. Thermoset polyimides are especially useful in engineering parts. Thermoplastic polyimides, due to their high elongation and toughness, are particularly advantageous in thin-film products (films, enamels, adhesives, fibers, and coatings).

Molded and machined parts fabricated from polyimides find applications in piston rings, bearings, valve seats, seals, electrical connectors, and coil shells, and for various components in the nuclear, automotive, aircraft, and space industries. Films are used in cable and wire insulation, as carrier in printed circuits, and in motor insulation. Solutions of polyimides or prepolymers are used as adhesives, and wire enamels, in the casting of films, impregnation of carriers for laminates, and in the production of solution-spun fibers.

Commercially available are fibers made from Polyimide 2080 of Upjohn and two types of poly(imide-amides) produced by Rhône-Poulenc under the trade name ®Kermel. The latter are prepared by polycondensation of trimellitic acid anhydride with 4,4′-diisocyanatodiphenylmethane or 4,4′-diisocyanatodiphenyl-ether, respectively, in DMAc or NMP 56. Market ability of these fibers appear to be limited in view of the competitive aramid fibers. Aramid fibers exhibit higher strength and retain their strength at elevated temperatures better than polyimide fibers.

TABLE 11-6
Properties of unfilled polyimides, polyamide-imides, and polyether-imides.

Physical property	Physical Unit	Property Values								
		Kamax 201	Kamax 301²⁾	Upjohn 2800	Envex 1000	Thermid MC-600	Meldrin PI	Vespel SPI	Torlon 4203	Ultem 1000
Processability[1]	—	I	I	C	C/E	C	I	I/E	I/E	I/E
Density	g/cm³	1.16	1.30	1.40	1.33	1.37	1.40	1.43	1.40	1.27
Tensile strength at										
23°C	MPa	51.7	43.4	118	88.3	82.7	80.0	86.2	190	105
260–280°C[3]	MPa	17.9	13.1	30	45	—	—	36	52	—
Modulus of elongation	GPa	2.10	1.83	—	2.59	3.93	—	—	—	—
Tensile strength at break	%	34	32	10	—	2	2.9	8	12	60
Flexural strength at										
23°C	MPa	83.4	64.8	200	165	131	—	117	212	—
260–280°C[3]	MPa	29.6	22.0	35	60	29	—	62	80	—
Flexural modulus at										
23°C	GPa	2.51	1.88	3.30	3.06	4.48	4.21	3.10	4.70	3.30
260–280°C[3]	GPa	1.38	1.31	1.10	1.0	—	—	1.79	3.00	—
Compression strength	MPa	55.1	48.9	210	204	172	427	>276	—	140
Impact strength with notch	J/m	267	134	240	—	—	30	53	—	—
Hardness (Rockwell)		L73	L58	E99	M122	—	—	E52	M119	M109
Continuous service temperature	°C	105	105	260	≥260	316	—	—	260	170
Heat distortion temperature (1.81 MPa)	°C	130	122	285	288	—	—	260	274	200
Vicat temperature	°C	148	147	—	—	—	—	—	—	219
Linear thermal expansion coefficient (×10⁵)	K⁻¹	9.3	9.6	5.0	6.8	4.4	4.3	5.6	3.6	—
Mold shrinkage	%	0.5–0.8	0.5–0.8	0.6	0.98	1	—	0.32	0.28	0.5–0.7
Water absorption (24 h)	%	0.43	0.36	3.4	—	—	—	—	—	0.25
Relative permittivity (1 kHz)		3.4	3.5	—	3.4	5.4	3.6	3.4	4.0	3.15
Dielectric strength	MV/m	29	32	—	12.3	—	—	22.1	23.6	—
Volume resistivity	Ohm × cm	10¹⁴	10¹⁴	—	10¹⁶	—	10¹⁶	10¹⁷	10¹⁷	—
Surface resistivity	Ohm	10¹⁴	10¹⁴	—	—	—	10¹⁵	10¹⁶	10¹⁷	—
Dissipation factor		0.010	0.009	0.006	0.005	0.0006	0.0025	0.0018	0.009	—
Arc resistance	s	117	83	—	—	—	—	—	125	218
Flammability (UL-94)		HB	V-0	—	—	—	—	—	—	—
Limiting oxygen index	%	18.6	25.2	44	—	—	—	—	—	47

[1] I = injection molding, E = extrusion, C = compression molding.
[2] with bromine containing flame protection agent.
[3] Kamax at 121°C, MC-600 at 316°C.

Polyimides are relatively high-priced materials. For some products, the following prices have been stated for the period 1979–1982:

Product	US-$/lb
Polyimide 2080	12.-
Kinel (depending on filler type)	7.- to 11.-
Torlon 4203	14.15
Torlon 4301 (graphite/PTFE filled)	13.05
Torlon 5030 (glass fiber reinforced)	12.17
Torlon 7030 (graphite fiber reinforced)	25.-
Torlon 9040 (high-heat grade)	5.40
Vespel, machined parts	40.- to 70.-
Kamax 201 and 301	1.07 to 2.45
Thermid MC-600	180.-
Thermid LR-600 (50% filler)	70.-
Thermid AL-600 (75% filler)	90.-
XU 218 general purpose grade	100.-
XU 218 high-purity grade	150.-
Envex, rods, Ø 1/4 to 2 inches	1.59 to 23.55/inch
Envex, tubing, Ø 1/2 to 1.1 inches	43.55 to 68.-/foot
Tribolon	10.-
NR-150 solution	75.-
Kerimid 601	11.-
Kerimid 711	18.-
Kerimid 353	29.-
Amoco AI polymer solutions	
AI 335, 37.5% solids	12.-
AI 830, 28% solids	2.45
AI 1035, 39% solids	10.-
AI 1130L, 30.5% solids	2.62
AI-Lite, 32.5% solids	2.88
Ultem 1000	4.25
Ultem 2300	3.40

11.10   FILLED AND REINFORCED POLYIMIDES

Polyimides, poly(amide-imides) and poly(ether-imides) can be filled with chopped glass fibers, graphite powder, molybdenum disulfide, PTFE powder, and, for the application as adhesives, aluminum

TABLE 11-7
Commercially available filled and reinforced polyimides.

Manufacturer/ Compounder	Trade name	Filler content in %				
		Graphite	MoS$_2$	PTFE	Glass fiber	Graphite fiber
Amoco Chemicals	Torlon 4301	×		×		
	Torlon 5030				30	
	Torlon 7030					30
	Torlon XG-549	×				
LNP	Torlon PDX-5200				30	
DuPont	Vespel SP-21	15				
	Vespel SP-211	15		10		
	Vespel SP-22	40				
	Vespel SP-31		15			
	Vespel SP-5				30	
General Electric	Ultem 2100				10	
	Ultem 2200				20	
	Ultem 2300				30	
Fiberite Corp.	PI 730				30	
	PI 740				40	
	PI 750				50	
Fluorocarbon	Tribolon					
	PI 550				40	
	PI 555	25				
	PI 558	40				
	PI 559			20		
	Xt 154			20		
Rhone-Poulenc	Kinel 4515				×	
	5504				×	
	5505	25				
	5508	40				
	5514				50	
	5515				50	
	5517	×				
	5518			×		
Rogers	Envex 1115		15			
	1228			×		
	1330	×				
	3540				×	
	1620			×		

* Only filler type but not filler concentration known.

powder and arsenic derivatives. Reinforcements include glass, aramid, graphite, quartz, and boron fibers and/or fabrics. A number of commercially available products are listed in Table 11-7.

Thermoplastic polyimides with graphite, MoS_2 or PTFE fillers are used for the fabrication of self-lubricating parts, e.g. for piston rings, valve seats, bearings, seals, and thrust washers. Wear characteristics of poly(amide-imide), filled with graphite and PTFE, are superior to bronze, polyacetals, fabric reinforced phenolic resins, and filled PTFE [57]. Such parts withstand temperatures of up to 260°C, compared to bronze with only 120°C. Injection moldable alloys of polyimides with PPS, unfilled or filled with graphite or PTFE, are also commercially available.

Coefficients of thermal expansion are considerably lowered by fillers or reinforcements. Creep is almost nonexistent even at high temperatures (Table 11-8). Deformation under load is below 0.05% for glas fiber reinforced moldings subjected to 28 MPa loads at 25°C for 24 h [69]. Wear resistance and coefficient of friction are further enhanced by use of fillers; PTFE-filled polyimide exhibits better wear and friction characteristics than are available in any filled PTFE-compound [69].

Filled and reinforced polyimides have found markets in the aircraft, aerospace, electric and electronic fields. Applications include jet engine and aircraft turbine engine parts, automotive parts that require electrical and thermal insulation, connectors, slot wedges, insulating sheets, and fire barriers. Laminates are used in printed circuits, honeycomb structures, and in jet engine parts. Laminates also demonstrate excellent strength at high temperatures (Table 11-9 and Table 11-10). The PMR technique has brought about great advantages by facilitating the fabrication of laminates [58].

11.11 POLYIMIDE ADHESIVES

Polyimides or their poly(amic acid) precursors as well as PMR systems are used in the form of solutions, pastes, or films on fabric carriers as thermally stable adhesives for bonding metals and graphite fiber reinforced laminates.

Older commercial products [2] include: Kerimid (Rhone-Poulenc), Quantad 159 (Quantum Inc.), FM34B and FM36 (American Cyanamid), Metlbond 840 (Narmco), DP 5505 (DuPont), IPA

TABLE 11-8
Properties of filled and reinforced polyimides.

Physical property	Physical unit	Kinel 5505	Kinel 5508	Envex 1115	Tribolon PI559	Fiberite PI 730	PI 740	PI 750	Kinel 5514	Envex 3540
						Property value				
Filler type	—	Graphite	Graphite	MoS₂	PTFE	Glass fiber	Glass fiber	Glass fiber	Glass fiber	Graphite fiber
Filler content	%	25	40	15	20	30	40	50	50	?
Density	g/cm³	1.45	1.55	1.5	1.42	1.85	1.90	1.95	1.70	—
Tensile strength	MPa	39.3	32.4	55.3	34.5	55.2	103.4	144.8	44.1	69.1
Modulus of elasticity	GPa	—	—	3.45	—	—	—	—	—	13.86
Elongation at break	%	1	1	1	2.4	<1.0	—	—	—	1.3
Flexural strength	MPa	129.6	78.6	110	49.0	69.0	248.3	248.3	146.9	124
Flexural modulus	GPa	6.21	7.24	4.63	2.69	13.79	—	—	13.65	14.90
Compression strength	MPa	137.9	108.3	193	137.9	241.4	—	—	234.5	244
Impact strength with notch	KJ/m	0.01	0.02	—	0.01	0.02	0.37	1.18	0.30	—
Hardness (Rockwell M)	—	110	100	111	115	117	118	118	118	—
Linear thermal expansion coefficient ($\times 10^5$)	K^{-1}	1.89	1.50	4.86	6.60	2.75	1.49	0.99	1.30	—
Continuous service temperature	°C	288	288	288	—	249	—	260	349	—
Heat distortion temperature	°C	288	288	288	—	>316	>316	>316	—	260
Dielectric strength	MV/m	—	—	—	—	15.8	17.7	17.7	17.7	—
Relative permittivity (1 GHz)	—	—	—	—	—	4.85	4.8	5.1	4.5	—
Dissipation factor (1 GHz)	—	—	—	—	—	0.0026	0.003	0.004	0.017	—
Water absorption (24 h)	%	0.6	0.6	0.67	0.30	0.1	0.1	0.2	0.7	—

TABLE 11-9
Properties of Polyimide Laminates [28, 35, 58].

Physical Property	Physical Unit	Kerimid 353			Thermid 600		PMR 15	PMR II	NR 150
		Glass	Graphite[1]	Aramid[2]	Glass	Graphite[3]	Graphite[3]	Graphite[4]	Graphite[4]
Reinforcing agent	—	Glass	Graphite[1]	Aramid[2]	Glass	Graphite[3]	Graphite[3]	Graphite[4]	Graphite[4]
Fiber content	Vol. %	60	60	60	35	35	38.4	55	57
Resin content	%	—	—	—	—	—	—	—	36
Density	g/cm³	2.0	1.57	1.39	—	—	—	—	1.59
Tensile strength									
at 23°C	MPa	1050	1100	1200	—	—	—	—	—
at 250°C	MPa	1020	1050	750	—	—	—	—	—
Modulus of elasticity									
at 23°C	GPa	42.2	170	62.1	—	—	—	—	—
at 250°C	GPa	38.0	162	50.0	—	—	—	—	—
Flexural strength									
at 23°C	MPa	1205	1450	645	479	1345	1475	—	1641
at 250°C	MPa	830	1150	310	—	—	—	—	841[5]
at 316°C	MPa	—	—	—	310	1021	890	570	—
Flexural modulus									
at 23°C	GPa	42	164	65	32.4	103.4	115	—	117.2
at 250°C	GPa	40	169	49	—	—	—	—	110.3[5]
at 316°C	GPa	—	—	—	20.7	82.7	112.6	53.1	—
Interlaminar shear strength									
at 23°C	MPa	78	75	27	64.1	83.4	108	—	115.8
at 250°C	MPa	49	49	21	—	—	—	—	—
at 316°C	MPa	—	—	—	45.5	55.2	59	40.7	44.1[5]

1) Modmor II S.
2) Kevlar 49.
3) Hercules HTS fiber.
4) Thornel T-300.
5) at 260°C.

TABLE 11-10
Flexural strength and interlaminar shear strength of polyimide/HT-S-graphite laminates [28, 58].

Physical Property	Physical Unit	Property value			
		NR-150 B 2	PMR 15	Thermid 600	LARC 160
Flexural strength					
at room temperature	GPa	1.43	1.43	1.38	2.13
at 260°C after 500 h at 260°C	GPa	1.30	1.84	1.17	1.52
at 288°C after 500 h at 288°C	GPa	1.21	1.24	0.99	1.44
at 316°C after 500 h at 316°C	GPa	1.14	0.97	1.04	—
Interlaminar shear strength					
at room temperature	MPa	70.3	111.0	83.4	95.8
at 260°C after 500 h at 260°C	MPa	40.7	53.8	60.0	53.9
at 288°C after 500 h at 288°C	MPa	39.3	47.6	51.0	46.9
at 316°C after 500 h at 316°C	MPa	34.5	17.9	41.4	—

380 (Rhone-Poulenc), Polyimide 2080 (Upjohn), and ST-1 (Soviet Union).

More recent products, of which some are still developmental materials, include Thermid LR-600 of Gulf Advanced Materials [59], NR-056X of DuPont [60], BXR 10314-151 C of American Cyanamid [70], LARC-13 [61] and LARC-TPI [71], both developed by the NASA Langley Research Center. Recently, NR-056X was withdrawn from the market by DuPont.

Thermid LR-600 is an acetylene end-capped poly(amic acid) supplied in NMP solution with 50% solids. NR-056X is a modification of NR-150 B2. It is prepared from 4,4'-hexafluoro-propylidene-bis-phthalic acid, diaminobenzene and diaminodiphenylether in a molar ratio of 1.0:0.75:0.25 in diglyme as solvent. LARC-13 with a molecular weight of about 1300 g/mole is prepared from the reaction of appropriate molar amounts of 3,3',4,4'-benzophenonetetracarboxylic acid dianhydride, 3,3'-diaminodiphenylmethane, and nadic anhydride in DMF. The nadic group undergoes a thermally induced reaction with the evolution of a small amount of cyclopentadiene to yield brittle resins. Silicone and NBR elastomers have been incorporated into LARC-13 for the purpose of improving fracture toughness. This improvement in toughness, however, is only accomplished at a sacrifice in the elevated temperature adhesive strength [72]. The addition polyimide adhesive type BXR10314-151C is reportedly also based on a nadic terminated imide system.

LARC-TPI (linear-aromatic-chain-thermoplastic-polyimide) is a multi-purpose polyimide and is prepared from 3,3',4,4'-benzo-phenonetetracarboxylic acid dianhydride and 3,3'-diaminobenzo-phenone in bis(2-methoxyethyl)ether (diglyme) as solvent. It can be processed as a thermoplast after imidization. The flexible, tough, thermo-oxidatively stable resin can be used for molding and as a matrix for fiber reinforced laminates. It also can be converted into fibers and films. Both the amic acid and the imidized form have been evaluated as high temperature adhesive. Experimental quantities of LARC-TPI are available from Gulf Research and Development Co.

The properties of these new adhesives have been compared with those of the older FM-34 and with a developmental [62] poly-(phenylquinoxaline) PPQ (Table 11-11) [16]. PPQ is supplied as monoether in m-cresole/xylene solution with 16% solids. LARC-13 and NR-056X contain aluminum powder as filler [15]. Thermid LR-600 contains 5 phr hydroquinone as inhibitor to prolong pot-life. The adhesives are used as tapes with glass fabric carriers.

Titanium/titanium (Ti/Ti) as well as laminate/laminate (C/C) bonded with polyimides exhibit high tensile lap shear strengths. The strength remaining after heat aging at temperatures of up to 316°C is still good. At 316°C, PPQ is no longer suitable for bonding Ti/Ti but is still good for bonding C/C. Compared with the older FM-34, the new products provide no significant increase in bonding strength, however, they demonstrate considerably prolonged shelf life (with the exception of NR-056X), and the evolution of volatiles during curing is less than 1%.

A more recent study investigated the performance of LARC-13 (NASA Langley), NR-150 A2, NR-150 B2, NR-056X (all of DuPont), HR-602 (Hughes Aircraft), LARC-13 with amide-imide modification (Boeing), LARC-13 with methylnadic capping and m-phenylenediamine modification (Boeing), PPQ (NASA Langley), PPQ with boron powder modification (Boeing), LARC-TPI (NASA Langley), and FM-34 (American Cyanamid) as baseline [73]. Lap shear bonds were tested at initial ambient and at 232°C after extended (to 3000 h) exposure to 232°C. In the next phase, the remaining PPQ and LARC-TPI systems were subjected to environmental durability tests. Unstressed lap shear coupons were exposed to 232°C for up to 10 000 h. The PPQ resin was pulling clear from

TABLE 11-11
Properties of temperature resistant adhesives [16, 59].

Physical properties	Physical unit	Property Value				
		FM 34	Thermid LR-600	NR-056X	LARC-13	PPQ
Tape						
Filler		yes	no	yes	yes	no
Tack		yes	yes	no	no	no
Shelf life		bad	good	bad	good	good
Volatiles	%	14	<1	11	<1	<1
Processing temperature	°C	343	343	343	343	399
Processing pressure	MPa	1.4	0.34	2.1	0.34	1.4
Glass transition temperature after curing	°C	322	—	328	306	318
Titanium/titanium bonding						
Shear strength						
at 23°C	MPa	22.7	26.2	24.8	13.8	30.3
at 288°C	MPa	—	14.5	—	—	17.2
at 316°C	MPa	11.7	11.4	11.9	11.5	2.1
at 316°C after 125 h	MPa	10.3	—	9.0	11.3	3.4
at 316°C after 750 h	MPa	—	10.7	—	—	—
Laminate/Laminate bonding						
Shear strength						
at 23°C	MPa	17.5	—	15.9	17.0	41.4
at 316°C	MPa	11.4	—	14.3	13.7	19.3
at 316°C after 125 h	MPa	10.1	—	9.2	10.4	17.2

the TiO_2 substrate. LARC-TPI exhibited stability to 9500 h exposure. The data for expected 20 000 h exposure had not been completed at the publication of the paper [73]. Lap shear coupons were subjected to −54°C, ambient and 232°C while stressed at 25, 30 and 50% of ultimate. At 232°C, the PPQ bond strengths after 5000 h had dropped to about 11 MPa which is below the structural 14.5 MPa requirement. For LARC-TPI at 232°C, the lap shear strengths at all stress levels remained significantly above 14.5 MPa. Both systems were subjected to 59°C/95% relative humidity for up to 2000 h, and neither system showed any shear strength degradation when tested at −54°C but did show moderate strength loss when tested at 232°C after high humidity exposure. Moreover, both systems were exposed to various fuels and fluids for up to 5000 h. Lap shear strength was relatively unaffected except by exposure to hot

Skydrol. The conclusion from all tests was that LARC-TPI demonstrated the best overall properties of all systems evaluated.

Principal areas for high temperature adhesives include advanced aircraft, space vehicles, missiles, engines, microelectronic and electrical components. In the microelectronic area, the adhesive must withstand processing temperatures (not use temperatures) of >400°C for a few hours in an inert atmosphere. Tens of thousands of hours at temperatures to 232°C will be required for use on supersonic aircraft. Application on missiles require mechanical performance for less than a minute at temperatures >538°C [70]. One example reported is the use of LARC-13 for bonding an experimental graphite/polyimide composite wing panel on the NASA YF-12 aircraft [72].

11.12 POLYIMIDE FOAMS

®Skybond [1] was the first polyimide foam; it was introduced by Monsanto in 1970. Recent products include the syntactic foams based on ®Polyimide 2080 introduced by Upjohn in 1978 [63–65], ®LMB 1907 marketed by Ciba-Geigy in 1980 [66], and ®Solimide presented by the Solar Division of International Harvester in 1981 [67, 68]. Solimide has been developed under NASA contracts; at present, licenses can not be obtained for this product.

Monsanto supplies a series of foamable polyimide powders with incorporated foaming agent that give rigid foams with densities ranging from 96 to 384 kg/m³ (Table 11-12). The density can be increased when the powders are compressed prior to foaming. Another Monsanto compound, commercially available, produces a flexible foam with density 16 kg/m³.

Upjohn has developed a technique for preparing syntactic foams. Hollow glass spheres (B-30 of 3 M Co.) are incorporated into the thermoplastic Polyimide 2080. Foams result with resin contents from 10 to 35% and densities from 160 to 592 kg/m³. The K-value is 0.20; compressive strength after 14 days heat aging at 232°C is 0.124 MPa. The upper continuous service temperature is 260°C.

LMB 1907 is commercially available from Ciba-Geigy as a one-component system supplied as pellets or tablets. It contains a chemically acting foaming agent. The foams are fabricated in heated closed molds. For objects with a wall thickness of 2 cm the following

TABLE 11-12
Properties of Polyimide Foams.

Physical Property	Physical Unit	Skybond 18	LMB 1907	Property value Solimide flexible	Structural foam with 40% glass fiber
Density	kg/m^3	328	450–480	19.2	1374
Flexural strength at					
23°C	MPa	—	11.9	—	—
180°C	MPa	—	12.4	—	—
Modulus of elasticity at					
23°C	MPa	—	465.5	—	—
180°C	MPa	—	303.3	—	—
Tensile strength at break	MPa	—	—	0.092	47.3
Elongation at break	%	—	—	39	1.1
Shear strength	MPa	—	7.3	—	—
Compression strength	MPa	2.5–7.6	18–24	—	—
Impact strength with notch	J/m	—	—	—	469
Compression set	%	—	6.2	—	—
Resilience	%	—	75	—	—
Relative Permittivity	—	1.41	1.7	—	—
Dielectric strength	kV/cm	—	127	—	—
Dissipation factor	—	6.2×10^{-4}	25	—	—
Volume resistivity	ohm × cm	—	7.3×10^{15}	—	—
Surface resistivity	ohm	—	2.7×10^{14}	—	—
Limiting oxygen index	%	—	—	45	60
Heat distortion temperature (1.81 MPa)	°C	—	—	>204	—
Weight loss (TGA)	°C (%)	—	—	204(0)	—

cycle times are required: 1 h at 160°C, then 1 h at 180–200°C, and subsequent annealing for 4–6 h at 180–200°C. The foam can be used at temperatures of up to 240°C. LMB 1907 costs approximately 40.-SFr/kg (≈8.65 US-$/lb).

The synthesis of Solimide is carried out by reacting BTDA, 2,6-diaminopyridine and 4,4'-diaminodiphenylmethane in a molar ratio of 1.0:0.4:0.6 in ethanol solution. Diaminopyridine provides foam flexibility and hydrolytic stability. The poly(amic acid) precursor is used in powdery form which loses ethanol and water on imidization. These compounds act as foaming agents. Foaming in a microwave oven gives a more homogeneous cell structure than foaming in an air circulating oven or with high-frequency heating. Surface active agents, e.g. Zonyl FB (DuPont) or FC-430 (3 M Co.), support the formation of homogeneous cell structures. Composite foams can be fabricated by incorporation of glass or graphite fibers and fabrics.

The open-cell flexible Solimide foam was developed to replace PUR foam seats used in aircrafts. Solimide does not ignite at temperatures of up to 430°C, it merely carbonizes, and almost no smoke is developed on decomposition. The gaseous decomposition products contain by far less HCN and CO than those obtained from PUR. It is hoped to gain approximately 3 additional minutes for the evacuation of passengers in accidents accompanied by burning fuel. The present specifications for aircraft interior materials request an evacuation time of only 2 min. Flexible seating for aircraft, auditoriums, and theaters is a major market for Solimide; additional uses can be found in noise abatement and thermal insulation applications. The foam retains its resilience at temperatures ranging from 149 to 260°C (300–500°F) [74]. International Harvester has received a contract from the Naval Research Laboratory to develop a closed-cell counterpart of its Solimide open-cell polyimide foam. The contract calls for a closed-cell foam with anti-sweat, accoustical, and fire resistant properties [75].

Reinforced rigid polyimide foams with honeycomb structure are used as components in the aircraft industry and for flame resistant noise abatement walls. Syntactic foams combine high compression strength at elevated temperatures, flame resistance, and accoustical and thermal insulation properties.

References

[1] H.-G. Elias, Neue polymere Werkstoffe 1969–1974, C. Hanser, München 1975;

New Commercial Polymers 1969–1975, Gordon and Breach, New York 1977.

[2] K. H. Bühler, Spezialplaste, Akademie-Verlag, Berlin 1978.

[3] P. E. Cassidy, Thermally Stable Polymers, M. Dekker, New York 1980.

[4] J. Preston, Heat Resistant Polymers, in Kirk-Othmer, Encycl. Chem. Technol. **12** (1980) 203.

[5] C. E. Scroog, Polyimides, Macromol. Revs. **11** (1976) 161.

[6] Anon., Kunststoffe **67**/1 (1977) 1.

[7] H.-G. Elias, Kunststoffe mit besonderen Eigenschaften, Ullmanns Encycl. Techn. Chem., 4th ed., vol. **15** (1978) 435.

[8] H. Domininghaus, Kunststoffe **69**/1 (1979) 2.

[9] Fr. P. 144 875 (1968), Rhône-Poulenc.

[10] Gulf BTDA, Gulf Oil Chemical Co., Technical Information, 1979 and 1981.

[11] A. K. St. Clair and T. L. St. Clair, SAMPE Symp. **26** (1981) 165.

[12] J. H. Bateman, W. Geresy and D. S. Neiditch, ACS Org. Coat. Plast. Prepr. **35**/2 (1975) 2.

[13] H. H. Gibbs, SAMPE Symp. **17** (1972) 17.

[14] H. H. Gibbs and C. V. Breder, High Temperature Laminating Resins based on Melt Fusible Polyimides, Adv. Chem. Ser. **142**: Copolymers, Polyblends and Composites, ACS, Washington, D.C., 1975.

[15] C. L. Hendricks and S. G. Hill, SAMPE Symp. **25** (1980) 39.

[16] P. M. Hergenrother and N. J. Johnston, in C. A. May (ed.): Resins for Aerospace, ACS Symp. Ser. **132**, Washington, DC, 1980.

[17] R. J. Jones, R. W. Vaughan and A. E. Samsonov, SAMPE Symp. **26** (1981) 470; ditto, USP 4 111 906 (1978), 4 196 277 (1980) and 4 203 922 (1980).

[18] Japan Plastics Age **3**/4 (1979) 17.

[19] Japan Chemical Week (June 19, 1980) 3.

[20] H. C. Fromuth, R. H. Weese and L. D. Trabert, SPE Nat. Techn. Conf., Detroit, November 1979.

[21] A. S. Wood, Modern Plast. Internat. **10** (July 1980) 40.

[22] P. S. Carleton, W. J. Farrissey, Jr., and J. S. Rose, J. Appl. Polym. Sci. **14** (1970) 1093; ditto, **16** (1972) 2983.

[23] J. T. Chapin, B. K. Onder and W. J. Farrissey, Jr., ACS Polym. Prepr. **21**/2 (1980) 130.

[24] W. J. Farrissey, Jr., L. M. Alberino and A. A. R. Sayigh, J. Elastom. Plast. **7** (1975) 285.

[25] A. L. Landis, N. Bilow, R. H. Boschan and R. E. Lawrence, ACS Polym. Prepr. **15**/2 (1974) 533.

[26] N. Bilow, A. L. Landis and L. J. Miller, USP. 3 845 018, 3 864 309, 3 879 349, 3 928 450, 4 075 111, 4 098 767 and 4 108 836.

[27] N. Bilow, Acetylene-substituted Polyimides as Potential High-Temperature Coatings, in C. A. May (ed.), Resins for Aerospace, ACS Symp. Ser. **132**, Washington, D.C., 1980.

[28] ®Thermid Polyimide Resins, Gulf Oil Chemical Co., Technical Information, 1977 and 1981.

[29] R. V. Subramarian and J. J. Jakubowski, in C. A. May (ed.), Resins for Aerospace, ACS Symp. Ser. **132**, Washington, D.C., 1980.

[30] H. R. Lubowitz (TRW), USP. 3 528 950 (1970).

[31] T. T. Serafini, Status Review of PMR Polyimides, in C. A. May (ed.), Resins for Aerospace, ACS Symp. Ser. **132**, Washington, D.C., 1980, 15.

[32] F. Grundschober and J. Sambeth, USP. 3 380 964 (1969) and 3 533 996 (1970).

[33] F. P. Darmory, in R. D. Deanin (ed.), New Industrial Polymers, ACS Symp. Series 4, Washington, D.C., 1972, 124.

[34] M. A. J. Mallet and F. P. Darmory, ACS Coat. Plast. Chem. Prepr. 34 (1974) 173.

[35] H. D. Stenzenberger, Appl. Polym. Symp. 31 (1977) 91.

[36] S. W. Street, SAMPE 25 (1980) 366.

[37] R. J. Jones et al. (TRW), USP. 3 927 027 (1975), 3 951 902 (1976) and 3 975 363 (1976).

[38] R. J. Jones, H. E. Green, O. F. Markles, Jr., and J. M. Ham, SAMPE 25 (1980) 520.

[39] ®Jeffamine Polyoxypropyleneamines, ®Jeffamine AP-22 Polyaromatic Polyamine Epoxy Curing Agents, Texaco Chemical Co., Technical Information 1978.

[40] Isochem Resins Co., Technical Bulletin.

[41] P. J. Dynes, R. M. Panos and C. L. Hamermesh, J. Appl. Polym. Sci. 25 (1980) 1059.

[42] T. T. Serafini, P. Delvigs and G. R. Lightsey, J. Appl. Polym. Sci. 16 (1972) 905; USP. 3 745 149 (1973; T. T. Serafini, P. Delvigs and W. B. Alston, Proc. of the 27th National SAMPE Symp. and Exhibition, May 1982.

[43] T. L. St. Clair and R. A. Jewell, SAMPE 23 (1978) 520.

[44] A. Wereta, Jr., and D. K. Hadad, in C. A. May (ed.), Resins in Aerospace, ACS Symp. Ser. 132, Washington, D.C., 1980.

[45] ®Ultem Poly(ether-imide), The Comprehensive Guide to Material Properties, Design, Processing, and Secondary Operations, General Electric Co., Technical Information 1982.

[45a] Anon., Kunststoffe 72 (1982) 288.

[45b] D. E. Floryan and I. W. Serfaty, Modern Plast. Intern. 12 (June 1982) 38; Mod. Plast. 59/6 (1982) 146.

[46] J. G. Wirth, R. D. Heath, E. G. Banucci, T. Takakoshi (General Electric), USP. 3 905 942 (1975), Brit. P. 1 392 649 (1975), Ger. Offen. 2 364 246 (1974), USP. 3 968 083 (1976), USP. 3 991 004 (1976), USP. 4 048 142 (1977), USP. 4 049 613 (1977), Can. P. 1 026 894 (1978), Can. P. 1 046 193 (1978), USP. 4 115 341 (1978), USP. 4 156 597 (1979) and USP. 4 157 996 (1979).

[47] ditto, USP. 3 803 085 (1974), USP. 3 875 116 (1975), USP. 3 989 670 (1976), USP. 4 011 198 (1977), USP. 4 024 110 (1977) and USP. 4 073 773 (1978).

[48] ditto, USP 3 983 093 (1976), USP. 3 998 840 (1976).

[49] ditto, USP. 3 998 840 (1976).

[50] D. E. Floryan and G. L. Nelson, J. Fire & Flammability 11 (October 1980) 284.

[51] A. Saiki, K. Mukai, S. Harada and Y. Miyadera, ACS Org. Coat. Plast. Prepr. 43 (1980) 459.

[52] A. M. Wilson, D. Lake and S. M. Davis, ACS Org. Coat. Plast. Prepr. 43 (1980) 470.

[53] P. Kästner, Halbleiter-Technologie, Vogel-Verlag, Würzburg 1980.

[54] G. A. Brown, ACS Org. Coat. Plast. Prepr. 43 (1980) 476.

[55] ®BT Resins, Mitsubishi Gas Chemical Co., Inc., Technical Information, Second edition, 1980.

[56] B. v. Falkai, Synthesefasern, Verlag Chemie, Weinheim 1981.

[57] ®Torlon Resins, Amoco Chemicals Corp., Technical Information, May 1974.

[58] J. DelMonte, Technology of Carbon and Graphite Fiber Composites, Van Nostrand Reinhold Co., New York 1981.

[59] Formulating Adhesives with ®Thermid 600, Gulf Advanced Materials, Technical Information, March 1981.

[60] P. S. Blatz, Adhesives Age **21** (1978) 39.

[61] T. L. St. Clair and D. J. Prograr, SAMPE Symp. **24**/1 (1979) 1081.

[62] P. M. Hergenrother and D. J. Prograr, Adhesives Age **20** (1977) 38.

[63] ®Polvimide 2080 Solution, Upjohn Chemical Co., Technical Information, and USP. 3 708 458.

[64] Procedure for the Preparation of Polyimide 2080 Syntactic Foams and Particulate Compounds, Upjohn Co., Technical Report.

[65] Anon., Modern Plast. Intern. **8** (March 1978) 50.

[66] Anon., Modern Plast. Intern. **10** (March 1980) 21.

[67] J. Gagliani and D. E. Supkis: Non-flammable Polyimide Materials for Aircraft and Spacecraft Applications; Manuscript received from J. S. Chinn, Solimide Venture Dept. Manager, International Harvester Co., June 17, 1981.

[68] Anon., Chem. Engng. News (January 5, 1981).

[69] N. A. Sullo, in Modern Plast. Encycl. 1980−1981, vol. **57**, 70.

[70] P. M. Hergenrother, ACS Org. Coat. Plast. Prepr. **48** (1983) 349.

[71] A. K. St. Clair and T. L. St. Clair, SAMPE Quarterly **13** (October 1981) 20; Anon., Modern Plast. Internat. (March 1982) 78.

[72] A. K. St. Clair and T. L. St. Clair, ACS Org. Coat. Plast. Prepr. **48** (1983) 354.

[73] C. L. Hendricks, S. G. Hill and P. D. Peters, ACS Org. Coat. Plast. Prepr. **48** (1983) 364.

[74] Anon., Modern Plast. Internat. **12** (April 1982) 68.

[75] Anon., Modern Plast. Internat. **13** (January 1983) 65.

[76] J. P. Critchley, G. J. Knight, W. W. Wright, Heat Resistant Polymers: Technologically Useful Materials, Plenum Press, New York 1983.

[77] R. O. Johnson, H. S. Burlhis, J. Polym. Sci.-Polym. Symp. **70** (1983) 129.

[78] H. D. Stenzenberger, M. Herzog, W. Römer, R. Scheiblich, and N. J. Reeves, Brit. Polym. J. **15** (1983) 2.

12 Other heterocyclic polymers

12.1 INTRODUCTION

The world-wide research activity has led to the synthesis of thousands of heterocylic polymers with five- and six-member rings, fused rings, and ladder polymers. In the beginning, the emphasis was on the synthesis of new thermally stable polymers [1–3, 18]. Recently, however, the synthesis effort has been directed toward the improvement of processability of the polymers. Only very few polymers have yet reached pilot plant or production scale, or are available as developmental products (see Table 12-1).

The older commercial products that had been developed before 1974 and which are illustrated by the structures I, IV, VI, X, XII, XIII, XIV, XV and XVIII (Table 12-1) have been reviewed in several books and articles [1–5].

Poly(benzoxazole) with structure I is produced in the Soviet Union under the designation "Metolon"; it is prepared from 3,3'-dimethoxy-4,4'-diaminodiphenylmethane and isophthalic acid. The soluble precursor is a poly(methoxamide) prepolymer which, during thermal processing as adhesive, film or molding material [1], is converted into crosslinked poly(benzoxazole).

Poly(benzimidazoles) with structure IV are prepared by polycondensation of aromatic dicarboxylic acid diphenylesters with aromatic tetraamines or salts thereof. Poly(2,2'-m-phenylene-5,5'-dibenzimidazole) is the only commercial product since the raw materials are commercially available. Syntactic products containing hollow SiO_2 or phenolic resin spheres have been available from Whittacker Co. since 1968 under the designations Imidite SA and Imidite PC, respectively [1, 2]. Oligomeric m-phenylenedibenzimidazoles found use as metal adhesive and laminating resin. They have been marketed by Narmco Materials Division under the designations Imidite 850 and 1850 [1]. A product similar to Imidite 850 is available in experimental quantities from Aerotherm Division, Acurex Corp. [16]; it is still under evaluation as an adhesive.

364

TABLE 12-1

Commercial and semi-commercial polymers with heterocyclic chain elements.

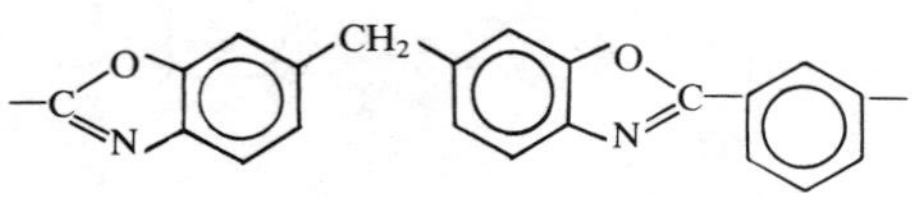	Poly(benzoxazole) "Metolon" UdSSR 1973
I	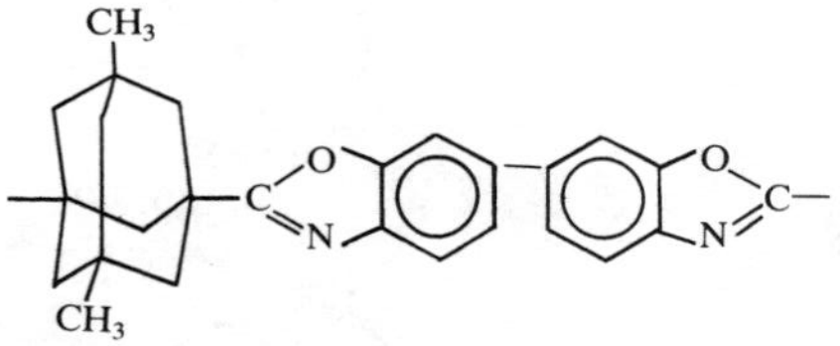 Poly(adamantyl-co-benzoxazole) Battelle (Genf) 1981
II	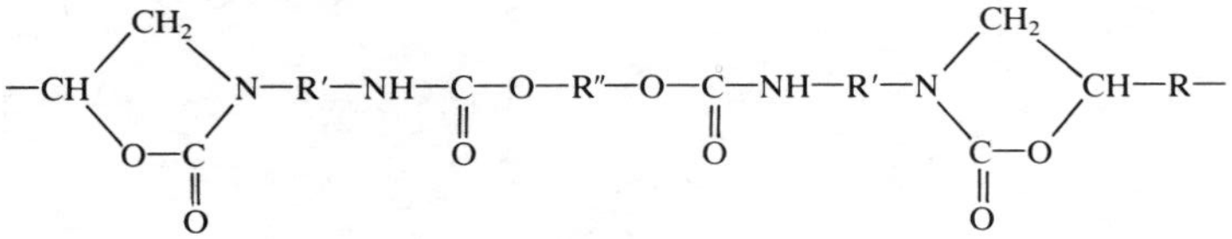 Poly(oxazolidone) Bridgestone Tire Co. 1979

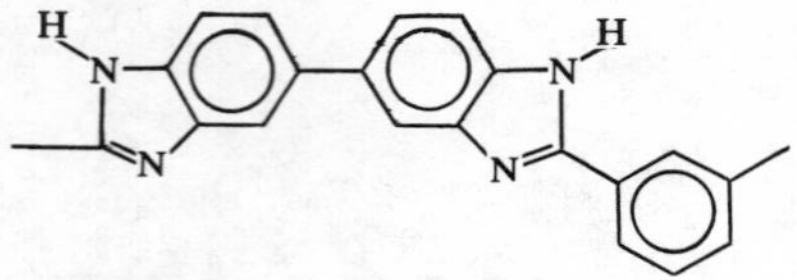

Poly-[2,2'-(m-phenylene)-5,5'
bibenzimidazole]

Imidite 850
Imidite 1850
Imidite SA, PC

Whittacker
Narmco
1964-1968

Pbi-Faser

Celanese
1980

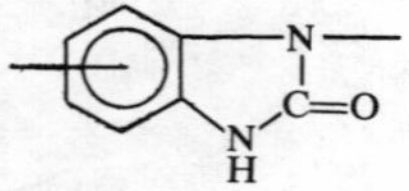

Poly(benzimidazolone)
Teijin
1978

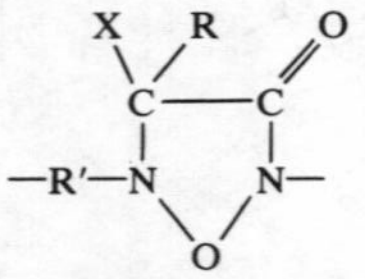

Poly(hydantoine)
Bayer
1970

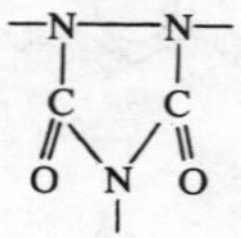

Poly(urazole)
Bayer

IV

V

VI

VII

VIII Poly(pyrrole)

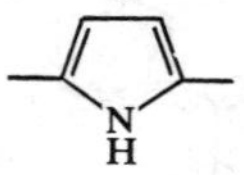

IX Poly(1,3-imidazolidine-2,4,5-trion); poly(parabanic acid) ®Tradlon, ®Tradlac, Exxon 1972 **Hoechst**

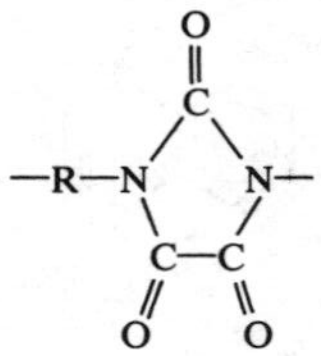

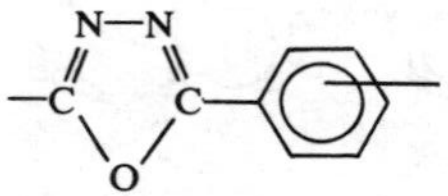

X Poly(phenylene-1,3,4-oxadiazole) ®POD-Film Furakawa Electric Co. 1974

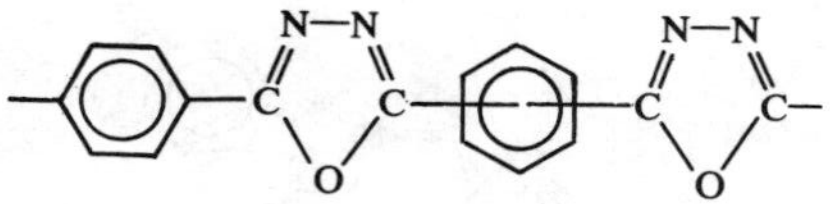

Oksalon, USSR 1971

XI Poly(p-phenylene-oxadiazole-co-N-methylhydrazide) Monsanto 1979

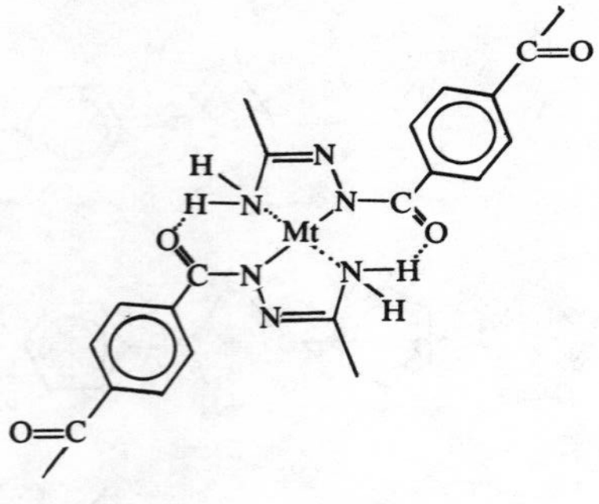

Poly(terephthaloyloxamidrazone)
Enkatherm
Akzo 1972

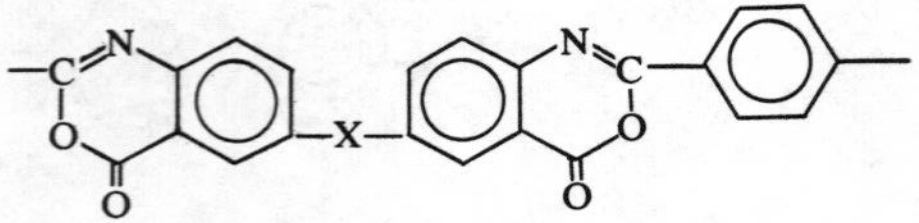

Poly(benz-3,1-oxazinone-4)
Toyo Rayon
1969

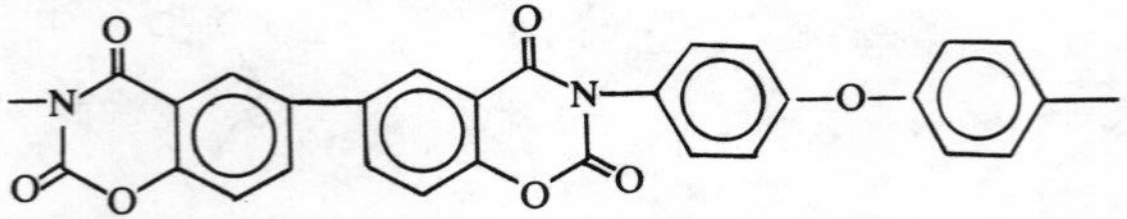

Poly(benzoxazindione)
Bayer
1971

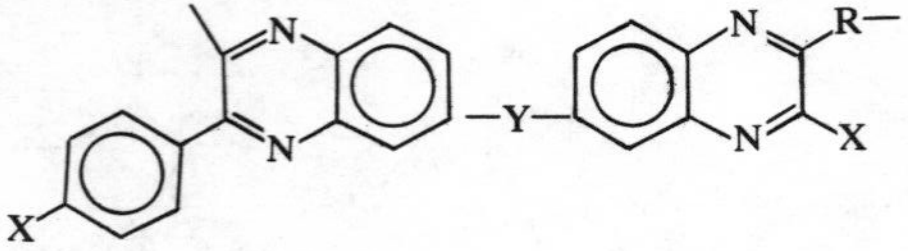

Poly(phenylquinoxaline)
Whittacker

XII

XIII

XIV

XV

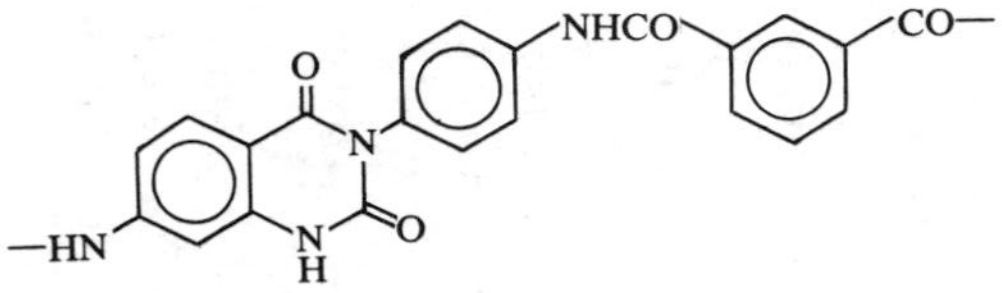

Poly(quinazolinedione)
AFT-2000
Bayer 1974

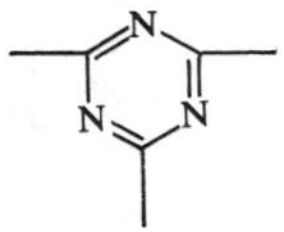

Poly(1,3,5-triazine)
NCNS-Polymers, Ciba-Geigy 1975
Triazin-A-Resins, Mobay 1976

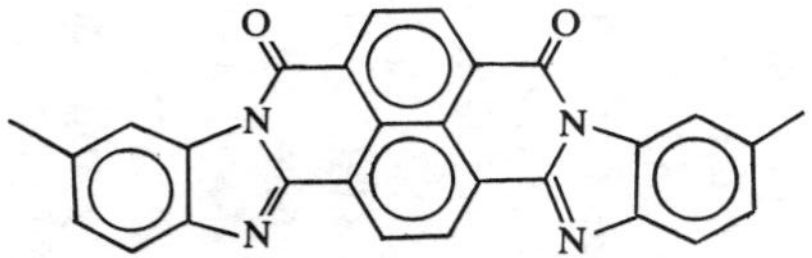

Poly(bisbenzimidazo-benzophenan-
throline)
BBB-Faser, Celanese 1969

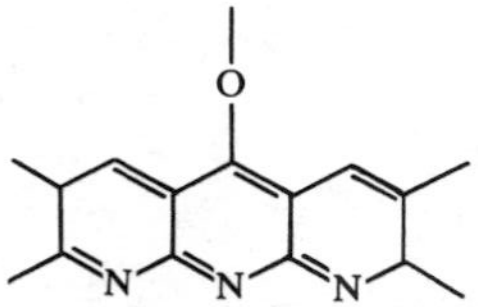

Cyclized and oxidized
poly(acrylonitrile)
Celiox
Toho Beslon/Celanese 1980

XVI

XVII

XVIII

XIX

Poly(parabanic acids) with structure IX have been prepared by Hoechst A. G. from oxamide esters and capped isocyanates. Exxon has developed a three-step reaction. First, HCN is added to the isocyanate to form a cyanoformamide. Then the cyanoformamide is added to another isocyanate to give the N-cyanoformyl urea which subsequently cyclizes to poly(imino-imidazolidinedione [4–6]. These imino polymers can be hydrolyzed to the trione analog, a poly(parabanic acid). Solutions of poly(parabanic acids) can be processed to give films, coatings and dielectric materials. Poly(parabanic acids) are 20 to 40% cheaper than polyimides. Nevertheless, Exxon decided to close down its production of ®Tradlon films and ®Tradlac resin solutions which had just started in 1979 [7].

Several routes exists for the synthesis of polyhydantoins with the general structure VI [1, 4, 5]. These products have been marketed for some time by Bayer A. G. for the use as insulation film and wire enamels. Of more recent origin are epoxides with hydantoine groups (see chapter 6.5).

A poly(oxadiazole) with structure X is produced in the Soviet Union under the designation "Oksalon", presumably by reacting a 70/30 mixture of isophthalic acid/terephthalic acid with hydrazine [8]. The intermediate polyhydrazide is cyclodehydrated to poly-(oxadiazole). "Oksalon" has been recommended as an economical, thermally stable fiber for uses in protective clothing and as tire cord [9].

Thermally stable fibers made of poly(terephthaloyloxamidrazone) with structure XII are prepared by interphase or solution polycondensation from oxalic acid bis-amidrazone and terephthalic acid dichloride and subsequent spinning of the alkaline solution into an acidic coagulation bath. The polymer is chelated by treatment with metal salt solutions [4, 10]. The fiber was produced only in experimental quantities from 1974 to 1976.

High-temperature insulating films fabricated from poly(benz-3,1-oxazinone-4) with structure XIII have been commercially available from Toyo Rayon [1]. They are prepared by reacting benzidine-3,3'-dicarboxylic acid with an aromatic diacid chloride. Films are casted from the tractable poly(amic acid) intermediate which is subsequently cyclodehydrated by heating under vacuum.

Poly(benzoxazinediones) with structure XIV were introduced as experimental products by Bayer A. G. in 1971 [1]. A one-step reaction between a diisocyanate and an aryl-bis(o-hydroxycarboxylic acid ester) yields an intermediate poly(ester urethane) which is not isolated prior to the cyclization [11]. The polymers are soluble in polar solvents and can be processed into films and fibers; they are said to be injection moldable, as well.

Several poly(phenylquinoxalines) or PPQ with the general structure XV have been available as developmental resins from Whittacker Corp. [5]. They are derived from the condensation of an aromatic tetramine with a bis(1,2-dicarbonyl) compound [1–3, 5]. They were first used as metal adhesives. The Boeing Co. evaluated PPQ compared to special polyimides and found that good quality structural bonds can be produced which exhibit durability to supersonic cruise vehicle environments [12, 17]. PPQ is unavailable as adhesive tapes although a PPQ solution is available in experimental quantities from King-Mar Laboratories [16]. End-capping of PPQ with acetylenic groups yields polymers (ATPQ) which exhibit shortcomings similar to those of polyimides with terminal acetylenic groups. The thermal reaction of the ethynyl group inhibits flow and the cured ATPQ is less thermooxidatively stable than a comparable PPQ [13, 16]. Crosslinking can be achieved also via terminal nitrile groups [14]. PPQ costs 990.-US-$/kg.

Poly(quinazolinedione) with amide groups in the backbone and structure XVI was introduced in experimental quantities by Bayer A. G. in 1974 under the designation AFT-2000 [4, 10]. It is prepared by polycondensation in dimethylacetamide solution from 3-(p-aminophenyl)-7-amino-2,4-(1H,3H)-quinazolinedione and isophthalic acid dichloride. Fibers made of AFT-2000 are said to exhibit better thermal stability and less flammability than aramide fibers.

Poly(bis-benzimidazobenzophenanthrolines) with structure XVIII were developed by the US Air Force and produced by Celanese under the designation BBB-fiber. The synthesis is carried out by polycondensation of 1,4,5,8-naphthalene-tetracarboxylic acid and 3,3′-diaminobenzidine [1–3]. BBB is a step-ladder polymer whereas the structurally related BBL is a true ladder. Thermal stabilities for the fibers are 450–550°C in air and 650–775°C in nitrogen.

Recent polymers which have been developed or became available since 1974 are illustrated by structures II, III, IV, VIII, X, XI, XVII and XIX (see Table 12-1). These products will be dealt with in the following chapters.

Also of recent origin is poly(benzimidazolone) with structure V [15]. The polymer containing phenylene sulfone and phenylene oxide groups was developed by Teijin and is being tested in an 100 million gallon per day Colorado River reverse osmosis desalination plant in Yuma, Arizona. Due to polymer structure, the membrane is said to exhibit good hydrophilic properties, stiffness, flexibility, and chemical resistance at pH 1 to 12 and at temperatures of up to 60°C. Details on the method of synthesis of V have not been disclosed.

Polymers with urazole groups VII have been developed by Bayer A. G. The products will not be available as pure polyurazoles but are combined with epoxides.

New polyimides which contain heterocyclic structures besides the imide group are dealt with in chapter 11 (polyimides). Other heterocyclic polymers derived from reactions with isocyanates are described in chapter 13.

References

[1] K. H. Bühler, Spezialplaste, Akademie-Verlag, Berlin 1978.

[2] P. E. Cassidy, Thermally Stable Polymers, M. Dekker, New York 1980.

[3] J. Preston, Heat Resistant Polymers, in Kirk-Othmer, Encycl. Chem. Technol., vol. 12 (1980) 203.

[4] H.-G. Elias, Neue polymere Werkstoffe 1969–1974, C. Hanser Verlag, München 1975; New Commercial Polymers 1969–1975, Gordon and Breach, New York 1977.

[5] H.-G. Elias, Kunststoffe mit besonderen Eigenschaften, in Ullmanns Enzykl. Techn. Chem., vol. 15 (1978) 421.

[6] T. L. Patton, ACS Polym. Prepr. 20/1 (1979) 183.

[7] Anon., Chem. Engng. News (April 20, 1981) 13.

[8] T. M. Gritsenko et al., Khim. Volokna 13/2 (1971) 64.

[9] Europ. Chem. News 21 (April 7, 1972) 28.

[10] B. v. Falkai, Synthesefasern, Verlag Chemie, Weinheim 1981.

[11] L. Bottenbruch, Angew. Makromol. Chem. 13 (1970) 109.

[12] C. L. Hendricks and S. G. Hill, SAMPE Symp. 25 (1980) 39.

[13] P. M. Hergenrother, Polym. Sci. & Eng. 21 (1981) 1072.

[14] W. B. Alston, SAMPE Symp. 21 (1976) 114.

[15] Anon., Modern Plast. Internat. 8 (March 1978) 65.

[16] P. M. Hergenrother, ACS Org. Coat. Plast. Prepr. 48 (1983) 349.

[17] C. L. Hendricks, S. G. Hill and P. D. Peters, ACS Org. Coat. Plast. Prepr. 48 (1983) 364.

[18] J. P. Critchley, G. J. Knight, and W. W. Wright, Heat-Resistant Polymers: Technologically Useful Materials, Plenum Press, New York 1983.

12.2   POLY(ADAMANTYL BENZOXAZOLE)

12.2.1   Structure and synthesis

The Battelle Memorial Institute in Geneva/Switzerland has developed a polymer named AD-PBO (adamantane-based polybenzoxazole) [1, 2]. The two-step synthesis involves polycondensation of adamantane dicarboxylic acid dichloride (I) with 4,4'-diamino-3,3'-dimethoxydiphenyl or o-dianisidine (II) at low temperatures to yield a prepolymer (III) which is then cyclodehydrated to AD-PBO (IV) [3, 4]:

(12-1)

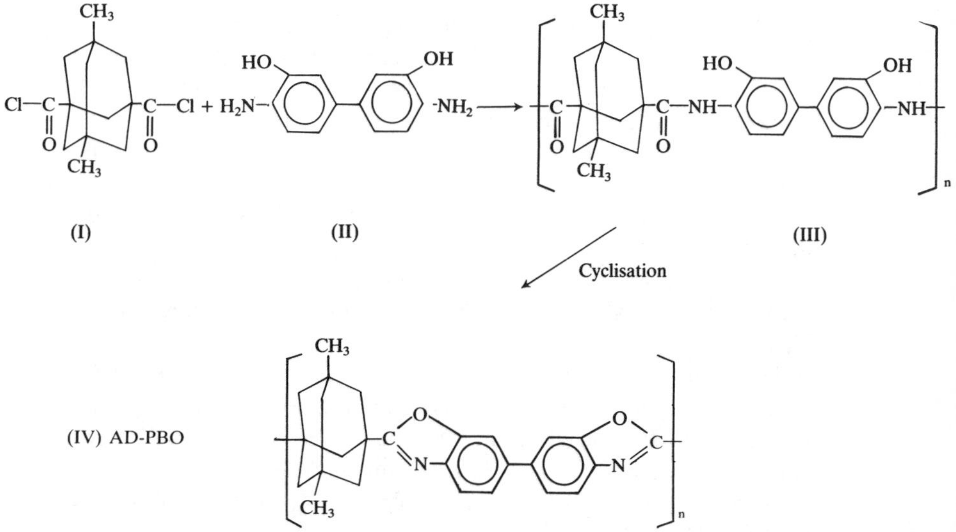

(I)   (II)   (III)

Cyclisation

(IV) AD-PBO

Starting material II is commercially available; I is obtained from acenaphthene or from the dimeric methylcyclopentadiene.

12.2.2   Properties and applications

So far only a few properties of the new polymer have been investigated [4]. AD-PBO films are fully transparent but are of slightly yellow color. Film tensile strength is 103 MPa, elongation at break

7.6%, and tensile modulus 3.0 GPa.

Thermogravimetric analysis in air shows that decomposition starts at 400°C with a weight loss of 10% at 465°C; in nitrogen the corresponding values are 400 and 500°C, respectively, with a weight loss of 40% at 600°C. No weight loss is observed at 300°C in air over 8 hours.

AD-PBO demonstrates excellent stability to sodium hydroxide and hydrochloric acid. Tensile strength and tensile modulus values remain constant on exposure to the following media: for 6 weeks in 30% NaOH at 90°C, 630 h in 30% NaOH at 150°C, and 4 weeks in 1 N hydrochloric acid at 40°C. AD-PBO is also resistant to oxidative attack. No visible change is observed on exposure for over 70 hours to 30% NaOH at 160°C under 30 bar oxygen pressure.

Optical and electrical properties and the degree of water absorption have not yet been determined. Development work is under way to introduce more hydrophilic characteristics to the hydrophobic AD-PBO for the purpose of creating ion-exchange capability. Such modification presumably means sacrifice of thermal stability [4].

Prepolymer III is soluble in DMAC, DMF, NMP and hexamethylphosphoramide. These solutions are processed into films, membranes and coatings, which are then thermally converted to AD-PBO.

Potential applications are electrical insulating materials, chemically and thermally resistant coatings, fibers, and membranes for electrolytic cells.

The estimated material costs for the preparation of AD-PBO are around 15 US-$/kg, and over-all production costs are in the range of 25 US-$/kg. Battelle offers non-exclusive license rights to interested companies. AD-PBO films (25−50 μm thickness, total surface area of three samples is 2 dm^2) can be obtained from Battelle at a price of US-$ 940.- [4].

References

[1] Battelle Today, no. 22 (April 1981) 1.
[2] Anon., Mater. Eng. **93**/6 (1981) 21.
[3] G. G. Bellmann, A. M. Groult and J. H. Arendt, USP. 3 852 239 (1974), Fr. P. 722 291, Brit. P. 1 379 357, Belg. P. 798 754, Swiss P. 566 358, and Ger. Offen. 2 330 452.
[4] AD-PBO, Information Note, Battelle Institute, Geneva 1981.

12.3 POLY(OXAZOLIDONE) ELASTOMERS

12.3.1 Structure and synthesis

Poly(2-oxazolidones) are prepared by catalytic 1,3-cyclo-addition of diglycidylethers to diisocyanates:

(12-2)

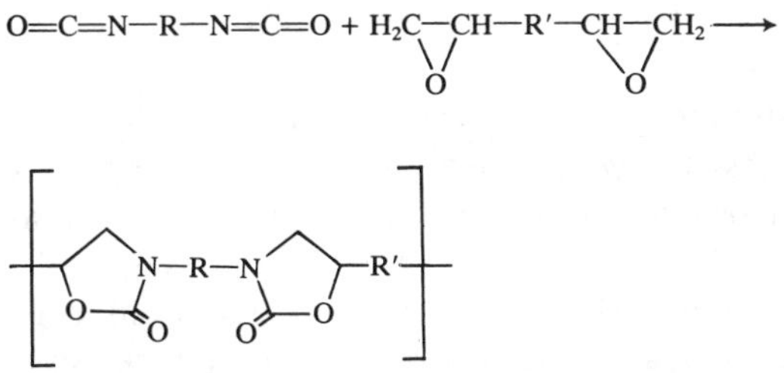

$$O{=}C{=}N{-}R{-}N{=}C{=}O + H_2C{-}CH{-}R'{-}CH{-}CH_2 \longrightarrow$$

This reaction has already been commercially used for the preparation of thermally stable casting resins from epoxy resin, MDI and styrene [1] as well as for the preparation of oxazolidone modified PIR foams [2].

The Bridgestone Tire Co. prepares poly(oxazolidone) elastomers by reaction of bisphenol A diglycidylether with isocyanate end-capped poly(oxytetramethylene glycol) [3]:

(12-3)

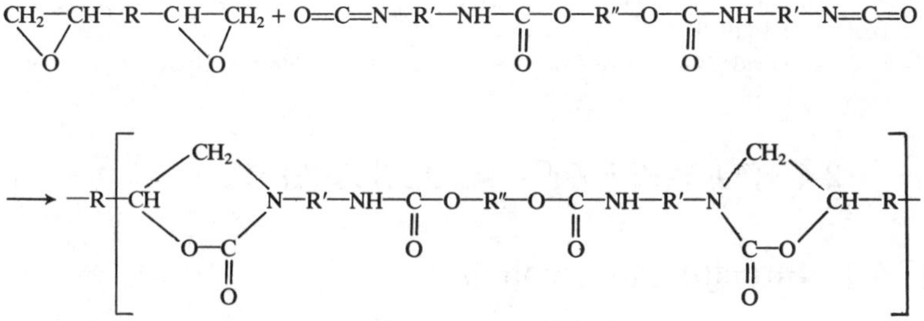

$$CH_2{-}CH{-}R{-}CH{-}CH_2 + O{=}C{=}N{-}R'{-}NH{-}\underset{\underset{O}{\|}}{C}{-}O{-}R''{-}O{-}\underset{\underset{O}{\|}}{C}{-}NH{-}R'{-}N{=}C{=}O$$

For this synthesis commercially available raw materials are used. The epoxy resin is Epikote 182 (Shell) with an epoxy equivalent of

188, the isocyanate is Adiprene L-100 (DuPont). The reaction is carried out at 150°C over 1 h without need of solvents. A catalyst composed of MgCl$_2$/hexamethylphosphortriamide yields polymers with low modulus and high tensile strength. When the reaction is conducted at temperatures below 120°C, the products contain unreacted isocyanate and isocyanurate polymers which lead to poor properties. Critical also is the epoxy/isocyanate equivalent ratio; best results are obtained with a 45/55 ratio.

12.3.2 Properties

Poly(oxazolidone) elastomers are comparable to the equally castable poly(ether urethanes) for which they have been developed as competitive materials. However, some significant differences exist in properties (Table 12-2).

Compared to polyurethane elastomers, poly(oxazolidone) elastomers exhibit a higher melt temperature range. Strength retained at 170°C is higher, however, curing rate and strengths at low temperatures are inferior. Development work is being undertaken to remedy these shortcomings [4]. Costs for poly(oxazolidone) are claimed to be lower than those for polyurethane elastomers because epoxy resins are cheaper than the amine hardeners commonly used for PUR, e.g., MOCA [3].

References

[1] R. Kubens, F. Ehrhard and H. Heine, Kunststoffe **69** (1979) 455.
[2] D. Brownbill, Modern Plast. Internat. (November 1980) 24.
[3] M. Kitayama, Y. Iseda, F. Odaka, S. Anzai and K. Irako, Rubber Chem. Technol. **53** (1980) 1.
[4] K. Irako (Bridgestone Tire Co. Ltd.), personal communication, April 30, 1980.

12.4 POLY(BENZIMIDAZOLE) FIBERS

12.4.1 Structure and synthesis

Poly(benzimidazoles) or PBI contain benzimidazole units in the backbone:

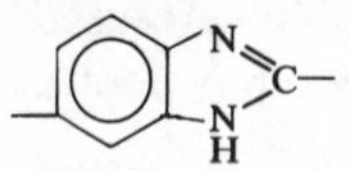

TABLE 12-2
Properties of Poly(oxazolidone) and Polyether-urethane Elastomers.

Physical property	Physical unit	Property Value	
		Poly-(oxazolidone)	Polyether-urethane[1]
Glass transition temperature	°C	−30/−20	−40/−30
Melting temperature	°C	235/245	200/220
Tensile strength at break			
at 23°C	MPa	26.3	40.8
at 23°C after 3 h at 100°C	MPa	—	27.8
at 23°C after 5 h at 150°C	MPa	31.2	—
at 23°C after 25 h at 170°C	MPa	25.6	26.7
at 100°C	MPa	2.1	10.6
Elongation at break			
at 23°C	%	527	443
at 23°C after 3 h at 100°C	%	—	492
at 23°C after 5 h at 150°C	%	540	—
at 23°C after 25 h at 170°C	%	553	426
at 100°C	%	158	372
Modulus of elasticity			
at 23°C	MPa	8.9	31.4
at 23°C after 3 h at 100°C	MPa	—	26.7
at 23°C after 5 h at 150°C	MPa	6.1	—
at 23°C after 25 h at 170°C	MPa	6.4	22.2
at 100°C	MPa	4.0	22.0
Stress relaxation (ratio of modul after 5 times drawing to 150% of initial modulus)	%	62	67
Deformation under creep (increase of elongation after 5 times loading with 4 MPa)	%	121	110

[1] From Adiprene L-100 with 4,4′-methylene-bis(2-chloroaniline) (MOCA), 3 h at 100°C.

PBI synthesis and properties have been extensively described [1–3]. PBI is not a new engineering polymer. Since 1964, oligomeric m-phenylenedibenzimidazoles have been commercially available as metal adhesives from Narmco Materials Division of Whittacker Corp. and were trade named Imidite 850 and Imidite 1850. Narmco has also supplied PBI laminating resins and, in 1968, introduced syntactic poly(m-phenylene-benzimidazole) foams under the designations Imidite SA and PC. A similar form of the Imidite 850 is available in experimental quantities and is still under evaluation as an adhesive by Aerotherm Div./Acurex Corp. [15].

PBI fiber became commercially available for the first time in 1980. Celanese produces "pbi Poly(benzimidazole)-High-Performance-Fiber" in developmental quantities [4, 5]. Large scale production is planned for the end of 1982 or beginning of 1983 with a plant capacity of 450 t/a [5, 6].

PBI fiber is produced by Celanese in a pilot plant in a two-step reaction. First, a melt polycondensation of 3,3′-diaminobenzidine and diphenylisophthalate is carried out at 250−300°C which, under foaming, yields a prepolymer:

(12-4)

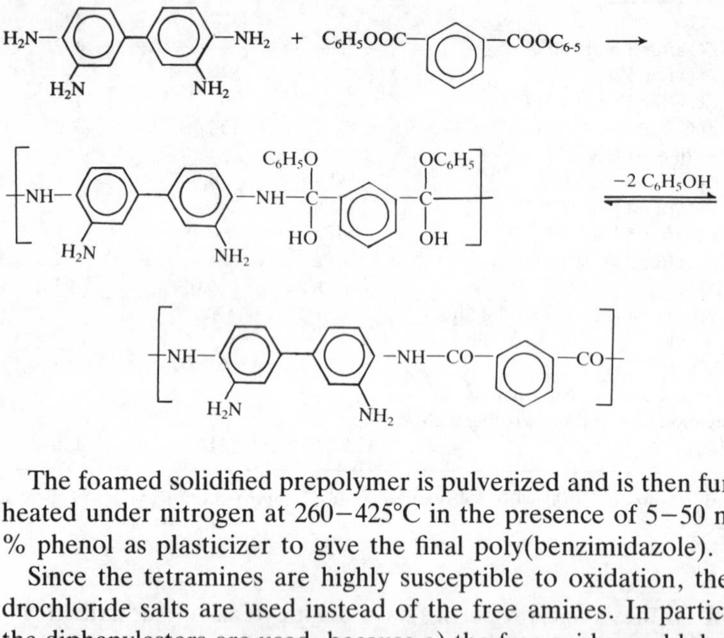

The foamed solidified prepolymer is pulverized and is then further heated under nitrogen at 260−425°C in the presence of 5−50 mass-% phenol as plasticizer to give the final poly(benzimidazole).

Since the tetramines are highly susceptible to oxidation, the hydrochloride salts are used instead of the free amines. In particular, the diphenylesters are used, because a) the free acids would decarboxylate under those reaction conditions, b) the acid chlorides would react too rapidly thus impeding ring closure, and c) the methyl esters would partially methylate the amino groups. During polycondensation, the exact stoichiometric amounts of reactants must be maintained: an excess of ester groups would cause gelation, an excess of amino groups would reduce the molecular weight.

(12-5)

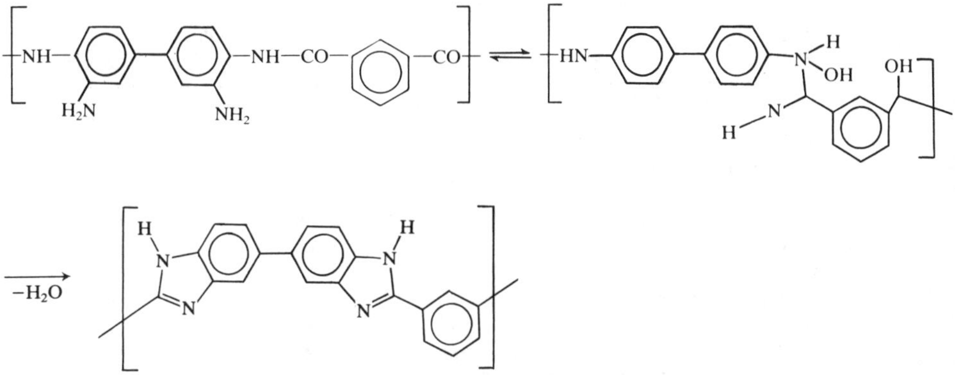

PBI fiber is dry-spun from DMAC solution [5, 7]. To this purpose, poly[2,2'-(m-phenylene)-5,5'-bisbenzimidazole] is dissolved
under pressure at 230°C. To avoid phase separation in the solution
during aging, 2% of lithium chloride is added; this is easily leached
out from the fibers after spinning. The fibers are dry-spun in an
atmosphere of nitrogen or superheated steam. The temperature of
the spinning solution is kept at 150°C, spinneret temperature is
180°C, and the inert gas is heated to 240°C. PBI fiber is drawn in a
two-step process. Predrawing is carried out at temperatures of 300—
650°C under a draw-ratio of 1:1.5—3.5; final drawing is done at
200—560°C with a draw-ratio of 1:1.05—1.5.

12.4.2 Properties and applications

Celanese supplies the gold colored PBI fiber as staple fiber in 5 cm
lengths which can be processed to fabric on conventional textile
machinery. The most obvious advantages of PBI fiber are thermal
stability, non-combustibility in air, resistance to chemicals, and good
textile properties (Table 12-3).

PBI fiber does not melt or ignite in air and no smoke is developed
on heating in air at temperatures of up to 560°C. Although PBI
fabric carbonizes at high temperatures, very little shrinkage occurs,
and the fabric still coheres as carbonized carcass. Celanese reports
that CO_2, H_2O and traces of SO_2 are liberated on decomposition

TABLE 12-3
Properties of the PBI Fiber [4].

Physical Property	Property Value	Phys. unit
Density	1.43	g/cm^3
Fineness	1.7	dtex
Tensile strength at break	3.1	g/den
	2.7	dN/tex
Elongation at break	30	%
Initial modulus of elasticity	45	g/den
	40	dN/tex
Crimp per length	0.5	mm^{-1}
Crimp	30	%
Water absorption (at 20°C and 65% R.H.)	15	%
Shrinkage in water at 100°C	0.5	%
Shrinkage in hot air at 204°C	1	%
Limiting oxygen index	38	%
Surface resistivity (at 21°C, 65% R.H.)	0.7×10^{10}	ohm

TABLE 12-4
Tensile strength and elongation of PBI fiber after storage in hot air [4].

Testing conditions	Remaining property values (in % of initial values)	
	Tensile strength	Elongation
Hot air of 302°C		
after 30 min	100	100
after 60 min	100	100
Hot air of 399°C		
after 30 min	95	50
after 60 min	80	30

above 560°C. The evolution of SO_2 is difficult to explain. Other sources report findings of CO_2, H_2O, HCN and NH_3 as off-gases during pyrolysis at temperatures of above 550°C]12].

PBI demonstrates excellent properties at elevated temperatures. Strength is retained on short-term heating to 400°C. Exposure to hot air has little effect on tensile strength and elongation (Table 12-4). For example, a PBI fabric exposed to air at 204°C over 24 h retains nearly 90% of its original strength and shrinkage is only 3% after 24 h at 316°C. Heating in vacuum at 316°C for 300 h causes no change in mechanical properties.

PBI fiber retains 100% of its strength after 168 h exposure at 30°C to acetic acid, methanol, perchloroethylene, DMAC, DMF, DMSO, kerosene, acetone and gasoline. Resistance to acids, bases and water vapor is excellent (Table 12-5).

TABLE 12-5
Tensile Strength of PBI Fiber after Aging in Various Solvents [4].

Solvent				Tensile strength in % of initial value
Name	%	Temperature in °C	Time in h	
Sulfuric acid	50	29	144	90
Sulfuric acid	50	71	24	90
Hydrochloric acid	35	29	144	95
Hydrochloric acid	10	71	20	90
Nitric acid	70	29	144	100
Nitric acid	10	71	48	90
Sodium hydroxide	10	29	144	95
Sodium hydroxide	10	93	2	65
Potassium hydroxide	10	25	24	88
Water vapor (462 kPa)	—	149	72	96

PBI fibers are relatively sensitive to UV radiation. PBI fibers tested in an Xenontest (ASTM 16E-1978) for 24, 120 and 192 h retained only 71, 53 and 53%, respectively, of their original strength [4].

Celanese foresees for PBI fibers a market potential of about 23 600 t, specifically 8200 t for replacement of asbestos, 5400 t for filters as flue gas dust collectors, 4100 t in paper manufacturing, and 1800 t in protective clothings [5]. PBI has equally good thermal stability as asbestos, but is superior to asbestos with regard to wear comfort, flexibility and wear resistance. Molten metal penetrates PBI fabrics of 253.5 g/m^2 basis weight after 17 s but penetrates similar asbestos fabrics after only 8 s. PBI felt is particularly suited in low-oxygen atmosphere for cleaning of hot flue and off-gases from coal firing plants, foundries, calcining furnaces, and combustion plants. Needle feltings provide better filtering performance than fabrics.

PBI fabrics as protective clothings are useful in high temperature applications and where exposure to open flames is involved. The cloth is said to be more comfortable than cotton. PBI absorbs 15%

moisture at 20°C and 65% relative humidity whereas cotton absorbs only 15% moisture under the same conditions. The oxygen index of 38% for PBI compares well to 17% for Nomex.

The U.S. Air Force has tested flight suits of PBI and found them superior to other materials [3]. During crossing of a flame zone of burning jet fuel at 980−1200°C unmodified PBI fabrics scorches to a degree of only 10%, and no blisters are formed beneath the cloth within 3 s [13].

Further, the PBI fibers show promise as reverse osmosis membranes and in graphization to high-strength, high-modulus fibers for composites [14]. More recently, the development of ultrafine fibers of PBI for use in fuel cell and battery separator applications has been undertaken by Celanese Research Co. [3].

At present, PBI fibers with a price of 30 US-$/lb cost considerably more than aramid fibers at 5 US-$/lb [5].

References

[1] H.-G. Elias, Kunststoffe mit besonderen Eigenschaften, in Ullmanns Enzykl. Techn. Chem., vol. **15** (1978) 440.

[2] K. U. Bühler, Spezialplaste, Akademie-Verlag, Berlin 1978.

[3] P. E. Cassidy, Thermally Stable Polymers, M. Dekker, New York 1980.

[4] pbi Polybenzimidazole High-Performance Fiber, Celanese Fibers Marketing Co., Technical Information 1980.

[5] Anon., Chem. Engng. News (September 29, 1980) 11.

[6] K. C. McAlister (Celanese), personal communication, May 13, 1981.

[7] A. Conciatori, J. Polym. Sci. **C 19** (1967) 49.

[8] Celanese, USP. 3 584 104 (1971).

[9] R. Singleton, ACS Polym. Prepr. **9**/2 (1968) 1158.

[10] R. Singleton, Appl. Polym. Symp. **9** (1969) 133.

[11] R. H. Jackson, Text. Res. J. **48** (1978) 314.

[12] G. Shulman, ACS Polym. Prepr. **6**/2 (1965) 773; J. Macromol. Sci. **A 1**/3 (1967) 413.

[13] Anon., Mater. Engng. **71**/4 (1970) 19.

[14] J. R. Leal, Modern Plastics **52** (August 1975) 60.

[15] P. M. Hergenrother, ACS Coat. Plast. Prepr. **48** (1983) 349.

12.5   POLY(PHENYLENE-1,3,4-OXADIAZOLES)

12.5.1   Structure and synthesis

The Furukawa Electric Co. Ltd. produces pilot plant quantities of poly(phenylene-1,3,4-oxadiazole) under the designation "POD-Film" [1]. Films are supplied with 50 μm thickness and maximum width of 400 mm [2]. The structure reportedly [3] is represented by

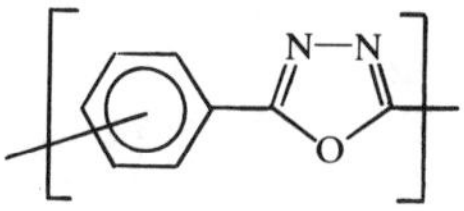

Five reaction schemes have been successful in producing aromatic poly(1,3,4-oxadiazoles) [4]. The reaction route to "POD-Film" has not been disclosed by Furukawa, however, the synthesis is said to be simple and economical and is directly linked to the fabrication of film [3, 5]. From these statements one can assume that "POD-Film" is obtained in a one-step, solution polycondensation by reaction of a dicarboxylic acid, e.g. terephthalic and/or isophthalic acid, with hydrazine or its salt in polyphosphoric acid or in oleum [4, 6]:

(12-6)

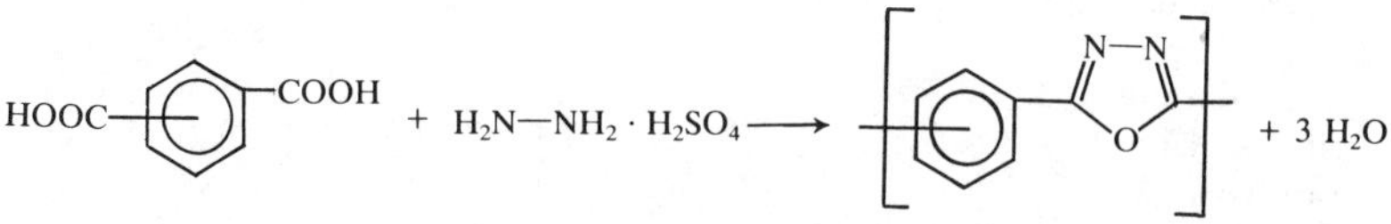

This reaction favors formation of very high molecular weight polymers; their solutions in oleum can be processed directly to tough films [6].

12.5.2 Properties and applications

The properties and uses of "POD-Film" have been extensively investigated, as evidenced by the 25 patents [4] issued to the Furukawa Electric Co. and the publication of a number of product bulletins [5]. The yellow transparent POD films are partially crystalline, hydrolytically and thermally stable, and they possess good mechanical properties (Table 12-6).

The mechanical properties can be enhanced by drawing to give a higher degree of orientation. POD-Film demonstrates good strength at temperatures of up to 250°C which gives it a position between aramides (®Nomex) and polyimides (®Kapton) (Figure 12-1). POD-

Film remains flexible even at the temperature of liquid nitrogen. An uncommon feature is the decrease of elongation-to-break with increasing temperature (Figure 12-2).

TABLE 12-6
Properties of Various Heat-Resistant Films [7].

Physical Property	Physical Unit	POD	Polyimid Kapton	Poly-ester	Aramid Nomex
Density	g/cm^3	1.40	1.42	1.40	1.35
Tensile strength					
at 25°C	MPa	137	177	245	93
at 180°C	MPa	93	142	80	46
Elongation at break					
at 25°C	%	100	70	117	10.6
at 180°C	%	70	102	140	17
Modulus of elasticity	MPa	29	29	39	—
Tear strength	kg/mm	35	15	35	11
Split tear strength	g/mm	600	700	800	110
Decomposition temperature	°C	440	450	300	350
Water absorption	%	7.0	2.9	0.4	—
Water absorption (25°C, 50% R.H.)	%	2.5	1.3	0.2	3.5
Shrinkage					
at 150°C	%	0.3	0.3	0.8	—
at 200°C	%	0.3	0.3	2.5	—
Dielectric strength					
at 25°C	kV/mm	210	275	235	17
at 150°C	kV/mm	195	230	210	—
Relative permittivity (50 Hz)					
at 25°C	—	3.3	3.5	2.3	2.7
at 180°C	—	3.0	3.0	—	—
Loss factor					
at 25°C	—	0.12	0.3	1.38	1.4
at 180°C	—	0.5	0.16	—	—
Volume resistivity					
at 25°C	Ohm × cm	10^{16}	10^{18}	10^{16}	10^{16}
at 180°C	Ohm × cm	10^{15}	10^{15}	—	—

Thermal decomposition of POD-Film starts at 440°C compared to films of polyimide, aramide, poly(parabanic acid), and poly(amide-imide) with temperatures of 470, 370, 360 and 390°C, respectively. Use times of POD-Film at 195 and 210°C are said to be 40 000 and

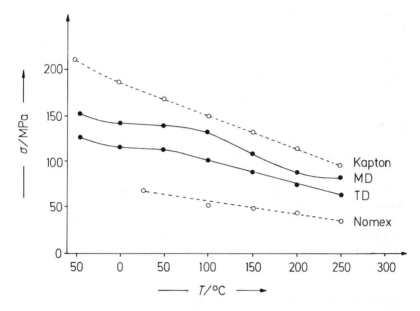

Figure 12-1   Temperature dependence of tensile strength of POD films (●——●) in machine direction (MD) and transverse to machine direction (TD) as compared to Kapton films and Nomex paper [7].

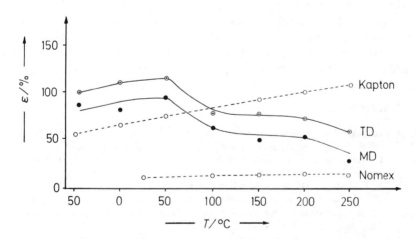

Figure 12-2   Temperature dependence of elongation at break of POD films in machine direction (●) and transverse to machine direction (○) as compared to Kapton films and Nomex paper [7].

20 000 h, respectively. No toxic gases are liberated on decomposition.

Disadvantageous is the relatively high absorption of moisture which causes a significant increase in elongation and a less pronounced decrease in strength. Approximately 95% of the absorbed water is removed by heating to 100°C, however, complete drying is only accomplished above 200°C. This rigorous drying procedure does not induce any hydrolytic degradation. The high diffusion rate for water vapor is attributed to the hygroscopic properties of POF-Film, but permeabilities for O_2 and N_2 are only modest.

The high values of dielectric strength and the low dielectric loss factor make POD an excellent insulating material. Dielectric strength is only slightly reduced at temperatures from 50 to 200°C. The electric properties of POD are less affected by thermooxidative degradation than are the mechanical properties.

POD-Film is completely insoluble in organic solvents; it is soluble only in concentrated sulfuric acid or polyphosphoric acid. Hydrolytic degradation occurs at pH values of <4 and >10; resistance to hydrolysis is superior to that of polyesters.

POD-Film upon weathering suffers a severe drop in tensile strength. It is surprising that the heterocyclic structure imparts no self-extinguishing properties to POD-Film; flammability is equivalent to that of polyesters.

POD-Film is used as electrical insulating material [3]. Other applications include laminates and carriers for thermally stable adhesive tapes [5], dyeable fibers, cation exchange resin, graphitized fibers, and weather-resistant, pigmented films [4].

References

[1] H. Sekiguchi, personal communication, May 19, 1981.

[2] "POD-Film", The Furukawa Electric Co., Ltd., Technical Information.

[3] H. Sekiguchi, J. Mura, K. Sadamitsu, J. Oda and T. Komiyama, World Electrotechnical Congress, Moscow, June 1977.

[4] P. E. Cassidy, Thermally Stable Polymers, M. Dekker, New York 1980; P. E. Cassidy and N. C. Fawcett, Thermally Stable Polymers: Polyoxadiazoles, Polyoxadiazole-N-oxides, Polythiazoles and Polyoxadiazoles, J. Macromol. Sci.—Rev. Macromol. Chem. C 17/2 (1979) 209.

[5] H. Sekiguchi and K. Sadamitsu, Japan Plastics (September/October 1974) 6.

[6] Y. Iwakura, K. Uno and S. Hara, J. Polym. Sci. A 3 (1965) 45.

[7] POD Technical Information No. 001 till 005 (1978) and Additional Techn. Data of POD Film (October 1978).

12.6 POLY(P-PHENYLENE-OXADIAZOLE-CO-N-METHYLHYDRAZIDE)

12.6.1 Structure and synthesis

Monsanto has developed a new high-modulus fiber [1, 2]. The three-step reaction starts from a mixture of terephthalic acid, dimethyl-terephthalate (DMT) and hydrazine sulfate with a molar ratio of terephthalic acid/DMT ranging from 50/50 to 75/125. Fuming sulfuric acid acts both as solvent and as dehydrating agent. The intermediates are not isolated:

A. Polycondensation
(12-7)

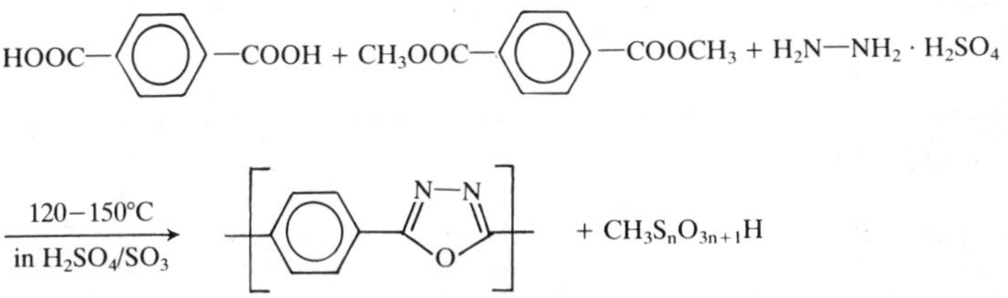

B. Methylation
(12-8)

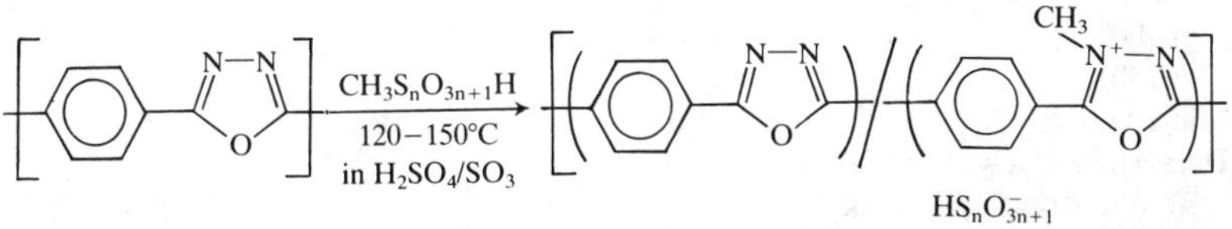

C. Hydrolysis
(12-9)

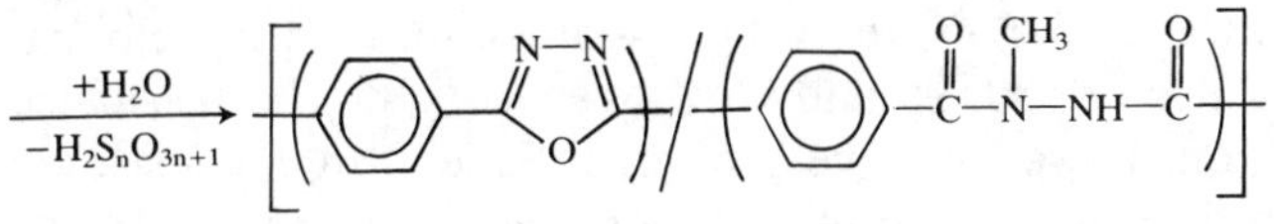

The resulting copolymer is soluble in fuming sulfuric acid and is directly wet-spun into an aqueous sulfuric acid coagulation bath. The copolymer overcame the intractability of the wholly aromatic homopolymer. The one-pot reaction circumvents the problems that arise from the classical route in the cyclodehydration step of the polyhydrazide intermediate [3, 4].

12.6.2 Properties and applications

The white to pale yellow poly(p-phenylene-oxadiazole-co-N-methyl-hydrazide) fibers exhibit high tenacities, high moduli and exceptionally high elongations (Table 12-7).

TABLE 12-7
Properties of Poly(p-phenyleneoxadiazole-co-N-methylhydrazide).

Physical property	Physical unit	Property values best	typical
Tensile strength at break	g/tex	19.0	14.4−16.2
Elongation	%	5.5	5.5−6.5
Modulus of elasticity	g/dtex	404	315−378
Fineness	tex	2.7	4.0−9.0
Density	g/cm^3		1.36
Decomposition temperature (N$_2$, DTA)	°C		460
Water absorption (25°C, 65% RH)	%		6.1−6.2
Shrinkage under hot air			
at 180°C	%		0.6
at 220°C	%		0.8
Retained tensile strength			
at 100°C	%		78
at 150°C	%		61
at 200°C	%		43
Heat aging (7 days at 150°C)			
Tensile strength at break	%		100
Elongation at break	%		105
Modulus of elasticity	%		104

The hydrolytic stability of the fiber is excellent; after 7 days immersion in water at 100°C the values retained for tenacity, elongation and modulus are 90, 96 and 96%, respectively. The resistance to UV radiation, however, is very poor; a Fade-O-Meter test over 96 h shows a drop in tenacity to 20% of original strength.

The new oxadiazole fibers were developed for the application as tire cord. They impart better wear characteristics to tires than glass or steel cords. In road tests, tires with oxadiazole fibers experienced failures after 13 300 and 21 600 miles whereas under the same conditions tires with Kevlar tire cord failed already after only 6000 and 12 500 miles. The new fiber is not yet commercially available.

References

[1] H. C. Bach, F. Dobinson, K. R. Lea and J. H. Saunders, J. Appl. Polym. Sci. **23** (1979) 2125.
[2] ditto, J. Appl. Polym. Sci. **23** (1979) 2189.
[3] Y. Iwakura, K. Uno and S. Hara, J. Polym. Sci. **A 3** (1965) 45.
[4] A. V. D'yachenko, V. V. Korshak and Ye. S. Krongauz, Vysokomol. Soedin. **A 9** (1967) 2231; J. Polym. Sci. USSR **9** (1967) 2523.

12.7   TRIAZIN POLYMERS

12.7.1   Structure and synthesis

Poly(s-triazines) with the structural unit

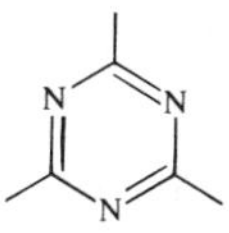

are present in the ®NCNS (N-cyanosulfonamide) polymers of Ciba-Geigy A. G. and in the ®Triazin-A-Resins of Mobay [1, 2].

In the preparation of the NCNS polymers, the reaction between primary and secondary biscyanamides in a molar ratio from 1:1 to 2:1, dissolved in lower alcohols or ketones, yields a soluble A-stage prepolymer:

(12-10)

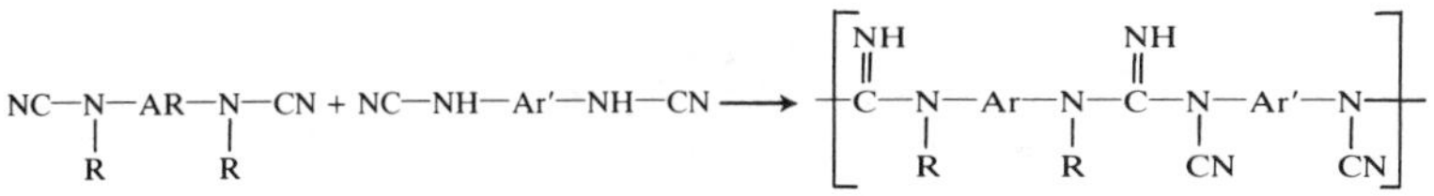

The thus obtained prepolymers are heated to 100−120°C for 30 min to increase the melt viscosity. Such polymers in the so-called B-stage are commercially available. The formation of triazin resins is accomplished by heating the B-stage polymers to 150−180°C:

(12-11)

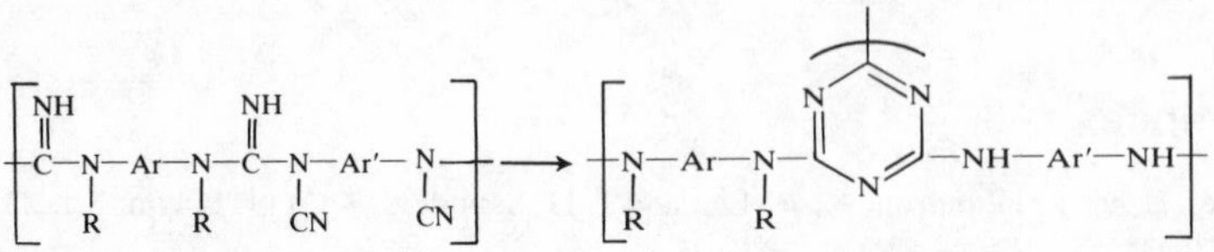

Ar and Ar′ are aromatic groups; R are electrophilic groups which impart enhanced processability, e.g. arylsulfonyl groups [3−5].

Mobay's process starts from bisphenol A which is reacted with cyanuric chloride to the corresponding cyanuric ester. Heating yields a prepolymer with molecular weight 2000 g/mole, solubility in ketones, and softening range 50−60°C. This prepolymer dissolved in methylethyl ketone with 70% solids is commercially available under the designation Triazin-A-Resin KL3-4000. The prepolymer is catalytically cured at 170−200°C to give the true Triazin-A-polymer:

(12-12)

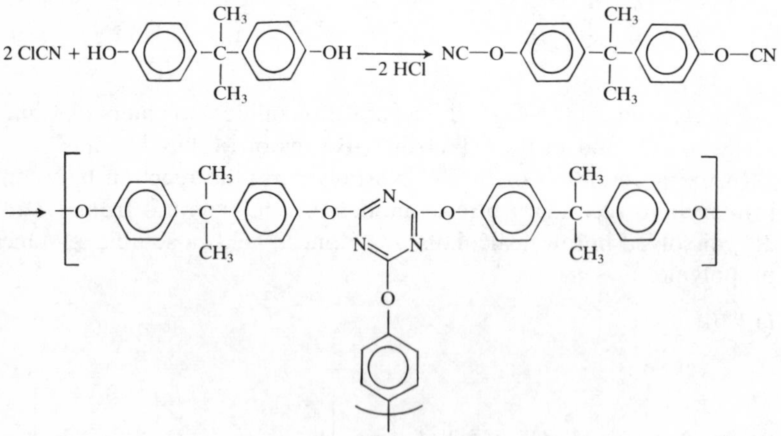

Zinc octoate, or a mixture of zinc octoate with catechole and tri-ethylenediamine are used as catalysts [6, 7].

The Mitsubishi Gas Chemical Co. has obtained the exclusive Triazin-A production and marketing rights for Asia; the same company also produces bismaleimide-triazin resins [8].

12.7.2 Properties and applications

The NCNS resins of Ciba-Geigy as well as the Triazin-A-Resins of Mobay and Mitsubishi Gas Chemical can be processed to molding materials or laminates (Table 12-8). Processability and properties of NCNS resins depend on the molar ratio of primary and secondary biscyanamides applied in the synthesis. The 1:1 resin is used for injection molding and transfer molding, the 2:1 resin for compression molding. The mechanical properties of the unfilled and glass fiber reinforced 2:1 resin are slightly better than those of the 1:1 resin. Composites and printed circuits can be manufactured from NCNS resins.

Triazine resins are recommended for all applications where performance of epoxy resins is no longer reliable. They show short-term resistance to 200°C, however, the long-term use temperature should not exceed 160°C. The advantages of triazine resins over epoxy resins are better dimensional stability, much better adhesion to copper-clad laminates at soldering temperatures, lower coefficient of thermal expansion along the z-axis, and improved dielectric properties. Electrical properties vary insignificantly over a wide range of frequencies, temperatures or water absorption levels.

References

[1] H.-G. Elias, Kunststoffe mit besonderen Eigenschaften, in Ullmanns Enzykl. Techn. Chem., vol. 15 (1978) 403.
[2] K. Demmler, Angew. Makromol. Chem. 76/77 (1979) 209.
[3] R. J. Kray, Soc. Plast. Ind. Proc., 30th Ann. Conf., Sect. 19-C (1975) 1.
[4] R. J. Kray, Modern Plast. 52/4 (1975) 72.
[5] R. J. Kray, R. Seltzer and R. A. E. Winter, ACS Coat. Plast. Prepr. 31/1 (1971) 469.
[6] K. K. Weihrauch, P. G. Gemeinhardt and A. L. Baron, 34th Meeting Soc. Plast. Engng. Techn. Paper 32 (1976) 317.
[7] ®Triazin A Resin, Mobay Chemical Corp., Technical Information 1976.
[8] Chem. Engng. (February 11, 1980) 111; ®BT Resin (Bismaleimide-Triazin), Mitsubishi Gas Chemical Co., Inc., June 1980.

TABLE 12-8

Properties of NCNS and Triazine A laminates. The NCNS Resins were 2:1 Resins, postcured for 2 h at 260°C.

Physical Property	Physical unit	NCNS non-reinforced	NCNS/ glass fiber	NCNS/ graphite fiber	NCNS/ Kevlar	Triazin-A/ glass fiber
Resin content	%	100	32.5	40.1	47.5	36–38
Density	g/cm³	1.31	1.90	—	1.40	—
Glass transition temperature	°C	300	—	—	—	>250
Tensile strength at break	MPa	49	—	—	—	—
Flexural modulus						
at 23°C	GPa	4.42	26.7	102.5	25.9	—
at 232°C	GPa	—	23.4	96.6	—	—
Flexural strength						
at 23°C	MPa	117	620	1400	370	614
at 150°C	MPa	—	—	—	—	483
at 200°C	MPa	—	—	—	—	297
at 232°C	MPa	—	525	940	—	—
at 250°C	MPa	—	—	—	—	172
Flexural strength after aging 4h at 180°C						
at 23°C	MPa	—	—	—	—	627
at 150°C	MPa	—	—	—	—	545
at 200°C	MPa	—	—	—	—	428
at 250°C	MPa	—	—	—	—	276
Impact strength with notch (Izod)	kJ/m	—	—	—	—	0.8
Water absorption	%	0.64	—	—	—	<0.5
Dielectric strength	MV/m	—	—	—	—	28
Volume resistivity	Ohm × cm	—	—	—	—	10^{15}
Surface resistivity	Ohm	—	—	—	—	10^{14}
Relative permittivity (50 Hz)	—	—	—	—	—	4.4
Arc resistance	s	—	—	—	—	130

12.8   CYCLIZED AND OXIDIZED POLY(ACRYLONITRILE)

12.8.1   Structure and synthesis

Poly(acrylonitrile) or PAN fibers have become the most important source for carbon/graphite fibers [1]. As a preliminary step, PAN fiber is stabilized by heating in an oxidizing atmosphere at temperatures ranging from 200 to 300°C for some hours to attain complete permeation of oxygen. The fiber is held under longitudinal tension to limit the shrinkage of fibers during heating. This oxidation of PAN fibers renders them infusible by oxidative crosslinking to enable them to undergo processing at higher temperatures. Further heat treatment of the fiber in an non-oxidizing atmosphere (argon or vacuum) results in carbonization at 1000−2000°C and then graphitization up to 3000°C. Fiber modifications include PAN copolymers with acrylic acid, acrylamide, or 1 to 5% methyl acrylate.

The preliminary reaction step yields a cyclized, oxidized, and crosslinked product with the idealized structure

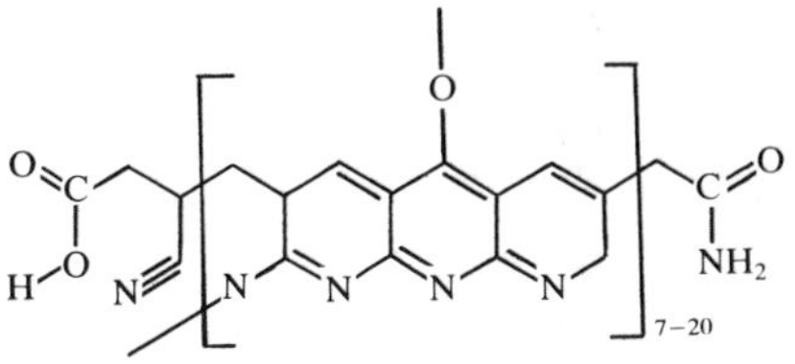

This reaction seems to be more complex. In one study, the theory of intramolecular crosslinking and cyclization was abandoned in favor of a series of elimination reactions, followed by intermolecular crosslinking and cyclization, induced by dipole interactions between pendant nitrile groups [2].

The ®Celiox fibers of Celanese Plastics and Specialties Co. are precursors in the production of ®Celion carbon fibers [3]. They are produced by Toho Beslon Co. in Japan and are marketed by Celanese in North America. The Gentex Corp. manufactures and processes fabrics from Celiox under the designation "Preox" [4].

More recently, similar products have been introduced by Hoechst A. G. [5] and by Universal Carbon Fibers, England [6]. Hoechst supplies the experimental fiber named "VF 1003" which is presumably derived from its ®Dolan 10 PAN homopolymer fiber. "Panotex" of Universal Carbon Fibers is also based on PAN fiber.

12.8.2 Properties and applications

Celiox is commercially available as filament and staple fiber. The fiber exhibits low density, low modulus, high elongation, and high moisture absorption (Table 12-9). The black fiber is non-fusible and non-flammable in air. At temperatures of over 360°C the fiber is slowly converted into carbon fiber which is accompanied by evolution of gaseous by-products.

The properties of fiber and fabrics are noticeably changed only above 200°C as illustrated by figures 12-3 and 12-4.

TABLE 12-9
Properties of the ®Celiox Fiber.

Physical Property	Physical unit	Property Value
Density	g/cm^3	1.4
Tensile strength àt break	GPa	2.14
Elongation at break	%	10
Modulus of elasticity	GPa	113.4
Surface resistivity	Ohm × cm	>10^{10}
Water absorption (20°C, 65% R.H.)	%	10
Limiting oxygen index	%	50

Celiox is targeted to replace asbestos in protective clothings for steel workers. Molten aluminum at 760°C and molten iron at 1560°C run off from cloth without causing any burns. Fire-protective clothing and heat resistant gloves are manufactured from Preox. The fabrics have poor abrasion resistance which can be improved by blending with ®Nomex or other thermally stable fibers. Gentex is investigating the substitution for asbestos in other applications, such as brake linings, filters, heat protective shields, and isolation of pipework.

Celiox yarn was priced at 13 US-$/lb and staple fiber at 10.-US-$/lb in 1980. Celanese plans to build its own large-scale plant in the U.S. from which price reductions for Celiox are expected.

The "VF 1003" fiber of Hoechst is also aimed at the protective

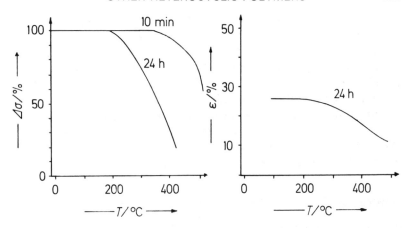

Figure 12-3 Residual tensile strength at break, $\Delta\sigma$, and elongation at break, ϵ, of Celiox fabrics after heat treatment in hot air at the temperatures indicated [1].

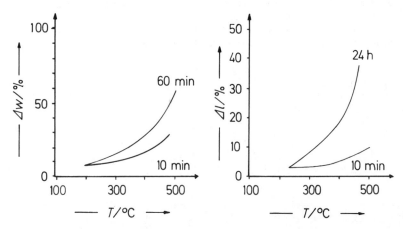

Figure 12-4 Weight loss, Δw, and shrinkage, Δl, of Celiox fabrics after treatment with hot air for various times [1].

clothing and friction lining markets. The non-fusible, thermally stable fiber is said to retain 2/3 of its original strength on heating at 160°C for 100 h. It is available as staple fiber, e.g., for brake linings, as filament for clutch linings. The fiber is said to have distinct price advantages over aramid fibers.

"Panotex" reportedly withstands temperatures of 1000°C for

several minutes. It is resistant to most acids and to strong bases. The cloth shows low thermal conductivity, good wear resistance and an air permeability like wool. The wear comfort is good. Fabrics can be laminated with aluminum or blended with glass fibers. "Panotex" is supplied as knitted or woven fabrics; it is also available in three different blends with aramide fibers. A rubber coated fabric is also manufactured. Applications include fire-protective coatings, special suits for SWAT teams, blends with wool for pilot suits, and double cloths for seats in aircrafts and cars to provide flame protection to the cellular plastic upholstery.

References

[1] J. Delmonte, Technology of Carbon and Graphite Fiber Composites, Van Nostrand Reinhold Co., New York 1981.
[2] C. A. Gaulin and W. R. Mcdonald (Aerospace Corp.), NTIS Report 755–350, May 1970.
[3] ®Celiox, Celanese Plastics and Specialties Co., Technical Information, 1980.
[4] Preox, Gentex Corp., Technical Information 1980.
[5] Anon., VDI Nachrichten **37**/13 (April 1, 1983) 17.
[6] Anon., VDI Nachrichten 37/15 (April 15, 1983) 28.

13 Polyurethanes and other polymers based on isocyanates

13.1 INTRODUCTION

Although polyurethanes are the most important reaction products derived from isocyanates with regard to scientific, technical and economical aspects, a series of other polymers based on isocyanate chemistry has been developed over recent years. The less common reactions of isocyanates have been described in several reviews [1–3].

Of commercial importance are the reactions with isocyanates for the production of the following polymers:

Cyclotrimerization:	Poly(isocyanurates)
Adduct formation	Poly(urethanes)
	Poly(ureas)
	Poly(oxazolidinones)
Condensation:	Poly(carbodiimides)
	Poly(amides)
	Poly(amide imides)
	Poly(imides)
Adduct formation/condensation:	Poly(parabanic acids)
	Poly(hydantoins)
	Poly(benzoxazinediones)

The polymers mentioned above with heterocyclic units are dealt with in chapters 11 and 12 with the exception of poly(isocyanurates) which will be described in this chapter.

References

[1] W. J. Farrissey, Jr., L. M. Alberino and A. A. R. Sayigh, J. Elast. Plast. 7 (1975) 285.
[2] H. Ulrich, J. Polym. Sci., Macromol. Revs. 11 (1976) 93.
[3] D. H. Chadwick and T. H. Cleveland, Isocyanates, in Kirk-Othmer, Encycl. Chem. Technol., 3th Edition, Vol. 13 (1981) 789.

13.2 POLYURETHANES WITH MULTI-PHASE SYSTEMS

Trends in modern polyurethane (PUR) technology aim at fully automated processing, further reduced cycle times and multi-phase systems with improved mechanical properties and/or processability [1–4].

The "ISP" (Instant Set Polymer) system of Dow Chemical Co. achieves molding cycles of less than 1 min [5]. One of the reactants contains up to 60% of an inert liquid with a boiling point over 150°C which is initially dissolved in the liquid phase but separates as fine droplets during the PUR formation and remains insoluble within the solid PUR phase. The inert liquid levels off temperature peaks and thus acts as so-called heat-sink, hence thick-walled parts can be produced without the risk of internal decomposition. Inert liquids mentioned in several patents include esters of phthalic, azelaic, sebacic and adipic acid [6, 7], organic phosphates, cyclic ethers and sulfones, carbonates, chlorinated hydrocarbons, hydrocarbons, and OH-free and estermodified poly(oxyalkylenes) [8].

ISP exhibits good mechanical properties (Table 13-1), however, due to the incorporated, non-reactive organic filler the heat deflection temperature is only 90°C. On the other hand, the filler reduces stress cracking and it absorbs and dissipates the heat created under long-term dynamic stress. Thus ISP demonstrates better fatigue resistance than ABS. ISP is a stiff engineering plastic with a tensile strength between that of thermoplastic polyesters and of polycarbonates. Weathering causes no loss of strength in ISP but a yellow surface discoloration is observed. ISP shows excellent resistance to aliphatic, aromatic and chlorinated hydrocarbons, to oils, acids and bases. ISP is decomposed by acetone, DMF, and ethyl acetate, whereas ethanol and concentrated hydrochloric acid cause strong swelling. ISP is processed by RIM or casting techniques. A number of applications have already been found to replace metals in corrosive environments [5].

Multi-phase PUR systems are obtained with a new generation of polyetherdiols which contain a solidly dispersed hard phase prepared in situ. Polymeric hard phases include styrene-acrylonitrile copolymers, urea-melamine-formaldehyde condensates, polyurethanes, and poly(hydrazodicarbonamides). For several years, dispersions of so-called grafted polyols have been commercially

TABLE 13-1
Properties of ISP and Analymer Multiphase—PUR.

Physical property	Physical unit	ISP	Analymers				Glass Fiber perpendicular to flow direction
			ISA 70	ISA 140	ISA 380	ISA 380/25% in flow direction	
Density	g/cm^3	1.16	1.10	1.23	1.16	1.29	1.29
Tensile strength							
at −40°C	MPa	93.0	—	—	—	—	—
at 23°C	MPa	58.6	21	106	30	32	27
at 60°C	MPa	50.3	—	—	—	—	—
Elongation at break							
at −40°C	%	6	—	—	—	—	—
at 23°C	%	15	100	24	55	28	25
at 60°C	%	20	—	—	—	—	—
Modulus of elasticity	GPa	2.16	—	—	—	—	—
Flexural strength	MPa	90.9	—	—	—	—	—
Flexural modulus							
at −23°C	MPa	—	770	5220	1033	1998	1309
at 25°C	MPa	1929	482	3347	861	1550	930
at 70°C	MPa	—	350	2900	829	1309	757
Impact strength with notch (Izod)	J/m	53	402	37.8	122	150	141
Linear thermal expansion coefficient (×10^5) between −23°/70°C	K^{-1}	—	1.02	0.56	1.01	0.61	0.68
Heat distortion temperature (1.81 MPa)	°C	88	—	—	—	—	—
Heat sag							
at 110°C/30 min	cm	—	0.5	0.025	—	—	—
at 149°C/30 min	cm	—	—	—	3.8	1.06	2.03
Relative permittivity (1 kHz)	—	4.22	—	—	—	—	—
Dissipation factor (1 kHz)	—	0.0052	—	—	—	—	—
Arc resistance	s	106	—	—	—	—	—
Dielectric strength	MV/m	13.8	—	—	—	—	—

available; they are prepared by free radical partial graft polymerization of alkenes onto polyols. A survey of the numerous patents has been published [1].

Of more recent origin are the "PHD" ("Polyharnstoff-Dispersion"-polyurea dispersion) polyols developed by Bayer and Mobay [9]. They contain finely dispersed particles of polyurea in polyetherpolyol. The preparation involves the formation of an emulsion by intensively stirring the incompatible liquid polyamines and polyols. On addition of diisocyanates competitive reactions occur within the polyol/isocyanate phase as well as at the interface of the emulsified polyamine/polyol droplets. The initially formed product from polyol and diisocyanate can react with additional polyol as well as with the polyamine:

(13-1)

$$OCN—R—NCO + HO—R'—OH \longrightarrow HO—R'—O—\underset{\overset{\|}{O}}{C}—NH—R—NCO$$

$$HO—R'—O—\underset{\overset{\|}{O}}{C}—NH—R—NCO \begin{cases} + HO—R'—OH \longrightarrow \text{OH-end-capped PUR prepolymer} \\ \\ + H_2N—R'—NH_{2-} \longrightarrow \text{Urea/urethane-copolymer} \end{cases}$$

The rate of both reactions is controlled by a variety of parameters, such as the ratio of primary to secondary OH-groups, functionality, molecular weight, aromatic/aliphatic isocyanates, aliphatic/aromatic amines, continuous or discontinuous feed, which all exert an influence on the differing reactivities of OH- and NH_2- groups. In practice, the continuous reaction is carried to a level of 20% polyurea. The amorphous polyurea particles have a broad particle size distribution with a medium diameter of 1 μm. Approximately 3 to 5% of the polyol is chemically bound to the polyurea.

Two-phase polyol prepolymers such as "PHD"-polyol increase the shear modulus in flexible foams. The resulting high-resilience (HR) foams have greater indentation load deflections than the conventional hot-cure foams. They are rated by the support-and-comfort (SAC) factor which is measured as the ratio of weight needed to depress a 50-sq. in., 4-in. thick foam by 65% to weight needed to depress it by 25%. Subjective comfort is greater as SAC is higher. The disperse phase acts like an active filler and increases crosslinking density; hence mechanical strength of HR foam is superior to that of hot-cure. Polyurea imparts greater temperature resistance and flame retardancy to foams prepared from PHD-polyol. The main application for PHD-polyols are molded foams for seattings in cars, for mattresses, and toys. PHD-polyols can be used also in slab flexible foam production. RIM processing and usability of PHD-polyols in coatings and adhesives are being studied [9].

A two-phased cast PUR elastomer was introduced by Dow Chemical Co. under the designation "Analymers" [10]. The polymers contain a in situ formed continuous elastomeric phase and a discontinuous hard phase which are mutually chemically bound. The chemical composition has not been disclosed. By variation of monomers and their molar ratio the proportion of both phases can be varied resulting in polymers which exhibit properties ranging from elastomers to high-modulus types (Table 13-1). They can be filled with glass fiber and are processable by the RIM technique. "Analymers" are still developmental products which are aimed at applications in the automotive industry. Compared to conventional RIM PUR types, "Analymers" are said to demonstrate greater stiffness, higher flexural moduli, improved impact resistance, and a lower coefficient of thermal expansion over a broad range of temperatures. Moreover, smooth and paintable surfaces are obtained on demolding.

The conventional methods which have so far been applied to increase flexural modulus of PUR bring about certain disadvantages. Increasing crosslinking density increases flexural modulus but sacrifices toughness. To increase flexural modulus through formation of crystalline domains leads to a trade-off in elongation and impact strength, and products with greater brittleness are obtained. In segmented PUR block copolymers, the extension of the hard domain results in greater flexural modulus at a sacrifice of elongation. Although the incorporation of reinforcements, e.g. glass or mineral fibers, increases flexural moduli, elongation and impact strength are simultaneously reduced. At high levels of incorporated fibers, the RIM process enforces an orientation of the fibers and anisotropic materials are obtained. In "Analymers", the coefficient of linear thermal expansion is 40% lower compared to that of similar PUR types. Reinforcement with 25% glass fiber gives products with less pronounced anisotropic properties (Table 13-1).

References

[1] K. C. Frisch, Recent Developments in Urethane Elastomers and Reaction Injection Molded (RIM) Elastomers, Rubber Chem. Technol. **53** (1980) 126.
[2] L. J. Lee, Polyurethane Reaction Injection Molding (Review), Rubber Chem. Technol. **53** (1980) 542.
[3] D. Dieterich, Angew. Makromol Chem. **76/77** (1979) 79.
[4] K. Uhlig and D. Dieterich, Polyurethane, in Ullmanns Enzykl. Techn. Chem. **19** (1980) 301.
[5] ISP, Instant Set Polymer, Dow Chemical Co., Technical Information, 1975.
[6] H. G. Newton (Dow Chemical Co.), USP 3 378 511 (1968).
[7] E. E. Jones, F. Olstowski and D. B. Parrish (Dow Chemical Co.), USP 3 726 827 (1973).
[8] F. Olstowski (Dow Chemical Co.), USP 4 000 105 (1976).
[9] K. G. Spindler and J. J. Lindsey, J. Cell. Plast. **17** (January/February 1981) 43.
[10] R. B. Turner, R. E. Morgan and J. H. Waibel, Internat. Conf. on Cellular and Non-cellular Polyurethanes, Straßburg; C. Hanser, München 1980, 865; J. Elast. Plast. **12** (1980) 155.

13.3 AQUEOUS POLYURETHANE SYSTEMS

Aqueous polyurethane (PUR) systems include dispersions, sols, xerosols and hydrogels. Dispersions are the most important products technically and economically. Of the sols, poly(carbamoylsulfonates) are already commercially available (see chapter 13.3.2). Xerosols are still under development; their market introduction is expected in the near future.

13.3.1 Polyurethane dispersions

13.3.1.1 Structure and synthesis

PUR dispersions have been commercially available since 1972. In the recent years, they have seen an unusually fast and worldwide development due to the fact that they allow a nearly problem-free application of isocyanate and polyurethane chemistry in aqueous medium. Bayer A. G. in particular has pioneered developments in this field. Reviews [1, 2] which also list the already numerous patents issued [3] have been published.

Among the variety of aqueous, ionic and non-ionic PUR dispersions the emulsifier-free anionic dispersions have gained the greatest share. They contain mostly sulfonate groups and less frequently carboxyl groups. Incorporation of sulfonate groups is accomplished by reaction of isocyanate or NCO-containing prepolymers with salts of hydrazine sulfonic acid, polyfunctional hydroxysulfonic acids, or with amino compounds, e.g. with N-(2-aminoethyl)-2-aminoethane sulfonate, whereas for the incorporation of carboxyl groups polyfunctional hydroxycarbonic acids are employed. Cationic PUR dispersions contain quaternary ammonium groups in the backbone; these are formed by reaction of the hydrogen in the NH-group of urethanes or ureas with chloracetamide.

For the commercial preparation of PUR dispersions a number of processes has been utilized [1–3] (Table 13-2). Shear force induced dispersing in the presence of emulsifiers can be applied to every PUR and led to the first products of technical importance. Stability of such two-phased systems, however, is in general not very good.

PUR ionomers are self-dispersable without the need of shear action and emulsifiers. In the so-called "acetone-process", a solution of high-molecular weight PUR or polyurea is mixed with water and the acetone is then distilled off. Depending on the level of ionic groups and solids, the ionomer dispersion is formed by precipitation of hydrophobic segments or by phase inversion of the initially formed reversed emulsion [4, 5]. The acetone process is also applicable to non-ionic dispersions when terminal or pendant polyethers based on ethylene oxide are built into PUR in place of the ionic centers.

Disadvantageous is the poor space/time yield of the acetone process. The so-called "prepolymer-ionomer-mix-process" operates without organic solvents. An ionically modified NCO-prepolymer is emulsified in water and is chain-extended in emulsion by addition of diamines. The diamine migrates from the aqueous phase into the polymer particles; the simultaneously proceeding chain propagation and crosslinking causes no viscosity increase of the dispersion. Although this process is easy to perform, it is applicable only to special NCO-prepolymers.

Of more general applicability is the so-called "melt-dispersion process". Ionically modified NCO-prepolymers are reacted with urea or ammonia and the thus formed terminal NH_2-groups are condensed with formaldehyde. Prior, during, or after the reaction with formaldehyde, the hot melt is mixed with water which results in spontaneous dispersion and subsequent chain extension through polycondensation.

TABLE 13-2
Processes for the Preparation of PUR Dispersions [3].

	Emulsifier/ shear force	Acetone	Prepolymer/ ionomer	Melt dispersion	Ketimine/ ketazine	Solids/ spontaneous dispersion
				Process		
Polyhydroxy compound	Liquid polyether	any	polyethers and some polyesters	any	any	any if softening temp. <40°C and molar mass <800 g/mol
Diisocyanate	TDI	any	TDI, IPDI	TDI, HDI, IPDI predominantly ionic	any	any
Glycols	small amounts only	any	dimethylol-propionic acid		any	any
Emulsifier	yes	no	no	no	no	no
Solvent	5–10% toluene	40–70% acetone	mostly 10–30% N-methyl-pyrrolidone	—	sometimes 5–30% acetone depending on glycol content	—
Shear force mixer	yes	no	no	no	no	no
Dispergation temperature (°C)	20	50	20–80	50–130	50–80	15–30
Product before dispergation	NCO prepolymers	PUR	NCO prepolymeric ionomers with COO⁻ groups	Biuret polymers	NCO prepolymers + ketimine and ketazine	Prepolymers
Reaction or process during dispergation	Amine extension	acetone distillation	amine extension	polycondensation	sometimes acetone distillation	addition of cross-linking agent
Final product	polyurethane urea	polyurethane, polyurethane/ urea	polyurethane/ urea isomer	polyurethane/ biuret	polyurethane/ urea	polyurethane
Solvent content of final dispersion	2–8% (toluene)	<0.5%	mostly 5–15% (N-methyl-pyrrolidone)	—	sometimes <2% acetone	—
Particle size (nm)	700–3000	30–100 000	100–500	30–10 000	30–1000	30–500
Postheating (°C)	>100	—	—	50–150	—	120

Advances in the acetone process led to the ketimine and ketazine processes. In the ketimine process, diamines are capped with ketones to yield bis-ketimines:

(13-2)

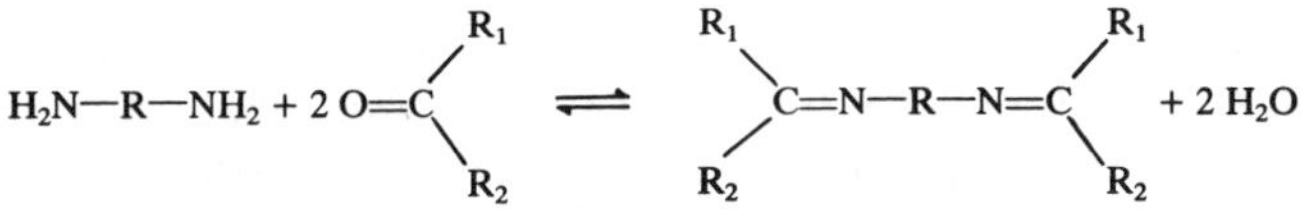

$$H_2N-R-NH_2 + 2\ O=C\overset{R_1}{\underset{R_2}{\diagdown}} \rightleftharpoons \overset{R_1}{\underset{R_2}{\diagup}}C=N-R-N=C\overset{R_1}{\underset{R_2}{\diagdown}} + 2\ H_2O$$

Bis-ketimines can be mixed with NCO-prepolymers without initiating a reaction. On addition of water, the prepolymer is dispersed and the ketimine is simultaneously hydrolyzed to the diamine in reversion of the reaction mentioned above; the liberated diamine then causes chain propagation.

In the ketazine process, NCO-prepolymers are mixed with a ketazine, aldazine or hydrazone followed by dispersion in water. Hydrolysis liberates hydrazine which in turn initiates chain propagation:

(13-3)

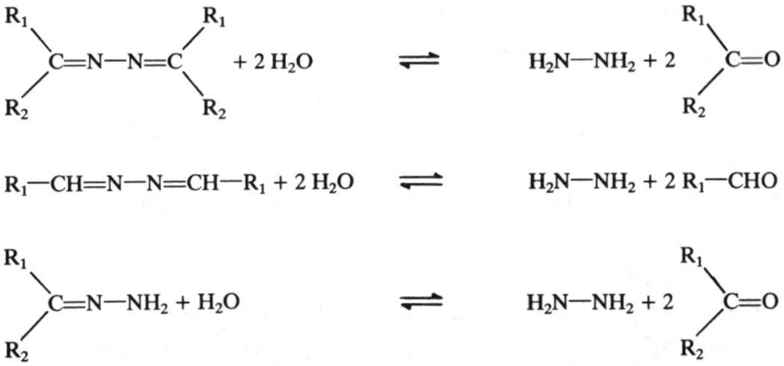

$$\overset{R_1}{\underset{R_2}{\diagdown}}C=N-N=C\overset{R_1}{\underset{R_2}{\diagup}} + 2\ H_2O \rightleftharpoons H_2N-NH_2 + 2\ \overset{R_1}{\underset{R_2}{\diagdown}}C=O$$

$$R_1-CH=N-N=CH-R_1 + 2\ H_2O \rightleftharpoons H_2N-NH_2 + 2\ R_1-CHO$$

$$\overset{R_1}{\underset{R_2}{\diagdown}}C=N-NH_2 + H_2O \rightleftharpoons H_2N-NH_2 + 2\ \overset{R_1}{\underset{R_2}{\diagdown}}C=O$$

Ketazines hydrolyze slower than ketimines; therefore, this process is also applicable to the particularly reactive aromatic isocyanates. For the first time, aromatic polyurethanes with a high degree of carbazide groups became thus available.

Characteristics and some parameters of the technical processes in use for the preparation of PUR dispersions are summarized in Table 13-2 [3].

13.3.1.2 Properties and applications

Emulsifier-free stable ionomer dispersions and suspensions are produced with particle sizes ranging from 30 to 800 nm and from 10^4 and 10^5 nm, respectively. They exhibit high mechanical and chemical stability, excellent film formation properties, good adherence, and a broad variety with regard to composition and properties. Like all ionomers, they are sensitive to freezing, electrolytes and changes in pH. Non-ionic PUR dispersions are less affected by those factors, however, due to their incorporated hydrophilic segments, they show increased sensitivity to water resulting in swelling, cloudiness, softening and finally hydrolytic degradation. By combining ionic and non-ionic groups, these disadvantages can be minimized and sufficient resistance to freezing and electrolytes is achieved [3].

Aqueous PUR dispersions can replace organic solvent based systems in adhesives and in coatings for textiles, leather and paper. Other applications are water-thinned paints [6, 7]. In floor coatings, the performance of aqueous PUR products is equivalent to that of conventional materials, including the solvent based PUR systems [8]. In glass fiber sizing, PUR dispersions outperform poly(vinylacetate). Ionomeric PUR types act as plasticizers in gelatin, casein, and PVC. Aqueous MDI-based PUR emulsions are used for bonding particle board and laminated hardboard.

Other products that are prepared from isocyanates or NCO-prepolymers via reactions in or from aqueous media, are PUR microcapsules, poromers, and highly hydrophilic foams. In the medical field, the conventional plaster of Paris support casts have been replaced by PUR casts. This is accomplished by impregnating bandages with a PUR prepolymer which on dipping into water hardens within a few minutes.

A variety of aqueous PUR dispersions, emulsions, and latices are already commercially available. In general, the exact composition of the products is not disclosed. Bayer A. G. has marketed a series of ionomer dispersions as product groups: ®Impranil dispersions for coating of textiles, ®Baybond for glass fiber sizing, ®Bayderm as primer, coating and top-coat in leather finish, ®Desmocoll 8065 and 8066 for adhesives, e.g. for bonding PVC and nonwovens, PU-Dispersion D and V for coating of paper, ®Baydur PU coupling agent for bonding metals or ceramics to self-skinning foams, and impregnating compound VP PU 3164 for hydrated alumina to impart flame-retardancy to flexible PUR foams.

W. R. Grace Co. has marketed two cationic PUR dispersions for coating of metals or flexible substrates [9]. Cationic and anionic PUR dispersions are commercially available from Lancro/England; they are used for coating of wood, concrete, leather and fabrics as well as in the formulation of adhesives. Films with a broad range of hardnesses can be obtained; these demonstrate high flexibility, abrasion resistance, and light fastness. The Ruco Division of Hooker Chemical Corp. has marketed two PUR latices which are said to be equally suitable for the preparation of paints, foams and adhesives. They are compatible with other water-based resins, take high filler loads, and are resistant to UV radiation. Two cationic PUR latices supplied by Witco Chem. Corp. yield fast drying antisoiling films having a good "feel". The materials are used in paper coatings and are said to be equally useful in binding fluorinated hydrocarbons. The PPG Industries Inc. has developed PUR latices for coatings which show setting to touch after 1 h and can be walked on after 12 h; full hardness is attained after one week.

References

[1] K. Uhlig and D. Dieterich, Polyurethane, in Ullmanns Enzykl. Techn. Chem., vol. **19** (1980) 301.
[2] D. Dieterich, Angew. Makromol. Chem. **76/77** (1979) 79.
[3] D. Dieterich, Angew. Makromol. Chem. **98** (1981) 133.
[4] O. Lorenz and H. Hick, Angew. Makromol. Chem. **72** (1978) 115.
[5] O. Lorenz, V. Budde and K. H. Reinmöller, Angew. Makromol. Chem. **87** (1980) 35.
[6] W. Brushwell, Farbe + Lack **87** (1981) 938.
[7] H. van de Viel and W. Zom, Polym. Paint Col. J. **170** (1980) 49.
[8] U.S. Government Report, National Bureau of Standards, Washington, December 1977; PB-275 390/3 WM.
[9] W. Brushwell, Farbe + Lack **87** (1981) 269.

13.3.2 Poly(carbamoylsulfonates)

13.3.2.1 Structure and synthesis

Poly(carbamoylsulfonates) (PCS) are water soluble thermoreactive bisulfite adducts of poly(isocyanates). They have been marketed by Bayer A. G. under the trade name ®Synthappret and by Dai-Ichi

Kogyo Seyaku Co. in Japan under the trade name ®Elastron [1].

PCS is obtained by reaction with bisulfite of NCO-end-capped prepolymers or poly(isocyanates) in an aqueous solution, which contains ethanol [2, 3, 4] or inert solvents [5]:

(13-4)

$R(OH)_n + n\,OCN—R'—NCO \longrightarrow R(OCONHR'—NCO)_n$

$R(OCONHR'—NCO)_n + n\,NaHSO_3 \longrightarrow R(OCONHR'—NHCOSO_3{}^-Na^+)$

$R''(NCO)_n + n\,NaHSO_3 \longrightarrow R''(NHCOSO_3{}^-Na^+)_n$

Prepolymers are prepared by a well-known process from poly-(propylene oxide triol) with a molecular weight of 3000 and aromatic or aliphatic diisocyanates. Depending on the type of diisocyanate, the reaction is conducted at temperatures of 60–120°C without solvents and without catalysts [4]. Dai-Ichi Kogyo Seyaku prefers using urethane prepolymers containing from 10 to 40 wt.-% oxyethylene groups which are prepared from polyether- or polyester glycols with diisocyanates [5]. In all procedures, isocyanates are used in excess of 5–10%.

When the reaction between the prepolymer or polyisocyanate and bisulfite is carried out in aqueous solution, only low molecular weight and water insoluble products are obtained; under these conditions such products hydrolyze rapidly to polyurea. The hydrolysis is prevented when water soluble solvents are added. The reaction of prepolymers with hexamethylenediisocyanate or TDI in water/ ethanol gives concentrated, gel-free solutions of PCS with yields of 90 and 80%, respectively. Such solutions containing 30% PCS (from hexamethylenediisocyanate) are stable for weeks at 60°C and for days at 70°C, when small amounts of sulfuric acid or H_2O_2 are added as stabilizers [4].

It is known that isocyanates react rapidly with water or alcohols, however, bisulfites react much faster. Model reactions between butylisocyanate and water, ethanol and bisulfite had reaction rates at a ratio of 1:1.5:500000.

13.3.2.2 Properties and applications

PCS is soluble [4] or can be emulsified in water [5]. The solutions reduce surface tension effectively. They are stable in acidic solutions for a fairly long time, but they undergo rapid hydrolysis in alkaline solution or on heating to 100°C.

At room temperature, reactivity of PCS to nucleophilic agents is by far less pronounced than that of isocyanates; this is also illustrated by the activation energy for the hydrolysis of butylcarbamoyl and butylisocyanate with values of 87 ± 5 and 62.3 kJ/mole, respectively. At 100°C, however, the hydrolysis rates of isocyanates and PCS are fairly equivalent, a feature which distinguishes PCS from other capped isocyanates.

The mechanism of PCS hydrolysis is not yet fully understood; both S_N2 and elimination reactions have been considered [4]. The hydrolysis of polyfunctional PCS compounds yields insoluble polyureas. For monofunctional PCS, the hydrolysis is formulated as follows:

(13-5)

$$2\,RNHCOSO_3^-Na^+ + H_2O \longrightarrow RNHCONHR + CO_2 + 2\,NaHSO_3$$

Besides with water, PCS reacts only with primary and secondary amines under formation of asymmetrical ureas:

(13-6)

$$RNHCOSO_3^-Na^+ + R'NH_2 \longrightarrow RNHCONHR' + NaHSO_3$$

The properties of PCS have been utilized by the commonwealth Scientific and Industrial Research Organization (CSIRO) in Australia for the development of a process to render woolen fabric creaseproof [2]. PCS can be applied as aqueous liquor, whereas polyisocyanates require organic solvents, e.g., perchloroethylene. The fabrics are wet rapidly by the strongly surface-active PCS solutions. Crosslinking conditions for PCS are suited to processing on conventional textile finishing machinery. To achieve crease-proofing, PCS in combination with dispersions or emulsions of polyurethanes or poly(acrylate esters) is applied to the fabric. Crosslinking is accomplished during drying of the fabric. The resulting film prevents felting when the fabric is washed. PCS enhances the durability of such films.

References

[1] G. B. Guise, Polymer News **7** (1981) 149.
[2] Commonwealth Scientific and Industrial Research Organisation, Australia, USP. 3 898 197 and 4 124 553.
[3] F. Reich and H. Schuster, Bayer Farben Review **30** (1978) 38.
[4] G. B. Guise, J. Appl. Polymer Sci. **21** (1977) 3427.
[5] T. Hamamura, S. Goto and K. Ishihara (Dai-Ichi Kogyo Seyaku Co., Ltd., Kyoto, Japan), USP. 4 039 517 (1977).

13.4 MODIFIED POLY(ISOCYANURATES)

13.4.1 Structure and synthesis

Poly(isocyanurates) or PIR are obtained by cyclotrimerization of aromatic, multi-functional isocyanates [1–3]:

(13-7)

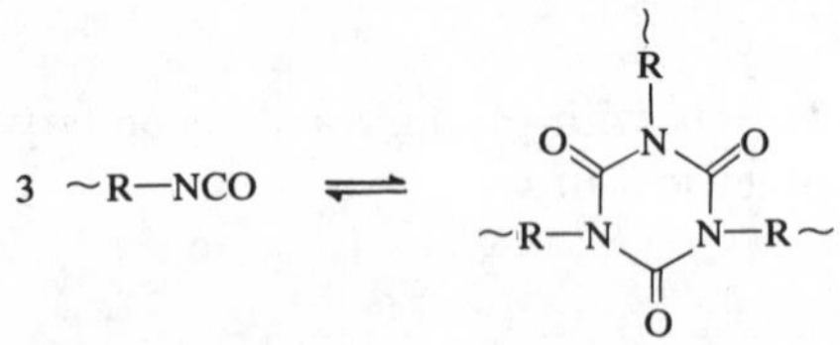

$$3 \ \sim\!\!R\!-\!NCO \ \rightleftharpoons$$

(isocyanurate ring structure)

Polymeric MDI is preferably used as the isocyanate component. Cyclotrimerization is catalyzed by phenolates, alcoholates and carboxylates of alkali metals, various organometallic compounds of silicon, tin, lead and zinc, tertiary amines as well as by combinations of amines with epoxides, alcohols, peroxides and alkylene carbonates. The reaction is exothermic, but less than in the formation of PUR, and it was utilized as early as 1968 for the preparation of PIR foams with chlorofluoroalkanes as foaming agents.

The earlier PIR foams demonstrated superior thermal stability and lower flammability than PUR foams, however, applications were hampered by brittleness due to the high crosslinking density. Attempt have been made to attain flexibility through modification without sacrificing flame retardancy. Modification with epoxides was suggested to give PIR with oxazolidine-2-one groups [4]. The Mitsubishi Chemical Co. produces polyoxazolidone and modified PIR foams from bisphenol A epoxides and polymeric isocyanates (PAPI) with CFCl$_3$ and complexed catalysts, such as AlCl$_3$/hexamethylphos-phortriamide, or complexes of AlCl$_3$, ZnCl$_2$, or FeCl$_3$ with THF, n-butylether or n-butylamine [5]. The Bendix Corp. developed imide-modified PUR/PIR foams by reaction of prepolymers from PAPI and 3,3′,4,4′-benzophenonetetraacid anhydride with polyols, a foaming agent and catalysts [6]. The products have not been yet commercialized.

The most practical and technically almost exclusively used modi-fication is trimerization of isocyanates in the presence of polyols to give PIR/PUR hybrid foams [2−5, 7−9]. This procedure requires the addition of halogen or phosphorus compounds in order to main-tain sufficient flame retardancy. Also, catalyst systems must be selective to initiate both trimerization and urethane formation.

Upjohn Co. formerly used chlorine containing polyols with the designations Isonol 36 or Isonol D 201 [2, 5], but is now marketing chlorine-free polyols which are said to give flame-resistant PIR foams according to specification DIN 4102 B2 [9, 10]. One typical catalyst for such PIR hybrid foams is sodium N-(2-hydroxy-5-nonylphenyl)methyl-N-methyl-glycinate [11]. Foams available from Upjohn are trade-named ®Trymer (low density), ®Kode-Panel and ®Isoderm (integral skin foam).

Bayer and Mobay have recommended a number of formulations for preparing modified PIR foams with low flammability and low smoke development [8]. The modifiers employed are aromatic poly-esterpolyols, either partially chlorinated or combined with phos-phorus compounds, e.g. tris-(chloropropyl)phosphate. A new copolyol is marketed by Bayer under the designation ®Baymer 1119 A.

Other polyols for the preparation of urethane-modified PIR foams are halogenated polyols trade named ®Ixol (Solvay), based on epichlorohydrin, and chlorine- and phosphorus-containing poly-ether-polyester polyols named ®Ugiflam 3700 (Ugine Kuhlmann). The Japanese company Inoue MTP supplies new catalysts for urethane-modified PIR foams; the metal salts of tertiary amino acids are said to exhibit better solubility in polyols and greater catalytic activity than conventional carboxylates such as potassium acetate or caprylate.

Vestrotex and ICI have jointly developed glass reinforced PIR foams and laminates [9]. Other PIR products of ICI are available under the trade names ®Hexafoam and ®Hexafroth.

Non-cellular, glass reinforced PIR can be prepared by the RIM technique [7, 12]. New thermally stable thermosets based on PIR are obtained from long-chained, linear or slightly branched polyols with molecular weights from 1000 to 4000 with a high excess of polymeric MDI [13]. The polymers are composed of hard, thermally stable PIR and elastic PUR structural units. Properties can be varied within a broad range by varying the ratio of reactants.

Thermally stable casting materials with PIR structures are obtained by a combined polyaddition/polymerization reaction. For example, 100 pt by weight of a MDI mix of 4,4'- and 2,4'-diphenyl-methanediisocyanate, 40 pt styrene, 5 pt bis-phenol-A-diepoxide (®Lekutherm X 20) are reacted with 1 pt dimethylbenzylamine as catalyst [14−16]. The Bayer products are named "KI-Duromere".

13.4.2 Properties and applications

Triphenylisocyanurate decomposes in the absence of oxygen and catalysts at temperatures above 480−500°C. Thermal stability is reduced by incorporation of flexible urethane groups. Properties of urethane-modified PIR foams not only depend on the urethane/isocyanurate ratio and the nature of polyol used, but also on the operating temperature of the foaming process. At temperatures ranging from 20 to 100°C, cure degree of the foam core, dimensional stability and adhesion to top coats all increase, and surface brittleness decreases, as the foaming temperature increases [8]. Glass fiber reinforcement reduces brittleness, and improves dimensional stability and some mechanical properties in continuously produces PIR foam laminates (Table 13-3). The laminates were tested at temperatures of up to 900°C. The carbonized layer, which is formed on the surface, protects the foam core against further combustion. Such laminates are suitable for insulation of flat building roofs; they withstand 250°C hot poured asphalt and bitumen.

The upper service temperature limit of non-reinforced, modified PIR foams is 150°C; they withstand short-term, i.e. up to 30 min, exposure to temperatures of up to 800°C [7]. Applications are insulation of pipelines and boilers. In tankers transporting liquified gas, PIR foams insulate at temperatures ranging from −180 to +150°C. In the rigid foam market, PIR foams have already gained a 20% share and are predicted to surpass PUR foams by 1985 [7].

Non-cellular PIR types (Table 13-4) are processed by RIM,

TABLE 13-3
Properties of PUR and PIR structural foams and PIR foam laminates [8–10].

Physical Property	Physical unit	Structural Foams: PIR (Isoderm)	Structural Foams: PUR	PIR Foam Laminates: with glass fiber	PIR Foam Laminates: non-reinforced
Thickness of laminate	mm	—	—	50	50
Density	kg/m^3	534	480	—	—
total	kg/m^3	—	—	37	34
core	kg/m^3	—	—	35	33.5
Glassfiber content	%	0	0	6.9	0
Closed cells	%	—	—	90	90
Tensile strength	MPa	10.3	9.7	—	—
Compression strength (10%)	MPa	11.7	10.3	—	—
in foaming direction	MPa	—	—	0.195	0.220
perpendicular to foaming direction	MPa	—	—	0.125	0.110
Compression modulus					
in foaming direction					
perpendicular to foaming direction	MPa	—	—	8.0	8.2
Flexural strength	MPa	—	—	3.15	1.80
Flexural modulus	MPa	57.2	48.3	0.265	0.180
Hardness (Shore D)	MPa	1273	690	5.8	3.0
Thermal conductivity	—	80	75	—	—
Mold shrinkage	W m^{-1} K^{-1}	—	—	0.0019	0.0019
Continuous service temperature	%	0.5	0.5	—	—
Change of length after 8 months	°C	150	<100	—	—
at −15°C	%	—	—	−0.01	0.03
at 23°C/50% RH	%	—	—	−0.01	0.08
at 70°C	%	—	—	−0.01	0.09
with cycle (24 h at −15°C, 24 h at 20°C, 24 h at 70°C, 24 h at 20°C)	%	—	—	−0.02	0.11

TABLE 13-4
Properties of non-cellular, urethane-modified PIR grades [7, 13, 14].

Physical property	Physical unit	Property Value of				
		PIR RIM [7]	PIR RRIM [7]	PIR[1] Baydur 120 [13]	KI-Duromere unfilled [14]	KI-Duromere filled [14]
Reinforcing agent						
type	—	—	Glass	—	—	Quartz powder
content	%	—	30	—	—	66
Tensile strength	MPa	—	—	—	54	77
Modulus of elasticity	MPa	—	—	—	3.2	9.4
Elongation at break	%	12.7	4.3	4	—	—
Flexural strength	MPa	—	—	75	95	106
Flexural modulus						
at 21°C	GPa	1.79	4.59	—	—	—
at 82°C	GPa	0.85	2.72	—	—	—
Impact strength	kJ/m^2	—	—	18	12.5	6.6
Impact strength with notch (Izod)	kJ/m	0.06	0.03	—	—	—
Hardness (sphere penetration)	MPa	—	—	—	227	583
Martens temperature	°C	—	—	245	208	>250
Continuous service temperature	°C	200	200	≥200	200	220
Heat conductivity	W m^{-1} K^{-1}	—	—	—	0.167	0.760
Density	g/cm^3	—	—	1.2	1.1	1.7

[1] prepared with conventional PUR foaming equipment; wt.-ratio polyol/PIR = 100:250.

RRIM and casting techniques. PIR thermosets see service in applications where high use temperatures are required and short cycle times must be sustained even at intricated mold designs and thick-walled parts. Disadvantageous is the subsequently required annealing to complete the cure. Properties of such non-cellular PIR thermosets are superior to those of polyacrylates, poly(styrenes), polyacetals and polyacetates, however, they are inferior to polyamides, polysulfones and polyamide-imides with regard to the combination of thermal stability and toughness.

References

[1] H. J. Diehr, K. J. Kraft and R. Wiedermann, Kunststoffe **62** (1972) 731.
[2] H. E. Reymore, Jr., P. S. Carleton, R. A. Kolakowski and A. A. Sayigh, J. Cell. Plast. **11** (November/December 1975) 328.
[3] H.-G. Elias, Kunststoffe mit besonderen Eigenschaften, in Ullmanns Enzykl. Techn. Chem., vol. **15** (1978) 442.
[4] K. Ashida and K. C. Frisch, J. Cell. Plast. **8** (1972) 160.
[5] D. Brownhill, Modern Plast. Internat. **10** (November 1980) 22.
[6] J. Salary and C. H. Smith, SPE ANTEC, Techn. Papers **21** (1975) 102.
[7] H. Ulrich, J. Cell. Plast. **17** (January/February 1981) 31.
[8] G. F. Baumann and W. Dietrich, J. Cell. Plast. **17** (May/June 1981) 144.
[9] D. Brownhill, Modern Plast. Internat. **11** (May 1981) 42.
[10] H. E. Reymore, Jr., R. J. Lockwood and H. Ulrich, J. Cell. Plast. **14** (1978) 332.
[11] R. A. Kolakowski (Upjohn Co.), USP. 3 986 991 (1976).
[12] P. S. Carleton, L. M. Alberino and D. J. Breidenbach, SPE Techn. Conf., Detroit, November 1979.
[13] H. Lüdke, Kunststoffe **67** (1977) 680.
[14] R. Kubens, F. Ehrhard and H. Heine, Kunststoffe **69** (1979) 455.
[15] R. Kubens, Report Internat. Polyurethane Conference, Strasbourg, June 1980; C. Hanser, Munich 1980, 285.
[16] D. Braun, Angew. Makromol. Chem. **78** (1979) 1.

13.5 MODIFIED UREA-FORMALDEHYDE AND POLYUREA FOAMS

13.5.1 Structure and synthesis

In situ produced urea-formaldehyde (UF) foams have been used in thermal insulation for over 30 years. The large-scale starting materials, i.e., urea and formaldehyde, are not based on petrochemical feedstocks. The foams are inherently flameproof without the need of additional flame retardants. Worldwide consumption of UF-foams was about 90 000 t in 1980. Besides sound and thermal insulation, a multitude of other applications have been developed, e.g., biologically degradeable moisture reservoirs for soil stabilization, medical powders, and oil absorbers. They are used also for sealing off mine damps and as inexpensive filler for padding cavities caused by rock thrust in mining operations [1−4].

In spite of easy processability, UF-foam is no do-it-yourself-product as are certain PUR foam formulations. For expertise processing, specification DIN 18 159, part 2, A1, and those of "Gütegemeinschaft Aminoplast Montageschaum" must be observed [5]. During cold setting of UF-foams traces of formaldehyde are released from the aqueous resin solution; these may possibly cause irritations. The Consumer Products Safety Commission has demanded a ban of UF-foams in the USA [6]. This demand is being challenged [7], since formaldehyde has been produced on a large scale for over 45 years, and the multitude of its applications has never led to any apparent detrimental health effects.

Independent of the outcome of this controversy, a number of US and European companies are actively engaged in developing urea-based foams from which generation of formaldehyde vapor is either minimized or completely absent. The Schaum-Chemie GmbH & Co. KG., Essen and Frankenthal, has developed so-called monomer-catalyst-foams [3]. Oligomeric alkylnaphthalene-sulfonic acid/ formaldehyde condensates are reacted with di- or polyhydroxy-benzenes at room temperature and at pH 1−2 to give solid polymers in an exothermic reaction within a few minutes. In alkaline media, the reaction requires heating and yields instantly porous materials. Depending on the molar ratio of the reactants gels, paste-like, or crosslinked polycondensates with various degrees of hardness are obtained. In some cases, fibrous materials are formed which retain their shape on drying. The monomer/catalyst solutions can be processed with the conventional UF foaming machines. The monomer-catalyst-foam prepared from single or multi-component systems is initially yellow, it turns pink and finally deep red as temperature and concentration increase during hardening [3].

Release of formaldehyde from UF-foams can be reduced when the concentration of formaldehyde in the prepolymer is increased and not, as is frequently attempted, decreased. If the formaldehyde concentration is not sufficient, excess formaldehyde can be added to the prepolymer solution. Furthermore, single or polynuclear phenols, phenolethers, or acenaphthene derivatives are added together with the catalyst or the catalyst/foaming agent solution. These additives bind formaldehyde which is present in the prepolymer solution, or which was added in excess, or is generated during the curing/drying process, under formation of cocondensates or crosslinks that are stable to hydrolysis [8]. If formaldehyde-binding amide groups are present in excess, the so-called UF-combination-foam shows enhanced flame retardancy. The foams obtained are stable, and exhibit no tendency to excessive shrinkage or formation of shakes and checks.

Rubicon Chemicals [9], Chemie Linz [10], and Schaum-Chemie [12] have developed new polyurea foams. Also, polyurea is used as filler in PUR foams (see chapter 13.2). Jefferson Chemical Co. has developed polyureas for coatings and for RIM processing [11].

Rubicon employs a two-component system, originally developed by ICI, consisting of polymeric diphenyl-4,4'-diisocyanate (A), a mixture of water, a "small amount" of polyol, catalyst, emulsifier and a flame retardant (B). The ratio of A to mixture B is 3–4:1 which is said to be less critical than the ratio of reactants for UF foams where deviations cause insufficient cure and generation of formaldehyde. For polyurea foams, a broad ratio of A/B ranging from 7:1 to 1:1 can be tolerated. The reaction is strongly exothermic and stock temperatures rise to 80–90°C [9]. The viscosity of the aqueous phase is too low to be handled by conventional UF and PUR foaming machines, therefore, the Gusmer Corp., Old Bridge, N.J., designed a foam generation machine specifically adapted to polyureas.

Chemie Linz also employs a two-component system composed of polymeric diphenyl-4,4'-diisocyanate (raw MDI with high levels of three- and polynuclear polyisocyanates) and aqueous urea-formaldehyde prepolymer solution [10]. The foam is generated by liberated CO_2. Like conventional UF foams, cure is accomplished by means of acidic catalysts. However, free acids can not be used because they would interfere with the primarily base-catalyzed reaction; thus, latent acids, e.g. benzoylchloride, which yield free acids only after hydrolysis are preferred. Combinations of tertiary amines with at least three hydroxy groups, in conjunction with tin-organic compounds or with another tertiary amine, catalyze the reaction between polyisocyanate and the methylol- and NH-groups of the UF prepolymer (eq. 13-8 to 13-10).

(13-8)   FOAMING REACTION:

$$2\,RNCO + HOH \longrightarrow RNHCONHR + CO_2$$

with R ⊨ —NCO

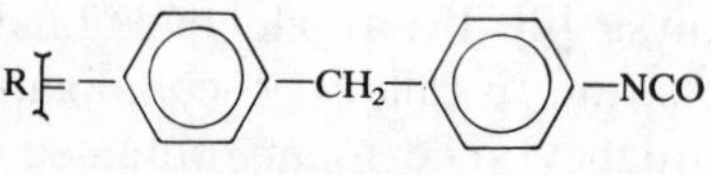

or with R = polynuclear polyisocyanate group

(13-9)   CURING REACTION:

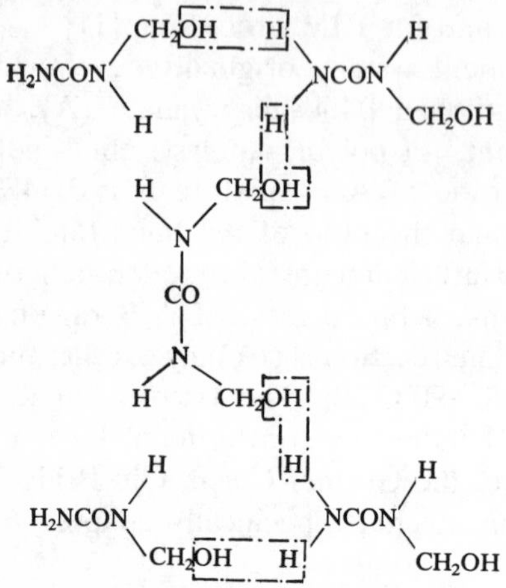

Toluylenediisocyanate is less suitable for the reaction mentioned above. The MDI used has the following specification: NCO-assay 31 ± 1%, viscosity 250 ± 50 MPa × s, density 1.22 g/cm³, hydrolytic chlorine <0.3%, total chlorine <0.8%, solids <1%, flash point >200°C. The UF prepolymer contains formaldehyde and urea in a molar ratio of 1.0–2.0:1, from 2 to 10% water; it has a density of 1.23–1.27 g/cm³ and viscosities ranging from 300 to 10000 mPa × s.

For example, a foamable mixture comprises 79.90 pt UF prepolymer solution, 15.98 pt raw MDI, 1.69 pt benzoylchloride, 0.19 pt

(13-10)

$$RNCO + HOH_2CNHCONH_2 \longrightarrow RNHCOOH_2CNHCONH_2$$

$$RNCO + H_2NCONHCH_2OH \longrightarrow RNHCONHCONHCH_2OH$$

$$\sim NHCONH \sim\ +\ RNCO \longrightarrow\ \sim NHCO\overset{\textstyle |}{\textstyle C}ONHR$$

$$RNCO + HO{-}\underset{\underset{\textstyle OH}{\textstyle |}}{R'}{-}OH \longrightarrow RNHCOO\underset{\underset{\textstyle OH}{\textstyle |}}{R'}{-}OH$$

dimethylethanolamine, 1.28 triethanolamine and 0.96 pt silicone oil. First, the amines are mixed into the polyisocyanate. After a very short residence time, to this mixture are simultaneously, but separately, added the prepolymer solution and the catalyst. Chemie Linz has developed a proprietary foaming machine for the two-step metering of the feed. The foam starts to develop after $8-15$ s, rises for another $60-150$ s and sets to touch after $60-150$ s.

Jefferson Chemical Co. has developed sprayable coating compounds based on monocyanoethylated poly(oxypropylenediamine) which is obtained by the addition of acrylonitrile to the amino groups:

(13-11)

$$2\ CH_2{=}CH{-}CN + H_2N{-}\underset{\underset{\textstyle CH_3}{\textstyle |}}{CH}{-}CH_2{-}\Big[\underset{\underset{\textstyle CH_3}{\textstyle |}}{CH}{-}CH_2{-}O\Big]_n CH_2{-}\underset{\underset{\textstyle CH_3}{\textstyle |}}{CH}{-}NH_2 \longrightarrow$$

$$H_2N{-}\underset{\underset{\textstyle CH_2{-}CH_2{-}CN}{\underset{\textstyle |}{\textstyle CH_3}}}{CH}{-}CH_2{-}\Big[\underset{\underset{\textstyle CH_3}{\textstyle |}}{CH}{-}CH_2{-}O\Big]_n CH_2{-}\underset{\underset{\textstyle CH_2{-}CH_2{-}CN}{\underset{\textstyle |}{\textstyle CH_3}}}{CH}{-}NH$$

The reactivity of the monocyanoethylated diamine (CNPA) is only 1/20 of that of a primary amine and hence allows a better control of the exothermic reaction. The reaction between CNPA and polyisocyanates yields polyureas, that between a mixture of CNPA and polyols with polyisocyanates gives crosslinked polyurea-urethanes [11].

13.5.2 Properties and applications

Polyurea foams (PUA) exhibit properties which are as anticipated between those of UF- and PUR-foams (Table 13-5). PUA foams can be produced with nearly the same low densities as UF foams and also show similar thermal insulation characteristics. Compressive strength of PUA foams is far superior to that of UF foams. Water permeability of PUA foams is lower than for PUR foams in spite of an equivalent degree of water absorption. PUA foams adhere better to surfaces than do UF foams. In this respect as well as in flammability PUA foams are similar to PUR foams. PUA foams are resistant to water, diluted acids and bases. Polar solvents cause swelling which, however, is reversible on drying. PUA foams are attacked by strong bases and acids. PUA foam can be cut, sawed and milled.

Recommended applications for PUA foams are acoustic and thermal insulation, e.g. in home refurbishing, thermal insulation of backwardly aerated facades, poured-in-place foam to insulate spaces between inner and outer walls, pipe insulation with prefabricated semi-shells, and footfall sound insulation under floating floors.

References

[1] H. Baumann, Plastoponik, Schaumkunststoffe in der Agrarwirtschaft, Dr. A. Hüthig Verlag, Heidelberg 1967.

[2] H. Baumann, Plastverarbeiter **27** (1967) 235.

[3] H. Baumann, Kunststoffe **69** (1979) 440.

[4] B. Meyer, Urea-Formaldehyde Resins, Addison-Wesley Publishing Co., Reading, Mass. 1979.

[5] GSM, Gütegemeinschaft Aminoplast Montageschaum, D-6000 Frankfurt, Mannheimer Str. 97.

[6] D. J. Manson, Chem. Engng. News (April 5, 1982) 34.

[7] H. Baumann, Kunststoffe **71** (1981) 835.

[8] H. Baumann, personal communication, June 1982; Technical Information Schaum-Chemie W. Bauer GmbH & Co. KG., Essen.

[9] C. J. Galbraith, M. J. Cartmell and R. Brown, Paper presented at the 6th Annual Urethane Foam Contractors Association, Orlando, Florida, January 22, 1981.

[10] L. Gosler, Paper presented at the Internat. Polyurethane Conference, Strasbourg, June 1980; Cellular and non-cellular polyurethanes, C. Hanser, München 1980, 865.

[11] R. L. Rowton, Rubber Chem. Technol. **50** (1977) 435.

[12] H. Baumann and F. Bayersdorf, Paper presented at the 12th Internat. Foamed Plastics Symp., Düsseldorf, May 25, 1983.

TABLE 13-5
Properties of polyurea foams as compared to UF, PUR and EPS foams [4, 9–11].

Physical Property	Physical unit	Polyurea foam (Chemie Linz)	Polyurea foam (Rubicon Chem.)	Urea/formaldehyde foam	Polyurethane hard foam	Poly(styrene) particle foam
Cell structure	—	open	open	open	closed	closed
Density	kg/m^3	12–60	8–12	5–15	20–100	15–30
Compression Strength	MPa	0.04–0.20	—	0.01–0.05	0.1–0.9	0.06–0.25
Heat conductivity	W m^{-1} K^{-1}	0.035	0.045	0.03	0.018–0.024	0.036
Water vapor diffusion resistance factor (DIN 53122)	µm	2	—	4–10	30–130	30–70
(ASTM C 355)	%	—	12–20	28–35	0.6–4.0	0.4–3.0
Flammability (ASTM E 84)						
Flame spreading	—	—	20–25	15–25	15–50	—
Smoke development	—	—	325–425	0–5	140–450	—

13.6 POLYESTER-URETHANE HYBRIDS

13.6.1 Structure and synthesis

Amoco Chemical Corp. produces pilot plant quantities of a poly-ester-urethane hybrid resin from its "Amoco Pilot Resin TG-89" [1, 2]. Starting material is an unsaturated polyester prepared from isophthalic acid, maleic anhydride and diethylene glycol in molar ratio of 1.0:1.0:2.64. The resin with acid number 4.9 and hydroxyl number 164.8 is dissolved in styrene to give a solution with 60% solids. When isocyanates and free radical initiators are added, chain propagation and crosslinking reactions occur simultaneously, where the propagation reaction rate at room temperature is greater than that of the crosslinking. The chain-extended polymer is formed according to

(13-11)

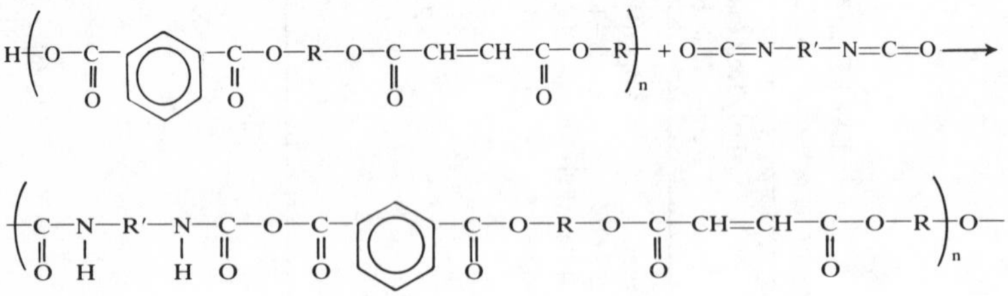

Bayer prepared similar products named "KI-Duromere" [3, 4]. Hydroxylic end-capped maleic/fumaric acid polyesters are reacted with diisocyanates in the presence of styrene and free radical initiator, where formation of urethane and copolymerization of styrene with the double bond of the unsaturated acid proceed in one step to yield crosslinked products. For example, a KI-Duromer casting resin requires: 100 pt by weight polyol component (i.e. Bayer experimental product KL 3-4503 from maleic anhydride, diethylene glycol and 1,2-propanediol, dissolved in styrene), 10 pt zeolite powder in castor oil (50%), 3 pt benzoylperoxide paste, 38 pt MDI-based isocyanate (®Baymidur K 88), and 60 pt quartz powder W 12 [4].

13.6.2   Properties and applications

Polyester-urethane hybrids have been developed as competitive products to RIM and RRIM-PUR as well as to modified SMC polyesters. All these polymers are aimed at applications in the automotive industry. Although parts fabricated from polyester demonstrate high dimensional stability, they are too brittle for many applications. On the other hand, RIM- and RRIM-PUR exhibit greater flexibility, however, their dimensional stability and heat deflection temperature are inferior to those of modified UP resins. Polyester-urethane hybrids such as BMC and SMC materials are designed to bridge this property gap [1]. The low viscosity of the hybrid resins allows high filler loads, as in the case of KI-Duromere.

Polyester-urethane hybrids show properties similar to those of some epoxy-urethane hybrids and cycloaliphatic polyurethane casting resins based on isophoronediisocyanate (IDPI) (Table 13-6).

KI-Duromere and cycloaliphatic PUR casting resins are entering fields of applications which so far have been served by epoxy resins. Their upper continuous service temperature is above 180°C. IPDI-PUR types are resistant to weathering and are used in insulation of overhead lines. Compared to cycloaliphatic epoxy resins, they exhibit better adhesion to quartz powder, improved resistance to hydrolysis, and better wet insulation characteristics [4].

References

[1] R. S. Rapp, Paper presented at the British Plastics Federation Congress, November 1980; R. S. Rapp (Amoco Chemical Corp.), personal communication, June 1, 1981.

[2] Anon., The Journal of Commerce (March 17, 1981) 5.

[3] R. Kubens, Kunststoffe **64** (1974) 666.

[4] R. Kubens, F. Ehrhard and H. Heine, Kunststoffe **69** (1979) 455.

13.7   OXAZOLIDINE-ISOCYANATE RESINS

13.7.1   Structure and synthesis

The Rohm & Haas Co. has marketed an oligomeric acrylic resin with oxazolidine groups under the trade name ®Acryloid AU-568 (previously Experimental Resin QR 568), which is a one-component

TABLE 13-6

Properties of polyester/urethane hybrids, KI thermosetting polymers and cycloaliphatic polyurethane molding resins [1, 4].

Physical property	Physical unit	Amoco Polyester-Urethane			Property values of KI-Thermosets		Cycloaliphatic PUR(IPDI) Resin	
		TG 89/ TDI	TG 89/ MDI	TG 89/ Poly-MDI	unfilled	with 60% quartz powder	unfilled	with 60% quartz powder
Tensile strength	MPa	91.7	82.9	85.5	66	68	85–90	70–80
Modulus of elasticity	GPa	3.25	3.12	3.10	3.40	10.60	—	—
Elongation at break	%	5.9	7.8	3.9	2	1	3–5	0.7–0.9
Flexural strength	MPa	155.8	141.7	141.3	165	113	160–175	132
Flexural modulus	GPa	3.98	3.44	3.31	—	—	—	—
Impact strength	kJ/m^2	—	—	—	15	7	20–30	13
Hardness (sphere penetr. H_{30})	MPa	—	—	—	185	390	226	500
Martens temperature	°C	—	—	—	117	130	130	133

system that cures by reaction of isocyanate with atmospheric moisture [1−4]. Acryloid AU-568 is an oligomeric oxazolidinylethyl methacrylate with a number average molecular weight of 1200 g/mol and oxazolidine equivalent weight of 265. It is supplied in 2-ethoxyethylacetate (ethylene glycol monoethylether acetate) solution with 85% solids and a viscosity of 4−7 Pa × s [4].

Oxazolidines are readily obtained by cyclodehydration of alkanolamines with aldehydes or ketones:

(13-12)

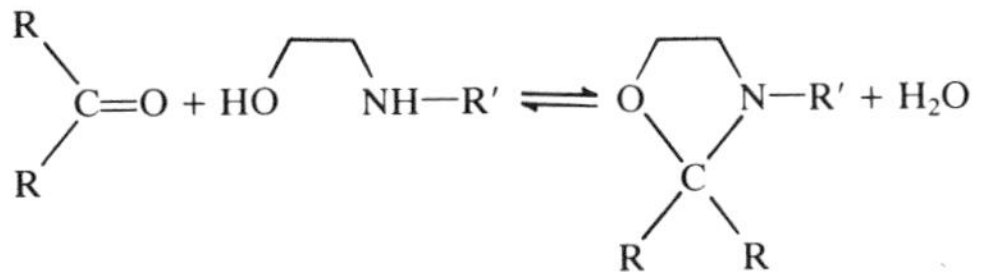

Formation of 2,2-dihydro- and 2-isopropyloxazolidines is particularly fast and quantitative yields are obtained [2]. Commercial production of polyfunctional oxazolidines is carried out by reaction of diethanolamine with formaldehyde or isobutyraldehyde followed by esterification with methacrylic acid methylester (eq. 13-13, eq. 13-14) or dialkyladipates [2, 5]. The methacrylicester is converted to oligomeric Acryloid AU-568 by free radical or anionic [6] polymerization in organic solvents (eq. 13-15).

Under anhydrous conditions, no reaction takes place between oxazolidines and aliphatic isocyanates. Presence of water causes ring opening of the oxazolidine with liberation of formaldehyde (eq. 13-16). At equilibrium, the ratio of compounds I, II and III is 99/0.5/0.5 [2]. Added isocyanates react with the active hydrogen in III and thus enforce the shift of the equilibrium entirely to the right-hand side of the equation (eq. 13-17). This reaction proceeds much faster than the reaction between hydroxyl groups and isocyanates, and no catalysts are needed.

13.7.2 Properties and applications

Oxazolidine groups act as blocking agents for hydroxyl groups in reactions with isocyanates. This method supplements the procedures known in polyurethane chemistry in formulating heat- or moisture activated one-component systems based on blocked isocyanates or

$$HO-\underset{H}{N}-CH_2CH_2OH + CH_2O \rightleftharpoons \underset{CH_2}{\overset{O}{N}}-CH_2CH_2OH + H_2O$$

(13-13)

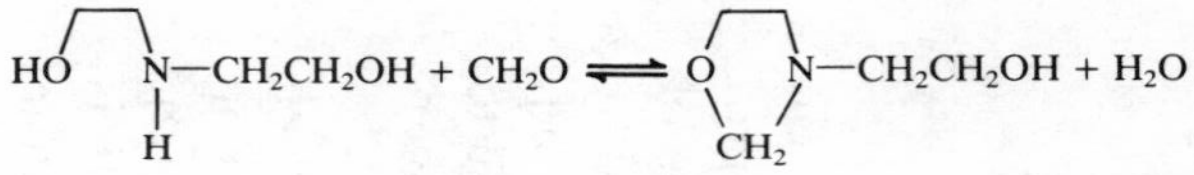

$$\underset{CH_2}{\overset{O}{N}}-CH_2CH_2OH + CH_2=\underset{CH_3}{C}-COOCH_3 \rightleftharpoons \underset{CH_2}{\overset{O}{N}}-CH_2CH_2OC\underset{O}{\overset{||}{}}-\underset{CH_3}{C}=CH_2 + CH_3OH$$

(13-14)

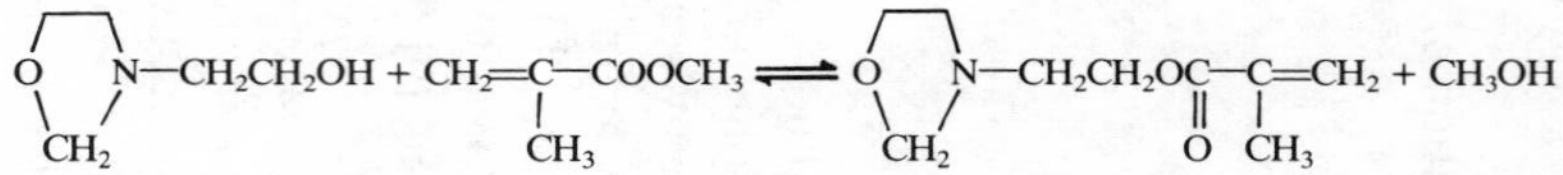

$$\underset{CH_2}{\overset{O}{N}}-CH_2CH_2OC\underset{O}{\overset{||}{}}-\underset{CH_3}{C}=CH_2 \longrightarrow \left[-CH_2-\underset{CH_3}{\overset{}{C}}-\underset{O}{\overset{||}{C}}-O-CH_2CH_2-\underset{CH_2}{\overset{O}{N}}\right]_n$$

(13-15)

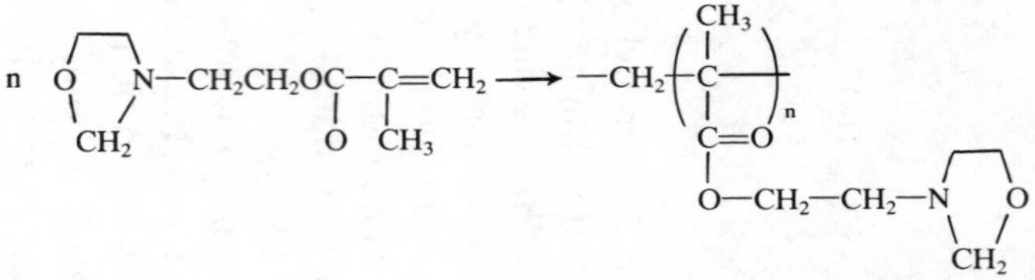

$$R-\overset{O}{N} + H_2O \rightleftharpoons \underset{OH}{R-N}\underset{OH}{} + CH_2O \rightleftharpoons \underset{H}{R-N}-OH$$

\hspace{2cm} I \hspace{3cm} II \hspace{3cm} III

(13-16)

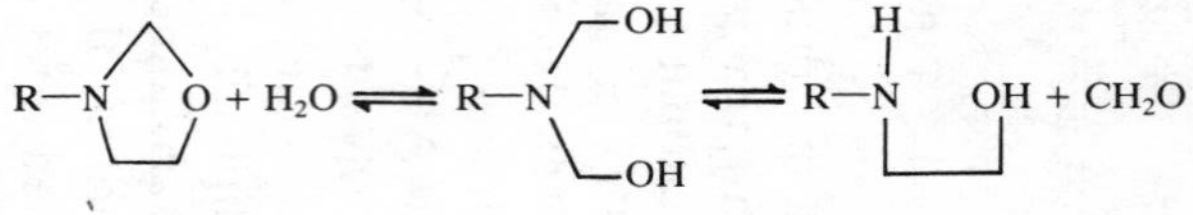

$$\underset{H}{R-N}-OH + 2R'-N=C=O \longrightarrow R'-\underset{H}{N}-\underset{O}{\overset{||}{C}}-\underset{R}{N}-O-\underset{O}{\overset{||}{C}}-\underset{H}{N}-R'$$

\hspace{6cm} III

(13-17)

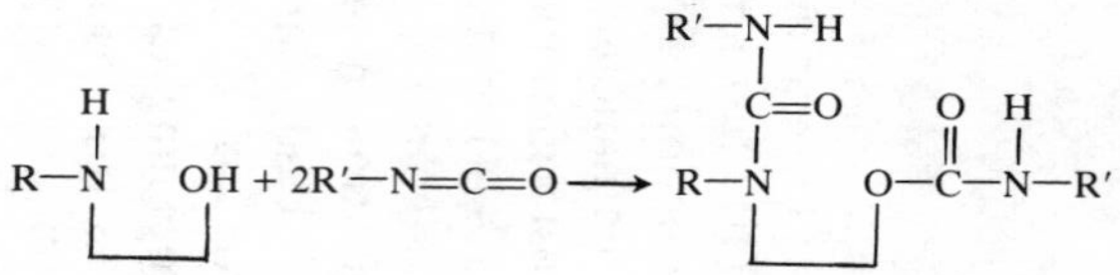

capped amines [7]. Also, oxazolidines can block isocyanate groups. N-hydroxyalkyloxazolidines and isocyanates give compounds that are resolved by hydrolysis [8].

The reaction of Acryloid AU-568 with aromatic diisocyanates proceeds too fast for practical applications. Aliphatic isocyanates, such as dimeryldiisocyanate (®DDI, General Mills) or preferably biuret-triisocyanate (®Desmodur N, Bayer), in mixtures with Acryloid AU-568 allow pot-lifes of more than 24 h, however, pot-life is reduced again by high filler loads, e.g. with TiO_2 [4]. Filled systems with up to 74% solids can be processed by conventional spraying techniques.

Clear lacquers dry at room temperature and set to touch after 3.5 h. They give weather resistant coatings from which mural "paintings" or graffiti can be removed without impairment of surfaces. For stripping of graffiti the following solvent mix is recommended: 46 pt by weight 2-ethoxyethylacetate, 42.1 methylethylketone, 4.7 pt dimethylaminoethanol, and 6.5 pt bentonite (as thickener). The mixture is allowed to penetrate for approximately 1 min, and is then brushed off and finally rinsed with water [4].

Clear or pigmented systems are used as original or repair laquers for cars, planes, and railway trains, and as metal coatings in such cases, where heat curing cannot be applied. The coatings demonstrate a good balance of hardness, flexibility, abrasion resistance, high resistance to chemicals and solvents, and excellent resistance to weathering.

References

[1] W. D. Emmons, A. Mercurio and S. N. Lewis, ACS Div. Org. Coat. Plast. Prepr. **34**/1 (1974) 731.
[2] W. D. Emmons, A. Mercurio and S. N. Lewis, J. Coat. Technol. **49** (1977) 65.
[3] W. D. Emmons et al. (Rohm & Haas), USP. 3 743 626 (1973), USP. 3 912 691 (1974), USP 4 032 686 (1976), Ger. Offen. 2 458 588 (1973).
[4] ®Acryloid AU-568, Rohm & Haas, Technical Information, May 1978.
[5] E. M. Hankins and W. D. Emmons (Rohm & Haas), USP. 3 037 006 (1962).
[6] S. N. Lewis and R. A. Haggard, Germ. P. 2 218 836 (1972).
[7] K. Uhlig and D. Dieterich, Polyurethane, in Ullmanns Enzykl. Techn. Chem., Vol. **19** (1980) 301.
[8] Brit. P. 1 463 944 (Bayer A. G.).

13.8 POLY(CARBODIIMIDES)

13.8.1 Structure and synthesis

Poly(carbodiimides) or PCD are prepared by polycondensation of multifunctional isocyanates under generation of CO_2 [1−4]:

(13-18)

$$n \sim R\text{—}N{=}C{=}O + n\, O{=}C{=}N\text{—}R\sim \longrightarrow \sim R(N{=}C{=}N\text{—}R)_n + n\, CO_2$$

No solvents are needed for the preparation of PCD foams; the liberated CO_2 acts as foaming agent. Polymeric MDI, which can already be partially carbodiimidized, is preferably used as the isocyanate component. Catalysts are phospholaneoxides or -sulfides, phospholineoxides or -sulfides, partially substituted with aromatic/aliphatic pendant groups and/or halogens. However, the reaction rate with starting times around 1 to 2 seconds is too fast for the catalyst to be homogeneously stirred into the rising foam. Addition of ligands prolongs potlife without extending cure time. Such ligands are mono-, di- or polyalcohols, protogenic acids, metal salts and acid chlorides [2]. PCD foam formation is less exothermic than reactions to PUR. A maximum temperature of only 70°C is measured in the core of large block of the foam.

Like PUR foams, also PCD foams are crosslinked. Crosslinks in PCD are formed by reaction of carbodiimide groups with free isocyanates to give polyfunctional uretoneimine structures:

(13-19)

$$\sim R(N{=}C{=}N\text{—}R)_n + O{=}C{=}N\text{—}R'\text{—}N{=}C{=}O \longrightarrow \sim R\text{—}N\underset{O}{\overset{\displaystyle C{=}N\text{—}R}{\underset{C\text{—}N\text{—}R'\text{—}N{=}C{=}O}{|}}{}_n$$

Raw and auxiliary materials for PCD foams are commercially available from Bayer A. G. under the trade name ®Baymid [5].

Soluble PCD types are obtained when the reaction is conducted at temperatures of 75−120°C in solvents like DMAC, benzene or xylene [1], or in 50/50 mixtures of chloroform and chlorobenzene [6]. Molecular weight is controlled by addition of mono-isocyanates [1] or p-bromoaniline [6]. From these solutions clear films can be cast; after annealing at 220°C, the films exhibit glass transition temperatures of 200–220°C, tensile strength 70 MPa, and elongation at break 10–15% [1]. Likewise, molded parts can be fabricated from PCD, but these and the films are still developmental products of the Upjohn Co.

13.8.2 Properties and applications

PCD foams combine high flame resistance (classification B 1) with good thermal insulation characteristics of rigid PUR foams and good sound absorption properties of flexible PUR foams. Disadvantages include obtainable densities are only in the range from 16 to 20 kg/m^3, and only open-cell slab foam can be produced [2, 7]. PCD foams are anisotropic, because the foam still rises to a small extent after setting, causing elliptic cells to be formed which are oriented in direction of flow [2]. Depending on the direction of orientation, compressive strengths range from 0.01 to 0.09 MPa, and thermal conductivity coefficients range from 0.0315 to 0.036 W/m × K. PCD foams carbonize readily when exposed to flames; smoke density is relatively low.

Applications for PCD foams include footfall sound insulation under floating floors, insulation of spaces between inner and outer walls, home refurbishing, thermal insulation of backwardly aerated facades, flame retardant insulations of vents, hot-air and ventilating ducts, and particularly pipe insulation using with semi-shell or flexible, wrappable foam sheets. Use temperature for pipe insulation is up to 150°C, and short term exposure to 250°C is tolerated.

References

[1] W. J. Farrissey, L. M. Alberino and A. A. Sayigh, J. Elast. Plast. **7** (1975) 285.
[2] M. Mann, Paper presented at the 5th Internat. Foamed Plastics Symp., Düsseldorf, May 26/27, 1975.
[3] K. Uhlig and J. Kohorst, Kunststoffe **66** (1976) 616.
[4] L. M. Alberino, Proc. of SPE 4th Internat. Cell. Plast. Conf., Montreal 1976, 1.
[5] Bayer-Polyurethane, Handbook, Bayer A. G., Leverkusen 1979.
[6] E. Dyer and P. A. Christie, J. Polym. Sci. [A-1] **6** (1968) 729.
[7] Anon., Modern Plast. Internat. **8** (May 1978) 21.

14 Electrically conductive polymers

14.1 INTRODUCTION

The field of electrically conductive polymers, a class of "Organic Metals" or "Synthetic Metals" or briefly "Synmetals", has attracted much attention over the last few years. Several books [1–3] and reviews [4–9] have been published.

Polymers suitable as synmetals contain conjugated double bonds. They are dielectric insulators in the ground state. Only by addition of oxidizing or reductive agents do the polymers experience a drastic increase in electric conductivity. With reference to the semiconductor technology, this procedure has been termed "doping", although no lattice sites are substituted; rather it involves a structural phase transition, caused by redox-reactions between a complexing agent and the conjugated polymeric system. Effective complexing agents include e.g., J_2, AsF_5, Br_2, $FeCl_3$, H_2SO_4, $HClO_4$, $XeOF_4$, XeF_2, JF_5, BF_3, BCl_3, SO_2, NO_2, O_2, $ClSO_3H$, NO_2X and NOX with X = SbF_6, $SbCl_6$ or BF_4, inorganic peroxides as FSO_2-O-O-SO_2F, and alkali metals e.g. in the form of sodium naphthalide.

Upon complexing, the polymers display conductivities equivalent to those of poorly conducting metals (Table 14-1). However, if the conductivity is related to unit weight, polymer conductivity comes close to those of the better conducting metals (Table 14-2). Conductivities of highly conductive metals such as copper are not attainable.

The mechanisms of polymer conductivity is still a matter of divergent discussions. For example, to date is has not been elucidated, whether the conductivity involves charge transfer along several chains or perpendicular to the direction of the backbone chain. A summary and critical evaluation of the models and mechanisms proposed for the electric conductivity of complexed polymers has been published [7].

The technical development of synmetals apparently is focussed on only a few polymers pioneered by a few companies.

TABLE 14-1

Electrical conductivity of "doped" polymers at room temperature ([7], and papers in ACS Polym. Preprints **23**/1 (1982) 73–138).

Polymer	Doping agent/ gegenion	Electrical conductivity in S/cm
Cis-Poly(acetylene)	J_2	1.6×10^2
	Br_2	0.5
	AsF_5	1.2×10^3
	H_2SO_4	1.0×10^3
	$(n\text{-}C_4H_9)_4NClO_4$	9.7×10^2
	$AgClO_4$	3.0
	Sodium naphthalide	80
Poly(p-phenylene)	J_2	$<10^{-5}$
	AsF_5	500
	HSO_3F^-	35
	BF_4^-	10
	PF_6^{2-}	45
	Sodium naphthalide	20
	Sodium naphthalide	5
Poly(m-phenylene)	AsF_5	10^{-3}
Poly(p-phenylensulfide)	AsF_5	1
	AsF_3/AsF_5	0.3–0.6
Poly(pyrrole)	BF_4^-	40–100
	$AgClO_4$	50
Poly(1,6-heptadiyne)	J_2	0.1
	AsF_5	0.1
Poly(aluminophthalocyaninefluoride)	$NOBF_4$	0.9
	$NOPF_6$	0.59

The University of Pennsylvania, Philadelphia, granted licenses to Allied, BASF, and Rohm & Haas [11] covering the process of doping poly(acetylene) which was discovered by A. G. MacDiarmid and A. J. Heeger [10]. Rohm & Haas supplies doped poly(acetylene) films as a developmental product.

At the present time, the conducting polymers such as doped poly-(p-phenylene sulfide), poly(p-phenylene), and poly(acetylene), are in the stage of preliminary applications evaluation and are not yet produced on a commercial scale at Allied [12]. The C & D Batteries Division of Eltra Corp. (a subsidiary of Allied Corp.) develops rechargeable batteries based on doped poly(p-phenylene) and poly-(acetylene) [11]. IBM is actively engaged in investigating poly-(acetylene), poly(p-phenylene sulfide), and poly(pyrrole). BASF

TABLE 14-2
Electrical conductivity of metals and doped poly(acetylene) [7]*.

	Density in g/cm^3	Electrical conductivity in S/cm	Electrical conductivity per density in cm^2 g^{-1} Ω^{-1} (S g^{-1} cm^2)
Copper	8.92	6.5×10^5	7.6×10^4
Gold	19.3	4.1×10^5	2.1×10^4
Aluminum	2.7	3.7×10^5	1.4×10^5
Iron	7.86	1.0×10^5	1.3×10^4
Lead	11.3	0.5×10^5	4.4×10^3
(SN)$_4$	2.3	3.7×10^5	1.6×10^3
Mercury	13.6	1.0×10^4	7.4×10^2
Germanium	5.36	10^3	1.9×10^2
cis-[CH(AsF$_5$)$_{0,14}$]$_n$	0.8	5.6×10^2	7.0×10^2
cis-[CH(AsF$_5$)$_{0,10}$]$_n$*	0.8	1.2×10^3	1.5×10^3

* Updated by data from A. G. MacDiarmid and A. J. Heeger, ACS Polym. Preprints 23/1, 73 (1982).

continuous its studies on poly(acetylene) but seems to focus recently more on the one-step synthesis of poly(pyrrole) [9]. BASF and Varta Batterie A. G. recently published their intention to jointly develop polymer batteries. Xerox Corp. is investigating poly(1,6-heptadiine) [13].

None of the conductive polymer systems is as yet commercially available. Unfortunately, all conductive polymers studied to date suffer from at least one of several undesirable characteristics, which include improcessability, poor mechanical integrity of the carrier materials, instability of the oxidized and/or reduced complexed polymer derivatives, sensitivity to oxygen, poor storage stability due to a gradual loss of conductivity, and the poor stability of polymer electrodes to electrolyte solutions.

A number of other polymers are being tested in the laboratory: poly(trimethylsilylacetaylene) [35], polyaniline [36], and various poly(thiophenes) [37].

14.2 POLY(ACETYLENE)

14.2.1 Structure and synthesis

For the synthesis of poly(acetylene) (PAZ) about 25 different metal containing initiator systems have been described in the literature [9]. The majority of the studies were based on PAZ that was obtained by the so-called "Shirikawa technique" [14, 15]. Here acetylene gas is contacted with the surface of a non-agitated, concentrated solution of a Ziegler type catalyst, e.g. $Ti(n-C_4H_9)_4/Al(C_2H_5)_3$. Lustrous, silvery, polycrystalline, thin layers of nonhomogeneous networks of low densities are immediately formed; these are sometimes, but incorrectly, described as "films". Cis-PAZ is formed at a polymerization temperature of $-78°C$, at temperatures above $100°C$, purely trans-PAZ is obtained [14, 15].

Another frequently applied, alkyl-free initiator system described by Green and Luttinger is based on cobalt or nickel salts in alcoholic solution containing $NaBH_4$ [16, 17]. Upon polymerization of acetylene on the surface of inert carriers between -80 and $-30°C$, crystalline cis-PAZ is obtained in the form of thin layers. They exhibit the same morphology as the PAZ "films" obtained by the procedure of Shirikawa. The degree of polymerization is nearly independent of the nature, concentration and conversion of the metal salt, and is only slightly affected by the polymerization temperature.

The poly(acetylenes) tend to undergo spontaneous crosslinking reactions which involve Diels-Alder type cycloadditions and additions under participation of catalyst fragments (from Ziegler catalysts) [7]:

(14-1)

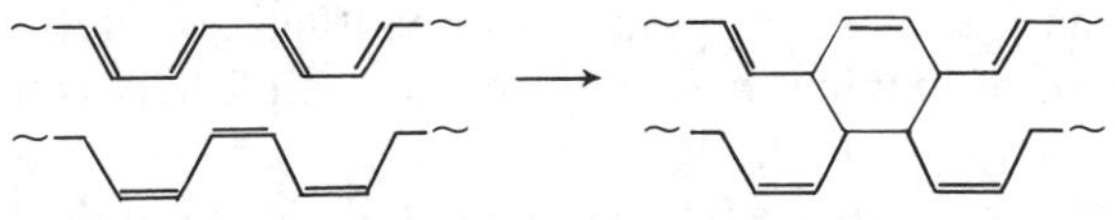

(14-2)

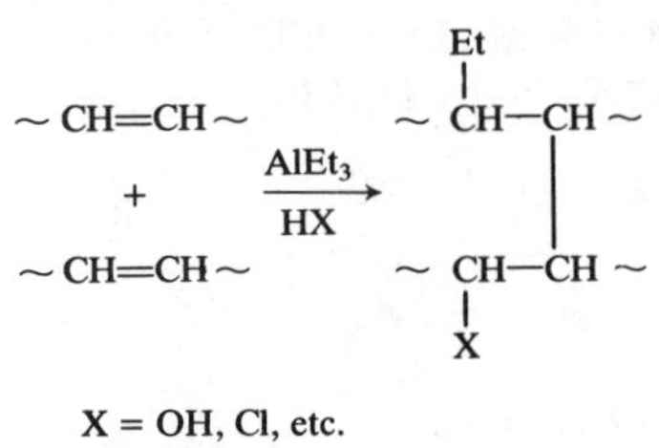

$X = OH, Cl$, etc.

14.2.2 Doping and properties

PAZ is nonfusible and insoluble in all solvents. The cis isomer is converted to the thermodynamically more stable trans isomer; the conversion is slow at 0°C, but fast and irreversible above 100°C. Conversion is also achieved by chemical or electrochemical doping and "undoping" (compensation) at room temperature. Electron microscopy studies show that the as-formed film consists of randomly oriented fibrils of 5–10 nm thickness and average diameter to 200 nm. All poly(acetylenes) are very sensitive to oxygen. Auto-oxidation at 30°C gives products with oxygen content from 4.9 to 22.4% depending on PAZ type. The maximum conductivity of complexed PAZ is only attained when the oxygen content is below 5% [18].

PAZ is an insulator with room temperature conductivities of about 10^{-9} S/cm for the trans isomer and 10^{-5} S/cm for the cis isomer. p-Type doping involves oxidation of PAZ; it occurs rapidly when either cis or trans film is exposed to a variety of reactants such as Br_2, J_2, AsF_5, H_2SO_4, etc. It may also be accomplished conveniently by electrochemical methods in which anions such as $(ClO_4)^-$, $(PF_6)^-$, etc., are introduced. Volatile compounds are reacted with PAZ under vacuum or in the gas phase for 1–2 days to give golden, lustrous, flexible films [10]. Non-volatile doping agents are dissolved in inert solvents, e.g., $FeCl_3$ in nitromethane, or $AgClO_4$ in toluene. Upon dipping the PAZ film into these solutions, doping is accomplished within minutes. n-Type doping, which involves reduction of PAZ, occurs rapidly when the film is immersed in a solution of lithium, sodium, or potassium naphthalide in tetrahydrofurane solution. n-Doping may also be accomplished conveniently by electrochemical methods.

Dopant concentrations as low as 0.01–0.05 mole-% cause a drastic increase of conductivity, the maximum is reached at dopant concentrations of 0.05–0.25 mole-% (see Figure 14-1) [19]. The electrical conductivity of doped PAZ and other polymers shows only a slight dependence on temperature (Figure 14-2) [6].

Doping can be formulated as redox reaction, e.g.:

(14-3)

$$2 (CH{=}CH)_n + 3\,AsF_5 \longrightarrow 2 (CH{=}CH)_n^+ + 2\,AsF_6^- + AsF_3$$

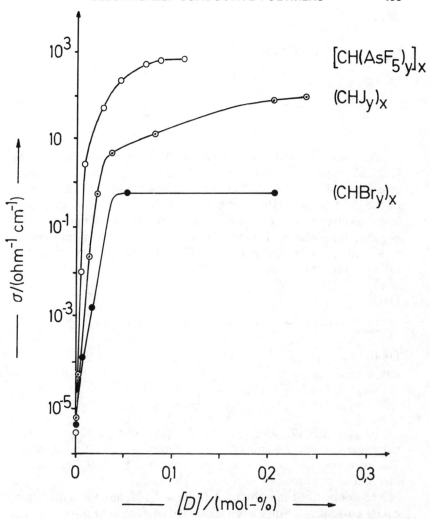

Figure 14-1 Change of "specific" electrical conductivity of poly(acetylene) $(CH)_x$ with type and concentration of doping agent [19].

Here AsF_5 serves as electron acceptor, whereas sodium naphthalide acts as electron donator:

(14-4)

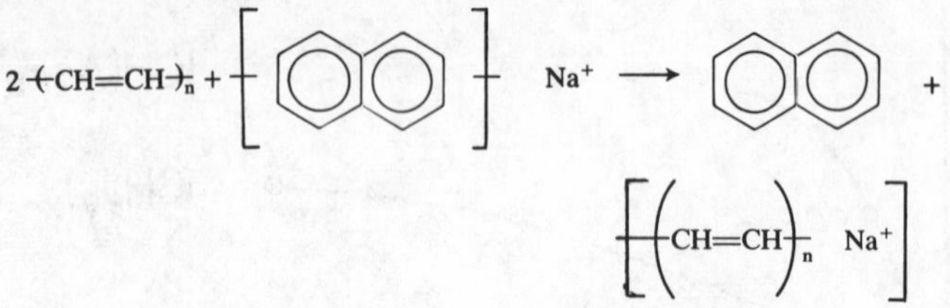

Much more elegant and more practical for technical applications is the electrochemical doping of PAZ in the presence of a suitable electrolyte. Two PAZ films are immersed in an electrolyte solution of 0.5 M tetra-n-butylammonium perchlorate dissolved in propylenecarbonate or CH_2Cl_2, and are then attached to the positive and negative terminals of a power source. Within 30 to 60 min and at 9 V and 40 mA, the polymer undergoes anodic oxidation and cathodic reduction, respectively [20]:

(14-5)

$$(CH)_x + 0{,}06x\ ClO_4^- \longrightarrow \left[(CH^{+0,06})(ClO_4)^{-0,06}\right]_x + 0{,}06\ x\ e^-$$

(14-6)

$$(CH)_x + 0{,}06x\ (n-Bu_4N)^+ + 0{,}06\ x\ e^- \longrightarrow \left[(n-Bu_4N)^{+0,06}(CH^{-0,06})\right]_x$$

The doped PAZ becomes electrically conductive. The conductivity increases by a factor of 10^{12}, i.e. from 10^{-8} to 10^2-10^3 S/cm. The reactions can be reversed and permit the electrolytic cell to function as galvanic cell or battery. The initial open circuit voltage of a 6% doped, 0.1 mm thick PAZ film is 2.5 V, and the initial short circuit current is 22 mA at an electrode surface area of only 1 cm^2 [20]. Other cells of this type employ 1 M $LiClO_4$ or $(Bu_4N)^+(PF_6)^-$, dissolved in THF or propylene carbonate, as electrolytes. They have an open circuit voltage of ca. 2.4 V for 6% doping and have a short circuit discharge current of ca. 100 mA per 1 cm^2 of PAZ. This type of p-doped PAZ cathode + n-doped PAZ anode would represent the ideal type of an all-polymer cell, but to date their shelf life has not been as good as the other three chief types of batteries employing PAZ electrodes [20a].

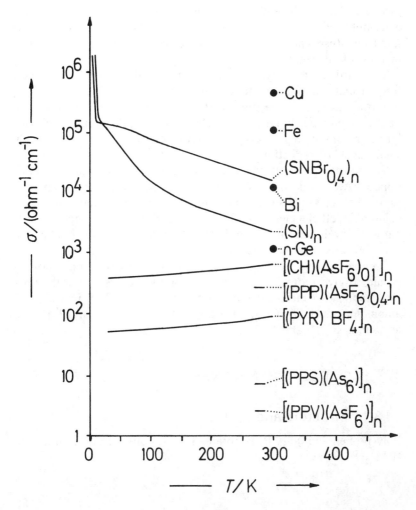

Figure 14-2 Temperature dependence of "specific" electrical conductivity of various complexed ("doped") polymers as compared to copper (Cu), iron (Fe), bismuth (Bi) and n-type germanium (n-Ge). $(SN)_n$ = poly(sulfazene), (CH) = poly(acetylene), PPP = poly(p-phenylene), PYR = poly(pyrrole), PPS = poly(phenylenesulfide), PPV = poly(p-phenylenevinylene) [6].

The second chief type consists of p-doped PAZ cathode and Li metal anode, immersed in an electrolyte of 1.0 M $LiClO_4$ in propylene carbonate. The electrochemical characteristics of this cell are extremely sensitive to the method of cell construction, presence of impurities (especially oxygen), relative ratio of electrolyte to PAZ, method of charging, etc.

The third chief cell type is composed of neutral PAZ cathode and lithium metal anode, immersed in an electrolyte of 1.0 M LiClO₄ in tetrahydrofuran. The present studies indicate that cells involving neutral and/or partly reduced PAZ have excellent stability and exhibit interestingly large energy and power densities even at relatively small levels of reduction of the PAZ.

The fourth chief cell type consists of neutral PAZ cathode and n-doped PAZ anode in an electrolyte of 1 M LiClO₄ in tetrahydrofuran. This cell shows excellent stability. It is fully rechargeable with coulombic efficiencies >99%. It is the first stable, rechargeable battery developed in which both the cathode and anode active materials are organic polymers [20a].

Compared to gas or liquid phase doping, electrochemical doping is said to give more homogeneously complexed systems with superior stability to moisture but not to oxygen [9]. Well investigated anions include BF_4^-, ClO_4^-, J^-, SbF_6^-, $SbCl_6^-$, AsF_6^-, and PF_6^- and as cations R_4N^+ and Li^+.

14.3   POLY(PHENYLENES)

14.3.1   Structure and synthesis

Poly(p-phenylene), PPP, is obtained by 1,4-dipolar addition of I to II [21]:

(14-7)

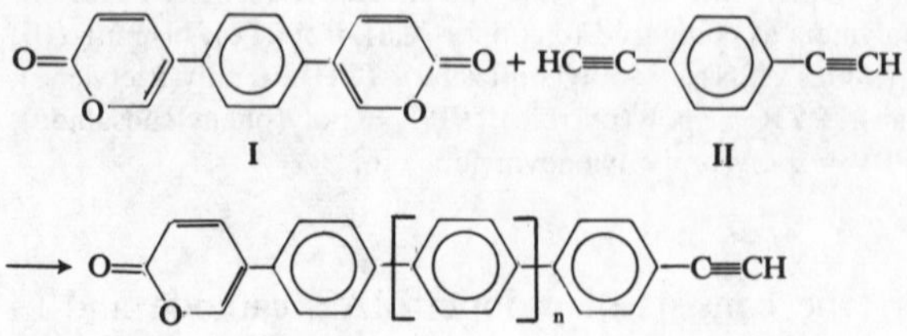

In another synthesis, cyclohexadiene-1,3 is polymerized by means of the catalyst system $Al(isoBu)_3/TiCl_4$ to poly(cyclohexadiene) which is then dehydrogenated, e.g. by chloranil, to PPP [22]:

(14-8)

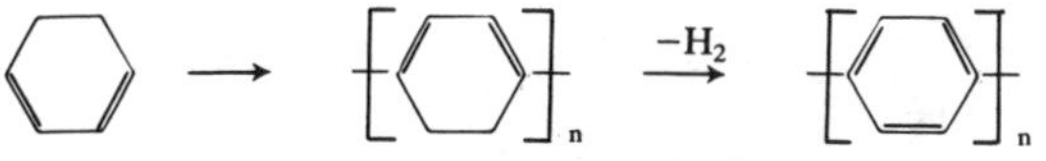

The synthesis which has been investigated best follows the procedure given by Kovacic; it involves oxidative coupling of benzene with $AlCl_3/CuCl_2$ as catalyst. This procedure has been recognized as a general principle applicable to the synthesis of polyaromatics, polyheterocyclics and polymetal complexes; see the literature reviewed in [9].

Allied Corp. has found a new way to fabricate conducting polymers. The one-step reaction uses films made from crystalline p-phenylene oligomers, such as biphenyl, terphenyl or quaterphenyl. They will complex with oxidizing dopants such as AsF_5 in the solid state, and then are polymerized into highly conducting doped poly-(p-phenylene) [23].

The PPP materials prepared to date are all nonfusible and insoluble. In November 1980, Allied announced the first melt- and solution processable conductor, doped poly(m-phenylene). The route of synthesis was not disclosed. It was stated, however, that the raw material is not commercially available [24].

PPP prepared by oxidative coupling of benzene and subsequently purified still contains 1−4% chlorine, <5% oxygen, 0.1−0.2% copper, and 0.3−0.6% aluminum [25]. The structural units are mainly linked in para position, but branching and crosslinks are likely to occur as well.

14.3.2   Doping and properties

Whereas PAZ can be obtained in the form of flexible films, the powdery PPP requires sinter techniques for shaping. Although PPP films show significantly less flexibility and strength than PAZ films, they withstand heating in air to nearly 400°C (see also chapter 2.10).

Like PAZ, PPP becomes electrically conductive upon doping,

except with bromine or iodine as dopants. Doped PPP exhibits somewhat lower conductivity than doped PAZ (Table 14-3).

TABLE 14-3
Electrical conductivities of doped PPP and PAZ [26].

Dopant	Doping method	Electrical conductivity in S/cm	
		PPP	PAZ
Acceptors			
J_2	Gas phase	$<10^{-5}$	500
AsF_5	Gas phase	500	1200
HSO_3F^-	Liquid phase	35	400
BF_4^-	electrochemical	10	100
PF_6^-	electrochemical	45	120
Donors			
K^+	potassium naphthalide in THF	20	50
Li^+	lithium naphthalide in THF	5	200

Like PAZ, PPP can be used as electrode in voltaic cells. A battery, composed of a PPP cathode complexed with acceptors, and a Li metal or Li/Al alloy anode, immersed in $LiAsF_6$ or $LiPF_6$ electrolyte dissolved in propylene carbonate, has a higher voltage than those obtainable from PAZ batteries (Table 14-4). With PPP electrodes current densities of up to 50 mA/cm^2 can be attained. When the Li anode is replaced by an alkali metal complexed PPP, the obtainable voltage is lowered; however, it is still higher than with the all-polymer PAZ batteries.

TABLE 14-4
Batteries from doped PPP and PAZ [26].

Battery Pair	Electrical Potential Difference in V	Theoretical Energy Density in W × h/kg
$Li/[(C_6H_4)^{+0.4} (AsF_6)^{-0.4}]_n$	4.4	285
$Li/[(C_6H_4)^{+0.4} (PF_6)^{-0.4}]_n$	4.4	320
$[(C_6H_4)^{-0.4} (K)^{+0.4}]_n/[(C_6H_4)^{+0.4} (AsF_6)^{-0.4}]_n$	3.3	150
$Li/[(CH)^{+0.06} (ClO_4)^{-0.06}]_n$	3.7	290
$Li/[(CH)^{+0.06} (PF_4)^{-0.06}]_n$	3.7	250
$[(CH)^{-0.06} (Li)^{+0.06}]_n/[(CH^{+0.06} (ClO_4)^{-0.06}]_n$	2.5	120

The redox reactions occurring at the anode and cathode can be formulated as follows [26]:

(14-9)

Anode

$$\left[(C_6H_4)^{-0,4}Li^{+0,4}\right] \underset{\substack{\text{charge} \\ +0,4\,ne^-}}{\overset{\substack{-0,4\,ne^- \\ \text{discharge}}}{\rightleftharpoons}} (C_6H_4)_n + 0,4\,n\,Li^+$$

(14-10)

Cathode

$$\left[(C_6H_4)^{+0,4}(AsF_6)^{-0,4}\right] \underset{\substack{\text{charge} \\ -0,4\,ne^-}}{\overset{\substack{+0,4\,ne^- \\ \text{discharge}}}{\rightleftharpoons}} (C_6H_4)_n + 0,4\,n\,AsF_6^-$$

In practice, the shelf life and coulombic effiency of the battery employing PPP at both electrodes have been low because a solvent has not been found which is stable to the alkali-metal doped anode and which has a voltage "window" wide enough to encompass the cathode as well.

14.4 POLY(P-PHENYLENE SULFIDE)

Poly(p-phenylene sulfide), PPS, has the advantages over PAZ and PPP of being commercially available and offering melt and solution processability. It is not susceptible to autoxidation. Its preparation and properties have been described and summarized [27].

Both Allied and IBM are investigating actively the doping of PPS with AsF_5 [24]. The reaction of PPS with AsF_5 is slow compared to the same reaction with PAZ. At a pressure of 250 torr, the reaction takes four days, during which the transparent PPS film becomes blue-green and finally blue-black. The maximum conductivity is attained at a 1:1 molar ratio of AsF_5 to polymer structural unit. The dopant molecules intercalate themselves between the polymer chains. It has been suggested that the doping may cause the formation of radical cations [28]:

(14-11)

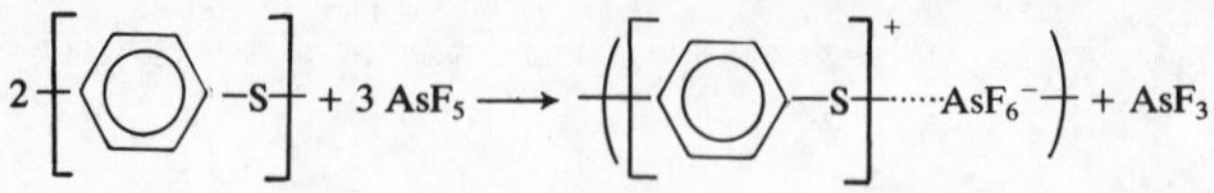

$$2\left[\underset{}{\bigcirc}-S\right] + 3\,AsF_5 \longrightarrow \left(\left[\underset{}{\bigcirc}-S\right]^{+}\cdots AsF_6^{-}\right) + AsF_3$$

Upon doping, the conductivity increases from $<10^{16}$ to about 1 S/cm, however, it is still a thousandfold lower than that of doped PAZ and PPP. Pretreatment of PPP with AsF_3 prior to doping with AsF_5 allows the reduction of time needed for doping from 4 days to 90 min. The maximum conductivity hereby obtained is 0.3–0.6 S/cm [29].

As with most other conducting polymers, the conductivity of PPS deteriorates when it is exposed to air. Another problem, also common in such materials, is that the doping process makes flexible PPS films turn brittle [24].

14.5 POLY(PYRROLE)

14.5.1 Structure and synthesis

Pyrrole is electrochemically polymerized to poly(pyrrole), PYR, where doping is simultaneously accomplished [30]. A solution of 0.06 M pyrrole and 0.1 M Et_4NBF_4 in acetonitrile containing 1% water is applied as electrolyte. Upon electrolysis at a current density of $0.5–1.5$ mA/cm^2, a blue-black film of an insoluble polymer is deposited on the anode. No film formation is observed in anhydrous or oxygen-free systems.

PYR films can also be obtained under anhydrous conditions, when an electrolytic cell with silver cathode, platinum or tin oxide anode, and carefully dried and purified $AgClO_4$ electrolyte dissolved in acetonitrile is employed [31].

The electrochemical procedure offers a fast and technically feasible synthesis route. At 0.5 V voltage and 1 mA current, films of 400–500 nm thickness are formed within 30 min [31]. BF_4^- doped PYR films can be electrochemically grown directly on n-type silicon semiconductors. They have been investigated as a potential new type of solar cell and photoanode [32].

The idealized structure of doped PYR has been suggested as

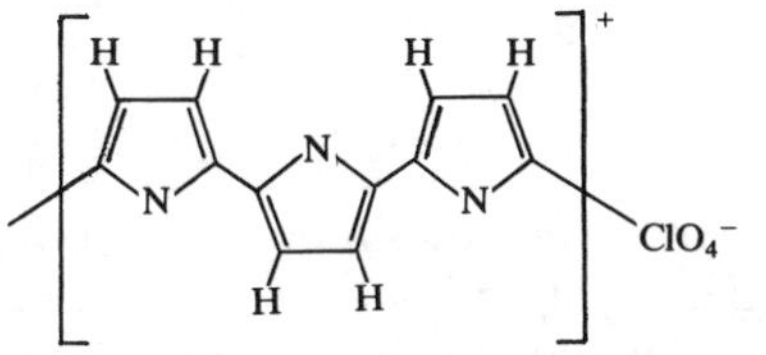

Relative to the structure given above, the films appear to be hydrogen rich suggesting that some of the pyrrole rings may have been hydrogenated. Isomerisation, branching and crosslinking are also likely to occur. ESR studies indicate the presence of unpaired electrons in PYR.

14.5.2   Properties

Electrochemically prepared thick PYR films are blue-black or copper-bronze colored, have a density of 1.48 g/cm^3, are amorphous by X-ray analysis, and are insoluble in organic solvents.

The conductivity of doped PYR films is in the range of 50−100 S/cm. Contrary to doped PAZ, PPP and PPS polymers, doped PYR retains its conductivity in air for a prolonged period of time; it starts to deteriorate only after several months [31]. PYR films can be heated to 200°C without an appreciable loss of electric properties.

The oxidized and ClO_4^- doped PYR films can be electrochemically reduced in n-Bu_4NClO_4 electrolyte dissolved in acetonitrile. The resulting neutral PYR is ClO_4^--free. The yellow-green films are very sensitive to oxygen and turn black within 15 min upon exposure to air. The oxygen uptake is 0.07 moles/mole of pyrrole after exposure to dry oxygen for about 12 hrs. The increase in oxygen uptake is accompanied by an increase in the conductivity of the neutral PYR from $<10^{-5}$ S/cm to around 10^{-2} S/cm. It has been shown that neutral PYR films can be oxidized by various metal solutions including Ag^+, Cu^{2+} and Fe^{3+}, as well as by bromine, iodine and $FeCl_3$ vapors to give conductivities in the range of 1−100 S/cm [31]. So far, high conductivities equivalent to those of doped PAZ have not been achieved with doped PYR.

14.6 POLY(1,6-HEPTADIYNE)

14.6.1 Structure and synthesis

Cyclopolymerization of 1,6-heptadiyne from the vapor state onto the surface of homogeneous Ziegler catalysts yields free-standing films of an insoluble polymer [33], for which the following structure has been suggested [13]:

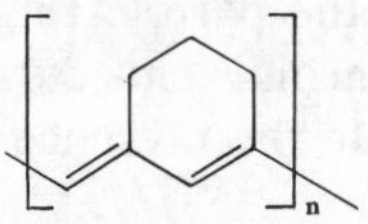

The polymerization with heterogeneous Ziegler catalysts yields soluble polymers.

14.6.2 Doping and properties

Poly(1,6-heptadiyne) films have a golden-green color with metallic appearance and density over 1 g/cm^3; they are amorphous by X-ray analysis. Scanning electron microscopy reveals no fibrillar open structure as observed with PAZ. Again in contrast to PAZ, the films do not stretch to any significant extent. They do possess some flexibility, however, dependent upon thickness. No glass transition temperature is observed between -40 and 330°C. DSC under N_2 reveals a broad exotherm at 100–120°C that is tentatively assigned to a cis-trans isomerization of the double bond, as seen at 145°C in PAZ. Poly(1,6-heptadiyne) is a substituted poly(acetylene) and thus shows many similarities. The films are found to be more sensitive to air or oxygen than is PAZ. Long exposure to air or oxygen leads to severe embrittlement and a yellow-orange coloration.

Films of poly(1,6-heptadiyne) have been doped with a number of dopants including iodine, AsF_5, and oxygen. Exposure to iodine gives a maximum conductivity of about 0.1 S/cm which is far below the conductivities of doped PAZ types. Doping with oxygen is followed by oxidation, which results in conductivity loss.

14.7 POTENTIAL APPLICATIONS OF ELECTRICALLY CONDUCTIVE POLYMERS

Polymers which combine low density, high and long term consistent conductivity, easy processability and free choice of shaping would indeed represent an entirely new class of materials. The new syn-metals already meet some of these requirements, but they are still far from complying with all of them.

A marketable polymer battery may not become a reality for at least five to ten years [11, 34]. Compared with conventional lead-acid batteries, polymer batteries have only one third the volume and one tenth the weight; and as much as 5/6 of their weight could be utilized as electro-active carrier due to the much greater power density. In other words: a polymer battery the same size as a lead battery could boost the power output 10-fold. Moreover, the poly-mer batteries' energy can be drawn off 10 times faster than is possible with lead batteries. The synthesis of the polymers is rela-tively straightforward and energy saving, in particular for the convenient electrochemical procedures. The materials presumably are non-toxic, and with nearly ash-free combustion the systems could be disposed off without causing pollution. The first commercial poly-mer batteries presumably will be rather expensive, therefore, first applications are likely to be in the aircraft and aerospace industry, in communication satellites and in the medical field, e.g. for heart pacemakers and hearing aids, as well as in watches. At an even later period, applications in the automotive industry are expected. Since conductive polymers are prepared in flexible sheets, batteries made from this material might be fabricated in a variety of novel shapes. Thus, an automobile's starter battery could be hidden away in a side panel. They seem ideal for powering electric vehicles. Polymer batteries are foreseen to serve as both power source and construc-tion material in the form of laminates.

Other potential applications include heating elements for wall coverings, floors, blankets, or clothing, antistatic coatings, and incineration-disposable electrostatic precipitator screens in air pollu-tion control. The most significant market in terms of volume for conductive polymers is likely to be in wire and cable applications. Polymeric electrical conductors have a potential weight saving of approximately 25% in comparison with aluminum and perhaps over

300% in comparison with copper. Also, conductive polymers are of interest in that they can offer a much more uniform level of electrical conductivity for power cable sheathing than can the currently used black loaded poly(ethylene) compounds. An additional large market could be in electric power load leveling stations.

The most immediate need for electrically conductive polymers is in electronic housings so that Federal Communications Commission (FCC) regulations regarding electromagnetic and radiofrequency interferences can be met [34]. However, conductive polymers would have to compete with other materials, e.g. metal foils, plated plastics, and conductive metal or graphite containing coatings.

Great expectations are placed in the development of conductive polymers for applications in photoelectrochemical cells, in particular for solar cells. For example, electrochemically grown thin films of highly conductive poly(pyrrole) have been shown to substantially prevent oxidation/corrosion of n-Si photoanodes and to allow higher effiency in conversion of light into electrically energy [32].

References

[1] L. Alcacer (Ed.), The Physics and Chemistry of Low Dimensional Solids, D. Reidel Publ. Co., Boston 1980.

[2] A. J. Epstein and E. M. Conwell, Molecular Crystals and Liquid Crystals, Gordon & Breach, New York 1981.

[3] R. B. Seymour (Ed.), Conductive Polymers, Plenum Press, New York 1981.

[4] M. M. Labes, P. Love and L. F. Nichols, Chem. Reviews **79** (1979) 1.

[5] H. Block, Adv. Polym. Sci. **33** (1979) 93.

[6] G. B. Street and T. C. Clarke, IBM J. R & D **25** (1981) 51.

[7] G. Wegner, Angew. Chem. **93** (1981) 352.

[8] K. Seeger, Angew. Makromol. Chem. **109/110** (1982) 227.

[9] H. Naarmann, Angew. Makromol. Chem. **109/110** (1982) 295.

[10] A. G. MacDiarmid and A. J. Heeger, Synth. Metals **1** (1980) 101.

[11] Anon., Chem. Engng. News (Sept. 28, 1981) 12; ditto, (Oct. 12, 1981) 34; ditto, Modern Plast. Internat. **11** (November 1981) 14, 16.

[12] R. H. Baughman (Allied Chemical, Corporate Technology, Morristown, N.J.), personal communication, May 29, 1981.

[13] H. W. Gibson (Xerox Corp., Rochester, N.Y.), manuscript and personal communication, December 10, 1980.

[14] H. Shirakawa and S. Ikeda, Polym. J. **2** (1971) 231.

[15] T. Ito, H. Shirakawa and S. Ikeda, J. Polym. Sci.-Polym. Chem. Ed. **12** (1974) 11.

[16] L. B. Luttinger, Chem. Ind. (London) (1960) 1135; J. Org. Chem. **27** (1962) 1591.

[17] M. L. H. Green, Chem. Ind. (London) (1960) 1136.

[18] H. Haberkorn et al., Synth. Metals, in press, cited in [9].

[19] C. K. Chiang, Y. W. Park, A. J. Heeger, H. Shirakawa, E. J. Luis and A. G. MacDiarmid, Phys. Rev. Lett. **39** (1977) 1098.

[20] A. G. MacDiarmid, A. J. Heeger et al., Org. Coat. Plast. Chem. **44** (1980) 372; ditto, J. Electrochem. Soc. **128** (1980) 1651; ditto, ACS Polym. Prepr. **23**/1 (1982) 241.

[20a] A. G. MacDiarmid, M. Aldissi, R. Kaner, M. Maxfield and R. J. Mammone, ACS Org. Coat. Plast. Chem. Prepr. **48** (1983) 531.

[21] J. K. Stille, Makromol. Chem. **154** (1972) 49.

[22] C. S. Marvel et al., J. Am. Chem. Soc. **81** (1959) 448.

[23] L. W. Shacklette, H. Eckhardt, R. R. Chance, G. G. Miller, D. M. Ivory and R. M. Baughman, J. Chem. Phys. **73** (1980) 4098; Chem. Engng. News (April 7, 1980) 42.

[24] cited in Chem. Engng. News (March 31, 1980) 36.

[25] P. Kovacic et al., J. Macromol. Sci. C **5** (1971) 295; ACS Polym. Prepr. **21**/2 (1980) 259.

[26] R. L. Elsenbaumer, L. W. Shacklette, J. M. Sowa, R. R. Chance, D. M. Ivory, G. G. Miller, and R. H. Baughman, ACS Polym. Prepr. **23**/1 (1982) 132.

[27] H.-G. Elias, Neue polymere Werkstoffe 1969−1974, C. Hanser, München 1975; New Commercial Polymers 1969−1975, Gordon and Breach, New York 1977.

[28] J. F. Rabolt, T. C. Clark, K. K. Kanazawa, J. R. Reynolds, and G. B. Street, J. Chem. Soc. Chem. Commun. (1980) 347.

[29] J. E. Frommer, R. L. Elsenbaumer, H. Eckhardt, L. W. Shacklette, and R. R. Chance, ACS Polym. Prepr. **23**/1 (1982) 107.

[30] A. F. Diaz et al., J. Chem. Soc. Chem. Commun. (1979) 635; ditto (1980) 397; Synth. Metals **1** (1979/1980) 329.

[31] G. B. Street, T. C. Clarke, R. Krounbi, P. Pfluger, J. F. Rabolt, and R. H. Geiss, ACS Polym. Prepr. **23**/1 (1982) 117.

[32] T. Skotheim, ACS Polym. Prepr. **23**/1 (1982) 136.

[33] H. W. Gibson, F. C. Bailey, A. J. Epstein, H. Rommelmann, and J. M. Pochan, J. Chem. Soc. Chem. Commun. (1980) 426; ACS Org. Coat. Plast. **42** (1980) 603; ditto **43** (1980) 860; ditto **44** (1981) 462.

[34] J. R. Ellis and R. S. Schotland, ACS Polym. Prepr. **23**/1 (1982) 134.

[35] Sandia National Laboratories, press release 1984.

[36] Anon., Chem. Engng. News (10 September 1984) 39.

[37] G. Tourillon and F. Garnier, J. Polym. Sci.-Polym. Phys. Ed. **22**, 33 (1984).

15 Silicon containing polymers

15.1 ORGANO-MODIFIED POLYSILOXANE ELASTOMERS

15.1.1 Structure and synthesis

Organo-modified polysiloxane elastomers are castable, thermoplastic-reinforced RTV (Room Temperature Vulcanizing) elastomers, where thermoplasts are at least partially grafted onto the silicone backbone. They were first developed by the Stauffer-Wacker-Silicones Corp. and have been commercially available since 1976 under the tradenames ®Silgan (Stauffer) and ®m-Polymer (Wacker) [1–5]. The grades supplied include ®Silgan J-500, J-501, H-621 and H-622, and ®m-Polymere 355 Z, 435 Z, 444 Z, 455 A, B, Z, SLM 91812, SLM 91813 and SLM 91910 [6, 7]. For a review of RTV's, see [8].

Patents covering the preparation of organo-modified polysiloxanes are listed in [3, 5]. To prepare these polymers, monomers such as styrene, acrylonitrile, butylacrylate or vinylacetate are polymerized with free radicals in the presence of α,ω-dihydroxy-poly-(dimethylsiloxane) (PDMS). The initiators used must be capable of abstracting hydrogen radicals from the methyl groups of PDMS. Suitable initiators are e.g. 1,1-di-tert.-butylperoxy-3,5,5-trimethyl-cyclohexane or tert.-butylperpivalate [3]. Starting material is a liquid PDMS with terminal hydroxyl groups and an average polymerization degree of 220–230. As the in situ polymerization proceeds, the solution yields a dispersion of thermoplast particles finely dispersed in the liquid PDMS phase. The dispersion is stabilized by grafted copolymers which are formed by free radical reactions between monomers and methyl groups of PDMS that act like nonionic tensides [3]. Under appropriate conditions of agitation, the precipitating thermoplast particles are obtained in rod-like shapes [1], which essentially act as reinforcing material.

Two-component systems are prepared by blending with esters of silicic acid; curing is accomplished with organotin catalysts at room temperature over 10 to 20 h. One-component systems are obtained by reactions of aminosilanes, acetoxy- or oxim−silanes, which are moisture-curable within one to ten days depending on the thickness of the layer.

15.1.2   Properties and applications

Vulcanizate properties depend on thermoplast composition as well as on the ratio of thermoplast/silicone-phases. For example, m-Polymer grades 455 Z and 435 Z differ with regard to the styrene/ butylacrylate ratio, containing 60 and 40 mole-%, respectively, of styrene [4] (Table 15-1). Vulcanizates exhibit greater Shore hardnesses than conventional RTV silicone elastomers and have superior resistance to tear propagation.

A trade-off must be made with thermal stability of vulcanizates due to the thermoplastic nature of one phase. Long-term exposure to temperatures above 120°C causes degradation and reduction of tear resistance, tensile strength and elongation at break. Vulcanizates based on styrene/butylacrylate moieties demonstrate good dielectric properties and are resistant to acids and bases. Organic solvents, in particular acetone, xylene, and trichloroethylene, cause various degrees of swelling.

Applications are alkali-resistant, self-releasing molds for precast concrete parts [9], fabrication of molds, casting and dip-coating materials in the electric and electronic fields, sprayable thixotropic compounds for corrosion-resistant coatings, non-dripping pourable materials for restoration work, low-viscosity one-component systems for formed-in-place gaskets. The casting compounds require no release agent. Properties in comparison with other casting resins are summarized in Table 15-2 [4].

References

[1] J. C. Getson and R. N. Lewis, Paper presented at the ACS Rubber Division Meeting, New Orleans, October 1975; Abstract published in Rubber Chem. Technol. **49** (1976) 402.
[2] B. Deubzer and F. H. Kreuzer, Kunststoffe **66**/10 (1976) 629.
[3] K. Marquardt, F. H. Kreuzer and M. Wick, Angew. Makromol Chem. **58/59** (1977) 243.
[4] K. Marquardt and W. Keil, Gummi, Asbest, Kunststoffe **30**/2 (1977) 76.
[5] F. H. Kreuzer, M. Wick, W. Keil and K. Marquardt, Kautschuk + Gummi, Kunststoffe **33**/8 (1980) 603.

TABLE 15-1
Properties of m-Polymers [5].

Physical property	Physical unit	Property Values for			
		355-Z	455-Z	444-Z	435-Z
Viscosity, (Brookfield at 23°C)	mPa × s	30 000 −60 000	8 000 −12 000	10 000 −20 000	9 000 −15 000
Properties of Vulcanizates[1]					
Density	g/cm^3	1.02	1.02	1.02	1.02
Hardness (Shore A)	—	85	75	55	23
Tensile strength	MPa	9	7	7	4
Elongation at break	%	300	300	300	350
Split tear strength	N/mm	70	30	19	5
Tracking resistance	—	KC >600	KC >600	KA 3C	KA 3C
Dielectric strength	kV/mm	23	27	27	26
Surface resistivity	Ohm	3×10^{13}	10^{13}	3×10^{13}	4×10^{13}
Volume resistivity	Ohm × cm	1.5×10^{15}	7×10^{14}	6×10^{14}	3×10^{14}
Relative permittivity (1 kHz)	—	2.7	2.7	2.8	3.1
Dissipation factor (1 kHz)	—	5×10^{-3}	6×10^{-3}	1.7×10^{-2}	3×10^{-2}

1) Vulcanization with 5% hardener T 935 and 9 days storage under standard clima.

TABLE 15-2

Properties of thermoplastics reinforced silicone rubbers and other materials for electric and electronic components [4, 6].

Physical Property	Physical Unit	Silgan H-622	m-Polymer	RTV Silicone rubber	Epoxides flexible	Epoxides rigid	Poly-urethane
				Property value			
Density	g/cm^3	1.0	1.0	1.2	1.1	1.6	1.1
Hardness (Shore A)	—	90	75	40	98	—	80
Tensile strength	MPa	10.3	7.5	2	17	41	24
Elongation at break	%	300	300	100	35	—	500
Split tear strength	N/mm	—	35	3	—	—	44
Continuous service temperature	°C	125	120	200	105	125	100
Brittleness temperature	°C	−90	—	−65	—	—	−60
Dielectric strength	kV/mm	24	27	30	13	14	16
Relative permittivity (100 Hz)	—	3.0	2.9	3.0	4.6	4.4	5.5
Dissipation factor (100 Hz)	—	0.008	0.001	0.005	0.02	0.03	0.05
Volume resistivity	ohm × cm	10^{15}	10^{15}	10^{15}	10^{13}	10^{13}	10^{11}
Exothermic vulcanization	—	no	no	no	yes	yes	yes

[6] ®Silgan Elastomers, SWS Silicones Corp., Technical Information, 1976–1978.

[7] ®Wacker m-Polymere, Wacker-Chemie GmbH, September 1978.

[8] E. L. Warrick, O. R. Pierce, K. E. Polmanteer, J. C. Saam, Silicone Elastomer Developments 1967–1977, Rubber Chem. Techn. **52** (1979) 438.

[9] W. Keil, K. Kollmannsberger, K. Marquardt, M. Wick and M. Gaschler, Betonwerk + Fertigteil-Technik **10** (1977) 512.

15.2   LIQUID SILICONE ELASTOMERS

15.2.1   Structure and synthesis

Liquid silicone elastomers are not entirely new products; they have been extensively described in a number of reviews [1–4]. New however, are liquid two-component systems for injection molding processing, which were introduced by Dow Corning Corp. in 1978 under the tradename ®Silastic LSR (standing for Liquid Silicone Rubber) [5–7].

The commercially available Silastic LSR grades are GP 590, QR-9591, Q3-9595 and X3-6596. They are compounds based on vinyl-functional dimethylsiloxane polymers, containing various reinforcements or fillers, catalyst, and a crosslinking agent for the addition reaction [8].

Injection moldable, liquid silicones are also supplied by the General Electric Silicones Co. under the designation ®LIM-Systems. In addition to the standard series, an electrically conductive material is also offered. Wacker has introduced a similar product [10]. In 1982 Bayer A.G. commercialized three materials, tradenamed ®Silopren LSR, with Shore A hardnesses of 35, 50 and 60. Of more recent origin are single component systems. New developments include Rhône-Poulenc's ®Rhodorsil PSE 31903 and 31904. Both use peroxide catalyzed crosslinking. GE Silicones have a self-lubricating material crosslinked by an addition reaction, designated ®LIM 1450 [10].

15.2.2   Properties and applications

Unlike the earlier silicone elastomers, the new materials require no working with rubber masticating equipment prior to fabrication. Silicone rubber liquids should not be confused with two-component RTVs. Liquid silicone rubbers cure only via heat. Blended liquid

TABLE 15-3

Comparison of LSR systems[1] with conventional, hot-vulcanizable silicone elastomers [6, 7].

Physical property	Physical Unit	Property value					
		GP 590	Q3-9591	Q3-9595	X3-6596	GP 30	GP 70
Density	g/cm³	1.22	1.55	1.12	1.35	—	—
Hardness (Shore A)	—	35	55	40	70	28	68
Tensile strength	MPa	5.5	3.8	6.9	5.2	5.2	7.1
Elongation at break	%	450	250	425	400	550	450
Tear strength	kN/m	14	14	35	26	—	—
Resilience	%	40	41	57	25	40	35
Compression Set							
after 22 h/177°C[2]	%	15	12	18	20	22	22
after 22 h/177°C[3]	%	10	10	12	20	—	—
Property values after 7 days at 265°C							
Hardness (Shore A)	—	54	64	51[4]	—	—	—
Tensile strength	MPa	3.9	4.5	3.1	—	—	—
Elongation at break	%	140	125	120	—	—	—
Property values after 60 days at 235°C							
Hardness (Shore A)	—	62	64	51	—	—	—
Tensile strength	MPa	5.0	5.0	4.1	—	—	—
Elongation at break	%	100	105	180	—	—	—

[1] LSR systems injection molded 30 s at 200°C; GP 30 and GP 70 vulcanized 10 min at 171°C.
[2] Annealed 4 h at 200°C.
[3] Annealed 8 h at 200°C.
[4] Values for LSR type Q3-9595 with 2% heat stabilizer.

silicone rubber components can be stored for 48 h at room temperature; shelflife of the individual components is approximately 12 months.

Injection molding of liquid silicone rubber is a completely automated operation [11]. Commercially available equipment includes pumps and meter-mixers to proportionally feed the components via static mixers to the injection molding machine. Injection pressures and clamping forces are lower than in processing of thermoplasts. Further, because of the need for less force, lighter and cheaper molds can be used; mold wear and mold maintenance are minimal, and deflashing of parts often is not needed. Curing requires only 15−30 s at mold temperatures of 175−200°C.

In addition to the excellent properties of the known silicone rubbers, LSR types exhibit better compression set values and enhanced flame resistance. Compression sets can be further improved by annealing, e.g. at 200°C for 4 to 8 h (Table 15-3).

The predominant application is mass production of small injection molded parts. For medical uses, LSR with properties of grade Q3-9591 is supplied; another grade is available for transparent articles. LSR types also can be extruded and vulcanized in hot air, e.g., for extrusion coating of wires, cables, and fabrics [9], however, fabrication of unsupported extruded profiles is not possible.

LSR types have distinct processing advantages over other elastomers and therefore can compete with other materials in spite of higher raw material costs. Liquid silicone elastomers cost about 6 US-$/lb versus 5 US-$/lb for conventional silicone rubbers, however, overall production costs for parts fabricated from liquid silicone rubber are significantly lower [8].

References

[1] M. G. Noble, Silicone Elastomers, in R. O. Babbit, The Vanderbilt Rubber Handbook, R. T. Vanderbilt Co. Inc., Norwalk, CT., 1978.

[2] W. Lynch, Handbook of Silicone Rubber Fabrication, Van Nostrand Reinhold Co., New York 1978.

[3] E. L. Warrick, O. R. Pierce, K. E. Polmanteer and J. C. Saam: Silicone Elastomer Developments 1967−1977, Rubber Chem. Technol. **52** (1979) 438.

[4] W. Hofmann, Kautschuk-Technologie, Gentner Verlag, Stuttgart 1980.

[5] J. L. Elias, M. T. Maxson and C. L. Lee, ACS Org. Coat. Plast. Chem. **39** (1979) 67.

[6] C. A. Romig and R. P. Sweet, Rubber World **184**/3 (June 1981) 28.

[7] R. Cush, Europ. Rubber Journal (December 1981) 19.
[8] J. Cooper, Modern Plast. Internat. **9** (February 9, 1979) 50.
[9] R. M. Fraleigh and G. P. Kehrer, Rubber World **183** (November 1980) 65.
[10] D. Brownhill, Modern Plast. Internat. **12** (November 1982) 41.
[11] Th. LaBue, Silicone, in Modern Plastics Encycl. 1980−1981, McGraw-Hill Inc., New York 1980, 110.

15.3   AQUEOUS RTV ELASTOMER DISPERSIONS

15.3.1   Structure and synthesis

In conventional two-component RTV (Room Temperature Vulcanizing) silicone elastomers, cure is accomplished either via vinyl groups by platinum-catalyzed addition, or via condensation reaction between silanol and alkoxy groups through elimination of alcohol. In one-component RTV elastomers, terminal acetoxy groups upon contact with the air's moisture eliminate acetic acid under formation of silanols, whhich subsequently undergo polycondensation to give siloxanes. Complete cure generally requires several days [1].

In 1980 for the first time, Dow Corning Corp. introduced aqueous RTV silicone elastomer dispersions, which yield elastic films when the water is evaporated at room temperature [2]. The experimental products are designated X3-5013, X3-5024 and X3-5025.

The binary systems are composed of colloidal silica or sodium silicate dispersed in an emulsion of poly(dimethylsiloxane) (PDMS) stabilized by anionic emulsifiers. Also, they contain small amounts of dialkyltincarboxylates [3−5]. The emulsion with 55% solids consists of PDMS with an average number molecular weight of $2.0 \cdot 10^5$ and is end-capped with silanols; particle size is about 0.3 μm. The dispersion is prepared by blending the PDMS emulsion with silica or sodium silicate.

Cure is catalyzed by organotin compounds, e.g. dioctyltindilaurate. The catalysts react in alkaline medium with water-soluble silicates:

(15-1)

$$(R'{-}COO)_2SnR_2 + OH^- \rightleftharpoons (R'{-}COO)\underset{\underset{\textstyle OH}{|}}{Sn}R_2 + RCOO^-$$

$$(R'{-}COO)\underset{\underset{\textstyle OH}{|}}{Sn}R_2 + Si(OH)_4 \rightleftharpoons (R'{-}COO)\underset{\underset{\textstyle OSi(OH)_3}{|}}{Sn}R_2 + H_2O$$

The formed silastannoxanes are more hydrophobic and thus migrate to the oil-water interface, where they condense with silanol groups present on the surface of the PDMS emulsion:

(15-2)

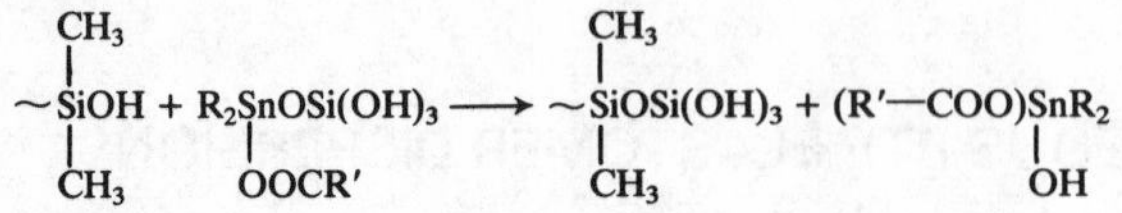

$$\underset{\underset{CH_3}{|}}{\overset{\overset{CH_3}{|}}{\sim SiOH}} + \underset{OOCR'}{\overset{}{R_2SnOSi(OH)_3}} \longrightarrow \underset{\underset{CH_3}{|}}{\overset{\overset{CH_3}{|}}{\sim SiOSi(OH)_3}} + (R'\!-\!COO)\underset{OH}{\overset{}{SnR_2}}$$

It is assumed that polymers with short, i.e. monomeric or dimeric, terminal silicate groups migrate into the interior of the PDMS emulsion where they cause crosslinking within the oil phase. Polymers with longer terminal silicate groups are solvated by water thereby causing steric and ionic stabilization of PDMS particles at the interface. "Vulcanization" is thus already terminated prior to the evaporation of water. The stabilizing effect is destroyed on evaporation of water, and silica and siloxane molecules come now in close contact with each other. The resulting hydrogen bonds are the last step in the formation of the elastomeric network [3].

15.3.2 Properties and applications

The dispersions can be pigmented and diluted with water. Dipping

TABLE 15-4
Properties of RTV dispersions and vulcanizates.

Physical Property	Physical Unit	Property value		
		X3-5013	X3-5024	X3-5025
Dispersion				
Color	—	white	opaque	opaque
Solids	%	44	40	40
Density	g/cm^3	1.07	1.02	1.02
pH	—	11.2	11.2	11.2
Viscosity (25°C)	10^3 cps	25	1	15
Vulcanizate (curing 7 days at 23°C)				
Hardness (Shore A)	—	35	40	40
Tensile strength	MPa	3.10	2.76	2.76
Elongation at break	%	450	400	400
Color	—	white	translucent	translucent

or spraying is used for coating of fibers or fabrics, whereby the thickness of the wet layer should not exceed 0.5 mm per pass. Cure is completed within 2 h at room temperature and 50% relative humidity. The resulting silicone elastomers are not thermoplastic and hence cannot be welded, but adhesive bonding is feasible.

Properties of dispersions and elastomers prepared therefrom are listed in Table 15-4.

Applications are coatings for fibers and fabrics as well as membranes. They are resistant to water and weathering. Use temperatures range from −45 to 150°C, for short-term exposure up to 200°C.

References

[1] W. Lynch, Handbook of Silicone Rubber Fabrication, Van Nostrand Reichhold Co., New York 1978.

[2] Silicone Water-Based Elastomers, Dow Corning Corp., Technical Information, June 1980.

[3] J. C. Saam, D. Graiver and M. Baile, Rubber Chem. Technol. **54**/5 (1981) 976.

[4] R. D. Robinson, J. C. Saam and C. M. Schmidt (Dow Corning Corp.), USP. 4 221 668 (1980).

[5] J. C. Saam (Dow Corning Corp.), USP. 4 244 849 (1981).

15.4 POLYCARBOSILANES

Polycarbosilanes are polymers with alternating Si and C atoms in the backbone:

$$+Si-C+_n$$

The synthesis involves dechlorinating polycondensation of dimethyldichlorosilane by means of sodium metal:

(15-3)

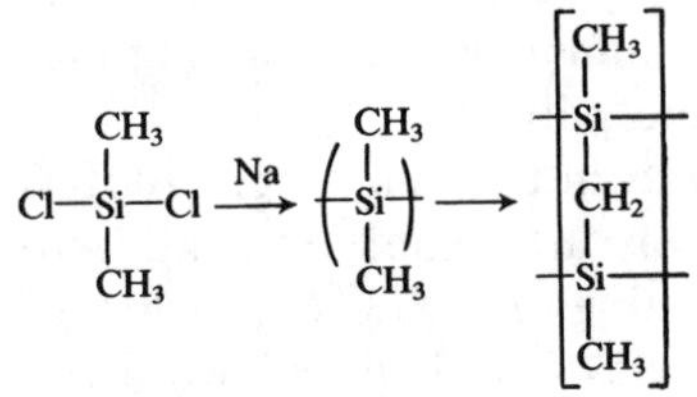

The initially formed poly(dimethylsilane) is converted to poly-carbosilane by heating to 500°C under pressures of 30 bar. The reaction is believed to occur via the homolytic cleavage of the Si-Si bond. The Si free radical abstracts a hydrogen atom to give a $-CH_2$ which subsequently couples with another Si radical [2−4].

The polycarbosilane has a molecular weight near 2000 g/mol. Its X-ray diffraction pattern is similar to that of β-SiC. The polycarbosilane is converted to ultrafine β-SiC (particle diameter 5−7 nm) at 1300°C under vacuum or inert gas. Filaments can be fabricated from polycarbosilane; after conversion to β-SiC fibers they exhibit tensile strength of 3400 MPa and tensile modulus of nearly 300 GPa. Fiber strength is retained up to 1250°C.

Polycarbosilane is used as the matrix material in composites of SiC or Si_3N_4 powders. The polymer and filler are mixed, shaped, and heated without need for pressure to 1300°C. The resulting composites have low densities but very high compressive strengths of about 1000 MPa. Furthermore, the polycarbosilane can be compounded with metal powders, such as Ti, Si, Mg, Mo, W, and others. A composite with Ti powder has a Vickers hardness of 2500 and shows no brittleness after 80 h heating at 1200°C [1].

A second route to polycarbosilanes utilizes poly(silastyrene) as intermediate:

```
        CH3 CH3 CH3 CH3
         |   |   |   |
       —Si—Si—Si—Si—
         |   |   |   |
        CH3 |  CH3 |
            ⬡    ⬡
```

It is synthesized by condensation of a mixture of dimethyldichlorosilane and phenylmethyldichlorosilane by finely dispersed sodium in toluene [5]. After 24 h heating at 50°C and further 24 h under reflux the reaction is terminated by addition of methanol. The solution is washed with saturated ammonium chloride solution to remove the sodium that is occluded in the polymer.

Poly(silastyrene) has molecular weights in the range from 100 000 to 400 000 g/mol. It is converted to polycarbosilane by heating to over 1200°C under inert gas; further heating to 1400°C causes loss of hydrogen, methane, and benzene to yield black β-SiC whiskers.

Filaments can be also fabricated from poly(silastyrene). Its lower melting range compared with poly(dimethylsilane) enhances processability. Shaped articles of silicon nitride fabricated by the conventional reaction sintering process have poor strength and, due to their porous structure, are sensitive to oxidative attack at high temperatures. Such sintered compacts are impregnated with molten poly(silastyrene). After heating to 1400°C, the material is resistant to oxidation and shows 21% greater strength.

So far, poly(silastyrene) has been used only by the U.S. Air Force. Applications are ceramic components for turbine blades, rocket nose cones, and radomes [6].

A third route to polycarbosilanes was developed by Union Carbide Corp. [7]. It involves reaction of a mixture of $ClCH_2Si(CH_3)_2Cl$, $CH_2=CH-Si(CH_3)Cl_2$ and $(CH_3)_2SiCl_2$ with potassium metal in THF at 66–68.5°C for 5 h. The one-step reaction yields branched, soluble polycarbosilanes, which can be converted by known procedures at high temperatures to SiC.

References

[1] S. Yajima, Ind. Eng. Chem., Prod. Res. Dev. **15**/3 (1976) 219.
[2] H. Sakurai, R. Kohi, A. Hosomi and M. Kumada, Bull. Chem. Soc. Japan **39** (1966) 2050.
[3] H. Sakurai, A. Hosomi and M. Kumada, Chem. Comm. (1969) 930.
[4] K. Shiina and M. Kumada, J. Org. Chem. **23** (1958) 139.
[5] W. Worthy, Chem. Engng. News (June 9, 1980) 20.
[6] Anon., Modern Plast. **57** (July 1980) 14.
[7] C. L. Schilling, Jr., J. P. Watson and T. C. Williams, Org. Coat. Plast. **46** (1982) 137.

16 Biopolymers and derivatives of natural products

16.1 POLY(β-D-HYDROXYBUTYRATE)

The bacterium Alcaligenes eutrophus, grown on a solution of glucose containing various nutrients, produces poly(β-D-hydroxybutyrate) with the structural unit

$$\left[CH\!-\!CH_2\!-\!COO \right]_n \qquad PHB$$
$$\big|$$
$$CH_3$$

PHB serves as a carbon reserve for bacteria, similar to the role of starch in higher-classified plants. PHB is found in the cytoplasm of bacteria as hydrophobic granules of approximately 500 nm diameter.

As early as the 60's, W. R. Grace became interested in the technical utilization of this biosynthesis. Recently, ICI has announced plans for an experimental production of PHB with an annual capacity of 500 kg [1]. Properties of PHB are similar to those of isotactic poly(propylene) (Table 16-1), although it is less flexible, its resistance to UV is better, and its resistance to solvents poorer. The relatively high degree of brittleness reportedly can be overcome by

TABLE 16-1
Properties of Poly(β-D-hydroxybutyrate) PHB and isotactic Poly(propylene) PP [2].

Physical property	Physical unit	Property value	
		PHB	PP
Density	g/cm^3	1.25	0.905
Crystallinity	%	80	70
Molar mass	g/mol	500 000	200 000
Melting temperature	°C	175	176
Glass transition temperature	°C	15	−10
Tensile strength	MPa	40	38
Elongation at break	%	6	400
Flexural modulus	GPa	4	1.7
UV resistance	—	good	bad
Solvent resistance	—	bad	good

improved work-up procedures and possibly by addition of plasticizers and fillers [1]. These property characteristics and the relatively high price (glucose obtained from corn is still three times as expensive as petrochemical raw materials) have led to the suggestion that PHB be used as chemical feedstock rather than as plastic [3]. For example, PHB is readily pyrolyzed to crotonic acid [4]. Also, degradation to oligomeric PHB with terminal ester groups is readily achieved by methanolic boron trifluoride [5]. However, it is too early to know if the polymer has a commercial future. Presently, it is being tested as biodegradable thread in surgery [6]. In this field, it has to compete with poly(lactide), poly(glycolide), and their copolymers.

ICI markets PHB under the tradename Biopol [7]. It plans to increase to several hundred tons per year the present production of a few tons.

References

[1] J. H. Mannon, Chem. Engng. (May 4, 1981) 41.
[2] Anon., Modern Plast. **58** (July 1981) 90.
[3] R. H. Marchessault, S. Coulombe, T. Hanai, and H. Morikawa, Trans. Techn. Section, Canadian Pulp and Paper Association **6**/2 (June 1980) 1 (TR52−TR56).
[4] M. Lemoigne, Ann. Inst. Pasteur **39** (1925) 144; J. N. Baptist and F. X. Werber, Soc. Plast. Eng. Trans. **4** (1964) 245.
[5] S. Coulombe, M. S. Thesis, cited in [3].
[6] Financial Times (London), April 23, 1982.
[7] Anon., Kautschuk + Gummi − Kunststoffe **37** (1984) 131.

16.2 CHITIN AND CHITOSAN

16.2.1 Chitin

16.2.1.1 Structure and synthesis

Chitin [1, 2] is poly(β(1 $\rightarrow$ 4)-N-acetyl-2-amino-2-desoxy-D-glucopyranose):

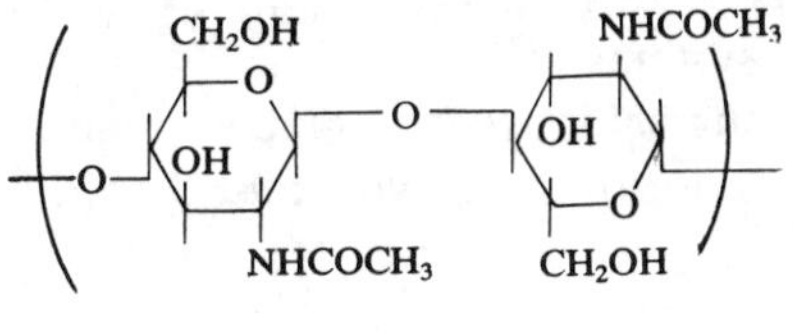

Chitin is also named 2-(N-acetylamino)-cellulose.

Chitin together with calcium carbonate forms the skeleton of arthopodes (crustacea, insects, spiders); it is also found in cell membranes of algae, yeast, fungi, and lichen. Skeleton materials contain proteins besides chitin. Chitin is one of the most abundant polysaccharides. The amount of chitin produced by only a single class of crustacea is estimated at several billion tons [1].

To recover chitin, first the calcium carbonate is dissolved by treatment with 5−6% hydrochloric acid for 24 h. The accompanying proteins are extracted with hot, 3−4% sodium hydroxide solution, or are removed by enzymes like pepsin or trypsin.

The Marine Commodities International Inc., Brownsville, Texas, recovers chitin that is then marketed by Hercules Inc. under the tradename ®KY-10 Chitin Marine Polymer [3]. The commercial product is of white to light brownish color, is supplied as flakes, and contains 10% moisture. Based on dry substance, it contains 5% ash, 6.4−6.8% N, 19.5−20.5% acetyl groups and 10 ppm heavy metals. A purified product with only 1% ash is also available.

Chitin is also commercially available from Velsicol Chemical Corp., Chicago, and from Nippon Tennen Gas Kogyo Co. Ltd., Japan. Velsicol plans a 30 000 t/a plant to recover chitin from lobster shells [4].

16.2.1.2 Properties and applications

Chitin is insoluble in water and in common organic solvents. A limited solubility is observed in anhydrous formic acid and in mixtures of chlorinated acids and methylene chloride [3]. It is dissolved under degradation by concentrated mineral acids and hot salt solutions.

A water-soluble, 50% de-acetylated product is obtained by careful hydrolysis in a homogeneous, alkaline medium [5]. Like cellulose, chitin is converted to the xanthate by reaction with carbon disulfide in strongly alkaline solution. From these solutions, regenerated chitin can be fabricated into films or fibers [6].

Chitin has good therapeutic properties in healing wounds. Due to its accessible amino groups, it is said to serve as an excellent substrate for immobilized enzyme systems [3].

The hydrolysis of chitin with strong acids gives D-glucosamine in good yields. D-Glucosamine hydrochloride react with diisocyanates to linear poly(urea-urethanes) [7].

16.2.2 Chitosan

16.2.2.1 Structure and synthesis

Strong alkali deacetylates chitin to various degrees to form poly-(2-amino-2-deoxyglucose) or chitosan. When chitin is treated with 40% NaOH at 135−140°C for 2 h, a product is obtained which is soluble in acids, whereas the reaction at 50−60°C for 24 h yields an insoluble material [1].

Chitosan is produced by Marine Commodities Inc., and is marketed by Hercules Inc. under the tradename ®Kytex [8].

16.2.2.2 Properties and applications

®Kytex contains only 10−15% of the acetylamino groups that were originally present in chitin. In solution, it behaves like a cationic polymer due to protonization.

Typical specification for Kytex indicates 10% moisture, 1% ash (as $CaCO_3$) and 0.5% insolubles. The granular, white to pale pink colored Kytex is available in three grades with different molecular weights. At pH 4 and 25°C (Brookfield), Kytex grades H and M in 1% acetic acid solution have viscosities of 100 and 10−25 mPa × s, respectively, and a 2% acetic acid solution of grade L has a viscosity of 25−50 mPa × s. Kytex is soluble in many organic and inorganic acids at pH values under 5.5.

Kytex has been recommended for numerous applications [8]: to improve washability and shrinkproof properties of wool, to enhance dyeability of cotton and synthetic fibers, and to improve paper wet and dry strengths; it has also been recommended as sizing agent and lubricant for glass fibers, as thickening agent and viscosity regulator, as binding resin in adhesives, as well as for cosmetics and pharmaceutical products.

Chitosan is used as chelating agent for cations in chromatography processes [9]. In aqueous solutions, the absorbance capacity per 1 g of chitosan for various metal cations (in mmoles) is as follows: Hg^{2+} 5.6; Cd 2.78; Pb^{2+} 3.97; Zn 3.70; Co^{2+} 2.47; Ni^{2+} 3.15; Cr^{3+} 0.46; Cu 3.12; Ag 3.26; Au 5.84; Pt 4.52; Pd 6.28; Mo 3.80, Fe^{2+} 1.18 and Mn^{2+} 1.44 [10]. The metal absorbance capacity of chitosan is greater than that of other cellulose derivatives or of poly(p-aminostyrene). Compared to chitin, chitosan absorbs considerably more $Pb^{2}+$ and Cr^{3+} [11].

Protonated chitosan is crosslinked by reaction with $K_4[Fe(CN)_6]$ or sodium polyphosphate to give an ionic network [12]. It is used as polymeric matrix for immobilizing enzymes.

References

[1] J. Conrad, Chitin, in Encycl. Polym. Sci. Technol., Vol. **3**, J. Wiley & Sons, New York 1965.

[2] R. A. A. Muzzarelli, Chitin, Pergamon, Oxford 1976.

[3] ®KY-10 Chitin Marine Polymer, Hercules Inc., Coatings & Specialty Products Department, Wilmington, Technical Information 1975.

[4] Anon., Chemical Week (June 7, 1978) 37.

[5] K. Kurita, T. Sannan and Y. Iwakura, Makromol. Chem. **177** (1976) 3589; ibid. **18** (1977) 3197; J. Appl. Polym. Sci. **23** (1979) 511.

[6] C. J. B. Thor et al. (Visking Corp.), USP. 2 168 374 (1939), USP. 2 168 375 (1939) and USP. 2 217 823 (1940).

[7] K. Kurita, N. Hirakawa and Y. Iwakura, Makromol. Chem. **178** (1977) 2939; ibid. **180** (1979) 2331.

[8] ®Kytex (Chitosan) Cationic Polymers, Hercules Inc., Coatings & Specialty Department, Wilmington, Technical Information 1975.

[9] R. A. A. Muzzarelli, Natural Chelating Polymers, Pergamon, New York 1973.

[10] M. Sid Masri, F. W. Reuter and M. Friedman, J. Appl. Polym. Sci. **18** (1974) 675.

[11] C. A. Eiden, C. A. Jewell and J. P. Wigthman, J. Appl. Polym. Sci. **25** (1980) 1587.

[12] K. D. Vorlop and J. Klein, Biotechnol. Lett. **3** (1981) 9.

16.3   PULLULAN

The attack of blackyeast Pullularia pullulans, also known as Aureo-basidium pullulans, on starch yields pullulan which consists of linear chains of poly[α-(1 → 6)-D-maltotriose [1]:

Every three D-glucose units are connected with regular alternation of two α-(1 → 4) and one α-(1 → 6) linkages. Pullulan also contains some α-(1 → 3) linkages and maltotetrose units [2, 3]. Since several years, pullulan has been produced by Hayashibara Biochemical Laboratories in semicommercial quantities with seven various molecular weights (from 10 000 to 400 000 g/mol) from hydrolyzed starch [2, 3]. Similar products are obtained from other species of yeast [4].

Pullulan is highly soluble in water. The white pullulan powder, after addition of a small amount of water, can be formed into biodegradable films, sheets and molded articles. The films display tensile strengths of ca. 70 MPa, elongations of 95%, folding endurance 750, light transparency 95% and show excellent gloss. Pullulan films are colorless, tasteless, odorless, oil and grease resistant, edible, and oxygen impermeable; they can be heat sealed.

Applications for pullulan films are coatings and packaging materials for food. They serve as a low calorie additive in diets, and are used in aqueous solution as adhesive. Dry-spun and drawn filaments reportedly have tenacities comparable to polyamides [2]. Esterified pullulan has found use for interior-use plywood coating or laminating, the surfaces of such treated plywood resemble melamine coatings.

Aqueous solutions of pullulan can be mixed with other water soluble polymers, e.g. amylose, gelatine and poly(vinylalcohol). Plasticizers such as sorbitol or glycerol improve the flexibility of pullulan films.

References

[1] H. Bender, H. Lehmann, and K. Wallenfels, Biochem. Biophys. Acta **36** (1959) 309.
[2] S. Yuen, Proc. Biochem. (November 1974) 7.
[3] S. Yuen, in R. D. Deanin (Ed.), New Industrial Polymers (ACS Symp. Ser. **4** (1974) 172), Amer. Chem. Coc., Washington, D.C.
[4] J. E. Zajic and A. La Duy, Encycl. Polym. Sci. Technol., Suppl. **2** (1977) 643.

16.4 XANTHAN

16.4.1 Structure and synthesis

The bacterium Xanthomonas campestris converts dextrose to the polysaccharide xanthan [1–3]. Xanthan is composed of irregularly linked structural units of D-glucuronic acid (I), D-mannose (II) and D-glucose (III):

The D-mannose units seem to be evenly distributed in the backbone and side chains. The latter consists of probably only one sugar unit, namely D-mannose (II) and 4,6-O-(1-carboxyethylidene)-D-glucose (IV) units. The exact position of linkage of this pyruvic acid ketal unit is not known; possibly it is linked at a non-reducing moiety. Approximately 8% of the hydroxyl groups are esterified with acetic acid. Normally, xanthan is obtained as potassium salt. The molar ratio of the individual structural units is D-glucose: D-mannose: D-glucuronic acid: pyruvic acid ketal: O-acetyl = 3.0:3.0: 2.0:0.6:1.7.

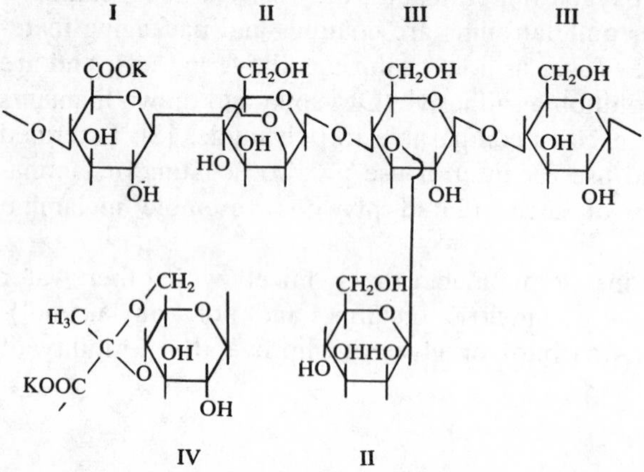

I   II   III   III

COOK   CH₂OH   CH₂OH   CH₂OH

IV   II

Xanthan has weight average molecular weights of about 5 000 000 g/mol, but it is open to question if this high value is not erroneous because of self-associations occuring in aqueous solution.

Technically useful properties of xanthan were soon recognized [4] and led to large-scale production [5]. Xanthan is now produced by several companies, such as General Mills (now Henkel Corp.), Kelco Division of Merck, Ceca (France) and Rhône-Poulenc. Kelco markets xanthan for food applications under the name Keltron, for technical applications under the name Kalzan, and for tertiary crude oil recovery under the name Xanflood [6]. Ceca's products are designated Actigum CX 9 and CX 12; they are still under development and not yet commercially available [7].

16.4.2 Properties and applications

The irregular chemical structure of xanthan contributes to the interesting rheological properties of this water-soluble polymer. Solution viscosities do not change significantly on addition of 0.01– 1% neutral salts, e.g., KCl. However, viscosity is lowered as small amounts of salts are added, and it increases on addition of larger amounts. Basic salts enhance the increase in viscosity; the polysaccharide itself is not precipitated even at high salt concentrations of 15–30%. Divalent cations precipitate xanthan at pH values over 9; trivalent cations cause precipitation in less alkaline or even acidic solutions, whereas at sufficient low concentrations of xanthan gelation occurs.

Aqueous xanthan solutions exhibit structural viscosity but very little thixotropy. Very small concentrations of xanthan in the 50– 100 ppm range cause considerable reduction of friction in turbulent flows.

These unusual rheological properties have been thoroughly investigated over the last few years; they have been extensively described in the scientific and technical literature, in particular with regard to the application of xanthan in tertiary oil recovery processes [8]. The solution characteristics of xanthan are presumably due to an ordered helical conformation which is not markedly altered on addition of salts [3, 9, 10]. Owing to the helical structure, the xanthan molecules assume extrinsic rod-like shapes [11]; they associate and form cholesteric mesophases at higher concentrations [14].

Besides the application in tertiary oil recovery, xanthan has found numerous other uses, e.g. in the fabrication of cellulosic spun yarn, for the stabilization of emulsions, as dispersing agent for liquid laundry starch, as carrier for agricultural chemicals, in gelling of explosives or detergents, as thickening agent for cosmetics, as adhesive and sizing agent [1], in the fabrication of inks, as processing aid for the extrusion of ceramic materials, and others [7].

Xanthan undergoes no metabolic degradation and hence can be used as a low calorie food additive for puddings, salad dressings, meat jelly, milk powder, and so on. Of particular interest to the food industry is that xanthans are said to intensify flavors and bring about a full-bodied taste in food.

References

[1] A. Jeanes, J. Polym. Sci. [Symp.] **C 45** (1974) 209.

NEW COMMERCIAL POLYMERS

[2] I. W. Cottrell, K. S. Kang and P. Kovacs, in R L Davidson (Ed.), Handbook of Water-Soluble Gums and Resins, McGraw-Hill, New York 1980.
[3] R. Moorhouse, M. D. Walkinshaw and S. Arnott, in P. A. Sandford and A. Laskin (Ed.), Extracellular Microbial Polysaccharides, ACS Symp. Ser. 45 (1977) 90; Amer. Chem. Soc., Washington, D.C.
[4] A. Jeanes, J. E. Pittsley and F. R. Senti, J. Appl. Polym. Sci. 5 (1961) 519.
[5] I. W. Cottrell and K. S. Kang, Dev. Ind. Microbiol. 19 (1978) 117.
[6] Xanthan Gum, Kelco, Product Information, Chicago 1976.
[7] Actigum, Ceca, Velizy-Villacoublay (France), Product Information; J. J. Piot (Ceca), personal communication (February 19, 1981).
[8] V. Martin and J. Klein, Erdöl-Erdgas 95/5 (1979) 164.
[9] J. S. Chen and E. W. Sheppard, J. Makromol. Sci.-Chem. A 13 (1979) 239.
[10] C. S. Chen and E. W. Sheppard, Polym. Engng. Sci. 20 (1980) 512.
[11] M. Rinaudo and M. Milan, Biopolym. 17 (1978) 2663.
[12] G. Holzworth and E. B. Prestridge, Science 197 (1977) 757.
[13] J. G. Southwick, H. Lee, A. M. Jamieson, and J. Blackwell, Carbohydr. Res. 84 (1980) 287.
[14] G. Maret, M. Milas, and M. Rinaudo, Polym. Bull. 4 (1981) 291.

16.5 ACRYLONITRILE/STARCH GRAFT COPOLYMERS

The chemical modification of graft copolymers of acrylonitrile onto starch yields materials that absorb water at amounts of more than two thousand times their weight [1]. The resulting polymer was appropriately nick-named "super slurper". The material was originally developed by the Northern Regional Research Laboratory of the U.S. Department of Agriculture, Peoria, Illinois. General Mills Chemicals (now Henkel Corp.) obtained a non-exclusive, royalty-free license and has marketed the polymer under the name SPG Absorbent Polymer [2].

The polymer is prepared by free radical graft polymerization of acrylonitrile onto starch. The resulting product is hydrolyzed in alkaline solution, leading to the formation of carboxylic as well as carboxamide groups. The acid derivative of the obtained polyelectrolyte is then dried to give a powder [1, 3].

A more recent process developed by NRRL Peoria starts from flour in lieu of starch; it employs a graft copolymerization with acrylonitrile, acrylic acid or other comonomers [4]. These products are said to absorb water at amounts of even greater than five thousand times their weight.

The dry powders dissolve in water and, after neutralization, give a thick, elastic paste [5]. It consists of dispersed, highly-swollen, narrowly packed gel particles with very little, if any, free water present between the individual gel particles. The thickening characteristics disappear at very large dilutions or at very high ion concentrations.

Self-supporting films can be fabricated from such solutions. The dispersions cannot be regenerated from the films, unless very strong mechanical comminution forces are applied. The cast films readily absorb water and hereby retain their film structure over a broad temperature and pH range.

Powerful agitation of a 1% dispersion reduces its viscosity from ca. 3000 mPa $\times$ s to 3 mPa $\times$ s. The resulting low-viscosity product yields true molecular solutions from which water-soluble films can be casted. These films become insoluble in water on heating, by ^{60}Co, radiation, or by aging at high relative humidities [7].

More than fifty different applications have so far been recommended for these products, ranging from throw-away diapers to soil conditioners to seed encapsulation.

References

[1] L. A. Gugliemelli, M. O. Weaver, C. R. Russell and C. E. Rist, J. Appl. Polym. Sci. **13** (1969) 2007.
[2] SPG 147 Absorbent Polymer Powder and SPG 101 Soft-Hand Film Laminates, Henkel Corp., Minneapolis, Product Information 1981.
[3] G. Fanta, R. C. Burr, W. M. Doane and C. R. Russell. J. Polym. Sci. [Symp.] **45** (1974) 89.
[4] Anon., Chem. Engng. (March 14, 1977) 74.
[5] N. W. Taylor and E. B. Bagley, J. Appl. Polym. Sci. **18** (1974) 2747.
[6] M. O. Weaver, E. B. Bagley and W. M. Doane, Appl. Polym. Symp. **25** (1974) 97.
[7] E. B. Bagley and N. W. Taylor, Ind. Engng. Chem.-Prod. Res. Dev. **14** (1975) 105.

16.6   ACRYLAMIDOMETHYL STARCH

The A. E. Staley Mfg. Co., Decatur, IL, has marketed acrylamido-methyl-groups containing starch with the general structure

Starch—O—CH_2—NH—CO—CH=CH_2

under the tradename ®Starpol 100 [1]. The acrylamido groups can be polymerized by suitable free radical initiators, photo-chemically by UV irradiation or by electron beam technique [2, 3]. Also, they can be added to the hydroxyl groups of starch by alkaline catalyzed reaction at temperatures above 60°C.

Only about 5% of the hydroxyl groups of starch are replaced by acrylamidomethyl groups. Starpol 100 is an odorless, white powder with an average granule particle size of 11 nm. Above 67°C, it forms a paste in water, whereas it dissolves completely in hot water, and is insoluble in the common organic solvents.

Starpol can be copolymerized with a number of other water-soluble monomers, e.g., with hydroxypropylacrylate, acrylic acid, acrylic amide, and dimethylaminoethylmethacrylate hydrochloride. The copolymerization with water-insoluble monomers, e.g. styrene, methylmethacrylate, and vinylacetate, is accomplished by suspension polymerization, in which simultaneous polymerization occurs in the continuous aqueous phase as well as in the dispersed monomer phase.

Starpol cured by free radical polymerization gives stiff polymer films with tensile strengths of 10−20 MPa. The films show good adhesion to cellulose, glass, metals and to some plastics. Therefore, Starpol is used as adhesive for laminates, abrasive paper, non-wovens, and foundry sand. The high degree of water absorption makes Starpol a candidate for applications in diapers, seed encapsulation, thickeners and encapsulating material for enzymes. Other recommended applications include pigmented paper, printing inks, lithographic printing plates, fillers for rubbers and plastics, protective colloid for emulsions, fire exstinguishing materials, and glass fiber sizing. Some of these applications are attainable when Starpol is copolymerized with other types of starch containing polymerizable groups, such as

$$-O-CH_2-CH(OH)-CH_2-O-CH_2-CH=CH_2,$$

$$-O-CO-CH=CH-COOH \quad \text{and}$$

$$-O-CH_2-NH-CO-CH=CH_2 \quad [2].$$

Chemically substituted starches are also supplied by American Maize-Products Co., Corn Processing Division, Hammond, Indiana.

References

[1] Anon., Industrial Research/Development (October 1978) 131.

[2] A. H. Young, F. Verbenac and T. F. Protzman, ACS Coat. Plast. Prepr. **36**/2 (1976) 411; ibid., in S. Labana (Ed.), Chemistry and Properties of Crosslinked Polymers, Academic Press, New York 1977, 191.
[3] ®Starpol 100, A. E. Staley Manufacturing Co., Technical Data TDS 311/871020.

16.7 POLYMERS FROM CARBON DIOXIDE

16.7.1 Poly(Alkylene Carbonates)

Although carbon dioxide, the anhydride of carbonic acid, has been used in the past for the monomer synthesis of urea (with ammonia) and hydroxybenzoic acid (with phenol), it was never utilized directly as monomer. Recently, Air Products and Chemicals, Allentown, PA, came to an agreement with two leading polymer producers under which poly(alkylenecarbonates) are to be prepared in pilot plant quantities.

The copolymerization is based on the discovery of Inoue et al. [1] who found that zinc alkoxides (I) react with carbon dioxide to form zinc alkylcarbonates (II):

(16-1)

$$\sim CHR-CHR'-O-Zn-X + CO_2 \longrightarrow \sim CHR-CHR'-O-CO-O-Zn-X$$

$$\text{I} \qquad\qquad\qquad\qquad\qquad \text{II}$$

which then react with epoxides under regeneration of zinc alkoxides:

(16-2)

$$\text{II} + \underset{\displaystyle \overset{\textstyle \backslash\,/}{O}}{HRC-CHR'} \longrightarrow \sim CHR-CHR'-O-CO-O-CHR-CHR'-O-Zn-X.$$

The most active catalysts are products obtained from the reaction of diethylzinc with equimolar amounts of compounds containing two or more active hydrogens, e.g. water, primary amines, di- and triphenols, dicarbonic acids, etc. Presumably, "polymeric" structures are formed in this reaction, for example (for a review see also [2]):

$C_2H_5(Zn-O)_xH$

$C_2H_5(Zn-O-\langle C_6H_4 \rangle-O)_xH$

The structure of the poly(alkylenecarbonates) that were recently licensed by Air Products and Chemicals has not been disclosed [3]. They are transparent elastomers with good barrier properties to oxygen (see Table 16-2). The tough polymers are resistant to fats; applications are aimed at food packaging, adhesives, coatings and possibly fibers. The polymers are biodegradable; on combustion, no products are formed that are detrimental to health.

TABLE 16-2
Properties of Polyalkylenecarbonates [2].

Physical Property	Physical Unit	Property values of copolymers of carbon dioxide with		
		Ethylene oxide	Propylene oxide	2-Butene oxide
Glass transition temperature	°C	5	30	66
Modulus of elasticity at strain rates of				
20 cm/min	MPa	2.1	—	—
1 cm/min	MPa	—	993	2190
Tensile strength	MPa	5.9	33	37
Hardness (Rockwell R)	—	—	—	112
Density	g/cm^3	1.429	1.275	1.18
Refractive index	—	1.470	1.463	1.470
Melt index (165°C)	g/10 min	7.0	68.5	—
Impact strength with notch	J/m	—	—	22
Heat of combustion	kJ/g	13.9	18.5	21.2
Relative permittivity	—	4.32	3.0	—
Dissipation factor (1 kHz)	—	0.031	0.007	—
Volume resistivity	ohm × cm	2.4×10^{14}	1.0×10^{16}	—
Water absorption (23°C)	%	0.41–0.69	0.40	—
Permeability for				
Water	cm^3 s g^{-1}	1.3×10^{-11}	9.8×10^{-11}	9×10^{-11}
Nitrogen	cm^3 s g^{-1}	5.2×10^{-16}	4.0×10^{-16}	32×10^{-16}
Oxygen	cm^3 s g^{-1}	11×10^{-6}	23×10^{-16}	150×10^{-16}

16.7.2 Other polymers

Carbon dioxide is also a monomer for the synthesis of a number of other polymers. The reactions have been reviewed in the literature [13, 14]. However, none of these polymers has been commercialized as yet. For example, carbon dioxide reacts with phenylethylene-imine [4, 5] and other imines [6] to form polyurethanes

(16-3)

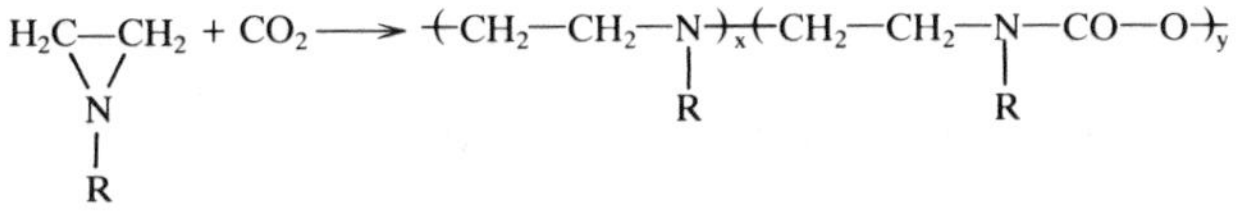

$$H_2C-CH_2 + CO_2 \longrightarrow +CH_2-CH_2-N\rightarrow_x+CH_2-CH_2-N-CO-O\rightarrow_y$$

with vinylethers to form low-molecular weight polyketoethers [7, 8]

(16-4)

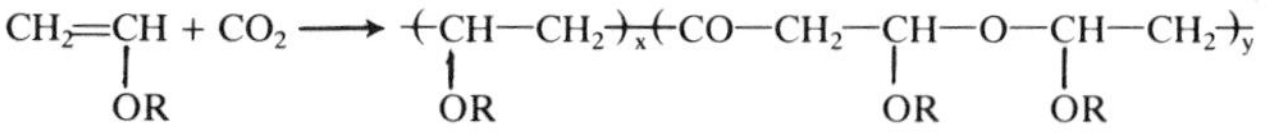

$$CH_2{=}CH + CO_2 \longrightarrow +CH{-}CH_2\rightarrow_x+CO{-}CH_2{-}CH{-}O{-}CH{-}CH_2\rightarrow_y$$

and with acrylic compounds and cyclic phosphonites to periodic terpolymers [9, 10]

(16-5)

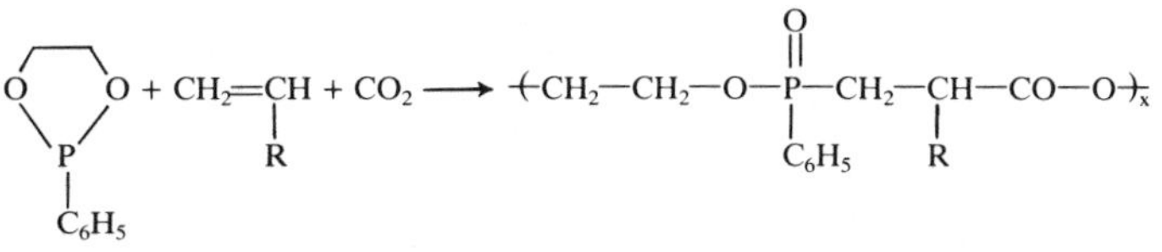

$$\text{cyclic} + CH_2{=}CH + CO_2 \longrightarrow +CH_2{-}CH_2{-}O{-}\overset{O}{\underset{C_6H_5}{P}}{-}CH_2{-}CH{-}CO{-}O\rightarrow_x$$

and with aromatic diamines to polyureas [11, 12]

(16-6)

$$H_2N-Ar-NH_2 + CO_2 \longrightarrow +NH-Ar-NH-CO\rightarrow_x$$

Each of these reactions requires a different catalyst. The formation of polyurethanes is catalyzed by phenol or manganesedichloride tetrahydrate [4, 5], that of polyketoethers by aluminum acetylace-tonate or aluminum alkoxides [7], and the synthesis of polyureas by diphenylphosphite/pyridine [11, 12].

References

[1] S. Inoue, H. Keinuma and T. Tsuruta, J. Polym. Sci.-Polym. Letters Ed. **7** (1969) 287; M. Takanashi, Y. Nomura, Y. Yoshida and S. Inoue, Makromol. Chem. **183** (1982) 2085.

[2] S. Inoue, Chemtech **6** (1976) 588.

[3] Anon., Chem. Engng. (December 28, 1981) 17.

[4] T. Kagiya and T. Matsuda, Polym. J. **2** (1971) 398.

[5] K. Soga, S. Hosoda, and S. Ikeda, J. Polym. Sci.-Polym. Chem Ed. **12** (1974) 729.

[6] K. Soga, W.-Y. Chang and S. Ikeda, J. Polym. Sci.-Polym. Chem. Ed. **12** (1974) 121.

[7] K. Soga, M. Sato, Y. Takazi, and S. Ikeda, J. Polym. Sci.-Polym. Letters Ed. **13** (1975) 543.

[8] K. Soga, M. Sato, S. Hosoda, and S. Ikeda, J. Polym. Sci.-Polym. Letters Ed. **13** (1975) 543.

[9] T. Saegusa, S. Kobayashi and Y. Kimura, cited in [2].

[10] T. Saegusa, Makromol. Chem. Suppl. **4** (1981) 73.

[11] N. Yamazaki, F. Higashi and T. Igushi, J. Polym. Sci.-Polym. Chem. Ed. **13** (1975) 785.

[12] N. Yamazaki, S. Nakahama and F. Higashi, Ind. Engng. Chem. Prod. Res. Dev. **18** (1979) 249.

[13] A. Rokicki and W. Kuran, J. Macromol. Sci.-Reviews in Macromol. Chem. **C 21** (1981) 135.

[14] N. Yamazaki, F. Higashi and S. Inoue, Org. Bio-Org. Chem. Carbon Dioxide 1982, 153, in S. Inoue and N. Yamazaki, eds., Kodansha, Tokyo, 1982.

17 Epilogue

Seven years ago, one of us (HGE) published a description [1] of ca. 60 polymers with new chemical structures, which had reached the commercial stage or had been commercialized in the years from 1969 to 1974. At one end of the spectrum of new polymers, there could be found new derivatives of the long-known cellulose, and at the other end polymers with completely new structures. In between were new polymers containing long-known functional groups such as aromatic polyamides and the many new copolymers. With regard to future development of new polymers the book came to the conclusion: "The possibilities for synthetic macromolecular chemistry are far from exhausted at this point. In all likelihood the next five years will witness the appearance of additional polymeric materials with new chemical structures, even if not in such abundance."

This optimistic view has not only been verified but was even surpassed by the actual developments. Where 11 new polymers had been turned out annually from 1969 to 1974, the number of new polymers per annum climbed to 19 for the years from 1975 to 1981 (see Table 1-5). Some of the new polymers have all prospects of becoming commodity polymers, e.g., the linear low-density poly-(ethylenes), p-methylstyrene based polymers, and the solution polymerized styrene/butadiene rubbers. Others will find niches in the specialties market, for which an annual growth of 11% (adjusted for inflation) in the U.S. is predicted [2]. The following turnovers in US dollars are estimated for 1986:

Industrial polymers with above average growth (860 million dollar sales):

poly(phenylene sulfide)
polyimides
ethylene/acrylate elastomers
polyamideimides
polyamides with low levels of water absorption (PA 11 and PA 12)
transparent polyamides
styrene/maleic anhydride copolymers
butadiene/styrene polymers
acrylic polymers of high impact strength

475

transparent styrene polymers
poly(1-butene)
crosslinked poly(ethylene) foams
poly(4-methylpentene-1)
barrier polymers based on acrylonitrile
low-cost styrene/acrylic polymers

Recently commercialized polymers (300 million dollars sales):
   ionic elastomers
   modified elastomers based on 1,3-butadiene
   elastomeric polyamides based on PA 11 and PA 12
   polyacrylates
   polyamides with high impact strength
   elastomers based on polyphenylene oxide
   styrene polymers resistant to weathering
   poly(ethylene terephthalate) engineering plastics
   triazine-bismaleimide polymers
Polymers in an advanced stage of development (90 million dollars sales):
   polyethersulfones
   polybenzimidazoles
   polyamides based on m-xylene
   polyestercarbonates
   sound dampening polymers (Keldex)
   aromatic polyesters
   polynorbornene

Newcomers (50 million dollars sales):
   polyether(ether)ketones
   polyetherimides
   self-reinforcing liquid-crystalline polymers
   elastomeric polyesteramides
   polyetheramide block copolymers

All these polymers are produced from petrochemical feedstocks; thus, at the presently prevailing over-supply of crude oil, it may be safely presumed, like 7 years ago [1], that the long-term provisioning of petrochemical feedstocks to the polymer industry seems to be secured. At the present time, the problem is no longer the supply of raw materials at increasing costs, but much more the world-wide recession hand in hand with falling oil prices. The receeding demand for crude oil is only in part a question of economy, it is due rather to the efforts undertaken by the industrialized countries to cut down on their energy consumption. Moreover, shifting from crude oil as the sole energy source is highly welcomed, since crude oil ought to be reserved for petrochemical operations.

For the next decade, crude oil will maintain its position as the predominant raw material source for the polymer industry. In this connection it is interesting to note that over the recent years not one polymer based on agricultural raw materials, such as soy beans or crambe, has been commercialized. With the present over-supply of corn in the U.S.A., starch could well develop into a raw material for the polymer industry (see chapters 16.3, 16.5 and 16.6). Under long-term aspects, however, starch is not likely to attain any greater importance, because the increasing soil erosion in the U.S. tends to reduce the agricultural production, which apparently is only delayed by greater consumption of fertilizers and new cultivations. It remains to be seen if poly(β-D-hydroxybutyrate) will ever gain significance as a raw material (chapter 16.1). Of greater potential interest is the introduction of carbon dioxide based polymers (chapter 16.7) which is the only true renewable source for raw materials [1].

A couple of years ago, biodegradable polymers were claimed to be needed for reducing environmental pollution. In the meantime, demand for these products has tapered off; nearly all biodegradable polymers were withdrawn from the market, and not one new product has been commercialized. Some U.S. states by law imposed returnable fees on beverage cans and bottles as effective means to keep streets and highways free from litter; money is still a very convincing argument.

Where are we bound? Human ingenuity combined with the pressure for technical innovation and profit orientation certainly will create new polymers over the forthcoming years. For example, the potential offered by polymers with heterocyclic structures has certainly not been fully exploited as yet. Also, application of RIM processing to polymers other than polyurethanes requires new structural materials. Furthermore, the search for electrically conductive polymers will stimulate the research in the field of polymers with "exotic" structures.

However, all these developments are largely empirical. They draw from experience that has been accumulated over some decades. This is illustrated by a simple example: it is known that p-phenylene moieties in the backbone result in stiff polymers whereas ether groups are known to impart flexibility. There are some indications that these strategies may be less successful in future. First, they are not very cost effective; secondly, they allow only crude estimates with regard to performance; and finally, they fail completely in predicting ultimate product properties.

Of crucial significance is the perception that polymer chain conformations control ultimate polymer properties [3]. Exemplary cases are liquid-crystalline polymers and ultra-drawn polymers. Both illustrate the attempt to attain better chain alignment of polymer chains; this effect is accomplished in liquid-crystalline polymers by their tendency to assume more or less parallel positions owing to the rod-like structure, and in ultra-drawn polymers by externally enforced tensile or shear strains.

What is now to be expected? For liquid-crystalline polymers, there is no theory that would allow predicting the maximum degree of chain orientation in relation to chemical structure. However, such a theory is needed before the rules of mixture known for composites can sensibly be applied to predetermine certain use properties. Even then, optimum properties will be obtained only if all the chains are uniformly and completely aligned.

For such completely aligned chains certain theoretical ultimate properties can be calculated. The calculated theoretical tensile moduli are largely confirmed through the so-called lattice moduli that are experimentally obtained from x-ray analysis at very low temperatures. In general, theoretical tensile moduli are much greater than the best moduli attainable from conventionally processed plastics or fibers (Table 17-1). Moreover, the data in Table 17-1 illustrate that the high tensile moduli of such super fibers as poly(p-benzamide) can be achieved also by less exotic chain structures. For example, ultra-drawn poly(ethylene) exhibits a tensile modulus of about 130 GPa and hence is equivalent to Kevlar and superior to aluminum.

Chain alignment is accomplished by a number of different processes: drawing, hydrostatic extrusion, and other drawing-type procedures, flow spinning, and gel spinning [4, 5]. A patent recently issued to the Michigan Molecular Institute covers a continuous high-pressure process to yield transparent, high-moduli polymers [6]. A pilot plant to produce ultra-drawn poly(ethylene) is under construc-

TABLE 17-1

Theoretical moduli of elasticity in chain direction, experimental moduli in chain direction, and experimental tensile moduli for extruded or injection molded plastics or fibers, resp.

Polymer	E/GPa				Remarks
	Theory	Lattice	Tensile Plastics	Tensile Fibers	
Poly(ethylene)	240–340	255	0.2–1.0	—	
it-Poly(propylene)	50	34	1.4	2.5	
st-Poly(vinyl chloride)	230	—	3.4	—	
Poly(oxymethylene),					
orthorhombic	220	105	—	—	
trigonal	48	54	3.7	—	
st-Poly(methylmethacrylate)	63	—	3.0	—	
Poly(ethylene terephthalate)	146	140	2.9	10	high strength
Polyamide 6.6	196	—	—	6.0	
Poly(p-benzamide)	203	—	—	77	
Poly(p-phenylenetere-					
phthalamide)	—	—	—	132	Kevlar 49
Cellulose I	129	130	—	105	Flax
Cellulose II	—	90	—	12	Rayon
Poly(oxyethylene)	9	10	—	—	
Graphite	1020	—	—	420	
Steel	270	—	—	210	
E-Glass				70	
Aluminum	—	—	70	—	

tion at the Celanese Corp.; the nature of the drawing process has not been disclosed.

The question is now which polymers will give the products with the highest possible tensile moduli upon ultra-drawing. The formulae to calculate the theoretical moduli are more or less complex depending on the degree of chain conformation that is obtained upon ultra-drawing. For approximate calculations two values are needed: the chain cross section area and the corresponding chain conformation (Figure 17-1). If the theoretical tensile moduli or lattice moduli are plotted versus chain cross section area, all polymers with zigzag chains (2_1-conformation) fall on one straight line, and for all polymers with 3_1-helix structures a second straight line is found. The 8_3-helices, e.g. poly(isobutylene), behave like

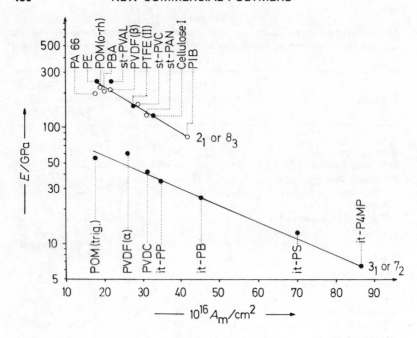

Figure 17-1 Logarithms of Young's modulus, E, as function of cross-sectional area, A_m, of polymer chains with zig-zag chains (2_1) or with 8_3, 3_1 or 7_2 helical conformation. (○) Theoretical moduli, (●) lattice moduli. PA 66 = polyamide 6,6; PE = poly(ethylene); POM = poly(oxymethylene); PBA = poly(p-benzamide); PVAL = poly(vinyl alcohol); PVDF = poly(vinylidenefluoride); PVC = poly(vinyl chloride); PAN = poly(acrylonitrile); PIB = poly(isobutene); PVDC = poly(vinylidenechloride); PP = poly(propylene); PB = poly(butene-1); PS = poly(styrene); P4MP = poly(4-methylpentene-1). it = isotactic; st = syndiotactic; o-rh = orthorhombic; trig = trigonal; α, β, II = various crystal modifications.

2_1-structures, and 7_2-helices, e.g. poly(4-methylpentene-1), like 3_1-helices, at least in first approximation.

The most important result is that polymers with closely related chemical structures exhibit quite different theoretical (i.e. maximal obtainable) tensile moduli: poly(ethylene) with E = 240−340 GPa is by far stiffer than is isotactic poly(propylene) with E = 34 GPa. The "natural" high stiffness of poly(benzamide) fibers is due to a small chain cross sectional area combined with a non-helical conformation and the tendency of self-alignment of chain segments, rather than to a peculiar chemical structure. Thus, not every polymer can be upgraded dramatically by ultra-drawing; a precise knowledge of the maximal obtainable theoretical moduli is essential. In turn, this requires information about the area of chain cross sections which can be easily calculated for simple chain structures from geometric models but only with difficulty for more or less angular chains. However, investigations have shown that the average cross-sectional area of chains may also be determined for amorphous polymers by x-ray analysis [7].

For a given polymer, tensile moduli go hand in hand with tensile strengths (Figure 17-2). As the obtainable tensile modulus increases upon ultra-drawing, the obtainable elongation at break also increases. Both values increase with polymer molecular weight, undoubtedly because the number of chain ends as potential defects decreases.

Ultra-drawn poly(ethylene) fibers are now semi-commercial. DSM (Geleen, The Netherlands) is using high-density poly(ethylene) with molecular weights of over 1 million to produce high-

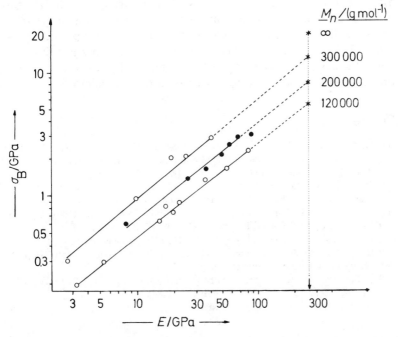

Figure 17-2 Tensile strength at break as function of Young's modulus for hot-drawn fibers from gel-spun filaments of poly(ethylene) (data of [8]). Numbers give number average molar masses.

modulus, high-strength polyethylene fibers by the gel-spinning technique. [9]. The fibers exhibit specific moduli (tensile modulus divided by density) of $50-130$ GPa cm^3 g^{-1} which compares favorably with steel fibers (25 GPa cm^3 g^{-1}), carbon fibers (125 GPa cm^3 g^{-1}), and aramid fibers ($40-90$ GPa cm^3 g^{-1}). Their specific strength (tensile strength at break divided by density) is up to 4 GPa cm^3 g^{-1} as compared to steel fibers (0.3 GPa cm^3 g^{-1}), carbon fibers (1.5 GPa cm^3 g^{-1}), and aramid fibers (less than 2 GPa cm^3 g^{-1}).

DSM has linked up with Toyobo, Japan, to develop this fiber. [10]. Another Japanese company, Mitsu Chemical Industries, also plans to introduce an ultradrawn poly(ethylene) fiber based on ultra-high molecular weight PE. A production of 36 tons per year is planned.

Ultra-drawn polymers as well as block polymers (chapter 4) present the first polymers which come fairly close to targeted synthesis and processing for the manufacture of polymers with predetermined use properties. At present, only few ultimate properties are theoretically accessible on a molecular basis. For many other properties, we are still far from being able to rely on theory-supported prognoses. Still, if the phrase "taylor-made" ever had any justification, then the new molecular engineering of ultradrawn polymers and preparation of block polymers is entitled to this claim now.

References

[1] H.-G. Elias, Neue polymere Werkstoffe 1969–1974, Hanser, Munich 1975; New Commercial Polymers 1969–1975, Gordon and Breach, New York 1977.
[2] Anon., J. of Commerce (1 October 1981) 10 A.
[3] H.-G. Elias, Makromoleküle, Hüthig and Wepf, Basle, 4th edition 1981; Macromolecules, Plenum, New York, 2nd ed. 1984.
[4] A. Keller and P. J. Barham, Plastics and Rubber Internat. 6/1 (1981) 19.
[5] I. M. Ward, ACS Polym. Preprints 23/2 (1982) 218.
[6] D. J. Meier and L. Jarecki, US-P. 4 348 350 (7 September 1982).
[7] R. L. Miller, R. F. Boyer, and K. Battjes, unpublished results, Annual Report, Michigan Molecular Institute, Midland, MI.
[8] P. Smith and P. J. Lemstra, J. Polym. Sci.-Polym. Phys. Ed. 19 (1981) 1007.
[9] Anon., Chem. Engng. News (20 February 1984) 37.
[10] Anon., Chem. Engng. News (19 March 1984) 10.

Appendix

TABLE A-1
SI Units.

Physical Symbol	Quantity Name	Physical unit Name	Symbol
Basic physical quantities and units			
l	Length	meter	m
m	Mass	kilogram	kg
t	Time	second	s
I	Electric current	ampere	A
T	Thermodynamic temperature	kelvin	K
l_v	Luminous intensity	candela	cd
n	Amount of substance	mole	mol
Supplemental quantities and units			
α, β, γ	Plane angle	radian	rad
ω, Ω	Solid angle	steradian	sr
Derived quantities and units			
F	Force	newton	$N = J\,m^{-1} = kg\,m\,s^{-2}$
E	Energy, work, heat	joule	$J = N\,m = kg\,m^2\,s^{-2}$
P	Power	watt	$W = V\,A = J\,s^{-1}$
p	Pressure	pascal	$Pa = N\,m^{-2} = J\,m^{-3}$
ν	Frequency	hertz	$Hz = s^{-1}$
Q	Electric charge	coulomb	$C = A\,s$
U	Electric potential difference	volt	$V = J\,C^{-1} = W\,A^{-1}$
R	Electric resistance	ohm	$\Omega = V\,A^{-1}$
G	Electric conductance	siemens	$S = A\,V^{-1}$
C	Electric capacitance	farad	$F = A\,s\,V^{-1}$
ε	Relative permittivity	—	1
Φ	Magnetic flux	weber	$Wb = V\,s$
L	Inductance	henry	$H = Wb\,A^{-1}$
B	Magnetic flux density	tesla	$T = Wb\,m^{-2}$
Φ_v	Luminous flux	lumen	$lm = cd\,sr$
E_v	Illumination	lux	$Lx = lm\,m^{-2}$
—	Radioactivity	becquerel	$Bq = s^{-1}$
—	Absorbed dose	gray	$Gy = J\,kg^{-1}$

TABLE A-2
Prefixes for SI Units.

Factor	Preflex	Symbol	Conventional Name Germany England	USA France
10^{18}	exa	E	Trillion	Quintillion
10^{15}	peta	P	Billiard	Quadrillion
10^{12}	tera	T	Billion	Trillion
10^{9}	giga	G	Milliard	Billion
10^{6}	mega	M	Million	Million
10^{3}	kilo	k	Thousand	Thousand
10^{2}	hecto	h	Hundred	Hundred
10^{1}	deca	da	Ten	Ten
10^{-1}	deci	d	Tenth	Tenth
10^{-2}	centi	c	Hundredth	Hundredth
10^{-3}	milli	m	Thousandth	Thousandth
10^{-6}	micro	μ	Millionth	Millionth
10^{-9}	nano	n	Milliardth	Billionth
10^{-12}	pico	p	Billionth	Trillionth
10^{-15}	femto	f	Billiardth	Quadrillionth
10^{-18}	atto	a	Trillionth	Quintillionth

TABLE A-3
Fundamental Constants.

Physical quantity	Symbol of Physical Quantity = Number × Physical Unit(s)
Speed of light in vacuum	$c = (2.997\,925 \pm 0.000\,003) \times 10^{8}\ \mathrm{m\ s^{-1}}$
Charge of proton	$e = (1.602\,10 \pm 0.000\,07) \times 10^{-19}\ \mathrm{C}$
Faraday constant	$F = (9.648\,70 \pm 0.000\,16) \times 10^{4}\ \mathrm{C\ mol^{-1}}$
Planck constant	$h = (6.625\,6 \pm 0.000\,5) \times 10^{-34}\ \mathrm{J\ s}$
Boltzmann constant	$k = (1.380\,54 \pm 0.000\,09) \times 10^{-23}\ \mathrm{J\ K^{-1}}$
Avogadro constant	$N_L = (6.022\,52 \pm 0.000\,28) \times 10^{23}\ \mathrm{mol^{-1}}$
(Molar) gas constant	$R = (83.143\,3 \pm 0.004\,4)\ \mathrm{bar\ cm^3\ K^{-1}\ mol^{-1}}$
	$= (8.314\,33 \pm 0.000\,44)\ \mathrm{J\ K^{-1}\ mol^{-1}}$
Permeability of vacuum	$\mu_o = 4\,\pi \times 10^{-7}\ \mathrm{J\ s^2\ C^{-2}\ m^{-1}}$
Permittivity of vacuum	$\varepsilon_o = \mu_o^{-1}c^{-2} = (8.854\,185 \pm 0.000\,18) \times 10^{-12}\ \mathrm{J^{-1}\ C^2\ m^{-1}}$

TABLE A-4
Conversion of Old and Anglo-Saxon Units into SI Units (* = permitted by ISO).

Name	Conversion Old Unit	= SI Unit(s)
1. Lengths		
Mile	1 mile	$= 1\,609.344$ m
Yard	1 yd	$= 0.9144$ m
Foot	1 ft = 1′	$= 0.3048$ m
Inch	1 in = 1″	$= 0.0254$ m
Mil	1 mil	$= 2.54 \times 10^{-5}$ m
Micron	1 μ	$= 10^{-6}$ m $= 1$ μm
Millimicron	1 mμ	$= 10^{-9}$ m $= 1$ nm
Ångstrøm	1 Å	$= 10^{-10}$ m $= 0.1$ nm
2. Areas (1 m² $= 10^4$ cm² $= 10^6$ mm²)		
Square mile	1 sq. mile	$= 2.589\,988\,11 \times 10^6$ m²
Hectare	1 ha	$= 10^4$ m²
Acre	1 acre	$= 4.047 \times 10^3$ m²
Ar	1 a	$= 100$ m²
Square yard	1 sq. yd.	$= 0.836\,127\,36$ m²
Square foot	1 sq. ft.	$= 9.290\,304 \times 10^{-2}$ m²
Square inch	1 sq. in.	$= 6.4516 \times 10^{-4}$ m²
Barn	1 b	$= 10^{-28}$ m²
3. Volumes (1 m³ $= 10^6$ cm³ $= 10^9$ mm³)		
Store	1 st	$= 1$ m³
Cubic yard	1 cu. yd	$= 0.764\,554\,857$ m³
Imperial barrel		$= 0.1636$ m³
US barrel petroleum	1 bbl	$= 0.158\,97$ m³
US barrel		$= 0.119$ m³
Cubic foot	1 CF	$= 2.831\,684\,659\,2 \cdot 10^{-2}$ m³

TABLE A-4

Conversion of Old and Anglo-Saxon Units into SI Units (* = permitted by ISO).

Name	Conversion Old Unit	= SI Unit(s)
Gallon (British or Imperial)	1 gal	$= 4.5459 \times 10^{-3} \text{ m}^3$
Gallon (US dry)	1 gal	$= 4.44 \times 10^{-3} \text{ m}^3$
Gallon (US liquid)	1 gal	$= 3.785412 \times 10^{-3} \text{ m}^3$
*Liter (cgs)	1 L	$= 1.000028 \times 10^{-3} \text{ m}^3$
*Liter	1 L	$= 1.000000 \times 10^{-3} \text{ m}^3$
Quart (US)	1 qt.	$= 9.463353 \times 10^{-4} \text{ m}^3$
Ounce (British liquid)	1 ounce	$= 2.8413 \times 10^{-5} \text{ m}^3$
Ounce (US liquid)	1 ounce	$= 2.9574 \times 10^{-5} \text{ m}^3$
Cubic inch	1 cu. in.	$= 1.6387064 \times 10^{-5} \text{ m}^3$
4. Masses		
Long ton (UK)	1 ton	$= 1016.046909 \text{ kg}$
*Ton	1 t	$= 1000 \text{ kg}$
Short ton (US)	1 ton	$= 907.18474 \text{ kg}$
Hundred weight (UK)	1 cwt	$= 50.8023 \text{ kg}$
Short hundred weight	1 sh cwt	$= 45.3592 \text{ kg}$
Slug	1 slug	$= 14.59 \text{ kg}$
Stone = 14 lb	1 stone	$= 6.35029318 \text{ kg}$
Pound (avoirdupois) = 16 drams	1 lb	$= 0.45359237 \text{ kg}$
Pound (apothecary)	1 lb	$= 0.373242 \text{ kg}$
Ounce (avoirdupois)	1 oz	$= 0.02834952 \text{ kg}$
Karat	1 ct	$= 2 \times 10^{-4} \text{ kg}$
Grain	1 gr	$= 6.48 \times 10^{-5} \text{ kg}$
5. Time		
Year	1 a	$= 3.15576 \times 10^7 \text{ s}$
Month	1 mo	$= 2.6298 \times 10^6 \text{ s}$

Day	1 d	= 86 400 s
*Hour	1 h	= 3600 s
*Minute	1 min	= 60 s

6. Temperatures

*Degree Celsius (= "centigrade")		= $y°C - 273.16°C = K$
Degree Fahrenheit		= $(x°F - 32°F)(5/9) = y°C$

7. Angles

*Degree	1°	= $(\pi/180)\text{rad} = 1.745\,329\,2 \times 10^{-2}$ rad
*Minute	1'	= $2.908\,882 \times 10^{-4}$ rad
*Second	1"	= $4.848\,136\,6 \times 10^{-6}$ rad

8. Densities ($1\ \text{kg m}^{-3} = 10^{-3}\ \text{g cm}^{-3}$)

1 lb/cu. in.		= $27.679\,904\,71\ \text{g cm}^{-3}$
1 oz/cu. in.		= $1.729\,993\,853\ \text{g cm}^{-3}$
1 lb/cu. ft.		= $1.601\,846\,337 \times 10^{-2}\ \text{g cm}^{-3}$
1 lb/gal. US		= $7.489\,150\,454 \times 10^{-3}\ \text{g cm}^{-3}$

9. Energies, work, heat ($1\ \text{J} = 1\ \text{Nm} = 1\ \text{Ws}$)

Kilowatt hour	1 kWh	= 3.6×10^6 J
Horse power hour	1 hph	= 2.685×10^6 J
Cubic foot-atmosphere	1 cu. ft. atm.	= $2.869\,205 \times 10^3$ J
British thermal unit	1 BTU	= $1.054\,350 \times 10^3$ J
—	1 therm	= $1.055\,056 \times 10^3$ J
—	1 ft^3 lb(wt)/in.^{2}	= $1.952\,378 \times 10^2$ J
Liter atmosphere	1 L atm	= $1.013\,250 \times 10^2$ J
—	1 m kgf	= $9.806\,65$ J
Calorie	1 cal$_{IT}$	= $4.186\,8$ J
Calorie	1 cal$_{th}$	= 4.184 J
—	1 ft-lbf	= $1.355\,818$ J
—	1 ft-pdl	= $4.215\,384$ J

TABLE A-4

Conversion of Old and Anglo-Saxon Units into SI Units (* = permitted by ISO).

Name	Conversion Old Unit	= SI Unit(s)
—	1 erg	$= 10^{-7}$ J
Electron-volt	1 eV	$= 1.6021 \times 10^{-19}$ J
10. Forces		
Notched impact strength	1 ft-lbf/in. notch	$= 53.37864$ N
Pound force	1 lbf	$= 4.448222$ N
Poundal	1 pdl	$= 0.1383$ N
Pond	1 p	$= 9.80665 \times 10^{-3}$ N
Gram force	1 gf	$= 9.80665 \times 10^{-3}$ N
Dyne	1 dyn	$= 10^{-5}$ N
11. Length related forces		
Impact strength	1 kp/cm	$= 980.665$ N m^{-1}
Impact strength	1 lbf/ft	$= 14.593898$ N m^{-1}
Surface tension	1 dyn/cm	$= 10^{-3}$ N m^{-1}
	1 ft. lb/in.^{2}	$= 2.103$ kJ/m^2
12. Area-related forces, pressures (1 MPa = 1 MN m^{-2} = 1 N mm^{-2})		
Phys. Atm. = 760 torr	1 atm	$= 0.101325$ MPa
	1 bar*	$= 0.1$ MPa
Techn. Atmosphere	1 at	$= 0.098065$ MPa
	1 kp/cm^2	$= 0.098065$ MPa
	1 kgf/cm^2	$= 0.098065$ MPa
	1 lbf/sq. in.	$= 6.89476 \times 10^{-3}$ MPa
	1 psi	$= 6.89476 \times 10^{-3}$ MPa
Inch mercury (32°F)	1 in. Hg	$= 3.386388 \times 10^{-3}$ MPa
Torr	1 torr	$= 1.333224 \times 10^{-4}$ MPa
Millimeter Mercury	1 mm Hg	$= 1.333224 \times 10^{-4}$ MPa
Millimeter water	1 mm H$_2$O	$= 9.80665 \times 10^{-6}$ MPa
—	1 pdl/sq. ft.	$= 1.488649 \times 10^{-6}$ MPa
—	1 dyn/cm^2	$= 10^{-5}$ MPa

13. Power ($1 \text{ W} = 1 \text{ J s}^{-1}$)

Horsepower (metric)	1 PS	$= 735.499 \text{ W}$
Horsepower (UK)	1 hp = 550 ft lbf/s	$= 745.700 \text{ W}$
—	1 BTU/h	$= 0.292\,875\,1 \text{ W}$
—	1 cal/h	$= 1.162\,222 \times 10^{-3} \text{ W}$
—	1 erg/s	$= 10^{-7} \text{ W}$

14. Heat conductivity

—	1 cal/(cm s °C)	$= 418.6 \text{ W m}^{-1} \text{ K}^{-1}$
—	1 BTU/(ft h °F)	$= 1.729\,577 \text{ W m}^{-1} \text{ K}^{-1}$
—	1 kcal/(m h °C)	$= 1.162\,78 \text{ W m}^{-1} \text{ K}^{-1}$

15. Heat transfer coefficients

—	1 cal/(cm² s °C)	$= 4.186\,8 \times 10^{4} \text{ W m}^{-2} \text{ K}^{-1}$
—	1 BTU/(ft² h °F)	$= 5.682\,215 \text{ W m}^{-2} \text{ K}^{-1}$
—	1 kcal/(m² h °C)	$= 1.163 \text{ W m}^{-2} \text{ K}^{-1}$

16. Length-related masses (= fineness = liter = "linear density")

*Tex	1 tex	$= 10^{-6} \text{ kg m}^{-1}$
Denier	1 den	$= 0.111 \times 10^{-6} \text{ kg m}^{-1}$

17. Tenacities

—	1 g/den	$= 9 \times 10^{3} \text{ m}$
—	1 g/den	$= 88.3 \times (\text{density in g/cm}^3) \text{ MPa}$
—	1 g/dtex	$= 98.06 \times (\text{density in g/cm}^3) \text{ MPa}$

18. Dynamic viscosities

Poise	1 P	$= 0.1 \text{ Pa s}$

19. Kinematic viscosities

Stokes	1 St	$= 10^{-4} \text{ m}^2 \text{ s}^{-1}$

20. Electrical quantities

Electric conductance	1 mho	$= 1 \text{ S}$
—	1 oersted	$= 79.577\,47 \text{ A m}^{-1}$
Electric field strength	1 V/mil	$= 3.937\,008 \times 10^{4} \text{ V m}^{-1}$

21. Radioactivity

Curie	1 Ci	$= 37 \text{ GBq}$

22. Absorbed dose

—	1 rem	$= 10^{-2} \text{ Gy}$

TABLE A-5

Abbreviations and acronyms for thermoplastics, thermosets, fibers, elastomers, and additives.

All abbreviations and acronyms used in this book were taken directly from the quoted literature. More complete lists of the more than 700 commonly used abbreviations and acronyms can be found in:

[1] H.-G. Elias et al., Nomenclature Committee, Division of Polymer Chemistry, Inc., American Chemical Society, Abbreviations for thermoplastics, thermosets, fibers, elastomers, and additives, Polymer News 9 (1983) 101.

[2] ditto, Part II, Polymer News 10 (1985) 169.

ABS	Thermoplastic terpolymer from acrylonitrile, butadiene and styrene
ACS	Graft polymer from styrene/acrylonitrile on chlorinated poly(ethylene)
AES	Thermoplastic polymer from acrylonitrile, ethylene, propylene, and styrene
AR	Elastomer from acrylic esters and olefins
BMC	Bulk molding compound
BR	Butadiene rubber
CDB	Conjugated diene butyl elastomer
CFK	1. Man-made fiber reinforced plastic
	2. Carbon-fiber reinforced plastic
CO	Elastomeric poly[(chloromethyl)oxirane]; poly(epichlorohydrin)
CR	Chloroprene rubber
CSM	Sulfochlorinated poly(ethylene)
DMAC	Dimethylacetamide
ECO	Elastomeric copolymer from ethylene oxide und epichlorohydrin
EP	1. Epoxide
	2. Ethylene/propylene Copolymer
EPDM	Elastomeric terpolymer from ethylene, propylene and a non-conjugated diene
EPR	Elastomeric copolymer from ethylene and propylene
EVA	Copolymer from ethylene and vinylacetate
EVAC	Copolymer from ethylene and vinylacetate
FE	Fluorine containing elastomer
FEF	Fast extruding furnace carbon black
FSI	Fluorinated silicone
GPF	General purpose furnace carbon black
HAF	High abrasion furnace carbon black
HDPE	Poly(ethylene) with high density
HIPS	High impact poly(styrene)
HPT	Hexamethylphosphortriamide
IIR	Elastomeric copolymer from isobutene and isoprene
LDPE	Low density poly(ethylene)
LLDPE	Linear low density poly(ethylene)
MDI	4,4'-Methylene-bis(phenylisocyanate)
MOCA	4,4'-Methylene-bis(2-chloraniline)
MT	Medium thermal carbon black

MTI	Methylenetri(p-phenyleneisocyanate)
NBR	Acrylonitrile/butadiene rubber
NMP	N-Methylpyrrolidone
NR	Natural rubber
PA	1. Polyamide
	2. Poly(acetylene)
	3. Phthalic acid anhydride
PAZ	Poly(acetylene)
PBT	1. Poly(butyleneterephthalate)
	2. Poly(1-butene)
PC	Polycarbonate
PCD	Polycarbodiimide
PEC	Chlorinated poly(ethylene)
PET	Poly(ethylene terephthalate)
PIR	Poly(isocyanurate)
PP	Poly(propylene)
PPO	Poly(phenyleneoxide) (trademark!)
PPP	Poly(p-phenylene)
PSU	Poly(phenylenesulfon)
PTFE	Poly(tetrafluoroethylene)
PUR	Polyurethane
PVAC	Poly(vinylacetate)
PVC	Poly(vinyl chloride)
PVDC	Poly(vinylidenechloride)
SAN	Thermoplastic from styrene and acrylonitrile
SBR	Elastomer from styrene and butadiene
SBS	Triblock polymer with styrene-butadiene-styrene-blocks
SEBS	Triblock polymer styrene-ethylene/butylene-styrene
SIR	1. Silicone rubber
	2. Elastomer from styrene and isoprene
	3. Standardized Indonesian rubber
SIS	Triblock polymer with styrene-isoprene-styrene blocks
SMA	Copolymer from styrene and maleic anhydride
SMC	Sheet molding compound
T	Polysulfide rubber
TMU	Tetramethyl urea
TOR	trans-Poly(octenamer)
TPE	Thermoplastic elastomer
TPO	Thermoplastic polyolefin elastomers
UP	Unsaturated polyester

TABLE A-6
Addresses of companies mentioned in this book.

Air Products and Chemicals, Inc., Plastics Div., Allentown, PA 18105, USA.
AKU Algemeene Kunstzijde Unie, Arnheim, Netherlands (now ENKA).
Akzo Plastics BV, P.O. Box 124, Zeist, Netherlands.
Albright & Wilson Ltd., Industrial Chemicals Div., P.O. Box 3, Oldbury, Warley, Worcs., England.
Allied Chemical Corp., Fibers Div., Plastics Film Dept., P.O. Box 1057R, Morristown, NJ 07960, USA.
Allied Chemical Corp., Fibers & Plastics Co., P.O. Box 2332, Morristown, NJ 07960, USA.
American Cyanamid Co., Industrial Chemicals Div., Wayne, NJ 07470, USA.
American Cyanamid Co., Elastomers Dept., Bound Brook, NJ 08805, USA.
Amoco Chemical Corp., Marketing Div., 200 East Randolph Drive, P.O. Box 8640 A, Chicago, IL 60601, USA.
Anhydrides & Chemicals, Inc., 377 Park Ave. South, New York, NY 10016, USA.
Arco Chemical Co., 1500 Market Street, Philadelphia, PA 19101, USA.
Asahi Chemical Industries Co., Ltd., 12, 1-Chome, Yurakucho, Chiyodaku-ku, Tokyo, Japan.
Asahi-Dow Ltd., c./o. Asahi Chemical Industries Co., Ltd., 12, 1-Chome, Yurakucho, Chiyodaku-ku, Tokyo, Japan.
Asahi Glass Co., Ltd., Hazawa-Cho, Kanagawa-Ku, Yokohama, Japan.
Asea Kabel AB, P.O. Box 42108, S-126-12 Stockholm, Schweden.
ATO Chimie, Tour Aquitaine Cedex no. 4, 92080 Paris La Défence, France.
Bakelite Xylonite Co., Ltd., London NW 1, England.
BASF AG, 6700 Ludwigshafen am Rhein 1, Federal Republic of Germany.
Battelle-Institut, 7 route de Drize, CH-1227 Carouge-Geneva, Switzerland.
Bayer AG, Bayerwerk, 509 Leverkusen, Federal Republic of Germany.
Dr. Beck & Co., now: BASF Farben + Fasern AG, Am Neumarkt 36, 2000 Hamburg 70, Federal Republic of Germany.
Bendix Corp., 20650 Civic Center Dr., Southfield, MI 48076, USA.
Borg-Warner Corp., 200 S. Michigan Ave., Chicago, IL 60604, USA.
Bridgestone Tire Co., Ltd., 10-1, Kyobashi 1-chome, Chuo-ku, Tokyo 104, Japan.
British Industrial Plastics Ltd., Oldbury, Birmingham, England.
Carborundum Co., Engineering Plastics Group, P.O. Box 337, Bldg. 1-2, Niagara Falls, NY 14302, USA.
C & D Batteries Division of Eltra Corp., Plymouth Meeting, PA, USA.
CdF Chimie, Cedex No. 5, F-92808 Paris Défence, France.
Ceca SA, 11 avenue Morane-Saulnier, F-78140 Velizy-Villacoublay, France.
Celanese Plastics & Specialties Co., 26 Main St., Chatham, NJ 07928, USA.
Celanese Corp., 522 Fifth Avenue, New York, NY 10036, USA.
Chemie Linz AG, Postfach 269, A-4021 Linz, Austria.
Chemische Werke Hüls AG, 437 Marl, Kreis Recklinghausen, Federal Republic of Germany.
Chemplex Co., 3100 Golf Rd., Rolling Meadows, IL 60008, USA.
Ciba-Geigy AG, Division Kunststoffe und Additive, CH-4002 Basel, Switzerland.

Ciba-Geigy Corp., Resins Department, Sawmill River Rd., Ardsley, NY 10502, USA.

Cities Service Co., Plastics Div., 3445 Peachtree Rd., N.E., Atlanta, GA 30326, USA.

Continental Gummi AG, Königsworther Platz, 3000 Hannover, Federal Republic of Germany.

Cooke Division of Reichhold Chemicals Inc., Thermoplastics & Elastomers Division, P.O. Box 400, Hackettstown, NJ 07840, USA.

Copolymer Rubber & Chemical Corp., P.O. Box 2591, Baton Rouge, LA 70801, USA.

Dai-Ichi Kogyo Seyaku Co., 55 Nishishichijyo, Higashikubocho, Shimokyo-ku, Kyoto 600, Japan.

Dainippon Ink & Chemicals America, Inc., 200 Park Ave., New York, NY 10017, USA.

Dart Industries, Inc., Fiberfil Division, Evansville, IN 47732, USA.

Disogrin Industries Corp., Grenier Field, Manchester, NH 03103, USA.

Dixon Industries Corp., Metacom Ave., Bristol, RI 02809, USA.

Dow Chemical Co., 2020 Dow Center, Midland, MI 48640, USA.

Dow Corning Corp., P.O. Box 1767, Midland, MI 48640, USA.

DSM, Postbus 7019, Utrecht, Netherlands.

DSM, Postbus 602, Geleen, Netherlands.

Dunlop Holdings, Polymer Engineering Div., Evington Valley Road, Leicester, England.

DuPont Inc., 1007 Market Street, Wilmington, DE 19898, USA.

DuPont of Canada Ltd., P.O. Box 26, Dominion Centre, Toronto, Canada.

Dynamit Nobel AG, Postfach 1209, 5210 Troisdorf, Federal Republic of Germany.

Eastman Chemical Products, Inc., P.O. Box 431, Kingsport, TN 37662, USA.

Eastman Chemical International AG, Baarerstr. 8, CH-6301 Zug, Switzerland.

Eitel-McCollough Inc., San Carlos, CA 94070, USA.

Ems-Chemie AG (formerly Emser Werke AG), Selnaustr. 16, CH-8039 Zürich, Switzerland.

Ems-Chemie AG, CH-7013 Domat/Ems, Switzerland.

Emery Industries, Inc., 1300 Caren Tower, Cincinnati, OH 45202, USA.

Enesco (Canada), Joffre, Alberta, Canada.

Exxon Chemical Co., P.O. Box 3272, Houston, TX 77001, USA.

Ferro Corp., Chemical Div., 7050 Krick Road, Bedford, OH 44146, USA.

Fiberfil Div., see Dart Industries, Inc.

Fiberite Corp., 515 West Third St., Winona, MN 55987, USA.

Fire Safe Products, Inc., 8411 Midcounty Industrial, St. Louis, MO 63114, USA.

Carl Freudenberg, 6940 Weinheim, Federal Republic of Germany.

Furukawa Electric Co., Ltd., 2-6-1 Marunouchi, Chioyoda-ku, Tokyo 100, Japan.

General Electric Company, 570 Lexington Ave., New York, NY 10022, USA.

General Electric Plastics, Plasticlaan 1, NL-4600 AC Bergen op Zoom, Netherlands.

General Mills Chemicals, Inc. (now: Henkel Corp.), 4620 W. 77th Street, Minneapolis, MN 55435, USA.

Gentex Corp., P.O. Box 315, Carbondale, PA 18407, USA.

The P. D. George Co., 5200 N. Second St., St. Louis, MO 63147, USA.

Gevaert Photo-Producten NV, 27 Septestraat, Mortsel/Antwerp, Belgium.

B.F. Goodrich Chemical Co., 6100 Oak Tree Blvd., Cleveland, OH 44131, USA.

The Goodyear Tire & Rubber Co., Plastics Dept., 1144 E. Market St. Akron, OH 44306, USA.

W. R. Grace & Co., 7 Hanover Square, New York, NY 10008, USA.

Gulf Oil Chemicals Co., Gulf Advanced Materials Dept., P.O. Box 10911, Overland Park, KS 66210, USA.

Gulf Oil Chemicals Co., US Plastics Div., P.O. Box 3766, Houston, TX 77001, USA.

The C. P. Hall Co., 7300 South Central Ave., Chicago, IL 60638, USA.

Hayashibara Biochemical Laboratories, Inc., Okayama, 2−3, 1-Chome, Shimoishi, Japan.

Heisler Compounding Div., Container Corp. of America, P.O. Box 1648, 1204 E. Twelfth St., Wilmington, DE 19899, USA.

Henkel Inc., Chemical Specialties Div., 1301 Jefferson St., Hoboken, NJ 07030, USA.

Herberts, Dr. Kurt, Lackfabrik, 5600 Wuppertal 2, Federal Republic of Germany.

Hercules Inc., 910 Market St., Wilmington, DE 19899, USA.

Hitachi Ltd., 1-5-1 Marunouchi, Chiyoda-ku, Tokyo 100, Japan.

Hitachi Chemical Co. of America, Ltd., 437 Madison Ave., New York, NY 10022, USA.

Hoechst AG, Sparte Kunststoffe, 6230 Frankfurt (M) 80, Federal Republic of Germany.

Hooker Chemical Co., Ruco Division, River Road, Burlington, NJ 08016, USA.

Hooker Chemicals & Plastics Corp., 1024−6 Iroquòis St., Niagara Falls, NY 14302, USA.

Hughes Aircraft Co., Centinela Ave. and Teale St., Culver City, CA 90230, USA.

The Humphrey Chemical Co., Devine St., North Haven, CT 06473, USA.

ICI, Plastics Division, P.O. Box 6, Bessemer Road, Welwyn Garden Cizy, Hertford, England.

Deutsche ICI GmbH, ICI Haus, Postfach 710330, 6000 Frankfurt (M) 71, Federal Republic of Germany.

ICI Americas Inc., Concord Pike & New Murphy Road, Wilmington, DE 19897, USA.

Idemitsu Petrochemicals Co., Ltd., 1-1, Marunouchi 3-chome, Chiyoda-ku, Tokyo 100, Japan.

Imhausen International Company mbH, Talacker 42, CH-8001 Zürich, Switzerland.

Isochem Resins Co., Cook St., Lincoln, RI 02865, USA.

ISR, The International Synthetic Rubber Co., Ltd., Brunswick House, Brunswick Place, Southampton S09 3AT, England.

Japan Synthetic Rubber Co., Ltd., 2−11−24 Tsukiji, Chuo-ku, Tokyo, Japan.

Jefferson Chemical Co., P.O. Box 4128, Austin, TX 78765, USA.

Kanegafuchi Chemical Industry Co., 3−2−4 Nakanoshima, Kita-ku, Osaka 530, Japan.

Kay-Fries Inc., Member Dynamit Nobel Group, 200 Summit Ave., Montvale, NJ 07645, USA.

Kelco, Division of Merck & Co. Inc., 20 N. Wacker Drive, Chicago, IL 60606, USA.

Kurary Co., 280 Park Ave., New York, N.Y. 10017.

LNP Corp., 412 King St., Malvern, PA 19355, USA.

Lonza-Werke GmbH, Postfach 1260, 7858 Weil am Rhein, Federal Republic of Germany.

Lord Corp., Biomedical Group, 1635 West 12th Street, Erie, PA 16514, USA.

Marine Commodities International Inc., Shrimp Harbor, Star Route Box 140, Brownsville, TX 78520, USA.

Maruzen Oil Co., Ltd., P.O. Box 293, Osaka Minami, Japan.

Meteo, Inc., 1101 Prospect Ave., Westbury, NY 11590.

Mitsubishi Chemical Industry, Ltd., Mitsubishi Building, 2−5−2 Marunouchi, Chiyoda-ku, Tokyo 100, Japan.

Mitsubishi Gas Chemical Co., Mitsubishi Building, 2−5−2 Marunouchi, Chiyoda-ku, Tokyo 100, Japan.

Mitsubishi Petrochemical Co., Mitsubishi Building, 2−5−2 Marunouchi, Chiyoda-ku, Tokyo 100, Japan.

Mitsui Petrochemical Industries Ltd., 2−5 Kasumigaseki 3-chome, Chiyoda-ku, Kasumigaseki P.O. Box 90, Tokyo 100, Japan.

Mobay Chemical Corp., Penn Lincoln Parkway, W., Pittsburgh, PA 15205, USA.

Mobil Chemical Co., Macedon, NY 14502, USA.

Monsanto Co., 800 N. Lindbergh Blvd., St. Louis, MO 63166, USA.

Montedison S.p.A., Foro Buonaparte 31, P.O. Box 3777, I-20121 Milano, Italy.

M & T Chemicals Inc., Woodbridge Ave., Rahway, NJ 07065, USA.

Naphthachimie S.A., B.P. 203, F-71708 Paris, France.

Narmco Division of Celanese Corp. (formerly Whittaker Corp.), 10880 Wilshire Blvd., Los Angeles, CA 90024, USA.

NASA National Aeronautics and Space Administration, Hampton, VA 23665, USA.

Nippon Petrochemicals Co., Ltd., 3−12, Nishishimbashi 1-chome, Minato-ku, Tokio 105, Japan.

Nippon Polyurethane Industries Co. Ltd., Toranomon Kotohira Kaikan Building, 1 Shiba Kotohira-Cho, Minato-Ku, Tokyo 105, Japan.

Nippon Synthetic Chemical Industry Co. Ltd., Higashi-umeda Building, 9−6 Nogakicho, Kita-ku, Osaka 530, Japan.

Nippon Tennen Gas Kogyo Co.

Nippon Unicar Co., Ltd., 6−1, Ote-machi 2 chome, Chiyoda-ku, Tokio 100, Japan.

Nippon Zeon Co. Ltd., Furukawa Sogo Building, 6−1 Marunouchi 2-chome, Chiyoda-ku, Tokyo, Japan.

Ohio Rubber Co., Orthane Division, P.O. Box 1398, 135 West, Denton, TX 76201, USA.

Pantasote Inc., Film/Compound Division, 26 Jefferson St., Passaic, NJ 07055, USA.

P. D. George Co., see George.

Phillips Chemical Co., 1501 Commerce Dr., Stow, OH 44224, USA.

Phillips Petroleum Co., Chemical Department, Plastics Division, Bartlesville, OK 74004, USA.

Polysar Ltd., 201 N. Front St., Sarnia, Ontario N7T 7VI, Canada.

PPG Industries Inc., Resin Products, 1 Gateway Center, Pittsburgh, PA 15222, USA.

Prolastomer Inc., 56 Eagle St., Waterbury, CT 06708, USA.

Quantum Inc., P.O. Box 748, Barnes Industrial Park, Wallingford, CT 06492, USA.

Quinn, K. J. & Co., Inc., 195 Canal St., Malden, MA 02148, USA.
Raychem Corp., 300 Constitution Drive, Menlo Park, CA 94025, USA.
Reichhold Chemie AG, Kettelerstr. 100, Postfach 38, 6050 Offenbach/Main 1, Federal Republic of Germany.
Ren Plastics, 5656 S. Cedar St., Lansing, MI 48909, USA.
Rhône-Poulenc, 21 Rue Jean-Goujon, 22 Ave. Montaigne, B.P. 75308, F-75360 Paris Cedex 08, France.
Rockwell International Corp., Rocketdyne Division, 6633 Canoga Ave., Canoga Park, CA 91304, USA.
Rogers Corp., Rogers, CT 06263, USA.
Rohm & Haas Co., Independence Mall W., Philadelphia, PA 19105, USA.
Rubicon Chemicals Inc., P.O. Box 751, Wilmington, DE 19897, USA.
Ruco Division, see Hooker.
SABIC (Saudi Basic Industries) Corp., Yanbu and Al Jubail, Saudi Arabia.
Schaum-Chemie Wilhelm Bauer GmbH & CO. KG, Kallenbergstr. 21, 4300 Essen 1, Federal Republic of Germany.
Schenectady Chemicals Inc., P.O. Box 1046, Schenectady, NY 12301, USA.
Schramm Lackfabriken, 6050 Offenbach/Main, Federal Republic of Germany.
A. Schulman Inc., 3550 West Market St., Akron, OH 44313, USA.
A. Schulman GmbH, Hüttenstraβe 211, 5014 Kerpen 3/Sindorf, Federal Republic of Germany.
Shell/Royal Dutch, Badhuisweg 3, Amsterdam-N., Netherlands.
Shell Chemical Co., 1 Shell Plaza, Houston, TX 77002, USA.
Showa Denko K. K., 1–13–9 Shibadaimon, Minato-ku, Tokyo 105, Japan.
Showa High Polymer Co. Ltd., Kanda Building, 3–20 Kanda-nishikicho, Chiyoda-ku, Tokyo 101, Japan.
Snia Viscosa S.p.A., Via Montebello 18, I-20121 Milano, Italy.
Solar Division, International Harvester Corp., 401 N. Michigan Ave., Chicago, IL 60611, USA.
Solvay & Cie., S. A., 33 rue du Prince Albert, B-1050 Bruxelles, Belgium.
Deutsche Solvay-Werke GmbH, Langhansstr., Postfach 110270, 5650 Solingen 11, Federal Republic of Germany.
A. E. Staley Mgf. Co., 2200 East Eldorado St., Decatur, IL 62525, USA.
Stauffer Chemical Co., 299 Park Ave., New York, NY 10017, USA.
Stauffer Chemical Co., Plastics Division, Westport, CT 06880, USA.
Stauffer-Wacker-Silicones Corp., Division Stauffer Chemical Co., P.O. Box 428, Adrian, MI 49221, USA.
Sumitomo Chemical Co. Ltd., 15–5 chome, Kitahama, Higashi-ku, Osaka, Japan.
Sumitomo Bakelite Co., 1–2–2 Ushisaiwai-cho, Chiyoda-ku, Tokyo Japan.
Sumitomo Rubber Industries, Ltd., 606 S. Olive St. 1014, Los Angeles, CA 90014, USA.
Sun Company, Inc., 100 Matsonford Rd., Radnor, PA 19087, USA.
Teijin Ltd., 1–11 Minamihonmachi, Higashi-ku, Osaka 541, Japan.
Teknor Apex Co., 505 Central Ave., Pawtucket, RI 02862, USA.
Texaco Chemical Co., Box 430, Bellaire, TX 77401, USA.
Toho Belson Co., c./o. Celanese Fibers Marketing Co., P.O. Box 32414, Charlotte, NC 28232, USA.

Toray Industries Co., Toray Building, 2−2 Nihonbashi-Muromachi, Chuo-ku, Tokyo 103, Japan.

Toyo Rayon, siehe Toray Industries Co.

Toyobo Co., Ltd., 2−8, Dojima Hama 2-chome, Kita-ku, Osaka 530, Japan.

TRW Inc., 35555 Euclid Ave., Cleveland, OH 44117, USA.

Ube Industries Ltd., 7−2 Kasumigaseki, 3-chome, Chiyoda-ku, Tokyo, Japan.

UCC, siehe Union Carbide Corp.

Ugine Kuhlmann S. A., Division Plastiques, Tour Manhattan, Cedex 21, F-92 087 Paris La Défense 2, France.

Unifos Kemie AB, Industrivägen, Postfach 44, S-44401 Stenungsund, Sweden.

Union Carbide Corp., 270 Park Ave., New York, NY 10017, USA.

Uniroyal Inc., Chemical Division, 58 Maple St., Naugatuck, CT 06770, USA.

Unitika Ltd., Plastics Division, 4−68, Kitakyutaro-Machi, Higashi-ku, Osaka, Japan.

Upjohn Co., Polymer Chemicals Division, P.O. Box 685, La Porte, TX 77571, USA.

Upjohn Polymer (Europa) AG, Poststr. 28, Postfach 91, CH-9001 St. Gallen, Switzerland.

U.S. Air Force Materials Laboratory, Wright-Patterson Air Force Base, OH 45433, USA.

U.S. Industrial Chemicals Co., 99 Park Ave., New York, NY 10016, USA.

U.S. Polymeric, Division Hitco, P.O. Box 2187, Santa Ana, CA 92707, USA.

U.S.S. Chemicals, Division U.S. Steel Corp., 600 Grant St., 28th Floor, Pittsburgh, PA 15230, USA.

Velsicol Chemical Corp., Chemicals & Resins Group, 341 E. Ohio St., Chicago, IL 60611, USA.

Wacker-Chemie GmbH, Prinzregentenstr. 22, 8000 München 22, Federal Republic of Germany.

Westinghouse Electric Corp., Insulating Materials Division, Manor, PA 15665, USA.

Whittaker, Corp., siehe Narmco Division of Celanese Corp.

Witco Chemical Corp., Organics Division, 277 Park Ave., New York, NY 10017, USA.

Wright Patterson Air Force Base, see U.S. Air Force Materials Laboratory.

Xenex Corp., 5822-T Parkersburg Dr., Houston, TX 77036, USA.

Yorkshire Chemicals Ltd., Kerkstall Road, Leeds, LS3 1LL, England.

SUBJECT INDEX